# Techniques and Technologies in Geoinformatics

# Techniques and Technologies in Geoinformatics

Edited by Chris Bevan

SYRAWOOD
PUBLISHING HOUSE
New York

Published by Syrawood Publishing House,
750 Third Avenue, 9th Floor,
New York, NY 10017, USA
www.syrawoodpublishinghouse.com

**Techniques and Technologies in Geoinformatics**
Edited by Chris Bevan



International Standard Book Number: 978-1-68286-607-8 (Hardback)



**Cataloging-in-Publication Data**

Techniques and technologies in geoinformatics / edited by Chris Bevan.
p. cm.
Includes bibliographical references and index.
ISBN 978-1-68286-607-8
1. Geoinformatics. 2. Geographic information systems. I. Bevan, Chris.
G70.212 .T43 2018
910.285--dc23

Printed in China.

# TABLE OF CONTENTS

# PREFACE

Geoinformatics is the study of geographical and geological processes and elements with the use of geospatial and geocomputational data. Satellites and remote sensing technologies gather information that can be sequentially monitored to calculate future availability of resources. Geoinformatics, as a field, combines various branches ranging from mathematics to computer science to architecture and conservation. This book elucidates the theories and innovative models around prospective developments with respect to geoinformatics. Scientists and students actively engaged in this field will find this book on geoinformatics full of crucial and unexplored concepts.

This book is the end result of constructive efforts and intensive research done by experts in this field. The aim of this book is to enlighten the readers with recent information in this area of research. The information provided in this profound book would serve as a valuable reference to students and researchers in this field.

At the end, I would like to thank all the authors for devoting their precious time and providing their valuable contribution to this book. I would also like to express my gratitude to my fellow colleagues who encouraged me throughout the process.

**Editor**

1

# Spatial Heterogeneity in Response of Male Greater Sage-Grouse Lek Attendance to Energy Development

**Andrew J. Gregory[1]*, Jeffrey L. Beck[2]**

**1** School of Earth, Environment and Society, Bowling Green State University, Bowling Green, Ohio, United States of America, **2** Department of Ecosystem Science and Management, University of Wyoming, Laramie, Wyoming, United States of America

## Abstract

Landscape modification due to rapidly expanding energy development, in particular oil and gas, in the westernUSA, have prompted concerns over how such developments may impact wildlife. One species of conservation concern across much of the Intermountain West is the greater sage-grouse (*Centrocercusurophasianus*). Sage-grouse have been petitioned for listing under provisions of the Endangered Species Act 7 times and the state of Wyoming alone represents 64% of the extant sage-grouse population in the eastern portion of their range. Consequently, the relationship between sage-grouse populations and oil and gas development in Wyoming is an important component to managing the long-term viability of this species. We used 814 leks from the Wyoming Game and Fish Department's lek survey database and well pad data from the Wyoming Oil and Gas Conservation Commission to evaluate changes in sage-grouse lek counts as a function of oil and gas development since 1991.From 1991–2011 we found that oil and gas well-pad density increased 3.6-fold across the state and was associated with a 24% decline in the number of male sage-grouse. Using a spatial and temporally structured analysis via Geographically Weighted Regression, we found a 1-to-4 year time lag between development density and lek decline. Sage-grouse also responded to development densities at multiple spatial neighborhoods surrounding leks, including broad scales of 10 km. However, sage-grouse lek counts do not always decline as a result of oil and gas development. We found similar development densities resulting in different sage-grouse lek count responses, suggesting that development density alone is insufficient to predict the impacts that oil and gas development have on sage-grouse. Finally, our analysis suggests a maximum development density of 1 well-pad within 2 km of leks to avoid measurable impacts within 1 year, and <6 well-pads within 10 km of leks to avoid delayed impacts.

**Editor:** Peter Shaw, Roehampton university, United Kingdom

**Funding:** The Wyoming Game and Fish Department provided funding for our study. Support for A. Gregory was provided by Northern Arizona University and a grant from the Gordon and Betty Moore Foundation. Support for J. Beck was provided by the University of Wyoming. The funders had no role in study design, data collection and analysis, decision to publish, or preparation of the manuscript.

**Competing Interests:** The authors have declared that no competing interests exist.

* E-mail: agregor@bgsu.edu

## Introduction

Anthropogenic impacts to nature are nearly omnipresent across Earth [1]. Today, human-caused fragmentation and land use are the main threats to biodiversity conservation [2,3]. Recent expansion of energy development throughout the Intermountain West has prompted concern about how such developments might impactprairie-grouse including greater sage-grouse (*Centrocercusur-ophasianus;* hereafter sage-grouse) populations [4]. The current distribution of sage-grouse encompasses 11 western states and 2 Canadian provinces and is tied to the presence of sagebrush (*Artemisia* spp.; [5]). Declines in the distribution and abundance of sage-grousehave been a concern since at least the early 1900s [6].However, this decline has increased across the range of sage-grouse over the last 50-years [7], prompting 7 separate petitions to listsage-grouse under provisions of the Endangered Species Act (ESA) of 1973. In the most recent attempt in 2010, the U. S. Fish and Wildlife Service (USFWS) found that listing greater sage-grousewas warranted, but precluded by higher priority listings for other species [8]. Concerns persist as current data suggest that over the last 50 years, human demand for energy has increased by >50% and a similar increase is anticipated by 2030 [9].

Several recent studies investigating the possible influence of oil and gas development on sage-grouse lek attendance have found that sage-grouse are negatively impacted by oil and gas development [4,10–15]. However, these studies evaluated the response of sage-grouse at local scales (e.g. basin or development site) or quantified the direction and magnitude of the impacts over broad geographic expanses that didnot account for possible spatial variation in response. It is possible that in some areas and under certain conditions, sage-grouse might be coping with oil and gas development, and so it is important to identify these areas and conditions so that we can better understand what characteristics of these locations are driving these trends.

A listing decision for the greater sage-grouse under the ESA would have severe economic impacts for thestates where sage-grouse are found. For example, Wyoming contains 64% of the sage-grouse in the eastern portion of the species range [16], and is an energy-rich state, with large reserves of coal, natural gas, oil, uranium, and high potential for wind energy development [17,18]. Sage-grouse distribution in Wyoming overlaps with high priority energy development sites, including those for oil and gas [19]. In recognition of this, the state of Wyoming instituted a regulatory approach to conserving sage-grouse through an executive order issued by the governor. This Sage-grouse Executive Order (SGEO) established sage-grouse core areas, limiting development in areas of high priority for sage-grouse [20–22].

Previous studies on the potential impacts of oil and gas development on sage-grouse lek attendance have typically modeled raw lek count data or a binomial characterization of lek occupancy as either occupied or unoccupied as a response variable [10,12,13,15].When dealing with data collected at spatially discrete locations or interpolated across spatially connected areas one expects there to be a high degree of spatial autocorrelation [23]. Moreover, because of some of the unique properties of spatially structured observations, near objects ought to be more related than distant objects [24]. Consequently, the assumption made by ordinary least squaresregression techniques that all observations are independent or nearly independent is violated and therefore the application of such techniques to spatially structured data is questionable [25]. Spatial regression controls for the lack of independence in observations by applying a spatial weights matrix to the response variables [26]. Whereas this approach does control for the effects of spatial autocorrelation, it masks possibly interesting and important local data structure that may exist and may average important local trends with a global average [27]. For example, relying on the average temperature for the United States would not be a helpful way to decide if one needs to wear a jacket on any given day in any given region of the United States. Geographically Weighted Regression (GWR) was designed for just this purpose, to control for global spatial autocorrelation in spatially structured data sets, while still maintaining local variation and patterns that might be important drivers of the system (i.e., spatial heterogeneity; [27].

Our primary objective was to explore the possibility of spatially varying relationships among oil and gas development density and sage-grouse lek attendance. Previous studies, working within important sage-grouse habitats in Wyoming, have suggested that such a relationship might exist. Doherty et al. [14] identified 17 leks in Wyoming that persisted despite a high development density of ≥40 wells within 32.2 $km^2$ of these leks, and via visual inspection of these leks suggested that on 11 of them oil and gas development was clustered leaving large open spaces. However, Doherty et al. 's analysis focused on lek persistence over a 4-year period, and over that period these 17 leks declined by ~55% [14]. Thus, our secondary objective was to investigate the possibility of similar development densities resulting in opposite trends in sage-grouse lek attendance, in an effort to understand if there is a sustainable development density of oil and gas well pads where lek attendance is not impacted. To obtain data to address our objectives we used a regression trend analysis of male sage-grouse lek count data for the state of Wyoming as an index of the statewide sage-grouse population trend. We modeled sage-grouse population trends for the most recent 3, 5, and 10 year time periods since 2011 as a function of oil and gas well-pad density at multiple spatial neighborhoods around leks. We also evaluated the effects of past well-pad densities on the current sage grouse lek count response to oil and gas development in an attempt to assess lag effects.

## Methods

### Terminology used throughout the manuscript

Throughout this article we refer to our response variable, based on Wyoming Game and Fish Department (WGFD) lek count data—as lek attendance. Previous studies using the same database have used the term sage-grouse population trends [13,19,28]. We hesitate to use the terms abundance or population trajectory as the lek-count data estimates we used do not account for detection probabilities ≠1.0, which is necessary to truly estimate abundance or a population trajectory [29]. In addition, we used lek count values for only peak male attendance so females were not included in our estimates. Thus, we prefer to use the term male lek attendance.

### Sage-Grouse Lek Data

We used WGFD annual sage-grouse lek survey count data to index sage-grouse population trend (T. Christiansen, WGFD Sage-Grouse Program Coordinator, personal communication, 2010). The WGFD lek database identifies a lek as a traditional courtship display area in or adjacent to sagebrush-dominated habitat attended by ≥2 male sage-grouse for two or more consecutive years (WGFD Sage Grouse definitions 2010). The database contains geographic coordinates and annual count data since 1948 for leks across Wyoming. However, the majority of these leks, 68%, do not have regular survey data (an average of ≥3 out of 5 years) until after 1973. Lek counts were conducted by WGFD,other natural resource agency personnel, or trained volunteersfollowing established WGFD protocol. Only ground counts were used in our analysis. In each given year, surveyed leks were categorized as active, inactive, or unknown. On all active leks numbers of male and female sage-grouse present on the lek at the time the survey was conducted wererecorded. What this means, is that both male and female sage-grouse were counted, but, due to the fact that adult males attend leks consistently throughout spring [30], and female lek attendance is inconsistent, we elected to use the count data for only peak male attendance, which is consistent with previous analyses [11,12].Leks were classified as occupied based on presence of strutting males at least once during the most recent 10 years. Leks that were surveyed only once were also included in our analysis because for analyses with >50 sage-grouse leks, repeated counts are not necessary to model population trend estimates [28]. Based on the annual lek survey data, leks were assigned a management status based on whether the lek site was observed to have had birds displaying on it sometime during the most recent 10-years as either occupied or unoccupied.

We used data from the WGFD sage-grouse lek survey database for 2002 to 2011. To be included in our analysis,sage-grouse leks had to have been surveyed in all years from 2009–2011, in at least 4 years from 2007–2011including all years from 2009–2011, and in at least six years from 2002–2011 including surveys in 2002 and in all years from 2009–2011. This restriction was necessary to allow us to characterize a regression of lek count data across each time period (most recent 3, 5, and 10 years, respectively). These restrictions left us with a database of 814 leks counted a total of 4,070 times in ten years (average of 5/lek) across Wyoming for our analysis (Figure 1).

### Well Pad Data

Oil and gas well-pad density data were summarized from the Wyoming Oil and Gas Conservation Commissions database of Wyoming Statewide Oil and Gas Drilling Activity to 2010 [31]. The database contains geographic coordinates and attribute data for all (114,246) oil and gas wells in the state of Wyoming from 1914–2010. All well-padswere assigned to 1 of 11 classifications and 1 of 29 statuses delineating the type of well pad and the amount of activity present at each well pad site [31]. Our well-pad density estimates are for active well-pads each year.

For the purposes of our study, well pads were classified according to the presence of physical structures on the landscape and of a status indicating that they were actively producing over the time period for which sage-grouse data were collected [31].

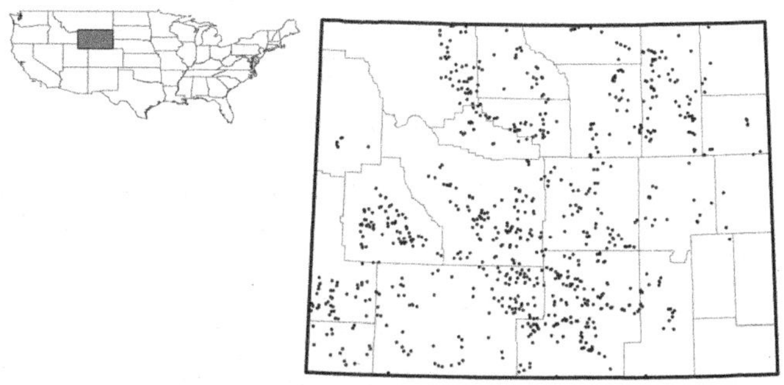

**Figure 1. Point locations of 814 sage-grouse lek sites used in the GWR analysis, from 2002–2011, Wyoming USA. Inset boundaries outline Wyoming's 23 county boundaries.**

For example, some well pads had a classification AP (Permit to Drill), but no indication that drilling had ever occurred; so these well-pads were not included in our analysis. In other instances a well pad might have a status of DH (dry hole) or DR (dormant) accompanied by a date for when activity at the site ceased. These well-pads were also not included in our estimate of well-pad densities. Eliminating these inactive or undeveloped well pads provided us with a geodatabase that included location data for 39,885 active well-pads for the state of Wyoming (Figure 2).

We used a roving window analysis to calculate the point density of well-pads within 0.5, 1, 2, 5, and 10 km areas surrounding leks in Arc Info 10.1 (Environmental Systems Research Inc., Redlands, California). Well pad densities were calculated for all well-pads present on the landscape for all years prior to 2009, 2007, 2006, 2004, 1996, and 1991 to test for possible existence of 1-, 3-, 5-, and 10-year time lags to population effects, respectively (e.g., Figure 3).

## Statistical Analysis

Sage-grouse lek count data shows periodicity [28]. While over 10- and 20-year periods the general trend (increasing, decreasing, or stable) is accurate, over 3- to 5-year periods within the 10- and 20-year periods, count data will suggest that populations at leks alternate from exhibiting an increasing trend to decreasing trend [28]. To control for this periodicity we used a logit function calculated from the 10-year LOWESS regression to back calculate lek attendance values for each year of the 10-year period [32]. The LOWESS function is a smoothing function used to detrend data and to dampen the effect of outliers on the data set [33]. We used a heuristic algorithm to choose a smoothing factor for the LOWESS function based on the $AIC_C$ Information Criterion [34]. We then applied a linear regression to the 10-year LOWESS predicted lek counts for the most recent 3, 5, and 10 year periods from 2011.We refer to these responses as the 3×10, 5×10, and 10×10 sage-grouse responses. We used parameter estimates from the 3-,5-and 10-year regressions possible sage-grouse responses trends to oil and gas development. Finally, we also used the raw LOWESS adjusted count data from 2011 and a binomial response variable of increasing (1) or decreasing (0) as response variables for each lek so that we could compare our analytical results with previously published studies.

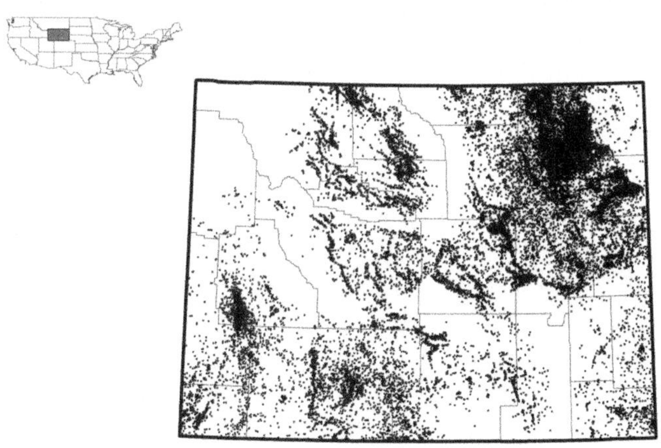

**Figure 2. Point locations of 39,885 active oil and gas well pads from 1991–2011 in Wyoming, USA used to create well pad density maps.**

We used GWR to assess the impact that oil and gas well-pad density had on male sage-grouse lek attendance and to test for possible spatial heterogeneity in the response of sage-grouse to oil and gas development [27].An advantage of the GWR approach is that it evaluates localized patterns in the response variable to localized patterns of predictor covariates. Thus, localized patterns in response are accounted for in a single analytical procedure [27]. This was an important component of this analysis as we expected different patterns in sage-grouse response to oil and gas development across our Wyoming study system, specifically when comparing the northeast to southwest portions of the state (T. Christiansen, WGFD Sage-Grouse Program Coordinator, personal communication, 2013).

We calculated separate GWR analyses for each response variable; thus, we had 5independent sets of GWR model results (3×10, 5×10, 10×10, 2011 Count, and Binomial). For predictor variables we used well pad densities at 0.5, 1, 2, 5, and 10 km surrounding leks, and tested for time lags of 1, 3, 5, and 10 years. These values were chosen based on observations that previous studies had found impacts at such temporal and spatial scales [11–14]. We used goodness-of-fit tests and model significance (Pseudo-$R^2$) values to rank suitability of response variables. A note about pseudo-$R^2$ values: pseudo-$R^2$ values are not the same as $R^2$, and cannot be directly interpreted as proportion of explained variance, their interpretation is actually more similar to an AIC value except that the larger the value the more support for that particular model; assuming that the data set from which the models are derived have similar magnitudes and variance [35].A second difference is that pseudo-$R^2$ values are not bounded 0-1, they can be negative and they can be >>1.0 [35]. This later issue is a particularly salient point to consider when dealing with pseudo-$R^2$ values generated via a Monte Carlo simulation procedure as was done for our analysis within GWR [27]. Thus, comparisons of pseudo-$R^2$ values to determine strength of model support are only appropriate for models using as a response variable the regression trends across LOWESS transformed point estimate (Table 1).

With the most supported modelwe used *post hoc*Monte Carlo procedures for goodness-of-fit tests to test for significant impacts of predictor variables that were driving model performance [36]. We generated parameter estimates and *Z*-scores for each predictor covariate at each lek point. We then mapped predicted trends in sage-grouse lek count values back onto the oil and gas development density maps of Wyoming for only those leks that had a significant response to the specified oil and gas development density. We then used inverse distance weighting spatial interpolation as implemented in Arc Info 10 (Environmental Systems Research Institute, Redlands, CA; [37]) to create a map of sage-grouse heterogeneic response to oil and gas development density by year that were identified as important in the most supported model.

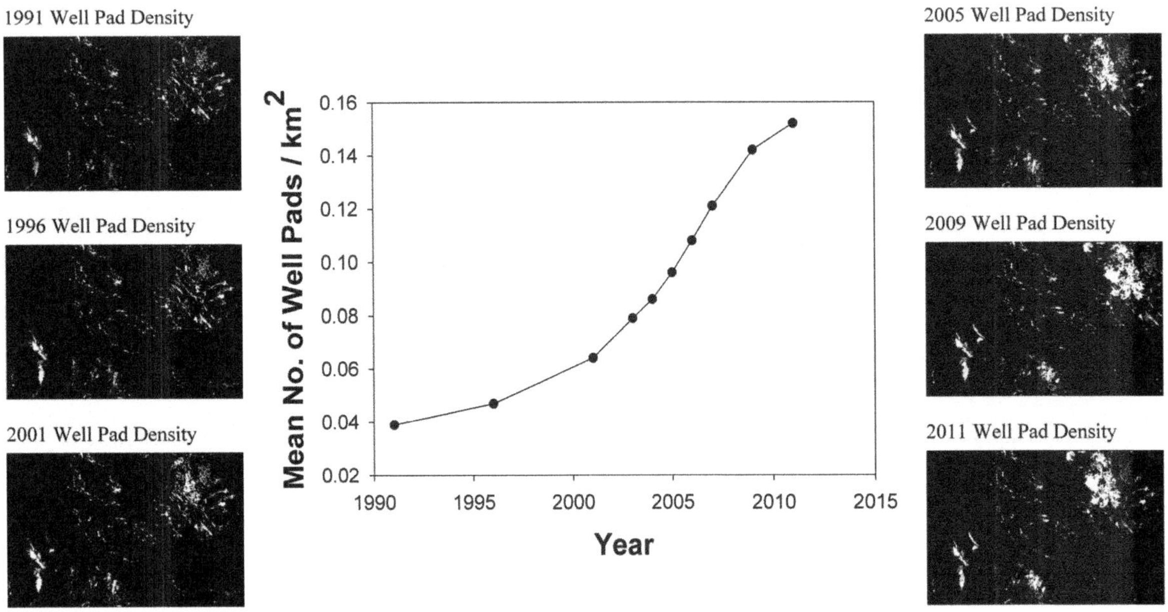


**Figure 3. Well pad density maps created from the WOGCC well pad data from 1991–2011, Wyoming, USA.** Maps depict relative well pad densities (number of well pads/km$^2$) across Wyoming from 1991–2011. White areas indicate areas of higher well pad density. Values range from 0 well pads/km$^2$ to 50.6 well pads/km$^2$, which was the highest observed density in 2011. Well pad densities plotted along the line plot, are the average well pad densities across Wyoming. Thus, the 3.6-fold increase in observed well pad density from 1991–2011 suggests not only an intensification of development within developed areas, but also an increase in area that was under development for oil and gas production.

Finally, we used the functional response curve of sage-grouse lek attendance to oil and gas development densities to forecast predictions about how sage-grousewill likely respond in the future to current development densities. This was done to aid in setting regional management recommendations for sage-grouse in light of ongoing oil and gas extraction within sage-grouse core areas in Wyoming. All statistical analyses were conducted in either Program R [38] or Program GWR [36].

## Results

From 1991–2011 we observed a 3.6-fold increase in the median well pad density across Wyoming (Figure 3). Over this sameperiod,at the 814 leks in our analysis,we observed a 23.8% decline (from 18,631 to 14,185 male sage-grouse) in male sage-grouse lek attendance. However, spatial analysis suggested that the rate of loss was not uniform across the state, and ranged from –4.2 males/lek/year average loss to +2.8 males/lek/year average gainacross the 20 years of our analysis at our 814 leks (Figure 4). It is important to note that while some leks were observed to increase, these were exceptions and not the norm.

We specifically used GWR to identify the appropriate response period and region over which a particular oil and gas development density had the strongest impact on male sage-grouse lek attendance. Analysis with GWR suggested that the5-year response period (adjusted for periodicity with LOWESS) form 2007–2011 showed the strongest response to oil and gas development ($F = 4.16$, $P = 0.042$; Table 1). The 5-year response period GWR analysis suggested that a model including spatial heterogeneity had 2.2-fold better performance than a spatial regression model that did not take spatial heterogeneity of response factors into consideration (spatial regression $AIC_C$ = 9,327; GWR $AIC_C$ = 4,238). *Post hoc* Monte Carlo significance tests indicated that there were 4 predictor covariates with a significant impact on the

**Table 1.** Results of GWR analyses with each response factor for the most parsimonious period over which the impacts of oil and gas development were observed onsage-grouse lek attendance,2002–2011, Wyoming, USA.

| Model | Leks | No. Parameters | Pseudo-R$^2$ |
|---|---|---|---|
| 5×10 | 814 | 47.8 | 0.32 |
| 10×10 | 814 | 30.8 | 0.07 |
| 3×10 | 814 | 65.9 | 0.17 |
| *Binary active/inactive | 1,298 | 31.3 | 0.13 |
| *2011 lek count | 2,012 | 1,276.0 | 451.20 |

*Analyses were conductedon data sets with different number of leks, data magnitudes, and unequal variance structures to LOWESS adjusted models; thus Psuedo-R$^2$ value comparison to LOWESS/regression models is not appropriate. Models are presented for comparison purposes only.

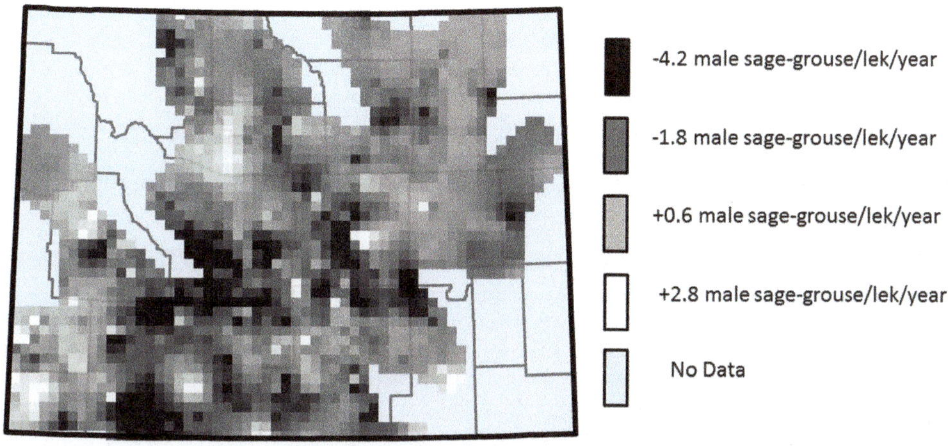


**Figure 4. Spatial pattern in male sage-grouse lek declines from 2007–2011 across Wyoming, USA.**

spatial heterogeneity of the GWR model performance (Table 2). Specifically, the Monte Carlo analysis suggested that heterogeneity in male sage-grouse lek attendance was most strongly associated with development density within 1, 2, and 10 km of the lek, 1 and 4 years (2006 and 2003, respectively) prior to the start of our 5-year response period.

Based on the results of the Monte Carlo analysis, we examined individual lek response to oil and gas development density at 1, 2, and 10 km. For each individual lek, GWR computes an independent *t*-test of significance. We used an inverse distance weighted function to characterize the spatial distribution over which development density at each significant parameter (1, 2, and 10 km and 1 and 4 year time lags) had a significant localized response (Figure 5). We then used the 5-Year LOWESS functional response for each lek and back calculated to determine the rate of change in terms of number of males per lek per year that were associated with development density in each region (Figure 5). Based on the observed rates of change in male sage-grouse lek attendance observed for each development density and time lags (Figure 5), we extrapolated what the expected response would be given current oil and gas well pad development density.

To extrapolate the effect of well pad density on male sage grouse lek attendance we created plots of observed lek attendance versus observed well pad density for each focal region surrounding leks and then calculated a trend line for each data plot. For development density within 1 km of leks with a 1-year time lag, the trend line was:*change in male lek attendance* = −1.45×well pad density −3.3. For development density within 2 km of leks with a 1-year time lag, the trend line was:*change in male lek attendance* = −5.98 ×well pad density −0.52. For development density within 1 km of leks with a 4-year time lag, the trend line was:*change in male lek attendance* = −7.7× well pad density +3.02. Finally, for development density within 10 km of leks with a 4-year time lag, the trend line was:*change in male lek attendance* = −2.1× well pad density −0.02. In2011, the average well pad density within 1 km of leks was 0.08 well pads/$km^2$, within 2 km of leks the average well pad density was 0.13 well pads/$km^2$, and within 10 km of leks the average well pad density was 0.17 well pads/$km^2$.

**Table 2.** GWR *post hoc* Monte Carlo analysis of covariate influence on GWR model performance, 1996–2007, Wyoming, USA.

| Time Lag x Well Pad Density | Parameter Estimate | Monte Carlo *P* |
|---|---|---|
| Intercept | −2.35 to −1.13 | 0.00 |
| 10 year; 10 km | −18.44 to −11.12 | 0.50 |
| 10 year; 5 km | −4.83 to −2.84 | 0.88 |
| 10 year; 1 km | −8.94 to −4.95 | 0.50 |
| 10 year, 500 m | −2.93 to −1.58 | 0.56 |
| *4 year; 10 km | −19.40 to −12.46 | 0.02 |
| 4 year; 5 km | −22.32 to −11.05 | 0.30 |
| *4 year; 1 km | −1.32 to −2.79 | 0.03 |
| 4 year; 500 m | −1.50 to −0.45 | 0.56 |
| 1 year; 10 km | −13.24 to −6.05 | 0.21 |
| 1 year; 5 km | −28.11 to −17.52 | 0.42 |
| *1 year; 1 km | −2.79 to −1.32 | 0.00 |
| *1 year; 2 km | −5.00 to −3.02 | 0.04 |

* Indicates parameters that had a significant influence on GWR model performance and thus where further analysis of the regional variation in covariate performance was informative.

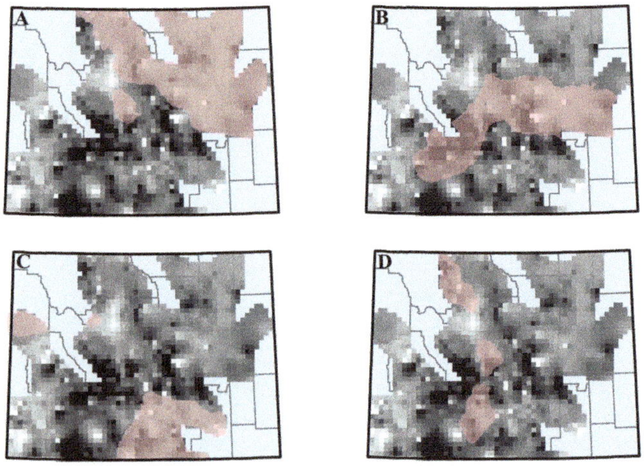


**Figure 5. Spatial variation in response of male sage-grouse lek counts from 2007–2011 to increases in oil and gas well pad development density from 1996–2011, across Wyoming, USA.** Lag effect in lek attendance was 1 year for the 2006 temporal scale and 4 years for the 2003 temporal scale. A = spatial region where 2007–2011 male sage-grouse lek attendance was significantly associated with well pad development density within 1 km of leks in 2006. B = spatial region where 2007–2011 male sage-grouse lek attendance was significantly associated with well pad development density within 2 km of leks in 2006. C = spatial region where 2007–2011 male sage-grouse lek attendance was significantly associated with well pad development density within 1 km of leks in 2003. D = spatial region where 2007–2011 male sage-grouse lek attendance was significantly associated with well pad development density within 10 km of leks in 2003. Average well pad density for A = 0.14 well pads/km$^2$. Average rate of lek attendance change for A = −1.89 males/lek/year. Average well pad density for B = 0.14 well pads/km$^2$. Average rate of lek attendance change for B = 1.63 males/lek/year. Average well pad density for C = 0.1 well pads/km$^2$. Average rate of lek attendance change for C = −0.82 males/lek/year. Average well pad density for D = 0.5 well pads/km$^2$. Average rate of lek attendance change for D = −2.29 males/lek/year.

## Discussion

From 1991–2011 we observed a 3.6-fold increase in the median well pad density across Wyoming, which was spatially associated with a 23.8% decline in the number of male sage-grouse counted at leks. During about this same time period (1997–2007) a range-wide sage-grouse lek count analysis of 3,679 leks in seven sectors indicated that 44% of leks declined, whereas 56% of leks increased [39]. Distance to nearest oil and gas development site was included as a possible explanatory covariate in this analysis and the two sectors in Wyoming had a distance to nearest oil and gas development an order of magnitude closer than in other sectors and also the largest declines in sage grouse lek count [39]. Our study in conjunction with the Johnson et al. [39] study suggests a dramatic decline in sage-grouse populations in Wyoming that is at least partially due to intense oil and gas development in that state.

Our findings indicated that male lek attendance across Wyoming varied spatially and temporally with the density of oil and gas well pads. We also identified similar development densities wherein male sage-grouse lek attendance increased or decreased, that is there were a few leks where male lek counts remained stable or actually increased slightly in spite of high density oil and gas development. As with other studies, our analysis indicated that a 4-to-5 year lag occurs between the time that oil and gas development reaches a particular density to when population-level sage-grouse responses are observed [12–14]. This lag period makes sense in terms of sage-grouse philopatry to lek sites and sage grouse biology. Although some males recruit to a lek at year one, in general it takes 2-3 years for male sage grouse to recruit to a lek and once they recruit they have relatively high adult survival as well as high philopatry to the lek site [7,12,13]. If oil and gas development leads to greatly reduced recruitment to a lek, but not complete reproductive failure, then philopatry to that lek and high annual survival of adult males ensures a relatively stable lek for several years into the future, and hence a lag to an observable impact on lek count. We also identified spatial regions where particular oil and gas development densities were having a significant impact on sage-grouse populations.

From 1991 to 2011 we observed a nearly 4-fold increase in oil and gas well-pad density across the state of Wyoming. In addition to the observed impacts on male sage-grouse lek attendance reported in this paper, such an increase in oil and gas development may also have incurred population-level impacts on other rangeland species. For example,oil and gas development in western Wyoming reduced mule deer (*Odocoileushemionus*) use of wintering habitat [40,41], and led to declines in abundance of sagebrush obligate songbirds [42]. Understanding and mitigating the effects of oil and gas development density for sage-grousemay also provide benefits for these other species, as sage-grouse are typically described as a landscape [7,39]or umbrella [43] species.

In our analysis, we observed that oil and gas development had both a different spatial and temporal scale of impact on male sage-grouse lek attendance. By evaluating the observed development density for those leks in our analysis with observed declines, we can better understand the influence that oil and gas development has on lek persistence. For example, our analysis suggests that leks in the northeastern portion of the state had an immediate (within 1 year) response to oil and gas development density within 1–2 km of the lek site (Figures 5A,B). One reason for this is likely due to lek size, in terms of number of males attending the leks in this region. Leks in this region were small, typically 4–8 birds (WGFD lek database data). Using the link function we were able to back calculate from our observed rates of decline in this region of 25–80% to an average decline of 1–4 males/lek/year (Figure 4). Conversely, in the southwestern portion of Wyoming, leks were typically larger ($12^+$ males/lek; WGFD lek database data). At small leks (4–8 birds), a 20–25% decline is likely a much more severe loss than is a similar rate of decline at large leks (12–20 birds). Moreover, at a small lek, a loss of a single bird will necessarily also account for a larger percent decline in lek attendance. Thus, high rates of decline noted in northeastern Wyoming, while representing only a small portion of the total state-wide sage-grouse population, may still be of significantly greater concern than observed rates of decline at other regions of this analysis because they could lead to more rapid extirpation of the species from that region. Specifically, we observed immediate effects of oil and gas development density on male sage-grouse lek attendance based on significant declines in attendance associated with development density occurring within 1 km and 2 km of leks in 2006, 1 year prior to our analysis period. Over this time frame, male sage-grouse lek attendance declined in areas where well-pad densities exceeded 0.19 well-pads/km$^2$ within 1 km of leks, and ≥2.13 well-pads/km$^2$ within 2 km of leks, particularly in the northeastern portion of Wyoming (Figures 5A,B). Consequently, our data suggest that to avoid immediate declines in male sage-grouse lek attendance, no active oil and gas well-pads should be placed within 2 km of leks. This is particularly true in the northeastern portion of Wyoming where lek sizes are typically small, and lek attendance responds rapidly to oil and gas development density. This has important management implications, as previous work has shown that reduction in lek attendance due to energy development in this region may increase sage-grouse susceptibility to other stressors such as West Nile Virus [44].

In addition, we also observed a temporal lag-affect in male sage-grouse lek attendance associated with oil and gas development. Specifically, we observed that the 5-year trend in male sage-grouse lek attendance was responding to oil and gas development that had occurred 4 years prior at two different spatial scales of 1 km and 10 km. What this means is that when we evaluated the trend in male sage-grouse lek attendance from 2007–2011 it responded strongly to the well-pad density that was present on the landscape in 2004. It also suggested that the spatial area over which sage-grouse were sensitive to oil and gas development was both local, within 1 km of leks, and broad, within 10 km of leks. Thus, leks with significant declines in male sage-grouse attendance had oil and gas development densities $\geq$0.06 well-pads/km$^2$ within 1 km of leks and $\geq$0.7 well-pads/km$^2$ within 10 km of leks.Leks with oil and gas development densities at or above these levels are expected to show declines in male lek attendance 4–9 years after development occurs. Previous studies have also noticed similar lag periods to effect [12,13], and presumably this lag effect is related to the time needed for juvenile sage-grouse to mature to reproductive age and recruit to a lek [13]. Because well-pad development density $\geq$0.7 well-pads/km$^2$ (7 well-pads within 10 km of the active lek site) resulted in significant declines in male lek attendance 9years after development, our data suggest that to avoid long-term declines, possibly associated with failed recruitment, no more than 6 well-pads should be placed within 10 km of active leks.

We also identified 51 lek locations where identical oil and gas well-pad development density resulted in opposing male lek attendance responses, such that some leks declined, whereas others leksremained stable or increased. The next step in identifying why lek attendance at these leks responded differently to equal intensities of energy development is to evaluate what differences exist in these areas such as development configuration, climate, and land management and evaluate what effect these characteristics have on male sage-grouse lek attendance [14,15].One possible explanation is that differences in well-pad development configuration may result in different sage-grouse lek response to similar overall development density. For example, over a 1,000 ha area, dispersed configurations may lead to high-localized development density, such that every lek is close to at least one and possibly several well pads. Over that same area, clustered configurations would likely lead to some leks being in an area of very high density (6–10 well pads), but most leks being far from any well-pads [14]. This is an intriguing possibility and the degree to which this is true and the consequences for sage-grouse needs to be examined in greater detail.

In addition, based on current development density and the lag to the effects observed for the 814 leks in our analysis, we predict an average of 5.09 male sage-grouse/lek/year decrease over the next 4 years (2012–2015) across Wyoming. Local weather patterns, management actions, and year-to-year variation in sagebrush habitat productivity will undoubtedly affect this predicted rate of loss up or down (*see* [45]). This rate of loss is still sobering, as it suggests that based only on the current development density as of 2011, up to 24% of the 2011 male sage-grouse population may be lost in a short time. However, even though the average response in all regions was negative, not all leks in a given region declined. Moreover, the same development density did not always result in the same functional response in male lek attendance. In 51 instances, identical development densities resulted in very different lek responses (Figure 6). However, it is important to note that development density at these leks may also have had a negative impact as it is possible that some of these leks which maintained stable lek attendance in light of current development density may have in the absence of development increased. While it is impossible with our analysis to directly assess this, it is an intriguing possibility.

Wyoming has undertaken efforts since 2007 under their Sage-Grouse Executive Order (SGEO) to develop regulatory mechanisms within "Core Areas" to provide protection and conservation forsage-grousewithin the State [20]. Then Governor Freudenthal formed a sage-grouse implementation team that recognized 31 sage-grouse"Core Areas" that encompassed approximately 24% of the surface area of the State and provided protections for 82% of the breeding population of sage-grouse (Wyoming Game and Fish Department, Cheyenne, unpublished data). Regulatory mechanisms that direct development within analysis areas established in Core Areas include (1) the number of surface disturbances is notto exceed a density of 1 per 2.6 km$^2$averaged across the disturbance analysis area, (2) total accumulated surface area affected (both existing and proposed) within an analysis area is not to exceed 5%, and (3) permanent surface disturbances may not occur within 1 kmof any active or occupied sage-grouse lek [22].Our finding that sage-grouse lek attendance was negatively impacted by as few as 1 well pad within 2 km of sage-grouse suggests that management for sage-grouse based on regulatory mechanisms provided in the Wyoming SGEOmay represent the absolute maximum sustainable development density.Moreover, secondary findings from our study indicate different spatial

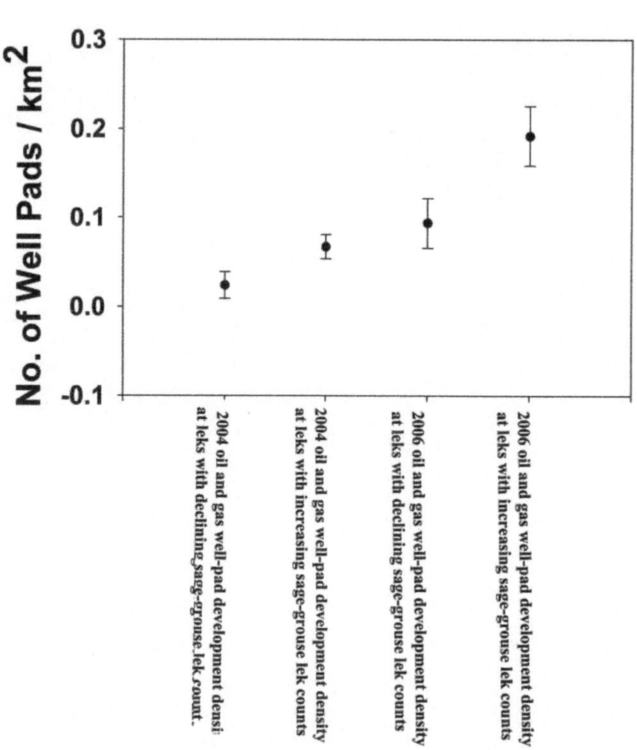


**Figure 6. Variation (mean ± SE) in male sage-grouse lek response from 2007–2011 to identical development density, in 2004 (4-year time lag) and 2006 (1-year time lag), Wyoming, USA.** In 2004 the average sage-grouse lek increase was +3.2 males/lek/year (n = 36 leks) and the average sage-grouse lek decline was −5.7 males/lek/year (n = 153 leks). In 2006 the average sage-grouse lek increase was +3.3 males/lek/year (n = 42 leks) and the average sage-grouse lek decline was −4.9 males/lek/year (n = 125 leks).

configurations of well pads that do not influence lek persistence need to be examined more fully to identify disturbance levels that may be more harmonious with leks inside and outside Core Areas.

## Acknowledgments

We thank M. Deutsch for compiling well pad density maps. T. Christiansen and S. Gamo provided helpful comments to an earlier draft. We also thank the Wyoming Game and Fish Department for their support of this research. E. Blomquist and C. Hagen provided helpful comments as reviewers that improved our manuscript.

## Author Contributions

Conceived and designed the experiments: AJG JLB. Performed the experiments: AJG JLB. Analyzed the data: AJG. Contributed reagents/materials/analysis tools: AJG JLB. Wrote the paper: JLB AJG.

## References


1. Ellis EC, Ramankutty N (2008)Anthropogenicbiomes of the World, Version 1. Palisades, NY: NASA Socioeconomic Data and Applications Center (SEDAC). Available: http://sedac.ciesin.columbia.edu/data/set/anthromes-anthropogenic-biomes-world-v1.Accessed 4 April 2013.
2. Noss RF, Cooperrider AY (1994) Saving natures legacy: Protecting and restoring biodiversity. Washington DC: Island Press. 416 p.
3. Fahrig L (2003) Effects of habitat fragmentation on biodiversity.Annu Rev EcolEvolSyst 34: 487–515.
4. Hagen CA (2010) Impacts of energy development on prairie grouse ecology: a research synthesis. Trans75th N AmWildlNatResourConf 75: 96–103.
5. Schroeder MA, Aldridge CL, ApaAD BohneJR, Braun CE, et al. (2004) Distribution of sage-grouse in North America. Condor 106: 363–376.
6. Hornady WT (1916) Save the sage grouse from extinction, a demand from civilization to the western states. New York ZoolPark Bull5: 179–219.
7. Connelly JW, Hagen CA, Schroeder MA(2011)Characteristics and dynamics of greater sage-grouse populations. Stud Avian Biol 38: 53–67.
8. USFWS (2010) Endangered and threatened wildlife and plants; 12 month findings for petitions to list the greater sage-grouse(*Centrocercusurophasianus*) as threatened or endangered. Federal Register 75: 13909–14014.
9. International Energy Agency, 2008World energy outlook 2008. Available: http://www.worldenergyoutlook.org.Accessed 4A April 2013.
10. Walker BLNaugle DE, Doherty KE (2007) Greater sage-grouse population response to energy development and habitat loss. JWildl Manage 71: 2644–2654.
11. Doherty KE (2008) Sage-grouse and energy development: Integrating science with conservation planning to reduce impacts [Dissertation]. Missoula: University of Montana.125 p.
12. Harju SM, Dzialak MR, Taylor RC, Hayden-Wing LD, WinsteadJB (2010) Thresholds and time lags in effects of energy development on greater sage-grouse populations. JWildlManage. 74: 437–448.
13. Holloran MJ, Kaiser RC, Hubert WA (2010) Yearling greater sage-grouse response to energy development in Wyoming. JWildlManage 74: 65–72.
14. Doherty KE, Naugle DE, Evans JS (2010) A currency for offsetting energy development impacts: horse-trading sage-grouse on the open market. PLoSONE 5(4): e10339. doi:10.1371/journal.pone.0010339
15. Hess JE, Beck JL (2012) Disturbance factors influencing greater sage-grouse lek abandonment in north-central, Wyoming. J WildlManage 76: 1625–1634.
16. Doherty KE, Naugle DE,Copeland HE, Pocewicz A, Kiesecker JM (2011) Energy development and conservation tradeoffs: Systematic planning for greater sage-grouse in their eastern range. Stud Avian Biol 38: 505–516.
17. American Gas Association (2013) State profiles in natural gas. Available: http://www.aga.org/Kc/analyses-and-statistics/state_profiles/Pages/default.aspx. Accessed17 February 2013.
18. American Wind Energy Association (2013) State fact sheets. Available: http://www.awea.org/learnabout/publications/factsheets/factsheets_state.cfm. Accessed 17 February 2013.
19. CopelandHE,Doherty KE, NaugleDE,PocewiczA,Kiesecker JM (2009) Mapping oil and gas development potential in the US Intermountain West and estimating impacts to species. PLoS ONE 4(10): e7400. doi:10.1371/journal.pone.0007400
20. State of Wyoming (2008) Office of Governor Freudenthal.State of Wyoming Executive Department Executive Order.Greater Sage Grouse Area Protection.2008-02. Available: http://will.state.wy.us/sis/wydocs/execorders/WS-GOV-EO-2008-02.pdf.Accessed 1 April 2014.
21. State of Wyoming (2010) Office of Governor Freudenthal.State of Wyoming Executive Department Executive Order.Greater Sage Grouse Area Protection.2010-04. Available: http://will.state.wy.us/sis/wydocs/execorders/WS-GOV-EO-2010-04.pdf.Accessed 1 April 2014.
22. State of Wyoming (2011) Office of Governor Mead.State of Wyoming Executive Department Executive Order.Greater Sage Grouse Area Protection.2011-05. Available: http://will.state.wy.us/sis/wydocs/execorders/EO2011-05.pdf.Accessed 1 April 2014.
23. Getis A, Ord JK (1992) The analysis of spatial association by use of distance statistics. Geogr Anal 24: 189–206.
24. Tobler WR (1970) A computer movie simulating urban growth in the Detroit region. EconGeogr 46: 234–240.
25. Charlton M, Fotheringham AS (2009) Geographically weighted regression white paper. National Centre for Geocomputation, National University of Ireland Maynooth.Maynooth, County Kildare, Ireland. Available: http://gwr.nuim.ie/downloads/GWR_WhitePaper.pdf.Accessed 1 April 2014.
26. Anselin L (1988) Spatial econometrics: Methods and models. Boston: Kluwer Academic Publishers.284 p.
27. Fotheringham AS, Brunsdon C, Charlton M (2002) Geographically weighted regression: The analysis of spatially varying relationships.Chichester, United Kingdom: John Wiley and Sons Incorporated. 284 p.
28. Fedy BC, Aldridge CL (2011) The importance of within-year repeated counts and the influence of scale on long-term monitoring of sage-grouse. JWildl-Manage 75: 1022–1033.
29. Schwarz CJ (2001) The Jolly-Seber model: more than just abundance. J AgricBiol Environ Stat 6: 195–205.
30. Jenni DA, Hartzler JE 1978. Attendance at a sage grouse lek: implications for spring censuses. J WildlManage 42: 46–52.46
31. Wyoming Oil and Gas Conservation Coalition (WOGCC) (2010) Available: http://wogcc.state.wy.us/.Accessed13August, 2011.
32. Cleveland WS, Grosse E, Shyu M-J (1996) A package of C and Fortranroutines for fitting local regression models: Loess user's manual. Berkeley Heights, New Jersey: *Bell Labs, Technical Report.* Available: http://stat.bell-labs.com/wsc/publish.html. Accessed 1 April 2014.
33. Cleveland WS (1979) Robust locally weighted regression and smoothing scatterplots, J AmStatAssoc 74(368): 829–836.
34. HurvichCM,Simonoff JS,Tsai C-L (1998) Smoothing parameter selection in nonparametric regression using an improved Akaikeinformation criterion.R Stat SocSeriesB Stat Methodol 60: 271–293.
35. Long JS, Freese J (2006) Regression models for categorical dependent variables using Stata, 2nd edition. College Station: Stata Press.527 p.
36. CharltonM,Fotheringham S,Brunsdon C (2003) GWR 3: software for geographically weighted regression.Spatial Ecology Research Group, Department of Geography, University of Newcastle upon Tyne, Newcastle upon Tyne, United Kingdom. Available: http://www.uvm.edu/rsenr/gradgis/GWR3Manual.pdf. Accessed 1 April 2014.
37. ZimmermanD,Pavlik C,Ruggles A,Armstrong MP (1999) An experimental comparison of ordinary and universal kriging and inverse distance weighting. MathGeol 31: 375–390.
38. R Core Team. (2012) R: A language and environment for statistical computing.R Foundation for Statistical Computing, Vienna, Austria. Available: http://www.R-project.org.Accessed June 20 2012.
39. JohnsonDH, Holloran MJ,Connelly JW,Amundson LA,Knick ST (2011) Influence of environmental and anthropogenic features on greater sage-grouse populations, 1997–2007. StudAvian Biol 38: 407–450.
40. Sawyer H, Nielson RM,Lindzey F, McDonald LL (2006) Winter habitat selection of mule deer before and during development of a natural gas field. JWildlManage 70: 396–403.
41. Sawyer H, Kauffman MJ, Nielson RM (2009) Influence of well pad activity on the winter habitat selection patterns of mule deer. JWildl Manage 73: 1052–1061.
42. Gilbert MM, Chalfoun AD (2011) Energy development affects populations of sagebrush songbirds in Wyoming. JWildlManage 75: 816–824.
43. Gamo RS, Carlisle JD,Beck JL,Bernard JAC,Herget ME (2013) Greater sage-grouse in Wyoming: an umbrella species for sagebrush dependent wildlife.Wildl-Profess7(1):56–59.http://www.uwyo.edu/esm/faculty-and-staff/beck/_files/docs/publications/spring-2013-rampage-pages 19–21.pdf.
44. Taylor RL, Tack JD, Naugle DE, Mills LS (2013) Combined effects of energy development and disease on greater sage-grouse. PLoSONE 8(8): e71256. doi:10.1371/journal.pone.0071256
45. Fedy BC, Doherty KE (2011) Population cycles are highly correlated over long time series and large spatial scales in two unrelated species: greater sage-grouse and cottontail rabbits. Oecologia 165: 915–924.

2

# Geographical and Temporal Body Size Variation in a Reptile: Roles of Sex, Ecology, Phylogeny and Ecology Structured in Phylogeny

**Pedro Aragón[1]*, Patrick S. Fitze[2,3,4,5]**

**1** Departamento de Biogeografía y Cambio Global, Museo Nacional de Ciencias Naturales, Madrid, Spain, **2** Departamento de Biodiversidad y Biología Evolutiva, Museo Nacional de Ciencias Naturales, Madrid, Spain, **3** Université de Lausanne, Department of Ecology and Evolution (DEE), Biophore, Lausanne, Switzerland, **4** Instituto Pirenaico de Ecología (IPE-CSIC), Avenida Regimiento de Galicia s/n, Jaca, Spain, **5** Fundación Araid, Edificio Pignatelli, Zaragoza, Spain

## Abstract

Geographical body size variation has long interested evolutionary biologists, and a range of mechanisms have been proposed to explain the observed patterns. It is considered to be more puzzling in ectotherms than in endotherms, and integrative approaches are necessary for testing non-exclusive alternative mechanisms. Using lacertid lizards as a model, we adopted an integrative approach, testing different hypotheses for both sexes while incorporating temporal, spatial, and phylogenetic autocorrelation at the individual level. We used data on the Spanish Sand Racer species group from a field survey to disentangle different sources of body size variation through environmental and individual genetic data, while accounting for temporal and spatial autocorrelation. A variation partitioning method was applied to separate independent and shared components of ecology and phylogeny, and estimated their significance. Then, we fed-back our models by controlling for relevant independent components. The pattern was consistent with the geographical Bergmann's cline and the experimental temperature-size rule: adults were larger at lower temperatures (and/or higher elevations). This result was confirmed with additional multi-year independent data-set derived from the literature. Variation partitioning showed no sex differences in phylogenetic inertia but showed sex differences in the independent component of ecology; primarily due to growth differences. Interestingly, only after controlling for independent components did primary productivity also emerge as an important predictor explaining size variation in both sexes. This study highlights the importance of integrating individual-based genetic information, relevant ecological parameters, and temporal and spatial autocorrelation in sex-specific models to detect potentially important hidden effects. Our individual-based approach devoted to extract and control for independent components was useful to reveal hidden effects linked with alternative non-exclusive hypothesis, such as those of primary productivity. Also, including measurement date allowed disentangling and controlling for short-term temporal autocorrelation reflecting sex-specific growth plasticity.

**Editor:** Marco Festa-Bianchet, Université de Sherbrooke, Canada

**Funding:** PA was supported by the Swiss National Science Foundation (SNF) and by a Ramón y Cajal contract (RYC-2011-07670). Financial support was provided by the SNF (PPOOP3_128375 to PSF) and by the Comunidad de Madrid, Spain (200530M090 to PSF). The funders had no role in study design, data collection and analysis, decision to publish, or preparation of the manuscript.

**Competing Interests:** The authors have declared that no competing interests exist.

* Email: paragon@mncn.csic.es

## Introduction

Spatial and temporal variation in phenotypic traits is a central issue in ecology and evolution. Geographical variation of body size along environmental gradients has been studied for more than 160 years [1]. This topic is currently the focus of intense interest, including in the context of the response of organisms to environmental change [2], and a major challenge is disentangling among (phylo)genetic and ecological factors [3, 4].

In endotherms, a positive association of body size with latitude and/or elevation, or a negative association with temperature, the so-called Bergmann's rule [1], has been frequently observed (e.g. [5]). An adaptive thermoregulatory explanation for this pattern was originally proposed by Bergmann [1] for which the area-volume ratio of larger sizes facilitates heat conservation in colder environments. However, it has been suggested that other non-exclusive mechanisms can be involved and even confounded with thermoregulatory mechanisms ([5–7] and references therein).

For ectotherms, there is a variety of results involving not only the underlying processes but also the observed patterns at the intraspecific and interspecific levels. In reptiles, different studies including different sources of variability reported contrasting results [8–13]. Moreover, body size differences between sexes along geographical gradients may exist [14], but have rarely been considered in reptiles and only over fragments of their ranges [15]. All of these sources of variability can yield contrasting results, and have contributed to the ongoing debate on the definition of Bergmann's rule [16–19].

Although latitude can explain geographical patterns at coarse scales, an increasing number of studies have encouraged the use of different environmental predictors directly linked with different mechanistic hypotheses [20]. Most of the potential processes proposed are related to physiology or biotic interactions. In ectotherms, the effects of climate and primary productivity on body size can be linked with thermoregulatory strategies, developmental dynamics, activity duration, and/or food availability. Several of these mechanisms are not mutually exclusive and might be acting jointly (e.g. [7]). Patterns and mechanisms commonly considered in previous studies are:

1) Bergmann's rule [8, 11]: Larger sizes facilitate heat conservation in colder environments. This rule predicts a negative association between temperature and size.
2) Converse-Bergmann's rule [8]: Smaller sizes facilitate higher heating rates in colder environments. It is predicted a positive association between temperature and size.
3) Temperature-size rule [21, 4]: Growth retardation during developmental stages in colder environments involves prolonged growth periods and a delay in maturation, which results in a larger size at maturity. This rule predicts a negative association between temperature and adult size.
4) Seasonality hypotheses: Potential effects of seasonality have been proposed also for opposite size trends in ectotherms. Higher temperatures and/or lower precipitation may reflect longer periods for growing due to an extended activity [22, 23]. This mechanism would predict a positive association of size with temperature and a negative association with precipitation. Alternatively, the opposite trend would be also plausible through the joint effects of seasonality and the temperature-size rule; where reduced activity is associated with higher survival, favouring in turn selection for a delayed maturation at larger sizes [24].
5) Resource rule [25, 26]: Higher abundance and/or quality of prey during the growing season allow attaining larger sizes, and predict a positive association between primary productivity and size. Also, potential joint effects of productivity and seasonality on body size have been considered for ectotherms [11, 27].

Whatever the mechanisms involved, several studies on the geographical body size variation suggested that different possibilities should be considered according to the idiosyncrasies of species [11,28].

On the other hand, while studies at the inter-specific level often provide analyses accounting for phylogenetic relationships [8], geographical studies of body size variation accounting for phylogenetic relationships among populations are scarce [9, 28] and are lacking at the individual level. Indeed, population-level studies in a phylogenetically explicit context may provide information on different evolutionary scenarios. Most phylogenetic studies use techniques to control for phylogenetic autocorrelation [29, 8, 9], whereas the application of variation partitioning methods to disentangle the independent contributions of ecology and phylogeny from phylogenetic niche conservatism has been only applied to higher taxonomic levels [6]. To our knowledge, there is no study of reptiles on geographical variation in body size that tests for different mechanistic hypotheses while simultaneously accounting for spatial, temporal and phylogenetic autocorrelation for each sex and at the individual level.

In this study, we collected field data and assessed the role of different sources of body size variation in males and females of the Spanish Sand Racer species group. A main challenge in studies of body size variation in vertebrates is accounting for the influences of past inter-annual fluctuations of environmental factors experienced through their entire lives. Thus, the longer their life span the more complex is their trajectory to attain such sizes. The rapid life cycle of sand racers makes them a suitable model to account for short-term temporal variation in body size because most of the environmental influences occur within one year, and age variability of adults is very small. Firstly, we performed a simple ecological model including climatic, topographic and primary productivity predictors that may be linked with previously proposed mechanisms involved in body size variation along gradients. Additional analyses were performed using available bibliographic data and two additional sampled populations, encompassing a time interval of 95 years, to assess the generality of our field-based data. Then, to distinguish between independent and shared effects of ecology and phylogeny [6], we applied a variation-partitioning methodology including ecological variables and pairwise genetic distances at the individual level. Our final models included relevant ecological predictors, while incorporating phylogenetic, spatial and temporal autocorrelation. The main objective of this study was to examine whether alternative sources of variation can emerge only after disentangling and accounting for the independent and shared components of ecology and phylogeny.

Spanish Sand Racers are thermophilic lizards whose distribution encompasses high environmental heterogeneity within the Mediterranean part of the Iberian Peninsula (absent from northern Spain) and southern France: (latitudinal range: 36°0′15″-44°32′6″, longitudinal range: 9°29′54″-6°51′00″). Within Lacertidae they belong to the lizards with smaller size, shorter life span (1–2 years), and earlier sexual maturity (5–9 months, no subadult stage) [30–32]. Females are larger than males and they have 1–2 clutches per year [30, 31, 33–35]. Until 2012, Spanish Sand Racers were considered as belonging to one single species with some morphological differentiation among populations. In 2012 three major lineages have been described based on phylogenetic analyses. These lineages diverged 4.8 and 8.3 Mya, and differ in morphology and realised ecological niche [36]. Recently they have been elevated to the species level (*Psammodromus hispanicus*, *P. occidentalis* and *P. edwardsianus*) [37, 38].

Considering the above-mentioned main hypotheses without joint effects, several predictions can be made for sand racers. Under Bergmann's and the temperature-size rules we predicted a negative association between temperature and adult body size. Under the seasonality hypothesis we predicted a positive association of body size with temperature, and a negative association with precipitation. Under the resource rule we predicted a positive association between primary productivity and body size.

## Methods

### Field sampling

Field captures throughout Spain were performed from April to June 2006. Sand racers exhibit a rapid life cycle. They reach sexual maturity in their first spring, there is almost no cohort overlap and juveniles appear mostly in August [31, 33–35]. During the field-sampling period only adults were observed and captured. Sampling was performed approximately each 250 km within the species distributions (Fig. 1). Environmental variation of sampled populations ranged from 0 to 1355 m a.s.l. for elevation, from 9.4 to 18°C for annual mean temperature, from 246 to 998 mm for annual precipitation and covering a maximum distance of 1005 km. Thus, our sampled populations were representative of the high environmental heterogeneity associated to the species distributions [33–35]. Lizard size (snout-to-vent length, SVL) was measured to the nearest 1 mm immediately after capture, and a small piece of the tail tip was collected and preserved in 70% ethanol at −20°C for DNA extraction. We obtained SVL and

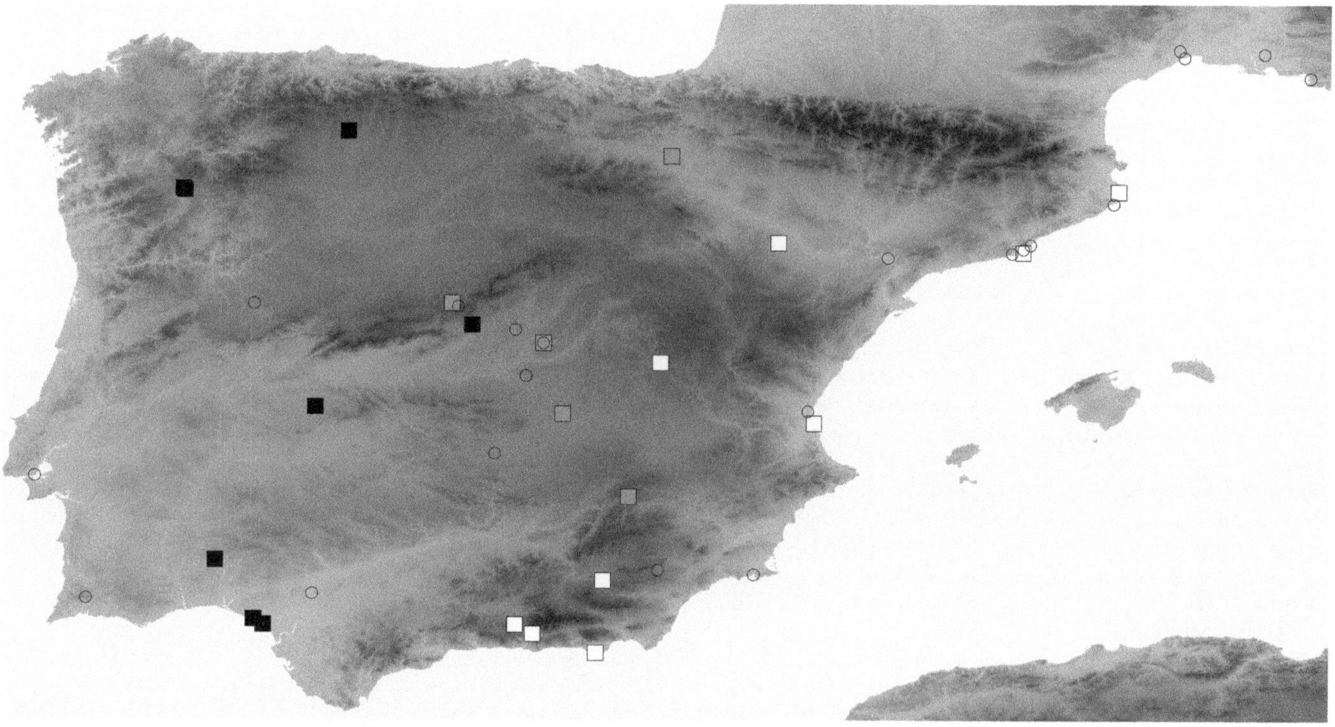

**Figure 1. Sampled localities for the Spanish Sand Racer species group corresponding to data-sets obtained from systematic field sampling of mt-DNA and SVL in 2006 (black squares: *Psammodromus occidentalis*, grey squares: *P. hispanicus*, white squares: *P. edwardsianus*) and a review of SVL data in the literature encompassing approximately 100 years (open circles).** Higher intensity in the grey scale denotes higher elevation.

genetic information for 240 captured adult individuals (130 males and 110 females) from 22 locations; for another 32 captured individuals only genetic information was obtained (see [36] for more details).

Captures and handling of sand racers were accomplished under the permits of Junta de Andalucía, Gobierno de Aragón, Junta de Castilla y León, Junta de Comunidades de Castilla-La Mancha, Generalitat de Catalunya, Junta de Extremadura, Xunta de Galicia, Comunidad de Madrid, Gobierno de Navarra, Generalitat Valenciana, Parque Natural de l'Albufera (Valencia), Parque Natural del Delta del Ebro (Cataluña), Parque Nacional de Doñana (Huelva), and Gobierno de España.

## Genetic distances among individuals

Previous phylogenetic and population genetic studies based on partial sequences of two mitochondrial genes (*cytb*, *nad4*), one nuclear gene (suppressor of SWI4 1) and an unknown nuclear region (clone17) showed congruency of derived phylogenetic trees, i.e., the cytochrome *b* (*cytb*) tree was congruent with the mitochondrial and the nuclear trees and the nuclear trees were as well congruent [36]. Mitochondrial *cytb* sequence data of a 249 bp alignment of partial *cytb* are available from GenBank for all sampled individuals, and sequence data for the other three markers are only available for a small subset [36]. For these reasons, we used the existing Maximum Likelihood phylogenetic tree of *cytb* [36], and calculated cophenetic genetic distances among individuals using the cophenetic function included in the ape package in R 2.7.0 software (Free Software Foundation, GNU Project, Boston, MA, USA). The derived pairwise genetic distance matrix was then used to investigate phylogenetic constraints on geographical body size variation. For additional details concerning the molecular methods used, see Fitze et al. [36].

## Ecological models

We used eight representative ecological predictors: two predictors were related to temperature, two predictors to precipitation, two to plant productivity, one to topography and one with the course of the season. One topographical and five bioclimatic predictors were obtained from the WorldClim source ([39], www.worldclim.org) at 1 $km^2$ resolution, and consisted of mean temperature of warmest quarter, mean temperature of coldest quarter, precipitation of wettest quarter, precipitation of driest quarter and mean elevation. As a measure of primary productivity, we used monthly maps of Enhanced Vegetation Indexes (EVI) generated from Moderate Resolution Imaging Spectroradiometer http://modis.gsfc.nasa.gov/(MODIS) satellite images available at the NASA-LP DAAC web page (https://lpdaac.usgs.gov/). We generated year-averaged monthly values of EVI for each grid-cell from the oldest period available to the sampled year (2000–2006). We then calculated the mean EVI for summer and winter for each cell. Thus, among the available temperature, precipitation and productivity predictors we used those encompassing the annual variation. Although the use of environmental predictors directly linked with processes is encouraged, at least one of the traditionally used pattern predictors (latitude or elevation) may be useful for comparisons with other studies (*s. l.* [20]). For this study elevation was included in models because preliminary analyses showed that it significantly correlated with SVL, which was not the case for latitude (see Appendix S1 for a more detailed justification). Finally, to estimate the effect of time on growth in these short-lived lizards with rapid growth, and to explicitly incorporate temporal autocorrelation, we used the capture date as an indicator of the course of the season.

To avoid multicollinearity or redundancy in our analyses, we performed a principal component analysis (PCA, [40]) and obtained eight orthogonal principal components as potential predictors (PC). Multiple regression analyses were performed for males and females with SVL as the dependent variable and the eight PCs as covariates. Significant PCs were considered relevant for subsequent analyses. To further understand the potential role of each ecological predictor, we performed regression analyses for each predictor having a factor loading >0.7 in relevant PCs. Lastly, to understand whether significant predictors reflected one-year or general trends, we used an independent inter-annual data set encompassing 95 years and correlated the environmental predictors that were significant in the intra-annual data set with body size (see Appendix S2).

## Phylogenetic models

To estimate potential phylogenetic inertia on body size, we applied Phylogenetic eigenVector Regression (PVR, [29]) on SVL. We used a double-centred phylogenetic pairwise distance matrix among individuals to perform a Principal Coordinate Analysis (PCoA, [41]) and obtain independent phylogenetic eigenvectors (PV). PCoA yields independent phylogenetic eigenvectors (PV) in an analogous manner as the PCA yields PCs for the ecological models. Thus, the obtained orthogonal PVs can be used as potential predictors of SVL in a regression frame work. This procedure has been shown to be a powerful tool for developing phylogenetically explicit models of body size gradients at higher taxonomic levels [29, 3]. An advantage of PVR is that PVs obtained from our individual-based genetic distance matrix can encompass genetic influences at different hierarchical levels (species, clades, sub-clades, populations and individuals), which is more appropriate than exclusively including the species level as a categorical factor in the analyses (see Appendix S3 for statistical tests). The number of derived PVs was large (240), and we used three previously considered selection criteria [6, 42]. First, a pre-selection of PVs was based on a descending sequential criterion regarding the cumulative variance explained up to 99%, which yielded the first 30 PVs. Then, multiple regression analyses were performed for males and females with SVL as the dependent variable, to test the significance of these remaining 30 PVs (full models and backward selection yielded same results). Lastly, to further examine phylogenetic autocorrelation, residuals of the final ecological, phylogenetic and combined models were regressed on the 30 retained PVs. This procedure is analogous to that proposed by Bini et al. [43]. Significant PVs in at least one of the analyses were considered relevant for subsequent analyses.

## Disentangling independent and shared components of ecology and phylogeny

To quantify the variance explained by independent components of ecology and phylogeny and by the component shared between them (the phylogenetically structured environmental variation), we used the variation-partitioning method proposed by Desdevises et al. [6], which is analogous to the widely applied general procedure of variation partitioning [44, 45]. The variation explained ($R^2$) by either the ecological or the phylogenetic model is subtracted from the combined model (including both relevant PCs and PVs) to obtain the independent components.

To estimate the significance of the independent components, we followed two steps: 1) extracting the structure not shared between phylogeny and ecology into independent components from the residuals resulting from regressing each relevant PC (dependent variable) on all relevant PVs (independent variables) and each relevant PV on all relevant PCs (hereafter ResPC and ResPV, respectively) and 2) estimating the significance of the independent components obtained in step 1 by performing multiple regressions of SVL on either all ResPC (independent components of ecology) or all ResPV (independent components of phylogeny). See Desdevises et al. [6] for a more detailed description of the overall procedure. Additionally, if a significant ResPC was obtained, residuals were also directly derived for ecological predictors with factor loadings >0.7 (Table 1) in the corresponding PC and tested for significance.

Although the shared component of phylogeny and ecology is potentially very important from an evolutionary perspective (reflecting phylogenetic niche conservatism), its significance cannot be estimated directly [6]. However, an approximation controlling for relevant independent effects should be at least informative. In previous studies it has been proposed that the pure effect of the environment can be obtained from the error term when regressing body size on the relevant PVs [29]. We followed the same philosophy but with a variant procedure to control for the relevant independent effects. We regressed SVL only on those independent components that were significant in the above-explained step 2 (*i.e.*, ResPCs and ResPVs expressing the relevant structure not shared between ecology and phylogeny). In this new regression ($SVL = ResPV_i + \cdots + ResPV_n + ResPC_i + \cdots + ResPC_n + e$), the error term $e$ represents the new resulting residuals. Thus, the residuals $e$ express the remaining variation of SVL that was not explained by the significant independent components. These new residuals $e$ were again regressed on all eight PCs and tested for significance ($e = PC\text{-}1 + \cdots + PC\text{-}8$). In other words, we estimated the effects of ecological factors on the remaining portion of SVL that was not explained by the unshared variation between ecology and phylogeny. In the same way as for the ecological models, we performed a regression analysis for each ecological predictor with a factor loading >0.7 in significant PCs. Overall, the potential roles of different ecological predictors on the pure and shared effects were inferred through changes in significance from models before and after disentangling the pure and shared effects.

SVL of lizards from field and bibliographic data did not differ significantly from a normal distribution, nor did residuals of phylogenetic and ecological models (Kolmogorov-Smirnov d range = 0.05–0.17, $P>0.05$ in all cases). Specific analyses to reduce dimensionality and assess autocorrelation were performed with R ([46], http://www.r-project.org/), and SAM ([47], http://www.ecoevol.ufg.br/sam/) software.

## Spatial autocorrelation in models

We assessed spatial autocorrelation by adding relevant spatial filters to the ecological, phylogenetic and combined models to estimate the extent to which our predictors absorbed the spatial structure of the SVL data ([47], see Appendix S4 for full details).

# Results

## Ecological models

The first four PCs obtained from the PCA explained 90.8% of the total variance (Table 1). In PC-1, four environmental predictors had factor loadings >0.7, where temperature predictors were positively associated, and elevation and EVI summer were negatively associated. In PC-2, PC-3 and PC-4, precipitation of wettest quarter, EVI winter and course of the season, respectively, had the highest factor loadings, all >0.7 and positive (Table 1).

When including all eight PCs in a multiple linear regression as predictors of SVL, PC-1 was significantly and negatively

**Table 1.** Principal components analysis for environmental predictors in Peninsular Spain at a 1 km$^2$-resolution.

| Ecological predictors | PC-1 | PC-2 | PC-3 | PC-4 | PC-5 | PC-6 | PC-7 | PC-8 |
|---|---|---|---|---|---|---|---|---|
| Course of Season | −0.55 | 0.22 | −0.25 | **0.72** | 0.24 | 0.09 | 0.01 | 0.02 |
| Mean Temperature of Warmest Quarter | **0.92** | 0.24 | 0.12 | 0.15 | −0.04 | 0.16 | −0.20 | −0.01 |
| Mean Temperature of Coldest Quarter | **0.83** | 0.51 | 0.16 | 0.03 | 0.02 | 0.01 | 0.09 | 0.13 |
| Precipitation of Wettest Quarter | −0.38 | **0.81** | −0.09 | 0.14 | −0.30 | −0.28 | −0.05 | −0.01 |
| Precipitation of Driest Quarter | −0.49 | 0.51 | −0.42 | −0.50 | 0.26 | 0.06 | −0.07 | 0.04 |
| Elevation | **−0.71** | −0.65 | 0.02 | 0.03 | −0.20 | −0.00 | −0.09 | 0.12 |
| Enhanced Vegetation Index of Winter | −0.57 | 0.17 | **0.74** | −0.03 | 0.28 | −0.11 | −0.05 | 0.00 |
| Enhanced Vegetation Index of Summer | **−0.74** | 0.46 | 0.25 | −0.07 | −0.26 | 0.33 | 0.05 | −0.01 |
| Eigenvalue | 3.59 | 1.95 | 0.90 | 0.82 | 0.40 | 0.24 | 0.07 | 0.03 |
| Variance explained (%) | 44.84 | 24.41 | 11.29 | 10.21 | 5.06 | 2.96 | 0.83 | 0.40 |
| Cumulative Variance (%) | 44.84 | 69.26 | 80.54 | 90.76 | 95.82 | 98.77 | 99.60 | 100 |

Bold lettering denotes factor loadings >0.7.

associated with SVL in both males and females (Table 2). Moreover, significance of PC-1 remained stable upon adding the relevant spatial filters (Table S1 in Appendix S4). However, we found sex differences regarding other PCs. PC-6 was only significant in males, and PC-4 and PC-7 were only significant in females (Table 2). For males, PC-6 remained significant after adding spatial filters with two selection criteria and marginally significant for a third selection criterion. For females, PC-7 was significant for one selection criterion (Table S1 in Appendix S4). Finally, when including only the first four PCs in the analyses, their significance did not change neither for females (PC-1: $b = -0.29$, $t_{105} = -3.18$, $P = 0.002$; PC-2: $b = -0.05$, $t_{105} = -0.60$, $P = 0.552$; PC-3: $b = -0.14$, $t_{105} = -1.54$, $P = 0.126$; PC-4: $b = 0.20$, $t_{105} = 2.20$, $P = 0.030$), nor for males (PC-1: $b = -0.26$, $t_{125} = -2.96$, $P = 0.004$; PC-2: $b = -0.04$, $t_{125} = -0.42$, $P = 0.675$; PC-3: $b = 0.002$, $t_{125} = 0.02$, $P = 0.984$; PC-4: $b = 0.03$, $t_{125} = 0.37$, $P = 0.716$).

When regressing separately those environmental predictors with factor loadings >0.7 in the significant PCs, we found similarities and differences between sexes (Table 3, Fig. 2). Temperature predictors were significantly and negatively associated with SVL of both males and females (Table 3, Fig. 2a, b). Both sexes also showed a significant positive association of their SVL with elevation (Table 3, Fig. 2c). There were no significant associations with primary productivity in either sex. Interestingly, the course of the season was significantly associated only for females (Table 3, Fig. 2d). Lastly, the analyses performed with the independent 95-year data-set confirmed the observed effects of temperature and elevation (see Appendix S2).

## Phylogenetic models

When including the first thirty PVs as predictors of SVL, only PV-2 was significant in males and females (Table 2), either in the full or backward eliminated models. When regressing residuals of ecological, phylogenetic and combined models on the 30 PVs, again only the PV-2 remained significant for both sexes. PV-2 as well remained significant in both sexes upon adding spatial filters, independently of the filter selection criteria (Table S1 in Appendix S4).

## Independent and shared components of ecology and phylogeny

Before the partition, the variation explained by phylogeny was similar in both sexes, whereas the variation explained by ecology was greater in females (Fig. 3a). The final combined models explained 19% and 22% of the variance for males and females, respectively. Variation partitioning again revealed similarities and differences between sexes. The independent component of phylogeny and the overlap between ecology and phylogeny were comparable in magnitude between sexes. The absolute difference between the independent phylogeny and the overlap was 1.74% and 1.13% for males and females, respectively (Fig. 3a). In contrast, the independent component of ecology was larger in females than in males (difference: 3.03%, Fig. 3a).

The independent contribution of phylogeny to SVL was significant in both sexes (males: $F_{1,128} = 11.29$, $P < 0.002$; females: $F_{1,108} = 7.65$, $P < 0.007$) and remained reasonably stable after adding spatial filters with different selection criteria (ResPV-2, Table S1 in Appendix S4). The independent effect of ecology was significant for females ($F_{3,106} = 2.78$, $P < 0.044$) but did not reach significance in males ($F_{2,127} = 2.83$, $P = 0.062$). For females, only ResPC-4 (reflecting the variance in PC-4 that could not be explained by phylogeny) showed significance; ResPC-1 did not (Table 2). ResPC-4 was also significant in a regression upon

**Table 2.** Multiple linear regression analyses of snout-to-vent-length on the ecological (PC) and phylogenetic (PV) factors, and on independent components, *i.e.* ecology independent of phylogenetic factors (ResPC) and phylogeny independent of ecological factors (ResPV).

| | Males | | | Females | | |
|---|---|---|---|---|---|---|
| **Ecological factors** | *b* | *t (121)* | *P* | *b* | *t (101)* | *P* |
| PC-1 | **−0.26** | **−3.04** | **0.002** | **−0.30** | **−3.27** | **0.001** |
| PC-2 | −0.03 | −0.32 | 0.747 | −0.05 | −0.56 | 0.572 |
| PC-3 | 0.01 | 0.09 | 0.926 | −0.14 | −1.55 | 0.123 |
| PC-4 | 0.01 | 0.11 | 0.908 | **0.22** | **2.44** | **0.016** |
| PC-5 | −0.03 | −0.36 | 0.720 | −0.09 | −0.95 | 0.340 |
| PC-6 | **−0.20** | **−2.37** | **0.019** | −0.09 | −1.02 | 0.308 |
| PC-7 | 0.12 | 1.42 | 0.159 | **−0.19** | **−2.06** | **0.042** |
| PC-8 | 0.11 | 1.29 | 0.197 | −0.01 | −0.12 | 0.900 |
| **Phylogenetic factors** | *b* | *t (99)* | *P* | *b* | *t (79)* | *P* |
| PV-1 | 0.06 | 0.58 | 0.559 | 0.15 | 1.76 | 0.081 |
| PV-2 | **0.30** | **2.00** | **0.048** | **0.32** | **3.32** | **0.001** |
| **Independent components** | *b* | *t (127)* | *P* | *b* | *t (106)* | *P* |
| ResPC-1 | −0.07 | −0.75 | 0.452 | −0.13 | −1.34 | 0.181 |
| ResPC-4 | - | - | - | **0.19** | **2.04** | **0.043** |
| ResPC-6 | **−0.20** | **−2.27** | **0.024** | - | - | - |
| ResPC-7 | - | - | - | −0.16 | −1.74 | 0.084 |
| | *b* | *t(128)* | *P* | *b* | *t(108)* | *P* |
| ResPV-2 | **0.28** | **3.36** | **0.001** | **0.26** | **2.77** | **0.006** |

Bold lettering denotes partial correlations significant at $P < 0.05$. Only the first two phylogenetic factors are shown for brevity, the remaining factors were not significant (See methods for details). *b*: denotes standardized regression coefficients.

adding the spatial filters with Moran's $I$ >0.1 (Table S1 in Appendix S4). The course of the season clearly showed the highest factor loading in PC-4 (Table 1). For this reason, we went directly to the season predictor, controlled for phylogeny (hereafter ResSeason) and included as a predictor of SVL. ResSeason was significantly associated with female SVL but not with male SVL (Table 3, Fig. 2e).

Additionally, the presence of only one relevant PV in the phylogenetic models enabled us to perform variance partitioning of each relevant ecological predictor separately against PV-2 in a balanced way. The independent proportion of explained variation by the course of the season (ResSeason) was much greater in females than in males, and greater than the contribution of the other predictors (Fig. 3b).

To extract the portion of SVL that was not explained by the overlap between ecology and phylogeny, we took the error term $e$ from the regressions including all the significant non-shared structures (males: SVL = ResPV2 + ResPC-6 + $e$; females:

**Table 3.** Linear regression analyses of snout-to-vent-length on each environmental predictor exhibiting factors loadings > 0.7 in significant PCs (see Table 2).

| | Males | | | Females | | |
|---|---|---|---|---|---|---|
| **Predictors** | *F (1,128)* | *b* | *P* | *F (1,108)* | *b* | *P* |
| Mean Temperature of Warmest Quarter | **12.31** | **−0.30** | **< 0.001** | **5.67** | **−0.22** | **0.02** |
| Mean Temperature of Coldest Quarter | **6.15** | **−0.21** | **0.01** | **9.82** | **−0.29** | **0.002** |
| Elevation | **6.78** | **0.22** | **0.01** | **8.15** | **0.26** | **<0.01** |
| Enhanced Vegetation Index of Summer | 1.78 | 0.12 | 0.18 | 1.40 | 0.11 | 0.24 |
| Course of the Season | 3.13 | 0.15 | 0.08 | **9.62** | **0.29** | **0.002** |
| **Phylogenetically independent predictors** | | | | | | |
| ResSeason | 0.24 | 0.04 | 0.62 | **4.19** | **0.19** | **0.04** |
| ResEVIsummer | 0.25 | 0.04 | 0.62 | 0.73 | 0.08 | 0.39 |

Bold lettering denotes *P*-values<0.05. *b*: denotes standardized regression coefficients.

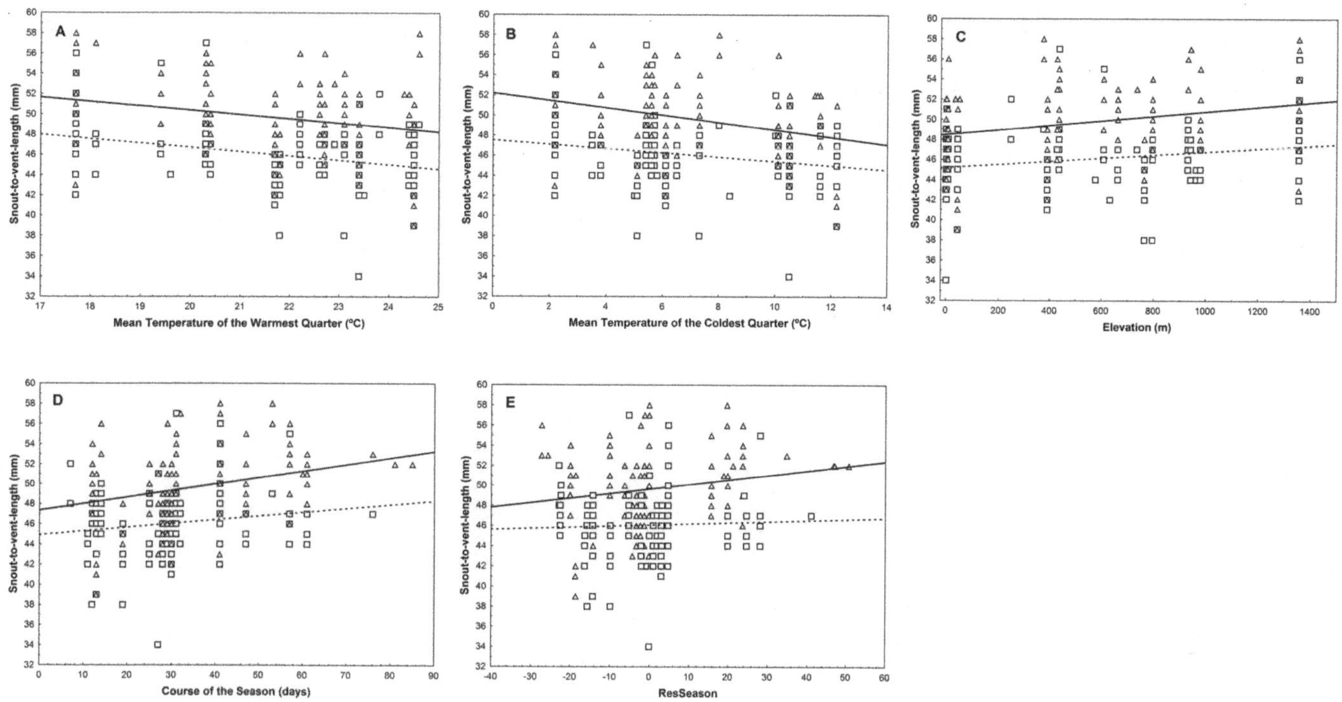


**Figure 2. Relationships between snout-to-vent-length of males (squares and dashed line) and females (triangles and solid line) and A) mean temperature of the warmest quarter, B) mean temperature of the coldest quarter, C) elevation, D) course of the season and E) the phylogenetically independent component of the course of the season (ResSeason).**

SVL = ResPV2 + ResPC-4 + $e$, see methods). These error terms $e$ (expressed as newly obtained residuals, hereafter residual SVL) were again regressed against all eight PCs. Estimations again revealed significant effects of PC-1 for males and females, remaining significant with the second and fourth filter selection criteria, respectively (Table 4). Also, there was a significant effect of the PC-7 only for females, which remained significant with three filter selection criteria (Table 4). In contrast to the analyses run before separating independent from shared components (Table 2), PC-4 and PC-6 were not significant (Table 4). When regressing separately those environmental predictors with factor loadings >0.7 in significant PCs, the mean temperature of the warmest quarter was negatively and significantly associated with the residual SVL regardless of sex (Table 5, Fig. 4a). Mean temperature of the coldest quarter was negative and significant in females, and a trend was found in males (Table 5, Fig. 4b). Elevation showed significant positive relationships only for females (Table 5, Fig. 4b). An interesting difference from the previous analyses using the entire variation (before separating independent from shared components, Table 3) is that there were positive significant associations of EVI of summer with the residual SVL for both sexes (Table 5, Fig. 4d).

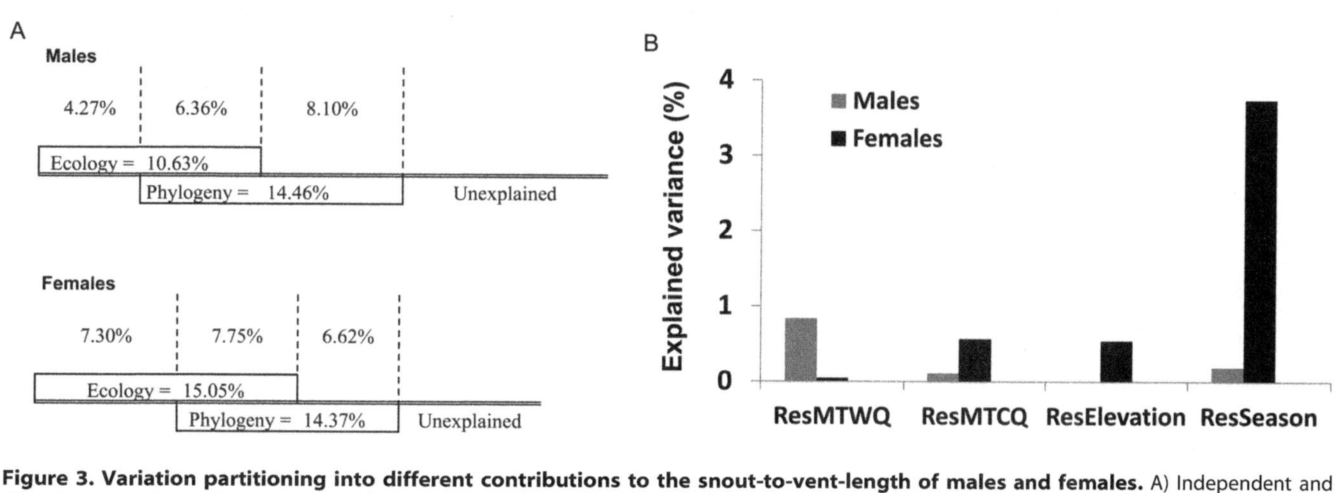


**Figure 3. Variation partitioning into different contributions to the snout-to-vent-length of males and females.** A) Independent and shared contributions of ecology and phylogeny and B) contributions of the phylogenetically independent component of relevant ecological predictors: mean temperature of the warmest quarter (ResMTWQ), mean temperature of the coldest quarter (ResMTCQ), elevation (ResElevation) and course of the season (ResSeason). Males are depicted in grey and females in black.

**Table 4.** Multiple regression analyses of the residuals of snout-to-vent-length on the eight ecological factors (PC).

| | Males | | | Females | | |
|---|---|---|---|---|---|---|
| **Ecological factors** | *b* | *t (121)* | *P* | *b* | *t (101)* | *P* |
| **PC-1** | **−0.28** | **−3.3** | **0.001**** | **−0.31** | **−3.43** | **<0.001******** |
| **PC-2** | 0.10 | 1.2 | 0.23 | 0.06 | 0.64 | 0.525 |
| **PC-3** | 0.02 | 0.2 | 0.83 | −0.12 | −1.27 | 0.206 |
| **PC-4** | 0.02 | 0.3 | 0.79 | 0.02 | 0.25 | 0.805 |
| **PC-5** | −0.09 | −1.0 | 0.32 | −0.16 | −1.77 | 0.079 |
| **PC-6** | −0.00 | −0.0 | 0.97 | −0.10 | −1.07 | 0.288 |
| **PC-7** | 0.15 | 1.8 | 0.08 | **−0.21** | **−2.24** | **0.027***** |
| **PC-8** | 0.10 | 1.2 | 0.23 | −0.01 | −0.15 | 0.878 |

Residuals of snout-to-vent-length (denoted as *e* in the formulation of methods) express the portion of SVL that was not explained by the unshared variation between ecology and phylogeny. Bold lettering denotes partial correlations significant at $P<0.05$. Asterisks denote the number of times that a PC was significant when applying the four selection criteria used to add spatial filters. See methods, Tables 2, 3 and Appendix S4 for details. *b*: denotes standardized regression coefficients.

## Spatial autocorrelation in combined models

The spatial structure not absorbed by the combined final models (including the relevant PCs and PVs) was small and did not affect the contribution of predictors in the combined models. Accordingly, the variance explained jointly by ecology and phylogeny independently of the spatial structure was much greater than that explained by the independent spatial component (see Appendix S4).

## Discussion

The two independent data sets analysed separately (the one-year field sampling and the inter-annual one) showed the same pattern and concomitant with Bergmann's cline: SVL increased with elevation and decreased with temperature both in males and females. Our procedures for separating the different sources of SVL variation revealed both similarities and differences between sexes. After separating the shared and independent components of phylogeny and ecology, we found a significant effect of phylogeny (independently of contemporary ecological factors). On the other hand, the effect of the independent component of ecology was much stronger in females than in males because females were much more influenced by the course of the season. Lastly, an effect of primary productivity on SVL in both sexes was only detected after controlling for the independent relevant components of ecology and phylogeny. Precipitation was not significant in any analyses. The fact that our models remained relatively stable after accounting for spatial autocorrelation shows that the predictors used are sufficiently informative to absorb most of the spatial structure of size variation at this scale.

The procedures used here to separate the independent and shared effects of phylogeny and ecology have been previously applied to higher taxonomic levels (the family level) in endotherms [6]. In this study, we applied this methodology using, for the first time, a matrix of genetic pair-wise distances at the individual level. The independent effect of ecology was stronger in females, primarily due to a differential effect of the course of the season, reflecting females grow faster. In agreement, longitudinal recapture data suggested that *P. hispanicus* adult females grew faster than males [30]. In organisms with rapid life cycles (early maturation and short adult lives), such as the species studied here [32], rapid female growth is advantageous, because reproductive success may depend on female body size (e.g., clutch size: [34, 35]; production of a second clutch: [31]; mass invested per clutch, or minimum size for producing a clutch). These sex differences could not be tested in the 95-year data-set because information on the course of the season through capture dates was not always available, indicating that sex-differences in developmental processes may be missed if a given dimension is lacking.

The independent effect of phylogeny was similar for males and females, reflecting a phylogenetic inertia that shaped part of the variation in lizard size as has been shown at other phylogenetic scales in endotherms (e.g. [29]). Phylogenetic inertia is the consequence of past evolutionary processes and may include

**Table 5.** Linear regression analyses of the residuals of snout-to-vent-length on each environmental predictor exhibiting factor loadings >0.7 in significant PCs (see Table 4).

| | Males | | | Females | | |
|---|---|---|---|---|---|---|
| **Environmental predictors** | *F (1,128)* | *b* | *P* | *F (1,108)* | *b* | *P* |
| **Mean Temperature of Warmest Quarter** | **9.20** | **−0.25** | **0.003** | **6.50** | **−0.24** | **0.012** |
| **Mean Temperature of Coldest Quarter** | 3.65 | −0.16 | 0.058 | **6.97** | **−0.24** | **0.009** |
| **Elevation** | 3.51 | 0.16 | 0.063 | **5.04** | **0.21** | **0.026** |
| **Enhanced Vegetation Index of Summer** | **10.32** | **0.27** | **0.002** | **4.94** | **0.21** | **0.028** |

Residuals of snout-to-vent-length (denoted as *e* in the formulation of methods) express the portion of SVL that was not explained by the unshared variation between ecology and phylogeny. Bold lettering denotes *P*-values<0.05. See methods for details. *b*: denotes standardized regression coefficients.

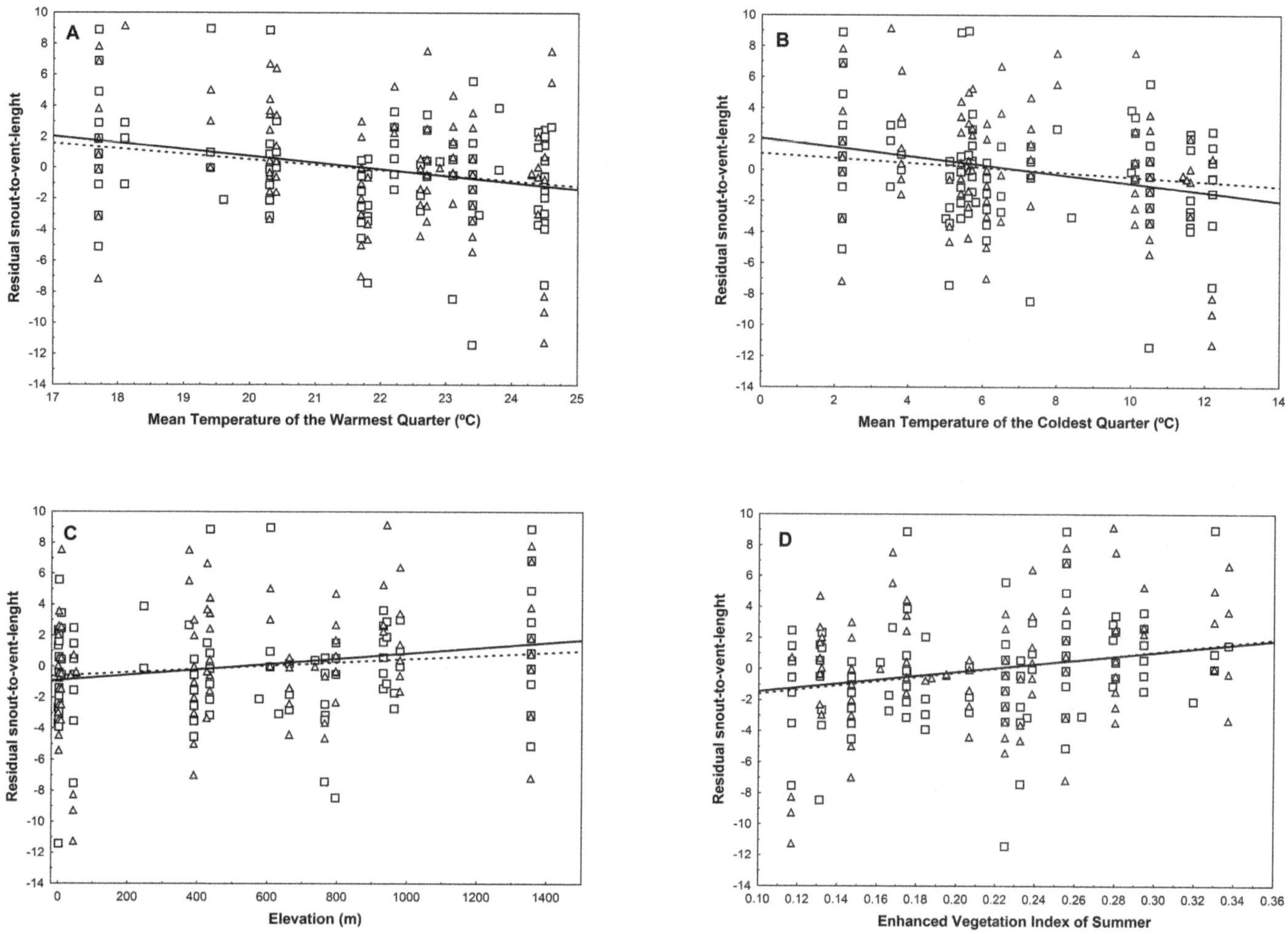


**Figure 4. Relationships between the residuals of snout-to-vent-length and A) mean temperature of the warmest quarter, B) mean temperature of the coldest quarter, C) elevation and D) Enhanced Vegetation Index of Summer.** Residuals of snout-to-vent-length (denoted as *e* in the formulation of methods) express the portion of SVL that was not explained by the unshared variation between ecology and phylogeny. Squares and dashed line: males, triangles and solid line: females.

geographic isolation, which has been proposed as an important driver of speciation in sand racers [36].

The proportion of variation explained by the overlap between ecology and phylogeny is similar in magnitude to that of the independent components (Fig. 3). This result is compatible with the existence of a link between contemporary environment and evolutionary mechanisms shaping body size. Once controlled for the relevant independent components, we found the same effect of temperature (Table 4) as when analyzing raw data (Table 1). Interestingly, an additional predictor that was previously overridden was then revealed as a factor with an important role in explaining this form of residual body size. Both sexes showed a significant positive relationship between the residual size and the vegetation index for summer, which reflects primary productivity during the activity period. The fact that PC-4 (in which the factor loading of the course of the season is larger) was no longer significant after controlling for the independent components supports the effectiveness of this procedure in controlling for short-term temporal effects.

Although the course of the season was mainly expressed by a PC factor different from that associated with other significant predictors, other predictors were correlated among them, as expressed in PC-1 (Table 1). Taken together several potential non-exclusive mechanisms on the partial role of the contemporary environment can be proposed in the light of previous studies and the species idiosyncrasy. While the thermoregulatory mechanism originally proposed to explain Bergmann's rule has been argued for endotherms, its role in ectotherm size is more controversial [16, 11]. For ectotherms, it is important to distinguish whether clinal variation may exist due to the benefits of larger sizes that are associated with higher thermal inertia or due to the benefits of smaller sizes that are associated with faster heat gain and loss. Although Bergmann emphasised the benefits of larger size in colder environments for endotherms, Kearney et al. [48] showed that the thermal challenge for terrestrial ectothermic animals may be to remain cool, depending on the inhabited climatic niche. From the available information on the climatic niche, morphology, behavior and physiology of sand racers it can be argued that, if a thermoregulatory mechanism was involved in the geographical pattern, it should be more related to heat loss in relatively warmer places. The Mediterranean climatic range of sand racers is warmer than that of other northern or montane distributions of European lacertids. Phylogenetically independent contrast of a lacertid clade reported that smaller species heat faster. Thus, sand racers gained heat most rapidly due to their relatively small size, which should be a constraint when operative ambient temperatures are above the preferred range [49]. Once corrected by size authors inferred an adaptive physiological mechanism to reduce overheating in Mediterranean lizards. Moreover, thermoregulatory behavioural strategies for gaining heat are less pronounced in Mediterranean lacertids (including *Psammodromus*) than in other northern/ montane European lacertids [49]. This scenario suggests that the thermal challenge for smaller individuals is more strongly associated with the avoidance of overheating. Therefore, the observed geographical pattern (smaller lizards in warmer places) would not match a thermoregulatory mechanism in a simple manner, but rather through a complex syndrome linking different aspects of the species biology as in other Mediterranean lacertids [50].

Several alternatives to thermoregulatory mechanisms have been proposed for ectothermic animals. The so-called temperature-size rule proposes that ectothermic animals in hotter environments grow faster but, paradoxically, mature at a smaller body size [21, 51]. A single explanation for the generality of this rule is lacking [52]. An explanation of the temperature-size rule for the lizard *Sceloporus undulatus* is that juveniles in colder environments have higher survival, mediated by biotic interactions, allowing them to delay maturation until they attain larger sizes, which provides a reproductive advantage [9]. Data on size at maturity are scarce in sand racers and restricted to females, but anecdotal evidence is consistent with the mechanism proposed by Angilletta et al. [9]: in Salamanca, the mean temperature of warmest quarter (21.1°C) is lower and the size at maturity (42 mm) and the mean adult size (48 mm) are higher [30] than in Barcelona (23.4°C, 38.7 mm and 42.8 mm, respectively, [53]). In *Z. vivipara*, a lacertid which inhabits a much larger geographic area, the temperature-size rule in combination with an effect of annual activity has been suggested to affect body size: reduced activity in colder environments has been suggested to increase survival and favour larger sizes at maturity [25]. Alternatively, it has been argued that in colder and seasonal environments larger sizes would be advantageous to survive periods of food shortage by means of fat storage [24]. In the Iberian Peninsula colder environments are often associated with a greater seasonality, and hence an effect of activity duration might partially shape sand racer sizes. However, this cannot be the single factor influencing body sizes since it is known that precipitation precludes activity in lacertid species [54], and we did not find any effect of precipitation in our analyses.

Once controlled by relevant independent effects, body size was positively correlated with primary productivity with the highest relative contribution (Table 5). The finding that the vegetation index of summer (when juveniles hatch), but not that of winter, predicted body size indicates that primary productivity during the activity period is an important determinant of size. Plant productivity is an indicator of invertebrate prey productivity [27]. An obvious direct effect of food availability (quantity or quality of prey) on lizard size is through fitness components, such as growth or survival, which is crucial at the juvenile stage in lacertids [55]. In this line, taking also into account additional analyses with a surrogate of lizard density (see Appendix S3) it appears that higher primary productivity allows a greater lizard biomass (greater number of lizards and larger sizes).

On the other hand, in this study the effect of primary productivity emerged only after correcting for phylogenetic inertia and short-term temporal variation within the season. This result invites to consider as plausible the role of contemporary environmental factors in evolutionary scenarios, besides the evident effect of food abundance on postnatal growth. A previous study suggested that prey size is an important driver of the evolution of size at birth in tiger snakes because it is correlated with gape size, which ultimately affects adult size in combination with the effect of resource availability during ontogeny [56]. A translocation of the lacertid, *Podarcis sicula*, into a novel environment revealed an evolutionary divergence in morphology due to strong changes in the diet [57]. Thus, it has been suggested that the temperature-size relationships may act through food intake during the growing season [27], in combination with the pulse of seasonality [7], or in combination with a direct effect of temperature [51]. The latter scenario might partially conciliate the puzzle in the temperature-size rule for ectothermic terrestrial vertebrates; in which many ectotherms grow more slowly but mature at a larger size at lower temperatures. Thus, higher juvenile survival due to higher food availability might outweigh the costs of delayed maturation [51], or higher growth rates due to a higher food intake might counteract growth retardation at lower temperatures.

## Conclusions

Our one year-field work and the multi-year data base converged showing that lizard size is negatively associated with temperature, concomitantly with Bergmann's and temperature-size rules. The applied procedures to disentangle among independent and shared components of ecology and phylogeny allowed detecting sources of variation connecting with different non-exclusive mechanisms. These procedures showed independent effects of phylogenetic inertia and sex-dependent short-term temporal autocorrelation. Primary productivity only emerged as an important predictor of size after controlling for phylogenetic inertia and sex-differences in growth, and consistently with the resource rule. This study highlights the importance of including individual-based genetic information, informative ecological parameters of different nature, and temporal and spatial autocorrelation in sex-specific models to detect potentially important hidden effects. Finally, besides the use of procedures to disentangle inter-correlations among environmental predictors, we encourage focusing on the species' idiosyncrasies to interpret patterns of body size geographical variation.

## Supporting Information

**Appendix S1 Additional analyses with latitude and initial considerations.**

**Appendix S2 Evaluating the generality of patterns with an independent inter-annual data-set.**

**Appendix S3 Additional analyses to further exploring potential mechanisms.**

**Appendix S4 Spatial autocorrelation in models.**

## Acknowledgments

We thank José María Delgado, Benet Pera Gresely, Guillem Pérez i de Lanuza, Tomás Ponce, Mario García-París, Jesús Mellado, Gustavo A. Llorente, Fernando Palacios, Juan Manuel Pleguezuelos Ana Cristina Andreu Rubio, Adolfo Marco Llorente, Isidro Román Maudo, and Luis M. San-José for field assistance, helping with logistics or indicating potential locations.

## Author Contributions

Conceived and designed the experiments: PA. Analyzed the data: PA. Wrote the paper: PA. Contributed data, generated the genetic distance matrix, discussed the results, and commented on the manuscript: PSF.

## References


1. Bergmann KGLC (1847) Über die Verhältnisse der wärmeokönomie der Thiere zu ihrer Grösse. Gött Stud 3: 595–708.
2. Gardner JL, Peters A, Kearney MR, Joseph L, Heinsohn R (2011) Declining body size: a third universal response to warming? Trends Ecol Evol 26: 285–291.
3. Diniz-Filho JAF, Bini LM, Rodríguez MA, Rangel TFLVB, Hawkins BA (2007) Seeing the forest for the trees: partitioning ecological and phylogenetic components of Bergmann's rule in European Carnivora. Ecography 30: 598–608.
4. Stillwell RC (2010) Are latitudinal clines in body size adaptive? Oikos 119: 1387–1390.
5. Olson VA, Davies RG, Orme CDL, Thomas GH, Meiri S, et al. (2009) Global biogeography and ecology of body size in birds. Ecol Lett 12: 249–259.
6. Desdevises Y, Legendre P, Azouzi L, Morand S (2003) Quantifying phylogenetically structured environmental variation. Evolution 57: 2647–2652.
7. Huston MA, Wolverton S (2011) Regulation of animal size by eNPP, Bergmann's rule, and related phenomena. Ecol Monogr 81: 349–405.
8. Ashton KG, Feldman CR (2003) Bergmann's rule in nonavian reptiles: turtles follow it, lizards and snakes reverse it. Evolution 57: 1151–1163.
9. Angilletta MJ, Niewiarowski PH, Dunham AE, Leaché AD, Porter WP (2004) Bergmann's clines in ectotherms: illustrating a life-historical perspective with sceloporine lizards. Am Nat 164: E168–E183.
10. Cruz FB, Fitzgerald LA, Espinoza RE, Schulte II, JA (2005) The importance of phylogenetic scale in tests of Bergmann's and Rapoport's rules: lessons from a clade of South American lizards. J Evol Biol 18:1559–1574.
11. Olalla-Tárraga MA, Rodríguez MA, Hawkins BA (2006) Broad-scale patterns of body size in squamate reptiles of Europe and North America. J Biogeogr 33: 781–793.
12. Pincheira-Donoso D, Hodgson DJ, Tregenza T (2008) The evolution of body size under environmental gradients in ectotherms: why should Bergmann's rule apply to lizards? BMC Evol Biol 8: 68.
13. Oufiero CE, Adolph SC, Gartner GEA, Garland TJr (2011) Latitudinal and climatic variation in body size and dorsal scale rows in *Sceloporus* lizards: a phylogenetic perspective. Evolution 65: 3590–3607.
14. Guillaumet A, Ferdy J-B, Desmarais E, Godelle B, Crochet P-A (2008) Testing Bergmann's rule in the presence of potentially confounding factors: a case study with three species of Galerida larks in Morocco. J Biogeogr 35: 579–591.
15. Amarello M, Nowak EM, Taylor EN, Schuett GW, Repp RA, et al. (2010) Potential environmental influences on variation in body size and sexual size dimorphism among Arizona populations of the western diamond-backed rattlesnake (*Crotalus atrox*). J Arid Environ 74: 1443–1449.
16. Watt C, Mitchell S, Salewski V (2010) Bergmann's rule; a concept cluster? Oikos 119: 89–100.
17. Pincheira-Donoso D (2010) The balance between predictions and evidence and the search for universal macroecological patterns: taking Bergmann's rule back to its endothermic origin. Theory Biosci 129: 247–253.
18. Olalla-Tárraga MA (2011) "*Nullius in Bergmann*" or the pluralistic approach to ecogeographical rules: a reply to Watt et al. (2010). Oikos 120: 1441–1444.
19. Meiri S (2011) Bergmann's Rule - what's in a name? Glob Ecol Biogeogr 20: 203–207.
20. Hawkins B, Diniz-Filho JAF (2004) 'Latitude' and geographic patterns in species richness. Ecography 2: 268–272.
21. Atkinson D (1996) Ectotherm life history responses to developmental temperature. In: Johnston IA, Bennett AF editors. Animals and temperature: phenotypic and evolutionary adaptation: Cambridge University Press. pp. 183–20.
22. Mousseau TA (1997) Ectotherms follow the converse to Bergmann's rule. Evolution 51: 630–632.
23. Chamaillé-Jammes S, Massot M, Aragón P, Clobert J (2006) Global warming and positive fitness response in mountain populations of common lizards *Lacerta vivipara*. Glob Chan Biol 12: 392–402.
24. Horváthová T, Cooney CR, Fitze PS, Oksanen TA, Jelić D, et al. (2013) Length of activity season drives geographic variation in body size of a widely distributed lizard. Ecol Evol 3: 2424–2442.
25. McNab BK (2010) Geographic and temporal correlations of mammalian size reconsidered: a resource rule. Oecologia 164:13–23.
26. Yom-Tov Y, Geffen E. (2011) Recent spatial and temporal changes in body size of terrestrial vertebrates: probable causes and pitfalls. Biol Rev 86: 531–541.
27. Cushman JH, Lawton JH, Manly BFJ (1993) Latitudinal patterns in European ant assemblages: variation in species richness and body size. Oecologia 95:30–37.
28. Ficetola GF, Scali S, Denoël M, Montinaro G, Tanja D, et al. (2010) Ecogeographical variation of body size in the newt *Triturus carnifex*: comparing the hypotheses using an information theoretic approach. Glob Ecol Biogeogr 19: 485–495.
29. Diniz-Filho JAF, De Sant'Ana CER, Bini LM (1998) An eigenvector method for estimating phylogenetic intertia. Evolution 52: 1247–1262.
30. Pascual JA, Pérez-Mellado V (1989) Datos sobre la reproducción y el crecimiento de *Psammodromus hispanicus* Fitzinger, 1826 en un medio adehesado de la España Central. Doñ Act Vertebr 16: 45–55.
31. Carretero MA, Llorente GA (1991) Reproducción de *Psammodromus hispanicus* en una arenal costero del nordeste ibérico. Amphibia-Reptilia 12: 395–408.
32. Bauwens D, Díaz-Uriarte R (1997) Covariation of Life-History Traits in Lacertid Lizards: A Comparative Study. Am Nat 149: 91–111.
33. Fitze PS (2012) Edward's Sand Racer – *Psammodromus edwardsianus*. In: Salvador A, Marco A editors. Enciclopedia Virtual de los Vertebrados Españoles: Museo Nacional de Ciencias Naturales (MNCN), Spain. pp. 18. http://www.vertebradosibericos.org/reptiles/pdf/psaedw2.pdf
34. Fitze PS (2012) Spanish Sand Racer – *Psammodromus hispanicus*. In: Salvador A, Marco A editors. Enciclopedia Virtual de los Vertebrados Españoles: Museo Nacional de Ciencias Naturales (MNCN), Spain. pp. 8. http://www.vertebradosibericos.org/reptiles/pdf/psahis2.pdf
35. Fitze PS (2012) Western Sand Racer – *Psammodromus occidentalis*. In: Salvador A, Marco A editors. Enciclopedia Virtual de los Vertebrados Españoles: Museo Nacional de Ciencias Naturales (MNCN), Spain. pp. 16. http://www.vertebradosibericos.org/reptiles/pdf/psaocc2.pdf
36. Fitze PS, González-Jimena V, San-José LM, San Mauro D, Aragón P, et al. (2011) Integrative analyses of speciation and divergence in *Psammodromus hispanicus* (Squamata: Lacertidae). BMC Evol Biol 11: 347.
37. Fitze PS, González-Jimena V, San-José LM, San Mauro D, Suarez T, et al. (2012) A new species of sand racer, *Psammodromus* (Squamata: Lacertidae), from the Western Iberian Peninsula. *Zootaxa* 3205: 41–52.
38. San-José LM, González-Jimena V, Fitze PS (2012) Patterns of phenotypic variation reveal substantial differentiation in sexual dimorphism of three *Psammodromus* (Squamata, Lacertidae) species. Contrib Zool 81: 181–197.
39. Hijmans RJ, Cameron SE, Parra JL, Jones PG, Jarvis A (2005) Very high resolution interpolated climate surfaces for global land areas. Int J Climatol 25: 1965–1978.
40. Quinn GP, Keough MJ (2002) Experimental Design and data Analysis for Biologist. Cambridge: Cambridge University Press. 537 p.
41. Gower JC (1966) Some distance properties of latent root and vector methods used in multivariate analysis. Biometrika 53: 325–338.
42. Diniz-Filho JAF, Bini LM, Rangel TFLVB, Morales-Castilla I, Olalla-Tárraga MÁ, et al. (2012) On the selection of phylogenetic eigenvectors for ecological analyses. Ecography 35: 239–249.
43. Bini LM, Diniz-Filho JAF, Rangel TFLVB, Akre TSB, Albaladejo RG, et al. (2009) Coefficient shifts in geographical ecology: an empirical evaluation of spatial and non-spatial regression. Ecography 32: 193–204.
44. Legendre P, Legendre L (1998) Numerical Ecology. 2d English ed. Amsterdam: Elsevier Science BV. 837 p.
45. Aubad J, Aragón P, Olalla-Tárraga MÁ, Rodríguez MÁ (2008) Illegal logging, landscape structure and the variation of tree species richness across North Andean forest remnants. For Ecol Manage 255: 1892–1899.
46. Ihaka R, Gentleman R (1996) R: a language for data analysis and graphics. J Comput Graph Stat 5: 299–314.
47. Rangel TFLVB, Diniz-Filho JAF, Bini LM (2006) Towards an integrated computational tool for spatial analysis in macroecology and biogeography. Glob Ecol Biogeogr 15: 321–327.
48. Kearney M, Shine R, Porter WP (2009) The potential for behavioral thermoregulation to buffer "cold-blooded" animals against climate warming. Proc Natl Acad Sci USA 106: 3835–3840.
49. Díaz JA, Bauwens D, Asensio B (1996) A Comparative Study of the Relation between Heating Rates and Ambient Temperatures in Lacertid Lizards. Physiol Zool 69: 1359–1383.
50. Belliure J, Carrascal LM, Díaz JA (1996) Covariation of thermal biology and foraging mode in two mediterranean lacertid lizards. Ecology 77: 1163–1173.
51. Angilletta MJ, Steury TD, Sears MW (2004) Temperature, growth rate, and body size in ectotherms: Fitting pieces of a life-history puzzle. Integr Comp Biol 44: 498–509.
52. Angilletta MJ, Dunham AE (2003) The temperature–size rule in ectotherms: simple evolutionary explanations may not be general. Am Nat 162: 332–342.
53. Carretero MA (1994) Algunes dades morfomètriques de *Psammodromus hispanicus* al Prat de Llobregat. Butll nat delt Llob 1: 13–19.
54. Aragón P, López P, Martín J (2001) Seasonal changes in the activity and spatial relationships of the Iberian rock-lizard, *Lacerta monticola*. Can J Zool 79: 1965–1971.
55. Massot M, Aragón P (2013) Phenotypic resonance from a single meal in an insectivorous lizard. Current Biol 23: 1320–1323.

56. Aubret F (2012) Body-size evolution on islands: Are adult size variation in tiger snakes a nonadaptive consequence of selection on birth size? Am Nat 6: 756–767.

57. Herrel A, Huyghe K, Vanhooydonck B, Backeljau T, Breugelmans K, et al. (2008) Rapid large-scale evolutionary divergence in morphology and performance associated with exploitation of a different dietary resource. Proc Natl Acad Sci USA 105: 4792–4795.

3

# Deciphering the Adjustment between Environment and Life History in Annuals: Lessons from a Geographically-Explicit Approach in *Arabidopsis thaliana*

**Esperanza Manzano-Piedras[1], Arnald Marcer[2,3], Carlos Alonso-Blanco[4], F. Xavier Picó[1]***

**1** Departamento de Ecología Integrativa, Estación Biológica de Doñana (EBD), Consejo Superior de Investigaciones Científicas (CSIC), Sevilla, Spain, **2** CREAF, Cerdanyola del Vallès, Spain, **3** Universitat Autònoma de Barcelona, Cerdanyola del Vallès, Spain, **4** Departamento de Genética Molecular de Plantas, Centro Nacional de Biotecnología (CNB), Consejo Superior de Investigaciones Científicas (CSIC), Madrid, Spain

## Abstract

The role that different life-history traits may have in the process of adaptation caused by divergent selection can be assessed by using extensive collections of geographically-explicit populations. This is because adaptive phenotypic variation shifts gradually across space as a result of the geographic patterns of variation in environmental selective pressures. Hence, large-scale experiments are needed to identify relevant adaptive life-history traits as well as their relationships with putative selective agents. We conducted a field experiment with 279 geo-referenced accessions of the annual plant *Arabidopsis thaliana* collected across a native region of its distribution range, the Iberian Peninsula. We quantified variation in life-history traits throughout the entire life cycle. We built a geographic information system to generate an environmental data set encompassing climate, vegetation and soil data. We analysed the spatial autocorrelation patterns of environmental variables and life-history traits, as well as the relationship between environmental and phenotypic data. Almost all environmental variables were significantly spatially autocorrelated. By contrast, only two life-history traits, seed weight and flowering time, exhibited significant spatial autocorrelation. Flowering time, and to a lower extent seed weight, were the life-history traits with the highest significant correlation coefficients with environmental factors, in particular with annual mean temperature. In general, individual fitness was higher for accessions with more vigorous seed germination, higher recruitment and later flowering times. Variation in flowering time mediated by temperature appears to be the main life-history trait by which *A. thaliana* adjusts its life history to the varying Iberian environmental conditions. The use of extensive geographically-explicit data sets obtained from field experiments represents a powerful approach to unravel adaptive patterns of variation. In a context of current global warming, geographically-explicit approaches, evaluating the match between organisms and the environments where they live, may contribute to better assess and predict the consequences of global warming.

**Editor:** Justin O. Borevitz, The Australian National University, Australia

**Funding:** This work was supported by grants from the Ministerio de Educación y Ciencia (CGL2009-07847/BOS) and the Ministerio de Economía y Competitividad (CGL2012-33220/BOS) of Spain to FXP. The funders had no role in study design, data collection and analysis, decision to publish, or preparation of the manuscript.

**Competing Interests:** The authors have declared that no competing interests exist.

* E-mail: xpico@ebd.csic.es

## Introduction

Annual plants exhibit contrasting life histories in different environments varying in predictability of the amount and timing of yearly precipitation. On the one hand, annuals occurring in highly unpredictable environments, e.g. deserts, are typically characterised by persistent seed banks and delayed seed germination. When environmental conditions become favourable for seed germination and seedling establishment, annuals can complete the life cycle in weeks. Hence, desert annuals reduce temporal variance in reproductive success and buffer the long-term risk of extinction in unfavourable years, representing a bet-hedging strategy in unpredictable environments [1–7]. On the other hand, annuals inhabiting humid temperate areas mostly behave as winter annuals because the environment is seasonal and far more predictable than in deserts. Winter annuals remain as seeds buried in the soil during summer, peak germination in autumn, overwinter as vegetative rosettes, and become reproductive in late winter or early spring [8–12]. Thus, bet-hedging and winter annual behaviour are regarded as adaptive life histories at the opposing margins of a gradient of environmental severity and predictability between deserts and temperate areas.

Obviously, environments are not homogeneous and annuals are expected to modify the architecture of life-history traits reflecting environmental and ecological differences among populations [13,14]. Two traits stand out above all when annuals adjust the life history to environmental conditions: the timing of germination and flowering. Due to the short duration of the life cycle in annuals, germination timing represents the first major developmental transition influencing all posterior life-history traits [15–19]. In addition, flowering time is also an important trait whose variation has been shaped by natural selection to maximise reproductive success [20–25]. Hence, annual plants are expected to fine-tune the timing of germination and/or flowering to succeed in a given environment. Indeed, recent empirical studies showed that some annuals tend to lower germination fractions [26–28] or flower earlier [21,29–32] as environmental stress increases, such as that caused by climate warming or increasing aridity.

The methodological approaches commonly employed to study the evolutionary consequences of variation in life-history traits of plants include common garden and/or reciprocal transplant experiments between contrasting environments. Such experiments are very powerful because they allow the assessment of the genetic differentiation among lineages from different populations for the life-history traits of concern [33–36]. However, an important drawback is that these experiments are very demanding and researchers generally opt for selecting an appropriate set of populations from representative environments [37–40]. Therefore, experiments have to sacrifice potentially relevant genetic variation, which may limit our understanding of how the adjustment between environmental factors and life-history traits works. Alternatively, correlation analysis between genetic variation in phenotypic traits and environmental variation has long represented a powerful method to detect selection along environmental gradients [41–45] because adaptive phenotypic variation shifts gradually across space as a result of the geographic patterns of variation in major selective pressures, e.g. climate. Given the fact that spatial patterns of variation exist in both environmental and phenotypic traits, understanding large-scale adaptive patterns of variation in life-history traits of annual plants demands geographically-explicit experimental approaches.

Here, we aim to analyse the adjustment between geographical variation in environmental factors and life-history traits of the annual plant *Arabidopsis thaliana*. We employ a collection of 279 geo-referenced populations from the Iberian Peninsula where *A. thaliana* has been studied for years due to its large genetic and environmental variation [12,46–49]. Study populations represent a wide array of Iberian natural environments, encompassing humanised and wild habitats as well as low- and high-elevation locations. Each population was characterised by climate, vegetation and soil features. Genetic variation in life-history traits was evaluated by replicating one individual from each population in a common garden experiment. Experiments in natural field settings are unavoidable if we aim to obtain realistic data on genetic variation in phenotypic traits. We specifically address two main questions. First, what is the spatial autocorrelation pattern of *A. thaliana*'s life-history traits in the Iberian Peninsula? If environmental factors have shaped variation in life-history traits, we would expect to detect it by the extent of spatial autocorrelation in life-history traits given the inherent spatial variation in environmental factors. Second, what is the relationship between variation in environmental factors and life-history traits in Iberian *A. thaliana* populations? The environmental factors and the life-history traits affected by them are expected to be correlated. The results are discussed in a context of adaptive geographic variation stressing the value of extensive sampling to identify ecologically and evolutionarily important life-history traits in plants.

## Materials and Methods

### Study Organism and Source Populations

*Arabidopsis thaliana* (L.) Heyhn. (Brassicaceae) is a cosmopolitan annual plant native to Europe and Central Asia [50]. In the Iberian Peninsula, the plant is widely distributed occurring in both humanised and wild habitats [46,48]. This self-compatible and self-fertile plant possesses a seed bank with an estimated half-life of approximately three years [12]. *Arabidopsis thaliana* can behave as a winter or spring annual [49,51,52] depending on the timing of germination (autumn *vs.* spring) and the state in which the plant overwinters (vegetative rosette *vs.* seed). Iberian *A. thaliana* populations are composed of individuals exhibiting both winter and spring annual habits, but the proportion of spring-germinated plants increases at high elevations as a result of high mortality rates of vegetative rosettes due to harsh winter conditions [49]. Flowering always occurs in late winter or early spring and fruits shed seed throughout spring and early summer [49].

We sampled 279 *A. thaliana* populations that were surveyed across the major part of the Iberian Peninsula (ca. 800×700 $km^2$; 36.00°N –43.48°N, 3.19°E –9.30°W; Fig. 1) during the period 2004 – 2009. Populations were geo-referenced for their latitude, longitude and altitude using a GPS (Garmin International, Inc., Olathe, KS, USA), with an estimated average positional error of 4–6 m. Populations were separated by 1–1,042 km and altitudes ranged 1–2,662 m.a.s.l. For each population, we collected seeds from several individuals per population (approximately 6–20 individuals) separated by a few metres from each other. For all individuals sampled, we produced a new generation of seeds in controlled conditions in the glasshouse facilities at the Centro Nacional de Biotecnología (CNB-CSIC) in Madrid. Individuals lacking a vernalization requirement to flower completed the life cycle in 1 – 5 months in glasshouse conditions at approximately 22°C and 16 h of light. In the case of individuals requiring a cold period to induce flowering, young vegetative rosettes grown in glasshouse conditions were transferred to growth chambers at 4°C with a short-day photoperiod (8 h of light/16 h of darkness) during 6 weeks. After that time, plants were placed back in glasshouse conditions to complete the life cycle. Seeds were stored in cellophane bags at room temperature and dry conditions in darkness.

In this study, we used one individual from each of the 279 *A. thaliana* populations. We selected one individual per population, which exhibited a common phenotype within its population based on average flowering time or vernalization requirement. Furthermore, all 279 selected individuals (accessions hereafter; [53]) were genetically different from each other, as shown by genetic data obtained from a set of 250 polymorphic genome-wide neutral SNPs previously analysed in Iberian *A. thaliana* accessions [47,48]. Genetically, the collection of 279 *A. thaliana* accessions exhibited a significant isolation-by-distance pattern ($N = 279$, $r = 0.35$, $P < 0.001$; Mantel test based on geographic and genetic distance matrices), which was highly consistent with previous results on the genetic structure of Iberian *A. thaliana* populations based on 100 [46] and 182 accessions [48].

*Arabidopsis thaliana* is not an endangered or protected species across its distribution range. Field sampling was carried out in locations where no permission was required, except at Doñana National Park (permission issued by Estación Biológica de Doñana) and Sierra de Grazalema Natural Park (permission issued by Red Andaluza de Jardines Botánicos de la Consejería de Medio Ambiente y Ordenación del Territorio de la Junta de Andalucía).

### The Field Experiment

In 2010, the 279 *A. thaliana* accessions were multiplied again by the single seed descent method as described above to discard possible effects of variable seed age on plant traits. For each accession, we prepared eight batches, each of them with 60 filled seeds, and stored them in 1.5 ml plastic tubes at room temperature in darkness until the sowing day. On October 6, 2010, we sowed the 60 seeds per batch in square plastic pots (12×12×12 $cm^3$) filled with standard soil mixture (Abonos Naturales Cejudo Baena S.L., Utrera, Spain) at the El Castillejo Botanical Garden in Sierra de Grazalema Natural Park in SW Spain (El Bosque, Cádiz province, 36.46°N, 5.30°W, 329 m.a.s.l.). This Botanical Garden

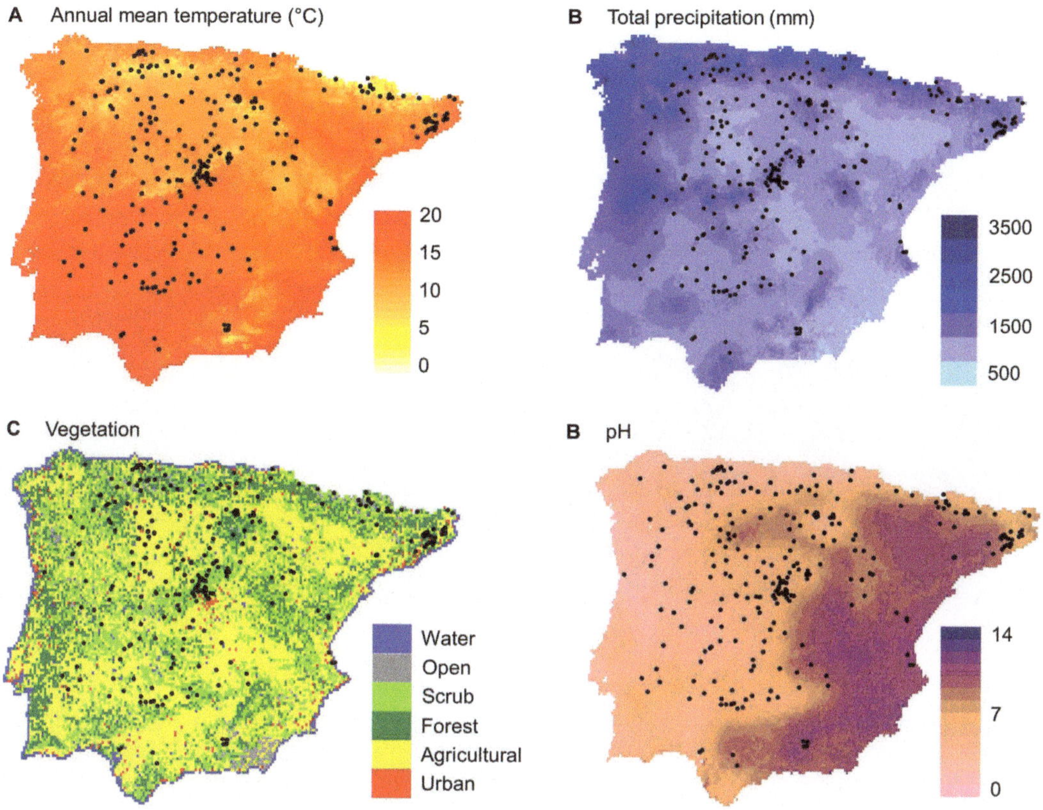


**Figure 1. Distribution of the 279 *A. thaliana* Iberian populations of study.** Maps show the geographical distribution of annual mean temperature (°C), total precipitation (mm), vegetation data, and pH.

is an appropriate environment for *A. thaliana* because several accessions from environmentally contrasted Iberian locations completed the life cycle successfully there [54]. During the experiment (from October 6, 2010, to May 5, 2011), mean monthly minimum and maximum temperatures were 7.2°C and 20.4°C, respectively, and precipitation totalled 810 mm (Figure S1). Temperature data were recorded with HOBO Pendant® UA-002-08 temperature loggers (Onset Computer Corporation, Inc., Bourne, MA, USA) and precipitation was recorded daily at the meteorological station of the experimental facility. The experiment had eight blocks and each block contained one replicate per accession that was randomly placed within the block (Fig. 2). Overall, this experiment included 2,232 pots (279 accessions × 8 blocks) and 133,920 seeds (2,232 pots × 60 seeds). Blocks were covered by 2-cm wire mesh to protect young plants from bird and rodent depredation (Fig. 2).

We surveyed the experiment every 15 days to record the maximum number of vegetative rosettes per pot. In total, we conducted five surveys between October 6 and December 17, 2010. In winter, mortality of vegetative rosettes increases and recruitment of new individuals is virtually nil, rendering winter surveys as unimportant for this trait [54]. A flowering date was assigned to each pot when the majority of the individuals in the pot developed flowering buds and had the first flower open (as in [54]). It must be emphasized that all individuals within each pot were sisters obtained by self-fertilization and their reproductive behaviour was quite homogeneous. Flowering time for each pot was estimated as the number of days between seed sowing, given that seed germination peaked within the first 15 days of the experiment for all accessions, and flowering date. We counted the number of fruiting individuals per pot and the total number of fruits per individual when all individuals finished flowering and fruiting. After recording data, plants were removed and incinerated, and the soil composted to avoid uncontrolled seed dispersal at the experimental facility.

In this experiment, we monitored a maximum of 57,638 vegetative rosettes during the autumn surveys (mean ± SD = 53,054.8±3,556.3 vegetative rosettes) and recorded a total of 46,638 reproductive individuals that overall produced 554,807 fruits (grand mean ± SD = 11.9±30.6; range = 1–982 fruits per plant). The extent of flower abortion, fruit depredation, insect damage and plant disease was rather low in this experiment; only 318 (0.68%) reproductive individuals were discarded due to these factors.

## Seed Weight and Seed Germination

We estimated seed weight and seed germination for the 279 *A. thaliana* accessions. For each accession, we weighted, to the nearest 0.1 mg, three batches of 60 filled seeds each using a Sartorious BP61S balance (Sartorius AG, Göttingen, Germany). Seed weight was obtained by averaging the three measurements. In this experiment, we used a total of 50,220 seeds (279 accessions × 60 seeds × 3 batches).

We also estimated seed germination under three temperatures (10, 22 and 30°C). 10°C represents the autumn mean temperature of several Iberian *A. thaliana* populations when seed germination peaks [12], 22°C is the temperature commonly used to germinate *A. thaliana* seeds in growth chamber conditions, and 30°C is known to induce secondary seed dormancy in *A. thaliana* [55]. For each accession and temperature treatment, we conducted two germi-

**Figure 2. Photographs of the experimental setting at the El Castillejo Botanical Garden.** Panels include a general view of the blocks and detailed views of pots with labels covered with the wire mesh, vegetative rosettes and fruiting plants of *Arabidopsis thaliana*.

nation assays that exhibited a very similar behaviour ($N = 279$; $r = 0.83$–$0.86$; Pearson's correlation). For each germination assay, one batch of 50 filled seeds per accession was placed in a moist Petri dish at 4°C in darkness for four days in an AGP-700-ESP incubator (Radiber S.A., Barcelona, Spain). After that, we transferred Petri dishes to a FITOCLIMA-10.000-EH growth chamber (ARALAB, Rio de Mouro, Portugal) for five days in light and in the corresponding temperature (10, 22 or 30°C). We counted the number of germinated seeds under a Stemi-2000-C stereomicroscope (Carl Zeiss Optical, Inc., Chester, VA, USA). Seeds were scored as germinated when the root tip protruded the seed coat. For each temperature treatment, seed germination was calculated by averaging the proportion of seeds germinated between the two assays. In this experiment, we used a total of 83,700 seeds (279 accessions × 50 seeds × 3 temperatures × 2 assays).

## Environmental Data

We developed a geographic information system to generate an environmental database for the 279 geo-referenced *A. thaliana* accessions (Fig. 1) at a resolution of 1 km$^2$. Weather data encompassed 19 bioclimatic variables [56], which consisted in a combination of annual trends, seasonality and extreme conditions relevant to species physiological tolerances. These variables were generated using monthly data from the Digital Climatic Atlas of the Iberian Peninsula (http://opengis.uab.es/wms/iberia/en_index.htm; [57]). Vegetation data were obtained from the CORINE Land Cover 2000 (http://www.eea.europa.eu/publications/COR0-landcover). We calculated the percentage of humanised habitat, including urban, crops and semi-natural grasslands, in a 78-ha circular area (500 m radius) around the GPS coordinates. The percentages of humanised and natural habitat, including primarily woody vegetation, were significantly negatively correlated ($N = 279$; $r = -0.95$; $P < 0.001$; Pearson's correlation) and nearly summed to 100%. Here, we only used the percentage of humanised habitat. Finally, soil data were characterised by pH, which was obtained from the Soil Geographical Database of Eurasia v.4 (http://eusoils.jrc.ec.europa.eu).

## Statistical Analyses

Life-history traits of *A. thaliana* accessions included in this study were seed weight, seed germination at three temperatures, maximum number of vegetative rosettes, flowering time, number of reproductive individuals, and mean number of fruits per individual. We also estimated a surrogate of individual fitness by multiplying the proportion of seeds becoming adult individuals times the mean number of fruits per individual. For all life-history traits, we used the mean among replicates of the different experiments. We also used replicates to estimate broad sense heritability for all life-history traits as $h^2 = V_G/(V_G + V_E)$, where $V_G$ is the among-accession variance component and $V_E$ is the residual variance.

Individuals within pots experienced asymmetric competition so that a few individuals performed much better, in terms of fruit production, than the rest of individuals in the pot. In order to avoid bias in the estimation of the reproductive output due to suppressed individuals bearing one fruit per plant, we estimated the number of fruits per individual by excluding the largest individual from each pot and averaging the number of fruits of the 15 largest individuals from all pots. Analyses conducted with different estimates for the mean number of fruits per individual (e.g. number of fruits of the largest individual, mean number of fruits of the 30 largest individuals) were highly consistent (results not shown). We considered that the 15 largest individuals captured the winners of the asymmetric competition process. In fact, the maximum number of individuals observed in the pots and the skewness of the number of fruits per individual were significantly positively correlated ($N = 279$; $r = 0.53$; $P < 0.001$; Dutilleul's *t*-test), indicating that asymmetric competition increased with density.

Previous experiments with an identical set-up also showed that density only affected the number of fruits per individual whilst the rest of life-history traits remained unaffected [54].

Spatial autocorrelation patterns of environmental variables (19 weather records, percentage of humanised habitat and pH) and life-history traits (seed weight, seed germination at three temperatures, maximum number of vegetative rosettes, flowering time, number of reproductive individuals, mean number of fruits per individual, and individual fitness) were analysed with PASSaGE v.2 [58]. For each environmental variable and life-history trait, we computed Moran's $I$ autocorrelation coefficients and their significance was estimated from 1000 permutations. Correlations between pairs of environmental variables and between pairs of life-history traits were tested with Dutilleul's modified $t$-test using SAM v.4 [59]. Dutilleul's $t$-test corrects the variance of the test statistic and the degrees of freedom according to the extent of spatial autocorrelation of each variable [60]. We also analysed the correlation between life-history traits, excluding the two traits used to compute fitness, and individual fitness with Dutilleul's modified $t$-test.

Simultaneous autoregressive models (SAR) were performed to test the effects of environmental variables on life-history traits using SAM. SAR is a regression technique, based on generalised least squares (GLS), to estimate regression parameters that takes spatial patterns of data into account by including an additional term for the autocorrelation matrix of the errors [61]. We did not transform any variable or trait because the lack of autocorrelation patterns in the residuals showed that the assumptions of the analyses were met. To check the robustness of patterns obtained with SAR models, we also conducted a complementary approach by performing a canonical correlation analysis (CCA) with SYSTAT v.13 (Systat Software, Inc., Chicago, IL, USA). CCA evaluates the relationship between two sets of variables and generates predicted values for each set of variables that have the highest linear correlation between them. In our case, CCA correlated the set of environmental variables and the set of life-history traits while taking among-environmental variable and among-life-history trait correlations into account. We only used weather variables showing pairwise correlation coefficients below 0.75 to avoid excessive collinearity between explicative variables [62]. As a result, we eventually included nine out of 19 weather variables in both SAR and CCA analyses. In the case of CCA, we also considered those life-history traits with pairwise correlation coefficients below 0.75 for the same reason.

Finally, for each of the nine weather variables selected, we estimated similarity matrices based on Euclidian distances. We also estimated similarity distances for life-history traits among accessions. Parametric partial Mantel tests were conducted with PASSaGE to analyse the relationship between climatic similarity and life-history trait similarity controlling for the geographic position of accessions given by the geographic distance matrix. In order to understand the relationship between phenotypic, genetic and environmental variation in this set of Iberian *A. thaliana* accessions, we also conducted partial Mantel tests to assess the relationship between genetic similarity and life-history trait similarity, and between genetic similarity and climatic similarity, always controlling for the geographic position of accessions.

## Results

### Patterns of Variation in Environmental Variables

The main environmental features of the 279 *A. thaliana* populations of study were as follows. Annual mean temperature ranged between 4.9 and 18.2°C (mean ± SD = 12.2±2.6°C; Fig. 1A), the minimum mean temperature of the coldest month varied from a low of −7.5°C to a high of 7.3°C (mean ± SD = 0.05±2.6°C), and the maximum mean temperature of the warmest month varied between 20.6°C and 35.9°C (mean ± SD = 29.1±3.4°C). Total annual precipitation ranged between 384.3 and 1,799.4 mm (mean ± SD = 771.3±281.6 mm; Fig. 1B), precipitation of the driest month varied between 0.6 and 90.3 mm (mean ± SD = 24.8±17.5 mm), and precipitation of the wettest month ranged from a low of 44.1 to a high of 288.4 mm (mean ± SD = 103.1±39.3 mm). The percentage of humanised habitat ranged between 0 and 100% (mean ± SD = 40.3±37.2%; Fig. 1C) and pH varied between the acidic 3.6 and the basic 7.5 (mean ± SD = 5.7±0.8; Fig. 1D). In general, correlations between pairs of environmental variables indicated that accessions from environments with warmer annual mean temperature were those with lower precipitations during the dry season, more seasonal precipitations throughout the year, and a higher percentage of humanised habitat (Table S1). Accessions from environments with basic soils were also those with lower annual precipitations (Table S1).

All 19 climatic variables were significantly spatially autocorrelated (range of Moran's $I = 0.19–0.36$; $P<0.001$ in all cases) in this set of *A. thaliana* accessions. The pH also exhibited a significant spatial autocorrelation (Moran's $I = 0.20$; $P<0.001$). The percentage of humanised habitat was not significantly spatially autocorrelated ($P = 0.18$). The distances between population pairs exhibiting significant autocorrelation patterns for environmental variables varied approximately between 130 and 350 km.

### Patterns of Variation in Life-history Traits

In our set of 279 *A. thaliana* accessions, seed weight (mean ± SD = $2.1±0.4×10^{-5}$; range = $1.3–3.6×10^{-5}$ g; Fig. 3A) was not significantly correlated with any life-history trait (Table 1). All seed germination proportions at three temperatures were significantly positively correlated among them (Table 1). Seed germination at 10°C (mean ± SD = 0.69±0.29; Fig. 3B) and 22°C (mean ± SD = 0.67±0.31; Fig. 3C) showed high proportions, whereas seed germination at 30°C (mean ± SD = 0.12±0.23; Fig. 3D) exhibited the lowest value. Seed germination at 10°C and 22°C were significantly positively correlated with number of vegetative rosettes (mean ± SD = 30.1±8.1; range = 8.9–46.3 rosettes per pot; Fig. 3E) and number of reproductive individuals (mean ± SD = 24.8±7.4; range = 4.8–43.3 individuals per pot; Table 1 and Fig. 3F). The number of vegetative rosettes and the number of reproductive individuals were strongly positively correlated (Table 1). Finally, flowering time (mean ± SD = 143.1±20.4; range = 85.6–187.5 days; Fig. 3G) was significantly positively correlated with mean number of fruits per individual (mean ± SD = 53.6±37.3; range = 10.3–242.1 fruits per individual; Table 1 and Fig. 3H).

Individual fitness was significantly positively correlated with seed germination at 10°C ($r = 0.17$; $P = 0.0004$), seed germination at 22°C ($r = 0.22$; $P<0.0001$), seed germination at 30°C ($r = 0.19$; $P = 0.001$), number of vegetative rosettes ($r = 0.41$; $P<0.0001$), and flowering time ($r = 0.33$; $P<0.0001$). In contrast, individual fitness and seed weight were not significantly correlated ($P = 0.67$).

Our estimates of heritability for all life-history traits indicated high values for seed weight, flowering time, and the three seed germination values at three temperatures (range $h^2 = 0.82–0.96$; Table 1). The number of vegetative rosettes, the number of reproductive individuals and the mean number of fruits per individual exhibited intermediate or low heritability values (range $h^2 = 0.08–0.41$; Table 1). Finally, heritability for individual fitness was also low in this set of *A. thaliana* accessions ($h^2 = 0.09$).

**Table 1.** Pearson's correlation coefficients among life-history traits and heritability values of 279 Iberian *A. thaliana* accessions.

| | Seed weight | Germination (10°C) | Germination (22°C) | Germination (30°C) | Number of rosettes | Flowering time | Number of adults | Fruit production |
|---|---|---|---|---|---|---|---|---|
| Seed weight | 0.96 | | | | | | | |
| Germination (10°C) | −0.116 *ns* | 0.87 | | | | | | |
| Germination (22°C) | −0.074 *ns* | 0.672*** | 0.82 | | | | | |
| Germination (30°C) | −0.053 *ns* | 0.319*** | 0.416*** | 0.86 | | | | |
| Number of rosettes | 0.096 *ns* | 0.344*** | 0.240*** | 0.185 *ns* | 0.41 | | | |
| Flowering time | 0.106 *ns* | −0.177 *ns* | −0.004 *ns* | 0.033 *ns* | −0.006 *ns* | 0.93 | | |
| Number of adults | 0.072 *ns* | 0.341*** | 0.211*** | 0.144 *ns* | 0.922*** | −0.107 *ns* | 0.24 | |
| Fruit production | −0.059 *ns* | 0.056 *ns* | 0.157 *ns* | 0.139 *ns* | 0.112 *ns* | 0.418*** | 0.048 *ns* | 0.08 |

Heritability values are given on the diagonal. Correlation coefficients were obtained from Dutilleul's modified *t*-tests. A new threshold significance value ($\alpha = 0.0018$) was set after applying the Dunn-Šidák correction $(1-[1-\alpha]^{1/n})$ for multiple comparisons. Significance: ***$P<0.0001$, *ns*; non-significant.

Only seed weight (Moran's $I = 0.33$; $P<0.001$) and flowering time (Moran's $I = 0.21$; $P<0.001$) showed significant spatial autocorrelation patterns whereas the rest of traits were not significantly spatially autocorrelated ($P>0.63$ in all cases). Individual fitness was not spatially autocorrelated either ($P = 0.91$). The distance between accession pairs with significant autocorrelation patterns was approximately of 500 km for seed weight and 300 km for flowering time.

## Relationship between Environmental Variables and Life-history Traits

Univariate SAR models showed that only seed weight and flowering time were significantly correlated with environmental variables (Table 2). Seed weight was significantly negatively correlated with annual mean temperature, mean temperature of the wettest quarter and the percentage of humanised habitat, whereas it was significantly positively correlated with annual precipitation (Table 2). Flowering time was significantly negatively correlated with annual mean temperature, mean temperature of the wettest quarter, mean temperature of the driest quarter, precipitation seasonality and the percentage of humanised habitat (Table 2). In contrast, flowering time was significantly positively correlated with precipitation of the driest month and pH. It is noteworthy that annual mean temperature exhibited the highest correlation coefficients with seed weight and flowering time (Table 2).

SAR models also indicated that individual fitness was significantly negatively correlated with annual mean temperature ($r = -0.23$; $P = 0.003$) and precipitation seasonality ($r = -0.23$; $P = 0.011$). This indicates that *A. thaliana* accessions from warmer environments and environments with higher seasonality in precipitation regimes, which were also positively correlated between them (Table S1), performed worse at the experimental facility (Fig. 4). Individual fitness was not significantly correlated with any other climatic variable ($P>0.07$ in all cases). Despite the significant negative linear trend between individual fitness and annual mean temperature, accessions from populations with annual mean temperatures below 7.5°C ($N = 45$ accessions; 16.1%) and above 15°C ($N = 20$ accessions; 7.2%) clearly showed reduced fitness (Fig. 4A).

CCA yielded two significant overall canonical correlation coefficients (Table 3), although the first canonical correlation coefficient was about two-fold higher than the second one (coefficients = 0.83 and 0.36, $P<0.008$ in both cases; Table 3). This indicated the preponderance of the first canonical correlation coefficient in this analysis. The most important canonical variate (i.e. a linear combination of environmental variables and life-history traits in this case) showed that only flowering time and annual mean temperature exhibited a negative relationship (Table 3). The second canonical variate showed the positive relationship between seed weight and number of vegetative rosettes, and the negative relationship of these two traits with mean number of fruits per individual (Table 3). Such life-history traits were in turn negatively related to mean diurnal range and the percentage of humanised habitat, whereas they were positively related to annual precipitation (Table 3). Overall, CCA results highlighted the strong negative relationship between flowering time and annual mean temperature in this set of Iberian *A. thaliana* accessions.

Partial Mantel tests correlating climatic similarity with similarity in life-history traits also showed that only seed weight and flowering time exhibited significant correlations (Table 4). Similarity in seed weight and flowering time was significantly positively correlated with similarity between environments for annual mean temperature (Table 4). Similarity in seed weight and flowering time was also significantly positively correlated with similarity between environments for mean temperature of the wettest quarter and precipitation seasonality, respectively (Table 4). These analyses provided complementary support to the relationship inferred from SAR and CCA models between flowering time and annual mean temperature, and to a lesser extent between seed weight and annual mean temperature. Finally, none of the partial Mantel tests correlating similarity between environments and similarity in individual fitness was significant ($P>0.12$ in all cases).

Additional partial Mantel tests between genetic similarity and environmental similarity yielded significant positive correlations for almost all environmental variables analysed in this study, although weaker than Mantel tests for life-history traits (range $r = 0.04$–$0.10$, $P<0.035$ in all cases). Only similarity in annual mean temperature, precipitation seasonality, and pH did not show significant correlations with genetic similarity ($P>0.08$ in all cases). The strong genetic structure of *A. thaliana* in the Iberian Peninsula, as reported elsewhere [46,48], accounts for the high similarity between genetic and environmental distances found in this study. Finally, genetic similarity was significantly positively correlated with similarity in seed weight, flowering time, mean number of fruits per individual and individual fitness (range $r = 0.06$–$0.11$, $P<0.001$ in all cases). However, these correlations were again

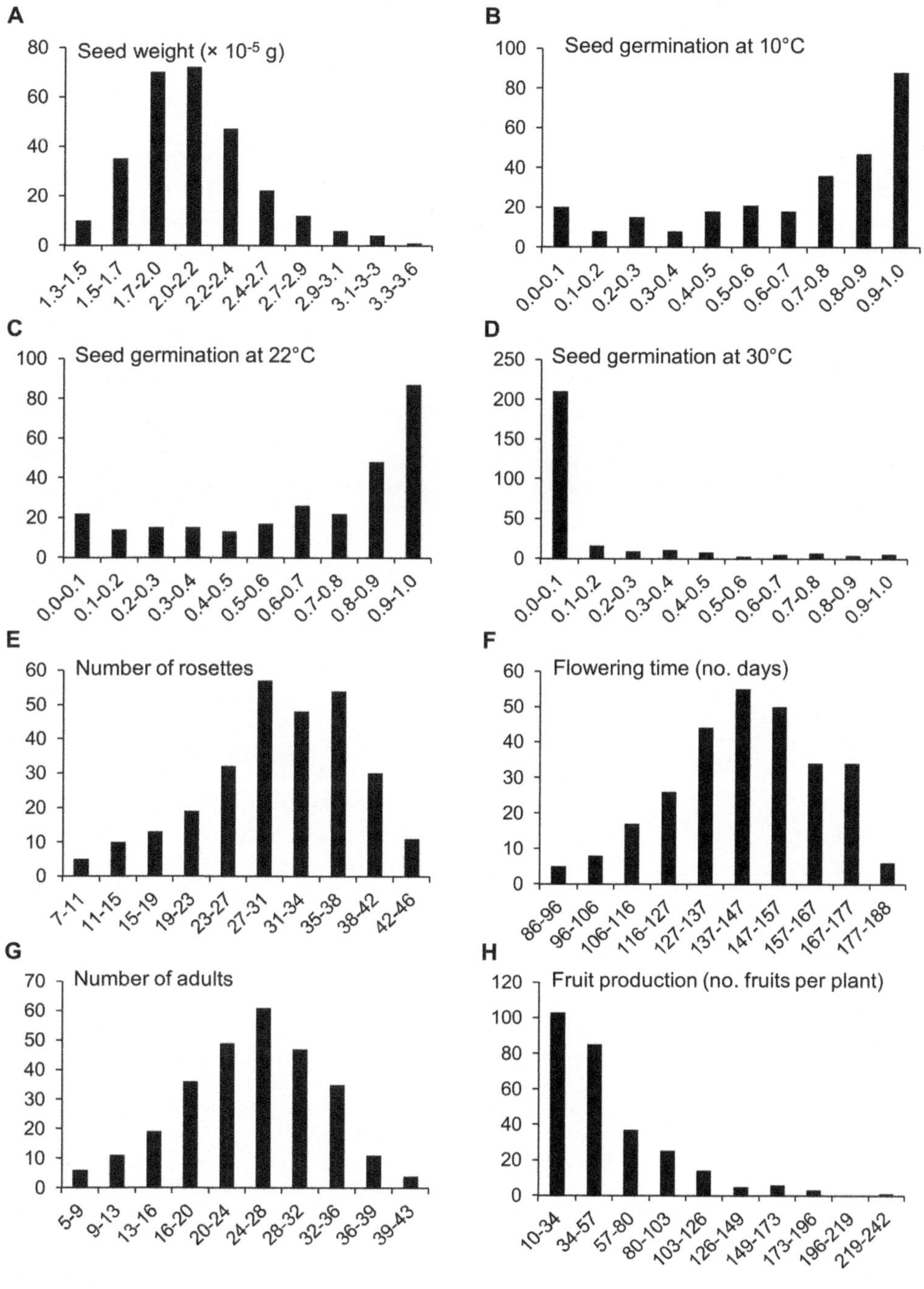


**Figure 3. Frequency distributions for life-history traits of *A. thaliana*.** Graphs depict the number of accessions within each interval. $N = 279$ accessions in all cases.

weaker than those between life-history traits and environmental variables.

## Discussion

Understanding large-scale adaptive variation requires the measurement of changes in phenotypic traits, the effects of such changes on fitness, and the identification of the environmental factors acting as proxies of selective agents. Overall, phenotypic variation is expected to move towards different phenotypes that best fit different environments, as developed by [63] and recently reviewed by [64], configuring the Wright's adaptive landscapes if phenotypes represent different gene combinations [65,66]. In this study, we have used the annual plant *Arabidopsis thaliana* to unravel the adjustment between environment and life history, highlighting the importance that the geographical component has to identify ecologically and evolutionarily relevant phenotypic traits and environmental factors. Geographically-explicit approaches evaluating the match between organisms and the environments where they live can be very important to assess and predict the consequences of dramatic environmental changes, such as global warming and increasing aridity.

**Table 2.** Correlation coefficients between life-history traits and environmental variables in *A. thaliana*.

| | Seed weight | Germination (10°C) | Germination (22°C) | Germination (30°C) | Number of rosettes | Flowering time | Number of adults | Fruit production |
|---|---|---|---|---|---|---|---|---|
| Annual temperature | −0.391*** | 0.064 *ns* | −0.135 *ns* | −0.051 *ns* | −0.194 *ns* | −0.699*** | −0.086 *ns* | −0.233 *ns* |
| Diurnal range | −0.137 *ns* | −0.006 *ns* | 0.095 *ns* | 0.053 *ns* | −0.086 *ns* | 0.205 *ns* | −0.089 *ns* | 0.133 *ns* |
| Temp. seasonality | −0.002 *ns* | −0.068 *ns* | 0.097 *ns* | −0.013 *ns* | −0.067 *ns* | 0.073 *ns* | −0.079 *ns* | 0.100 *ns* |
| Temp. wettest quarter | −0.287*** | −0.032 *ns* | −0.141 *ns* | 0.049 *ns* | −0.125 *ns* | −0.385*** | −0.071 *ns* | −0.073 *ns* |
| Temp. driest quarter | −0.220 *ns* | 0.099 *ns* | −0.076 *ns* | −0.070 *ns* | −0.160 *ns* | −0.427*** | −0.133 *ns* | −0.118 *ns* |
| Annual precipitation | 0.225*** | 0.045 *ns* | 0.018 *ns* | −0.018 *ns* | 0.155 *ns* | 0.070 *ns* | 0.132 *ns* | −0.012 *ns* |
| Prec. driest month | 0.278 *ns* | −0.023 *ns* | −0.014 *ns* | 0.086 *ns* | 0.131 *ns* | 0.307*** | 0.041 *ns* | 0.117 *ns* |
| Prec. seasonality | −0.092 *ns* | 0.091 *ns* | −0.030 *ns* | −0.097 *ns* | 0.035 *ns* | −0.506*** | 0.056 *ns* | −0.258 *ns* |
| Humanised habitat | −0.241*** | 0.141 *ns* | 0.061 *ns* | 0.087 *ns* | 0.041 *ns* | −0.262*** | 0.084 *ns* | 0.018 *ns* |
| pH | 0.098 *ns* | −0.149 *ns* | −0.056 *ns* | −0.097 *ns* | −0.038 *ns* | 0.225*** | −0.055 *ns* | 0.049 *ns* |

Correlation coefficients were obtained from SAR models. A new threshold significance value ($\alpha = 0.0006$) was set after applying the Dunn-Šidák correction ($1-[1-\alpha]^{1/n}$) for multiple comparisons. Significance: ***; $P<0.0001$, *ns*; non-significant.

## Variation in Environmental Variables and Life-history Traits

The analyses performed in this study indicated that the main adjustment between *A. thaliana*'s life history and the environmental conditions found across the Iberian Peninsula was mainly carried out through variation in flowering time. In addition, geographical variation in flowering time appeared to be related to geographical variation in annual mean temperature. These patterns of variation in flowering time are in agreement with recent empirical evidence indicating that flowering in annuals [29,30,32] and perennials [29,67–72] takes place earlier as environments get warmer. Such generalised advance of reproductive phenology with increasing temperature occurs because reductions in the life cycle duration allow plants to escape from unfavourable environmental conditions for growth and reproduction [73,74]. It must be emphasized, however, that other traits correlated with flowering time, but not measured in this study, could also represent targets of selection, e.g. rosette size at bolting or leaf production rate during vegetative growth [75]. Further field experiments are needed to take into account architectural plant traits to be compared with life-history traits.

Our results showed that *A. thaliana* accessions with more vigorous seed germination, higher recruitment and later flowering times exhibited higher individual fitness at the experimental facility. We have recently shown that the environmental conditions at the El Castillejo Botanical Garden correspond quite well to environments with mild and moderately cold winters [54]. Such environments activate the vernalization pathway, i.e. the flowering inductive effect of low temperature during winter months on vegetative rosettes, which is in turn associated to late-flowering behaviours in *A. thaliana* [48,54]. Hence, the better fitness performance of late-flowering *A. thaliana* accessions at the experimental facility could reflect their adaptation to the environments where they come from and the good match between local and experimental climatic characteristics for these accessions. The view that the patterns of variation in flowering time found in this study are adaptive is supported by the results of the partial Mantel test, indicating that a higher climatic similarity between local environments was significantly positively correlated with higher similarity in flowering time. Recent findings also indicated that Iberian *A. thaliana* populations were more genetically differentiated for flowering time than for neutral molecular markers, suggesting that flowering time is likely to be under divergent selection in *A. thaliana* [54]. It must be noted, however, that environmental adaptations of *A. thaliana* mediated by changes in flowering time must also be influenced, and probably enhanced, by the demographic history of accessions in the Iberian Peninsula, as indicated by the significant correlations between combinations of environmental, genetic, and flowering time similarity matrices.

Seed weight was the other trait showing a significant spatial autocorrelation pattern as well as significant relationships with environmental variables. In particular, accessions from warmer environments produced lighter seeds. As shown by previous observations in *A. thaliana* [14,76], variation in seed weight might partly be determined by variation in flowering time, since early flowering reduces the length of vegetative growth and subsequently the amount of resources that mother plants allocate to seeds. On top of that, it has also been shown that inflorescences are more effective than rosettes fixing carbon and increasing instantaneous water use efficiency in *A. thaliana* [77]. As a result, late-flowering individuals can possess the physiological means to increase carbon acquisition during the reproductive phase and eventually produce heavier seeds. Nonetheless, the lack of relationship between flowering time and seed weight found in this study suggests the existence of other factors, such as the known trade-off between seed size and seed number in *A. thaliana* [14,78], which need to be taken into account in future experiments to better assess the relationship between life-history traits and seed production in field environments.

Finally, individual fitness at the experimental facility decreased in *A. thaliana* accessions from environments with warmer annual mean temperatures and higher seasonal rains. Interestingly, genetic similarity and individual fitness similarity were also significantly positively correlated, suggesting that the genetic background of accessions also influenced their performance in the field experiment. Despite such relationships, it is noteworthy that accessions from environments with more extreme annual mean temperatures (i.e. 15°C< annual mean temperature <7.5°C) exhibited a trend toward reduced fitness. We have already seen the expected effects of warm environments on flowering time, but an explanation is due for the lower fitness performance found for accessions from cold environments, which are generally located at high altitudes in the Iberian Peninsula [46,48]. It has recently been shown that high-altitude *A. thaliana*

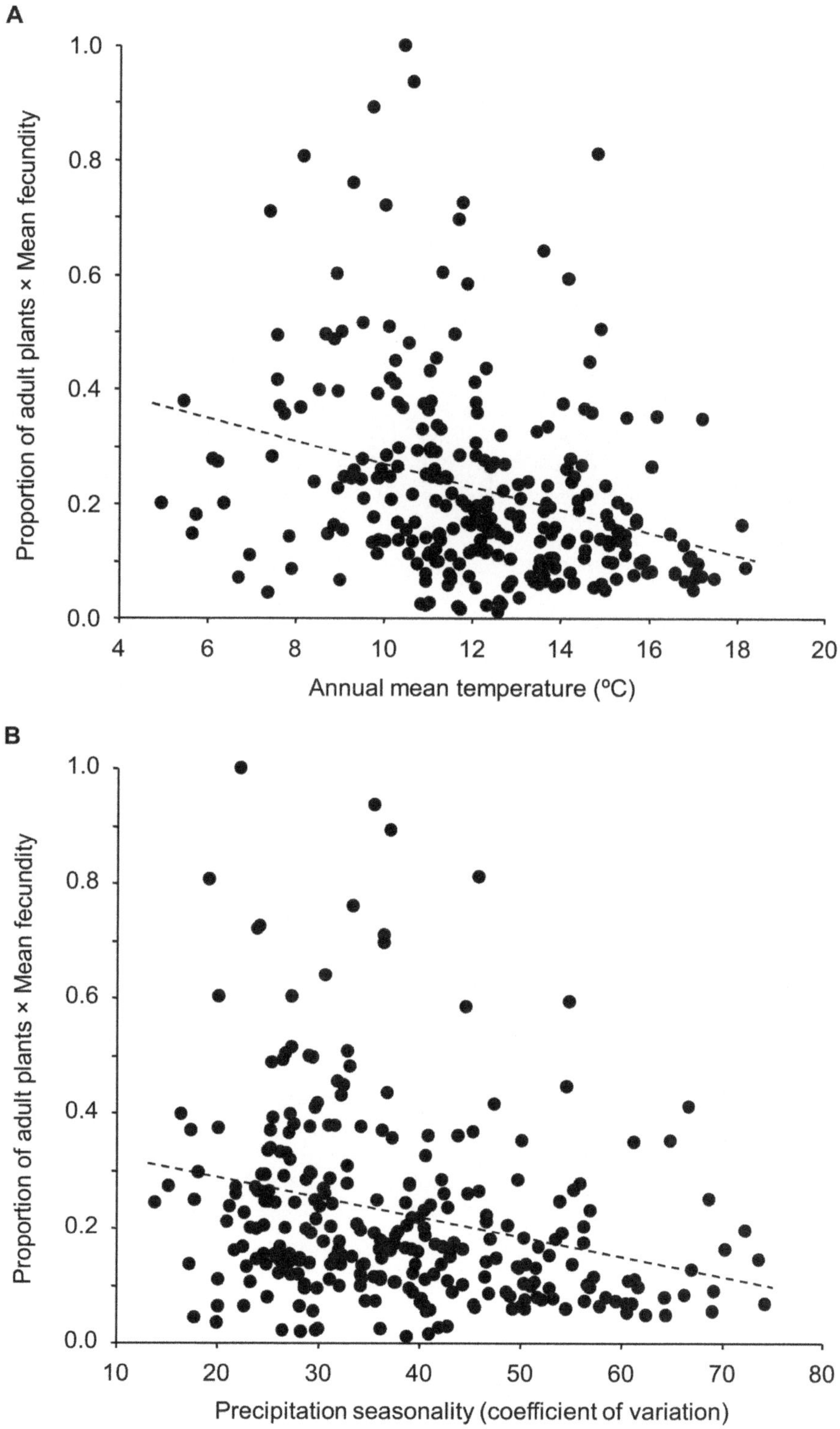


**Figure 4. Relationship between individual fitness and annual mean temperature or precipitation seasonality in *A. thaliana*.**

populations are predominantly composed of spring-germinated plants that overwinter as seeds in the soil seed bank. This has been interpreted as an escape strategy to mitigate the effects of massive mortality of rosettes during the cold harsh winter on population viability [49]. Thus, the lower fitness performance of accessions from cold environments at the experimental facility might reflect the extent of local adaptation of such accessions to their spring annual behaviour.

**Table 3.** Canonical correlations between environmental variables and life-history traits in *A. thaliana*.

| Group | Variables | 1st Canonical variate correlation | 2nd Canonical variate correlation |
|---|---|---|---|
| Life history | Seed weight | 0.35 | **0.83** |
| | Germination (10°C) | −0.10 | −0.08 |
| | Number of rosettes | 0.19 | **0.38** |
| | Flowering time | **0.95** | −0.28 |
| | Fruit production | 0.36 | **−0.40** |
| Environment | Annual temperature | **−0.93** | −0.10 |
| | Diurnal range | 0.10 | **−0.56** |
| | Temp. seasonality | 0.17 | −0.24 |
| | Temp. wettest quarter | −0.46 | −0.12 |
| | Temp. driest quarter | −0.60 | −0.21 |
| | Annual precipitation | 0.17 | **0.59** |
| | Prec. driest month | 0.42 | 0.24 |
| | Prec. seasonality | −0.54 | 0.26 |
| | Humanised habitat | −0.46 | **−0.36** |
| | pH | 0.23 | 0.12 |

Results for the first two significant canonical variates are given. Overall canonical correlations for the first and second canonical variates are 0.83 and 0.36, respectively. Canonical correlations of environmental variables and life-history traits above the overall canonical correlation for each canonical variate are in bold face.

## Expected Effects of Global Warming on A. Thaliana's Life History

We want to discuss the results of this study in a context of global warming given the current great concern on this issue [79–81]. Based on our findings and the evidence about the effects of increased temperature on flowering time, we expect *A. thaliana* to shift its life history toward early-flowering behaviour with smaller seed sizes with increasing warming. However, it must be emphasized that *A. thaliana* is very rare in vast areas across the driest and hottest regions of the Iberian Peninsula, including lowlands from southern and eastern Spain (Fig. 1). Hence, we hypothesize a more pessimistic scenario in which *A. thaliana* might go extinct in those regions that resulted more affected by increasing warming and drought. In dry hot environments, advancing flowering time might not be enough to maintain viable populations and *A. thaliana* should acquire other traits, such as delayed seed germination, as shown in desert annuals along gradients of aridity and climate predictability [27]. Additional experiments are also needed to assess realistically the adaptive potential and the developmental constraints of the species in dry environments.

Recently, we have shown the effect of loss- and change-of-function alleles of some well-known flowering genes, such as *FRIGIDA*, *FLOWERING LOCUS C*, and *PHYTOCHROME C* on flowering time variation of Iberian *A. thaliana* accessions [48,82]. It is becoming increasingly evident that nucleotide polymorphisms of genes involved in *A. thaliana* natural variation for flowering generally have a low frequency and are strongly geographically structured [39,48,53,82,83]. These results suggest that the role that regulatory pathways play to account for flowering time variation also vary geographically across the Iberian Peninsula. For example, accessions carrying specific alleles of *FLOWERING LOCUS C* (i.e. vernalization pathway for flowering) or *CRYPTOCHROME2* (i.e. photoperiod pathway for flowering) mainly occur in northern wet or south-western dry areas of the Iberian Peninsula, respectively [48]. Therefore, the adaptive potential of flowering time in *A. thaliana* across different Iberian regions might strongly be constrained by the differences in the genetic basis of the regulatory pathways accounting for flowering in each region. The effect of increased warming and drought on flowering time variation regulated by different genetic pathways seems difficult to predict and represents an important future research line.

## Conclusions

We stress the need to conduct experiments with extensive collections of geo-referenced accessions to evaluate the interplay between environmental factors and phenotypic variation. The reasons are manifold. First, environmental main selective pressures, although with some exceptions (e.g. pathogens or competition), vary geographically in a continuous fashion. Hence, experiments ought to capture such continuous adaptive landscape in order to be realistic and conclusive. Second, the increasing evidence that the genetic regulatory pathways underlying phenotypic variation are also geographically structured should also warn us against the use of a limited number of populations in experimental designs. Third, given the inherent spatial patterns of variation in environmental variables and life-history traits,

**Table 4.** Correlation coefficients between life-history trait and environmental similarity in *A. thaliana*.

| | Seed Weight | Germination (10°C) | Germination (22°C) | Germination (30°C) | Number of rosettes | Flowering time | Number of adults | Fruit production |
|---|---|---|---|---|---|---|---|---|
| Annual temperature | 0.185** | 0.009 *ns* | −0.011 *ns* | −0.007 *ns* | 0.010 *ns* | 0.429** | −0.040 *ns* | 0.044 *ns* |
| Diurnal range | 0.055 *ns* | 0.018 *ns* | 0.029 *ns* | −0.068 *ns* | 0.024 *ns* | −0.011 *ns* | 0.047 *ns* | −0.033 *ns* |
| Temp. seasonality | −0.031 *ns* | 0.030 *ns* | 0.077 *ns* | −0.048 *ns* | 0.039 *ns* | 0.051 *ns* | 0.045 *ns* | −0.011 *ns* |
| Temp. wettest quarter | 0.155** | 0.007 *ns* | −0.023 *ns* | −0.047 *ns* | −0.048 *ns* | 0.070 *ns* | −0.054 *ns* | 0.018 *ns* |
| Temp. driest quarter | −0.015 *ns* | 0.018 *ns* | −0.048 *ns* | 0.089 *ns* | 0.113 *ns* | 0.096 *ns* | 0.066 *ns* | 0.032 *ns* |
| Annual precipitation | 0.092 *ns* | −0.037 *ns* | −0.019 *ns* | −0.034 *ns* | 0.027 *ns* | −0.036 *ns* | 0.021 *ns* | −0.033 *ns* |
| Prec. driest month | 0.019 *ns* | 0.005 *ns* | 0.035 *ns* | 0.024 *ns* | 0.087 *ns* | 0.045 *ns* | 0.066 *ns* | −0.027 *ns* |
| Prec. seasonality | −0.007 *ns* | −0.011 *ns* | 0.031 *ns* | −0.004 *ns* | −0.021 *ns* | 0.181** | −0.034 *ns* | 0.033 *ns* |

Correlation coefficients were obtained from partial Mantel tests. A new threshold significance value ($\alpha = 0.001$) was set after applying the Dunn-Šidák correction ($1-[1-\alpha]^{1/n}$) for multiple comparisons. Significance: **; $P<0.001$, *ns*; non-significant.

statistical analyses need to take spatial autocorrelation into account, and only extensive spatially-explicit data sets can cope with spatial autocorrelation properly [84]. Finally, although the ideal experimental design should include the replication of the experiment in different environments [37,85], the combination of spatially-explicit statistical tools proved to be a useful method to depict patterns of adaptive variation. Overall, merging environmental, phenotypic and molecular data using extensive geo-referenced collections of natural populations can strongly accelerate the pace with which we obtain comprehensive insights into adaptive variation in plants.

## Supporting Information

**Figure S1** Daily weather records at the experimental facility during the experiment.

**Table S1**

## Acknowledgments

We are grateful to R. Gómez, J.C. López, M. Ramiro, J. Pozas, B. Méndez-Vigo, and the staff of the El Castillejo Botanical Garden for field and lab assistance. We thank M. Expósito for comments on an earlier draft of the manuscript. An editor and two reviewers also contributed to improve the manuscript. The geographic information system was developed by V. Garcia and A. Escobar. We also thank the administration of the Sierra de Grazalema Natural Park for permission to work at the El Castillejo Botanical Garden.

## Author Contributions

Conceived and designed the experiments: CAB FXP. Performed the experiments: EMP FXP. Analyzed the data: FXP. Contributed reagents/materials/analysis tools: CAB AM FXP. Wrote the paper: CAB AM FXP.

## References

1. Freas KE, Kemp PR (1983) Some relationships between environmental reliability and seed dormancy in desert annual plants. J Ecol 71: 211–217.
2. Philippi T (1993) Bet-hedging germination of desert annuals: variation among populations and maternal effects in *Lepidium lasiocarpum*. Am Nat 142: 488–507.
3. Pake CE, Venable DL (1996) Seed banks in desert annuals: implications for persistence and coexistence in variable environments. Ecology 77: 1427–1435.
4. Venable DL (2007) Bet hedging in a guild of desert annuals. Ecology 88: 1086–1090.
5. Childs DZ, Metcalf CJE, Rees M (2010) Evolutionary bet-hedging in the real world: empirical evidence and challenges revealed by plants. Proc R Soc Lond B Biol Sci 277: 3055–3064.
6. Simons AM (2011) Modes of response to environmental change and the elusive empirical evidence for bet hedging. Proc R Soc Lond B Biol Sci 278: 1601–1609.
7. Tielbörger K, Petrů M, Lampei C (2012) Bet-hedging germination in annual plants: a sound empirical test of the theoretical foundations. Oikos 121: 1860–1868.
8. Ratcliffe D (1959) *Hornungia petraea* (L.) Rchb. J Ecol 47: 241–247.
9. Baskin CC, Baskin JM (1988) Germination ecophysiology of herbaceous plant species in a temperate region. Am J Bot 75: 286–305.
10. Kalisz S (1991) Experimental determination of seed bank age structure in the winter annual *Collinsia verna*. Ecology 72: 575–585.
11. Kluth C, Bruelheide H (2005) Central and peripheral *Hornungia petraea* populations: patterns and dynamics. J Ecol 93: 584–595.
12. Montesinos A, Tonsor SJ, Alonso-Blanco C, Picó FX (2009) Demographic and genetic patterns of variation among populations of *Arabidopsis thaliana* from contrasting native environments. PLoS One 4: e7213.
13. Pigliucci M, Schlichting CD (1998) Reaction norms of *Arabidopsis*. V. Flowering time controls phenotypic architecture in response to nutrient stress. J Evol Biol 11: 285–301.
14. Montesinos-Navarro A, Wig J, Picó FX, Tonsor SJ (2011) *Arabidopsis thaliana* populations show clinal variation in a climatic gradient associated with altitude. New Phytol 189: 282–294.
15. Rathcke B, Lacey EP (1985) Phenological patterns of terrestrial plants. Annu Rev Ecol Evol Syst 16: 179–214.
16. Donohue K (2005) Niche construction through phenological plasticity: life history dynamics and ecological consequences. New Phytol 166: 83–92.
17. Finch-Savage WE, Leubner-Metzger G (2006) Seed dormancy and the control of germination. New Phytol 171: 501–523.
18. Donohue K, Rubio de Casas R, Burghardt L, Kovach K, Willis CG (2010) Germination, postgermination adaptation, and species ecological ranges. Annu Rev Ecol Evol Syst 41: 293–319.
19. Rubio de Casas R, Kovach K, Dittmar E, Barua D, Barco B, et al. (2012) Seed after-ripening and dormancy determine adult life history independently of germination timing. New Phytol 194: 868–879.
20. Roux F, Touzet P, Cuguen J, Le Corre V (2006) How to be early flowering: an evolutionary perspective. Trends Plant Sci 11: 375–381.
21. Franks SJ, Sim S, Weis AE (2007) Rapid evolution of flowering time by an annual plant in response to a climate fluctuation. Proc Natl Acad Sci USA 104: 1278–1282.
22. Amasino R (2010) Seasonal and developmental timing of flowering. Plant J 61: 1001–1013.
23. Anderson JT, Lee CR, Mitchell-Olds T (2011) Life-history QTLs and natural selection on flowering time in *Boechera stricta*, a perennial relative of *Arabidopsis*. Evolution 65: 771–787.
24. Andrés F, Coupland G (2012) The genetic basis of flowering responses to seasonal cues. Nat Rev Genet 13: 627–639.
25. Simpson GG, Dean C (2002) *Arabidopsis*, the rosetta stone of flowering time? Science 296: 285–289.
26. Clauss MJ, Venable DL (2000) Seed germination in desert annuals: an empirical test of adaptive bet hedging. Am Nat 155: 168–186.
27. Volis S, Mendlinger S, Ward D (2002) Differentiation in populations of *Hordeum spontaneum* along a gradient of environmental productivity and predictability: life history and local adaptation. Biol J Linn Soc 77: 479–490.
28. Lebrija-Trejos E, Cardiel-Lobato ML, Sternberg M (2011) Reproductive traits and seed dynamics at two environmentally contrasting annual plant communities: from fieldwork to theoretical expectations. Isr J Ecol Evol 57: 73–90.
29. Fitter AH, Fitter RSR (2002) Rapid changes in flowering time in British plants. Nature 296: 1689–1691.
30. McKay JK, Richards JH, Mitchell-Olds T (2003) Genetics of drought adaptation in *Arabidopsis thaliana*: I. Pleiotropy contributes to genetic correlations among ecological traits. Mol Ecol 12: 1137–1151.
31. Petrů M, Tielbörger K, Belkin R, Sternberg M, Jeltsch F (2006) Life history variation in an annual plant under two opposing environmental constraints along an aridity gradient. Ecography 29: 66–74.
32. Craufurd PQ, Wheeler TR (2009) Climate change and the flowering time of annual crops. J Exp Bot 60: 2529–2539.
33. van Tienderen PH, van der Toorn J (1991) Genetic differentiation between populations of *Plantago lanceolata*. I. Local adaptation in three contrasting habitats. J Ecol 79: 27–42.
34. Linhart YB, Grant MC (1996) Evolutionary significance of local genetic differentiation in plants. Annu Rev Ecol Evol Syst 27: 237–277.
35. Hufford KM, Mazer SJ (2003) Plant ecotypes: genetic differentiation in the age of ecological restoration. Trends Ecol Evol 18: 147–155.
36. Kawecki TJ, Ebert D (2004) Conceptual issues in local adaptation. Ecol Lett 7: 1225–1241.
37. Fournier-Level A, Korte A, Cooper MD, Nordborg M, Schmitt J, et al. (2011) A map of local adaptation in *Arabidopsis thaliana*. Science 334: 86–89.
38. Wagmann K, Hautekèete NC, Piquot Y, Meunier C, Schmitt SE, et al. (2012) Seed dormancy distribution: explanatory ecological factors. Ann Bot 110: 1205–1219.
39. Brachi B, Villoutreix R, Faure N, Hautekèete N, Piquot Y, et al. (2013) Investigation of the geographical scale of adaptive phenological variation and its underlying genetics in *Arabidopsis thaliana*. Mol Ecol 22: 4222–4240.
40. Debieu M, Tang C, Stich B, Sikosek T, Effgen S, et al. (2013) Co-variation between seed dormancy, growth rate and flowering time changes with latitude in *Arabidopsis thaliana*. PLoS One 8: e61075.
41. Haldane JBS (1948) The theory of a cline. J Genet 48: 277–284.
42. Mayr E (1956) Geographical character gradients and climatic adaptation. Evolution 10: 105–108.
43. Endler JA (1977) Geographic variation, speciation, and clines. Princeton: Princeton University Press. 262 p.
44. Mitchell-Olds T, Shaw RG (1987) Regression analysis of natural selection: statistical inference and biological interpretation. Evolution 41: 1149–1161.
45. Rausher MD (1992) The measurement of selection on quantitative traits: biases due to environmental covariances between traits and fitness. Evolution 46: 616–626.
46. Picó FX, Méndez-Vigo B, Martínez-Zapater JM, Alonso-Blanco C (2008) Natural genetic variation of *Arabidopsis thaliana* is geographically structured in the Iberian Peninsula. Genetics 180: 1009–1021.

47. Gomaa NH, Montesinos-Navarro A, Alonso-Blanco C, Picó FX (2011) Temporal variation in genetic diversity and effective population size of Mediterranean and subalpine *Arabidopsis thaliana* populations. Mol Ecol 20: 3540–3554.
48. Méndez-Vigo B, Picó FX, Ramiro M, Martínez-Zapater JM, Alonso-Blanco C (2011) Altitudinal and climatic adaptation is mediated by flowering traits and *FRI, FLC* and *PhyC* genes in *Arabidopsis thaliana*. Plant Physiol 157: 1942–1955.
49. Picó FX (2012) Demographic fate of *Arabidopsis thaliana* cohorts of autumn- and spring-germinated plants along an altitudinal gradient. J Ecol 100: 1009–1018.
50. Platt A, Horton M, Huang YS, Li Y, Anastasio AE, et al. (2010) The scale of population structure in *Arabidopsis thaliana*. PLoS Genet 6: e1000843.
51. Donohue K (2002) Germination timing influences natural selection and life-history characters in *Arabidopsis thaliana*. Ecology 83: 1006–1016.
52. Wilczek AM, Roe JL, Knapp MC, Cooper MD, Lopez-Gallego C, et al. (2009) Effects of genetic perturbation on seasonal life history plasticity. Science 323: 930–934.
53. Weigel D (2012) Natural variation in *Arabidopsis*: from molecular genetics to ecological genomics. Plant Physiol 158: 2–22.
54. Méndez-Vigo B, Gomaa NH, Alonso-Blanco C, Picó FX (2013) Among- and within-population variation in flowering time of Iberian *Arabidopsis thaliana* estimated in field and glasshouse conditions. New Phytol 197: 1332–1343.
55. Donohue K, Heschel MS, Butler CM, Barua D, Sharrock DA, et al. (2008) Diversification of phytochrome contributions to germination as a function of seed-maturation environment. New Phytol 177: 367–379.
56. Hijmans RJ, Cameron SE, Parra JL, Jones PG, Jarvis A (2005) Very high resolution interpolated climate surfaces for global land areas. Int J Climatol 25: 1965–1978.
57. Ninyerola M, Pons X, Roure JM (2007) Monthly precipitation mapping of the Iberian Peninsula using spatial interpolation tools implemented in a Geographic Information System. Theor Appl Climatol 89: 195–209.
58. Rosenberg MS, Anderson CD (2011) PASSaGE: pattern analysis, spatial statistics and geographic exegesis. Version 2. Methods Ecol Evol 2: 229–232.
59. Rangel TF, Diniz-Filho JAF, Bini LM (2010) SAM: a comprehensive application for spatial analysis in macroecology. Ecography 33: 46–50.
60. Dutilleul P, Clifford P, Richardson S, Hemon D (1993) Modifying the *t* test for assessing the correlation between two spatial processes. Biometrics 49: 305–314.
61. Beale CM, Lennon JJ, Yearsley JM, Brewer MJ, Elston DA (2010) Regression analysis of spatial data. Ecol Lett 13: 246–264.
62. Dormann CF, Elith J, Bacher S, Buchmann C, Carl G, et al. (2013) Collinearity: a review of methods to deal with it and a simulation study evaluating their performance. Ecography 36: 27–46.
63. Fisher RA (1930) The genetical theory of natural selection. Oxford: Oxford University Press. 272 p.
64. Orr HA (2005) The genetic theory of adaptation: a brief history. Nat Rev Genet 6: 119–127.
65. Wright S (1932) The roles of mutation, inbreeding, crossbreeding, and selection in evolution. Proc 6th Int Congr Genet 1: 356–366.
66. Kauffman S, Levin S (1987) Towards a general theory of adaptive walks on rugged landscapes. J Theor Biol 128: 11–45.
67. Parmesan C, Yohe G (2003) A globally coherent fingerprint of climate change impacts across natural systems. Nature 421: 37–42.
68. Peñuelas J, Filella I, Comas P (2002) Changed plant and animal life cycles from 1952 to 2000 in the Mediterranean region. Global Chang Biol 8: 531–544.
69. Inouye DW (2008) Effects of climate change on phenology, frost damage, and floral abundance of mountain wildflowers. Ecology 89: 353–362.
70. Gordo O, Sanz JJ (2009) Long-term temporal changes in plant phenology in the Western Mediterranean. Global Chang Biol 15: 1930–1948.
71. Amano T, Smithers RJ, Sparks TH, Sutherland WJ (2010) A 250-year index of first flowering dates and its response to temperature changes. Proc R Soc Lond B Biol Sci 277: 2451–2457.
72. Anderson JT, Inouye DW, McKinney AM, Colautti RI, Mitchell-Olds T (2012) Phenotypic plasticity and adaptive evolution contribute to advancing flowering phenology in response to climate change. Proc R Soc Lond B Biol Sci 279: 3843–3852.
73. Heschel MS, Riginos C (2005) Mechanisms of selection for drought stress tolerance and avoidance in *Impatiens capensis* (Balsaminaceae). Am J Bot 92: 37–44.
74. Brachi B, Aime C, Glorieux C, Cuguen J, Roux F (2012) Adaptive value of phenological traits in stressful environments: predictions based on seed production and laboratory natural selection. PLoS One 7: e32069.
75. Méndez-Vigo B, de Andrés MT, Ramiro M, Martínez-Zapater JM, Alonso-Blanco C (2010) Temporal analysis of natural variation for the rate of leaf production and its relationship with flowering initiation in *Arabidopsis thaliana*. J Exp Bot 61: 1611–1623.
76. Paul-Victor C, Turnbull LA (2009) The effect of growth conditions on the seed size/number trade-off. PLoS One 4: e6917.
77. Early EJ, Ingland B, Winkler J, Tonsor SJ (2009) Inflorescences contribute more than rosettes to lifetime carbon gain in *Arabidopsis thaliana* (Brassicaceae). Am J Bot 96: 786–792.
78. Alonso-Blanco C, Blankestijn-de Vries H, Hanhart CJ, Koornneef M (1999) Natural allelic variation at seed size loci in relation to other life history traits of *Arabidopsis thaliana*. Proc Natl Acad Sci USA 96: 4710–4717.
79. Vitousek PM (1994) Beyond global warming: ecology and global change. Ecology 75: 1861–1876.
80. Cox PM, Betts RA, Jones CD, Spall SA, Totterdell IJ (2000) Acceleration of global warming due to carbon-cycle feedbacks in a coupled climate model. Nature 408: 184–187.
81. Menzel A, Sparks TH, Estrella N, Koch E, Aasa A, et al. (2006) European phenological response to climate change matches the warming pattern. Global Chang Biol 12: 1969–1976.
82. Sánchez-Bermejo E, Méndez-Vigo B, Picó FX, Martínez-Zapater JM, Alonso-Blanco C (2012) Novel natural alleles at *FLC* and *LVR* loci account for enhanced vernalization responses in *Arabidopsis thaliana*. Plant Cell Environ 35: 1672–1684.
83. Nordborg M, Hu TT, Ishino Y, Jhaveri J, Toomajian C, et al. (2005) The pattern of polymorphism in *Arabidopsis thaliana*. PLoS Biol 3: e196.
84. Legendre P, Dale MRT, Fortin MJ, Gurevitch J, Hohn M, et al. (2002) The consequences of spatial structure for the design and analysis of ecological field surveys. Ecography 25: 601–615.
85. Fournier-Level A, Wilczek AM, Cooper MD, Roe JL, Anderson JT, et al. (2013) Paths to selection on life history loci in different natural environments across the native range of *Arabidopsis thaliana*. Mol Ecol 22: 3552–3566.

# 4

# Resilience to Disturbance Despite Limited Dispersal and Self-Recruitment in Tropical Barrel Sponges: Implications for Conservation and Management

**James J. Bell[1]*, David Smith[2], Danielle Hannan[1], Abdul Haris[3], Jamaludin Jompa[3], Luke Thomas[4]**

**1** School of Biological Sciences, Victoria University of Wellington, Wellington, New Zealand, **2** Coral Reef Research Unit, University of Essex, Colchester, United Kingdom, **3** Research and Development Center on Marine, Coastal and Small Islands, Hasanuddin University, Makassar, Indonesia, **4** The Oceans Institute, School of Plant Biology, University of Western Australia, Australia

**Abstract**

While estimates of connectivity are important for effective management, few such estimates are available for reef invertebrates other than for corals. Barrel sponges are one of the largest and most conspicuous members of the coral reef fauna across the Indo-Pacific and given their large size, longevity and ability to process large volumes of water, they have a major role in reef functioning. Here we used a panel of microsatellite markers to characterise the genetic structure of two barrel sponge species, *Xestospongia testudinaria* and a currently undescribed *Xestospongia* species. We sampled across seven populations in the Wakatobi Marine National Park, SE Sulawesi (Indonesia) spanning a spatial scale of approximately 2 to 70 km, and present the first estimates of demographic connectivity for coral reef sponges. Genetic analyses showed high levels of genetic differentiation between all populations for both species, but contrasting patterns of genetic structuring for the two species. Autocorrelation analyses showed the likely dispersal distances of both species to be in the order of 60 and 140 m for *Xestopongia* sp. and *Xestospongia testudinaria*, respectively, which was supported by assignment tests that showed high levels of self-recruitment (>80%). We also found consistently high inbreeding coefficients across all populations for both species. Our study highlights the potential susceptibility of barrel sponges to environmental perturbations because they are generally long-lived, slow growing, have small population sizes and are likely to be reliant on self-recruitment. Surprisingly, despite these features we actually found the highest abundance of both barrel sponge species (although they were generally smaller) at a site that has been severely impacted by humans over the last fifty years. This suggests that barrel sponges exhibit environmental adaptation to declining environmental quality and has important implications for the management and conservation of these important reef species.

**Editor:** Roberto Pronzato, University of Genova, Italy, Italy

**Funding:** This work was funded by the Victoria University of Wellington Research Fund. The funders has no role in the study design, data collection and analysis, decision to publish, or preparation of the manuscript.

* E-mail: james.bell@vuw.ac.nz

## Introduction

Coral reefs across the world have been seriously impacted by human activities to the extent that many reefs have been severely degraded [1]. Furthermore, local-scale impacts such as increases in sedimentation, habitat destruction and overfishing, coupled with global scale threats such as global warming and ocean acidification provide an uncertain future for coral reefs ecosystems [2]. However, while there has been considerable focus on how corals and fish respond to such degradation and might be impacted by future climate change scenarios, far less is known about other important reefs organisms, and little information is available to support their conservation and management. Connectivity is widely recognised as a key component of reef resilience and population viability [3–4]. Connectivity is a broad term that can be further split in to: 1) genetic connectivity, which refers to the degree to which gene flow affects evolutionary processes within populations; and 2) demographic connectivity, which refers to the degree to which population growth and vital rates are affected by dispersal and recruitment [5]. A key step in measuring demographic connectivity is to determine the actual numbers of individuals that are exchanged between populations, but this is very difficult for marine species as most rely on a pelagic larval phase to link populations. Understanding patterns of exchange between populations is important because its controls a population's buffering potential from local catastrophes, a population's potential as a source of new individuals to other populations and the level of genetic mixing between populations [6–7]. Generally, environmental managers are most interested in demographic connectivity as it controls processes over the temporal scales that management typically operates (years to decades).

Genetic techniques have been widely applied to estimate patterns of genetic connectivity, and more recently demographic connectivity of tropical marine species, but the focus has primarily been on corals and fish [8–11], with little information on connectivity patterns of other reef organisms. In order to understand broader patterns of connectivity for biologically complex reef systems it is important to understand the exchange

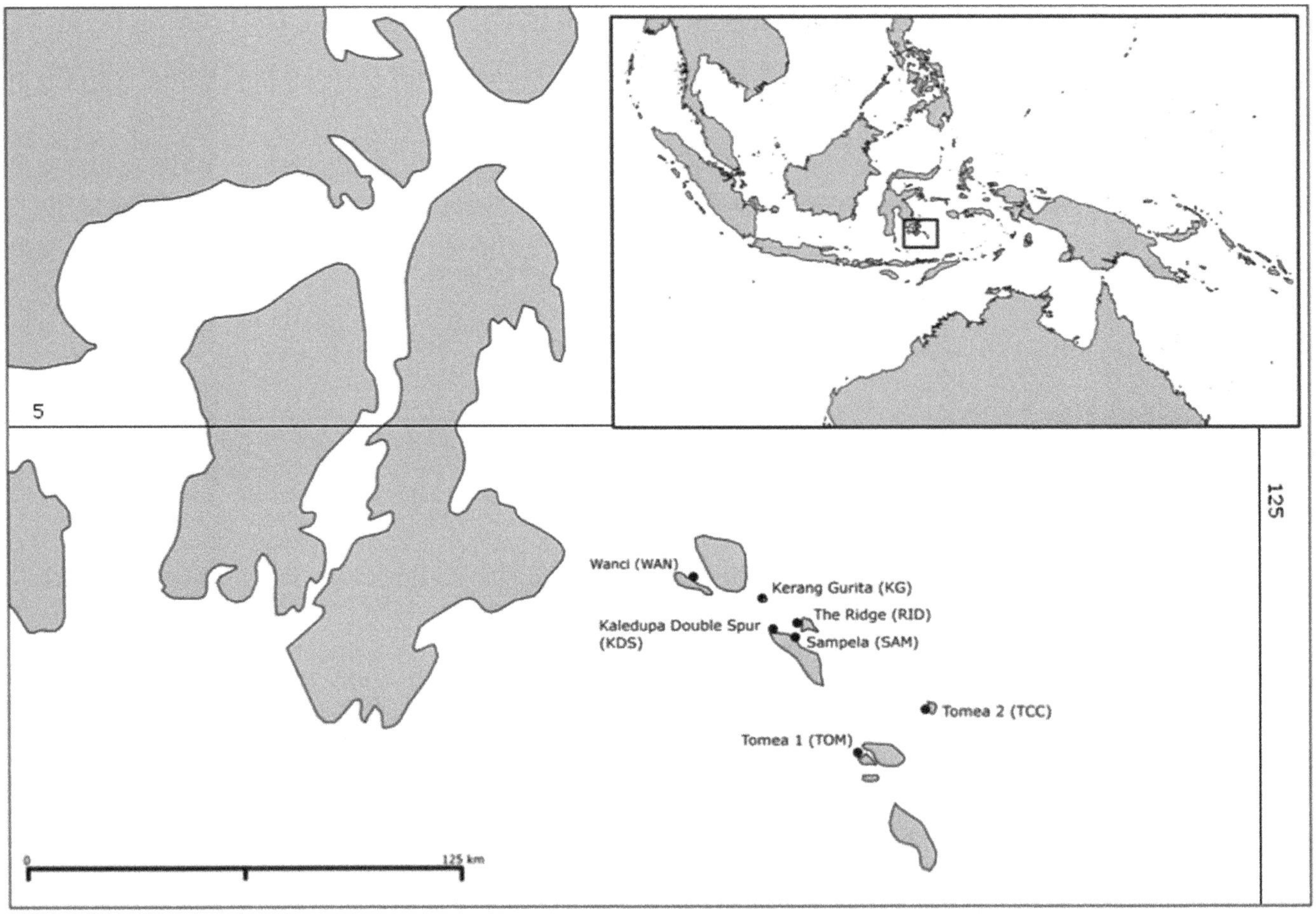


**Figure 1. Sampling locations of *Xestospongia* spp. in the Wakatobi Marine National Park.**

patterns of other major components of tropical reef systems. Without such information it will not be possible to effectively manage entire reef systems.

Sponges are common and important coral reef organisms across the world with a range of important roles in ecosystem functioning [12], key roles include the ability of sponges to process vast quantities of water consuming picoplankton, Dissolved Organic Carbon and potentially viruses, bioerosion of carbonate substrate, and supporting rich macro and microbial communities [12–14]. Furthermore, recent research suggests that sponges may be one of the few taxonomic groups to benefit from the long-term trends declines coral reefs [15–17]. Little is known about the spatial scales at which sponge populations are interconnected across reef systems; this represents a major gap in our knowledge of the functional ecology and dynamics of both healthy and degraded systems, and the role sponges play in these systems. In fact there are comparatively few estimates of genetic connectivity for sponges [18–20] and no estimates of demographic connectivity for coral reef sponges (but see [21] for a sub-tropical species).

Of the 10,000+ species of known sponges, one of the most conspicuous and charismatic are the giant barrel sponges of the genus *Xestospongia*, which can reach several meters in diameter and have been suggested to reach several thousand years in age [22]. Given their size and common occurrence, they are likely to be one of the more important sponge species on Indo-Pacific reefs [12]. While *Xestospongia muta* from the Caribbean has been one of the most intensively studied sponge species (e.g., [21–24]), far less is known about the *Xestospongia* species from the Indo-Pacific (but see [25–27] despite barrel sponges being commonly found across the region. Currently, two barrel sponge species have been described from the Indo-Pacific, (*X. testudinaria* and *X. bergquistia*), which have been separated on the basis of morphology [25] and sterol chemistry [26]. While the reproductive biology has been extensively studied in *X. testudinaria* and *X. bergquistia*, with both species being gonochoristic and oviparous, the length of the larval period is unknown, but is likely to be short (hours to days) based on studies of other sponge species [28]. Until recently *X. testudinaria* has been considered a single species, but a recent study by Swierts et al. [29] found evidence for a species complex within what has been considered *X. testudinaria.* With a combination of mitochondrial and nuclear markers and a morphological analysis, these authors proposed that there are at least two distinct *Xestospongia* species around Lembeh Island, Indonesia, each with different habitat preferences.

Given the slow growth and small census population sizes of barrel sponges, they would appear susceptible to environmental degradation and disturbance. However, large, old individuals are common on reefs suggesting that other species traits allow populations to persist. This may be explained by their ability to exist across a range of environmental conditions, their role as strong spatial competitors (through upward rather than highly competitive horizontal growth), their potential to tolerate environmental disturbances, or as a result of high levels of population connectivity. Identifying these traits will be increasingly important

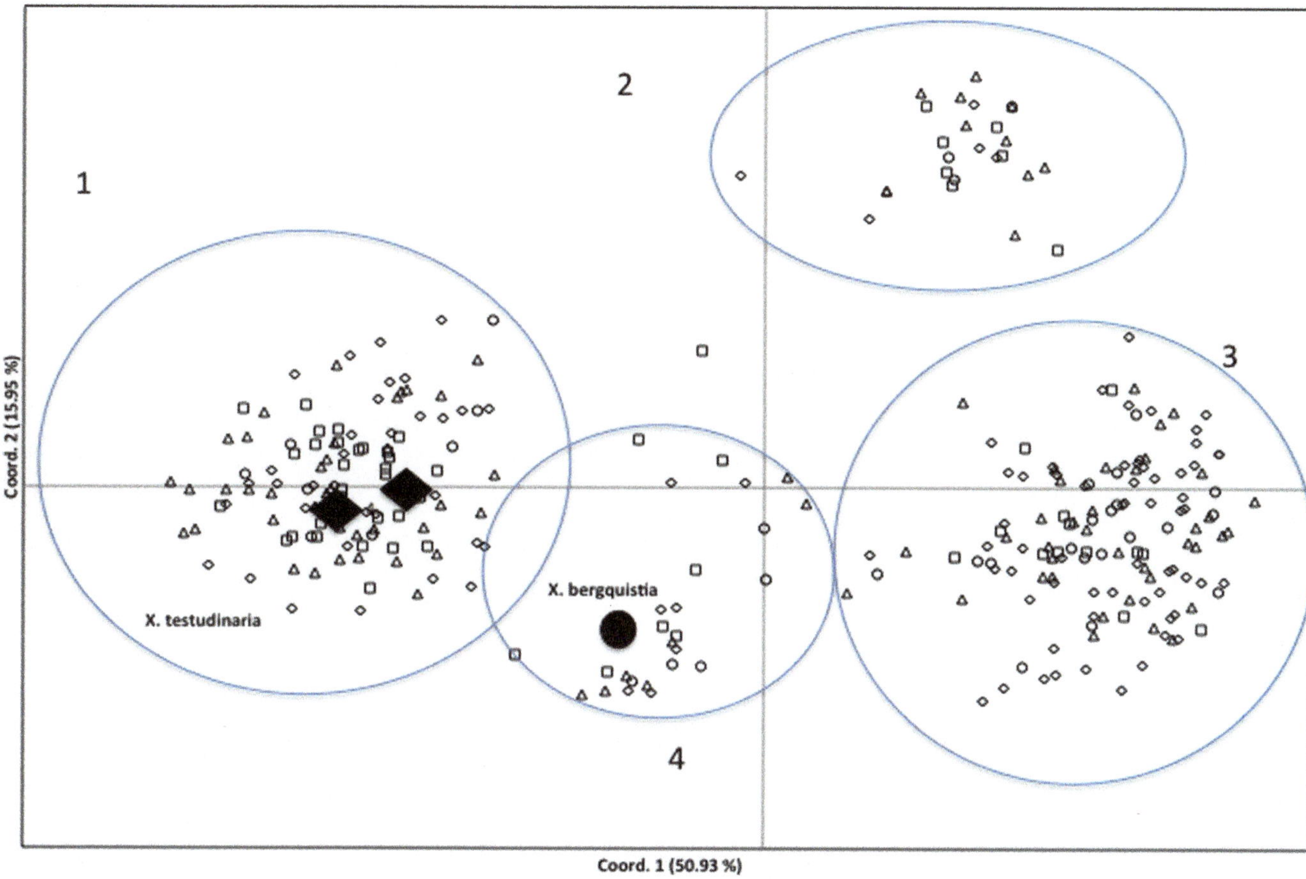


**Figure 2. Principle Coordinate Analysis (PCA) implemented in GENEALEX v 6.3 of pairwise (A) Nei's genetic distance for all samples from the Wakatobi.** The museum specimens of *Xestospongia testudinara* and *X. bergquistia* are shown as large symbols.

as predicted future reef impacts provides more opportunities for benthic species other than coral to dominate reef systems. We recently described the development of a panel of microsatellite markers for *Xestospongia* spp. [30], with the initial results from these markers providing evidence to support the division of barrel sponges into genetically distinct groups and likely species. Based on this earlier analysis and Swierts et al. [29], here we explore data across a great geographic range within an Indonesian archipelago and characterise the genetic structure of two putative *Xestospongia* species. We aimed to: 1) measure and compare the levels of genetic and demographic connectivity between populations of the two species of *Xestospongia* at a range of spatial scales (2 km to 70 km); 2) estimate dispersal distances using an spatial autocorrelation analysis and assignment testing; 3) examine differences in the level of genetic diversity and inbreeding at sites with different environmental conditions; and 4) compare the abundance and size distributions at sites with different environmental and biological characteristics to consider the impacts of disturbance on population structure. The life-history characteristics and population demography of barrel sponges would appear to make sensitive to disturbance. Therefore we hypothesise that barrel sponges will show low levels of dispersal, high levels of self-recruitment and therefore must have evolved physiological traits enabling them to tolerate degraded and disturbed environments. Understanding levels of connectivity will provide insights into how these functionally important reef organisms should be managed and conserved.

## Materials and Methods

### Sample Sites

Seven sites (see Fig. 1) were sampled in the Wakatobi Marine National Park (WMNP) in southeast Sulawesi in Indonesia in June 2012 under permit number 0212/SIP/FRP/SM/V11/2011 from the Ministry of Research and Technology (RISTEK). The WMNP was gazetted in 1996 and is the third largest marine national park in Indonesia [31]. The park is located in the coral triangle and supports some of the world's most diverse marine communities, but is also inhabited by over 90,000 people who are heavily reliant on reef resources [32]. There is a large variation in the coral cover, fish abundance and environmental conditions at sites across the park, and many areas have suffered major declines in coral cover over the last 10 years [33]. Sites were classified based on observations of live coral cover, water clarity, distance to major human populations, fish abundance, evidence of bomb fishing and degree of coral bleaching/disease. Based on our observations (at 6–18 m depth), the sites in Wanci Harbour and at Samplea were considered to be the lowest quality sites based on being close to a large population (<1 km), experiencing heavy sedimentation and having very low coral cover (5–10%) and fish abundance. The sites at Kaladupa Double Spur, the Ridge and Tomea 1, were considered to have moderate site quality with good water clarity, moderate coral cover (25–35%) and some distance from large human populations (3–5 km), while the sites at Karang Gurita and Tomea 2 where considered to be the highest quality sites with high water clarity, high coral cover (>35%) and the longest distance (5–

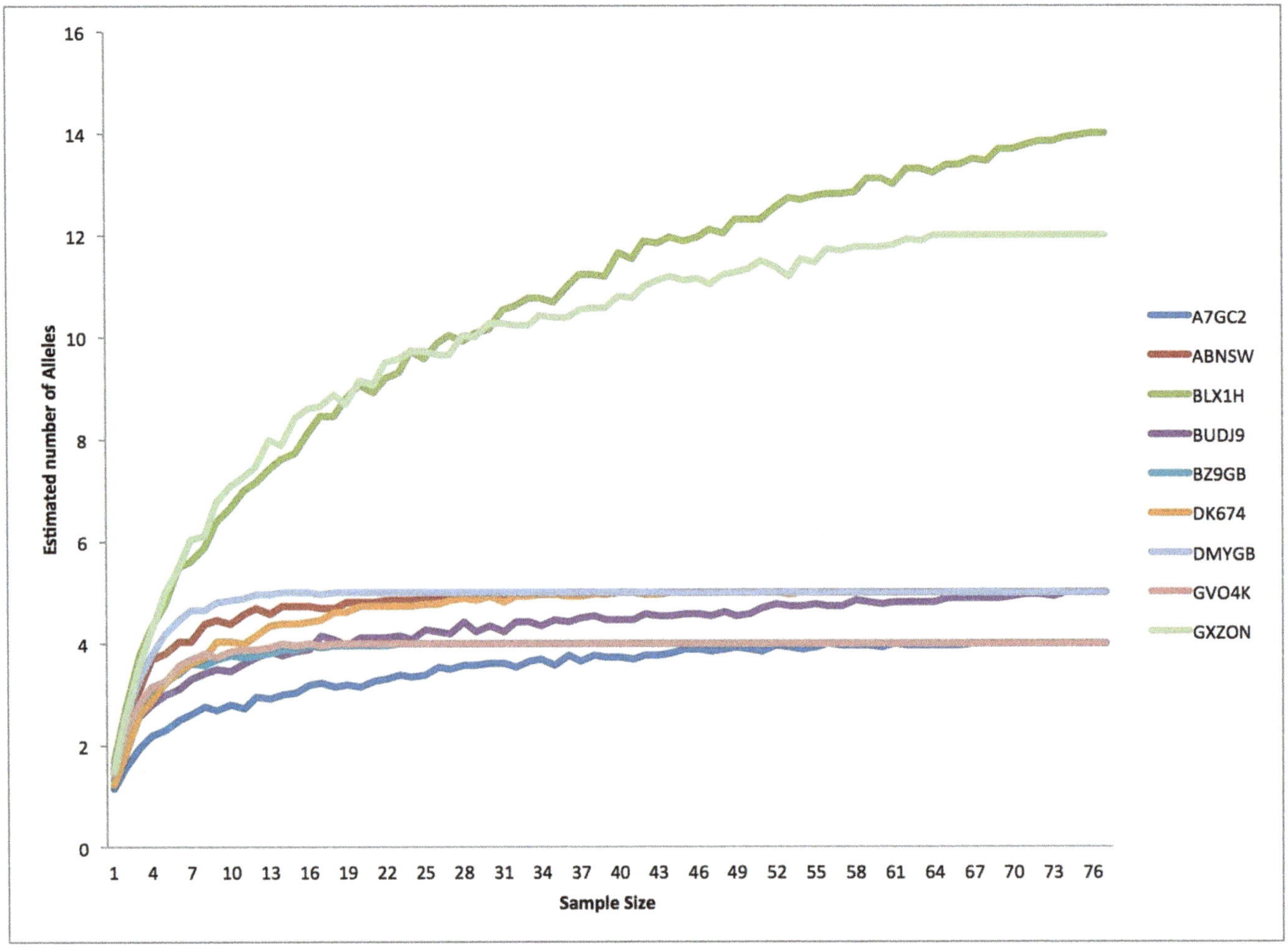


**Figure 3. Allele discovery curves for each locus using samples from the Sampela site, which as the largest population size.**

10 km) to the nearest large population. Observations and site categorisation are consistent with the findings of the long term monitoring programme within the study area [33].

All barrel sponges were sampled in an area approximately 250 m long between 6 and 21 m at each site, except Sampela, where the maximum depth of the reef extends to 12 m, giving a total sampling area (and population size for the site) of 2400 $m^2$ at each site except Sampela where 1200 $m^2$ was surveyed. Sampling was depth restricted in line with safe diving practises (<21 m) but some barrel sponges were observed below this depth threshold. One $cm^3$ tissue samples were collected from each sponge and preserved in 99% ethanol. Photographs and measurements were also taken of each sponge.

The restricted reef formation reef at Sampela (depth restricted to 12 m) allowed us to sample the entire population along a 250 m section of the reef. This survey was used for a spatial autocorrelation analysis (see below), where we mapped the location of each barrel sponge on the reef relative to each other sponge by running a transect along the base of the reef at 12 m and determining x,y coordinates (in metres) along the reef for each sponge.

## DNA Extraction and Microsatellite Amplification

Genomic DNA was extracted from the ethanol preserved *Xestospongia* tissue samples using an Isolate Genomic DNA Mini Kit (Bioline) following the manufacturer's protocols. We used the nine of the twelve markers developed Bell et al. [31] for *Xestospongia* (three markers were discarded based on initial screening that detected null alleles at two loci and evidence of selection acting on one locus).

Microsatellite loci were amplified for each sponge sample on a GeneAmp 2700 (Life Technologies) thermocycler in a final reaction volume of 12 μl containing: ~50 ng template DNA, 1X Bioline MyTaq Red Mix (0.11 units/μl TaqDNA polymerase, 82.5 mM Tris–HCl pH 8.5, 22 mM $(NH_4)_2SO_4$, 1.65 mM $MgCl_2$, 0.22 mM dNTPs), equal amounts of forward and reverse primer (see 31), and $ddH_20$ to volume. The following cycle conditions were used: 94°C for 5 minutes; followed by 40 cycles of 94°C for 30 seconds, 60°C for 45 seconds, 72°C for 60 seconds; followed by a final extension at 72°C for 10 minutes. PCR products were visualised on a 1.5% agarose gel using ethidium bromide staining and if successfully amplified were genotyped on an ABI3730XL (Life Technologies) automated capillary sequencer. The 5′ ends of the forward primers were tagged with the fluorescent labels FAM, VIC, NED or PET. Multiplex Manager 1.2 [34] was used to arrange loci into two multiplex PCR panels.

## Preliminary Analyses

Because de Swierts et al. [19] have proposed a species complex for *Xestospongia testudinaria*, we first conducted an exploratory

**Table 1.** Sampling locations within the Wakatobi Marine National Park with standard genetic diversity indices.

| Population | Coordinates | N | $N_A$ | $H_O$ | $H_E$ | $F_{IS}$ | $P_{HWE}$ |
|---|---|---|---|---|---|---|---|
| **a) Group 1** | | | | | | | |
| Sampela (SA) | 36° 34′ S 174° 46′ E | 28 | 4.00 | 0.47 | 0.57 | 0.131 | 0.001 |
| Wanci (WAN) | 46° 38′ S 167° 37′ E | 10 | 3.00 | 0.42 | 0.53 | 0.186 | 0.030 |
| Karang Gurita (KG) | 43° 55′ S 176° 43′ W | 21 | 3.78 | 0.45 | 0.56 | 0.192 | 0.001 |
| Kaledupa Double Spur (KDS) | 41° 20′ S 174° 48′ E | 8 | 3.00 | 0.31 | 0.54 | 0.431 | 0.001 |
| Ridge (RID) | 42° 24′ S 173° 40′ E | 5 | 3.89 | 0.51 | 0.67 | 0.248 | 0.080 |
| Tomea 1 (TCC) | 43° 53′ S 166° 09′ E | 19 | 3.89 | 0.43 | 0.58 | 0.229 | 0.001 |
| Tomea 2 (TOM) | 36° 49′ S 139° 49′ E | 17 | 3.56 | 0.35 | 0.48 | 0.271 | 0.001 |
| **b) Group 2** | | N | $N_A$ | $H_O$ | $H_E$ | $F_{IS}$ | $P_{HWE}$ |
| Sampela (SA) | 36° 34′ S 174° 46′ E | 46 | 4.11 | 0.375 | 0.49 | 0.212 | 0.001 |
| Wanci (WAN) | 46° 38′ S 167° 37′ E | 7 | 3.44 | 0.515 | 0.54 | 0.054 | 0.116 |
| Karang Gurita (KG) | 43° 55′ S 176° 43′ W | 20 | 3.33 | 0.369 | 0.45 | 0.186 | 0.001 |
| Kaledupa Double Spur (KDS) | 41° 20′ S 174° 48′ E | 19 | 4.33 | 0.449 | 0.53 | 0.167 | 0.001 |
| Ridge (RID) | 42° 24′ S 173° 40′ E | 14 | 4.33 | 0.498 | 0.51 | 0.067 | 0.06 |
| Tomea 1 (TCC) | 43° 53′ S 166° 09′ E | 5 | 2.77 | 0.289 | 0.46 | 0.364 | 0.006 |
| Tomea 2 (TOM) | 36° 49′ S 139° 49′ E | 17 | 3.55 | 0.398 | 0.49 | 0.149 | 0.002 |

(N) number of individuals sampled; ($N_A$) mean number of alleles per locus; ($H_O$) observed and ($H_E$) expected heterozygosity; and ($F_{IS}$) inbreeding coefficients with corresponding significant values ($P_{HWE}$).

analysis of our entire dataset to determine whether or not there was support for dividing our data, representing two or more species. In addition, we genotyped archived specimens from the Queensland Museum (Australia) representing *X. testudinaria* (G320630, G322086, G322647, G321612, G321769, G324859, G25009, G25010, G25011, G25012) and *X. bergquistia* (G25018, G25019) from the Indo-Pacific. To visualise the genetic relationships between all samples from all populations, a Principle Coordinate Analysis (PCA) was conducted by GENEALEX 6.3 using the pairwise matrices of Nei's genetic distance [35]. Based on the results of this plot, and subsequent examination of the morphologies of each sponge sampled (from photographs), we conducted all analyses separately for the two genetically distinct groups previously identified by Bell et al. [31]. To support this, we used a hierarchal analysis of molecular variance (AMOVA, n = 10 000 permutations) to determine the proportion of genetic variation that could be attributed to differences between the two main groups identified from the PCA using GENEALEX v 6.3.

## Genetic Diversity

Allele discovery curves were generated using PopGenKit Package in the software R 2.13.1 for all Samples from Sampela. Curves that reach an asymptote indicate that the allelic diversity would not be increased by analysing additional samples. We quantified genetic diversity within each of the sampled populations for each species based on the mean number of alleles per locus and total number of alleles using FSTAT v 2.9.3 [36] and Nei's unbiased heterozygosity [37] with GENEALEX 6.3 [38]. We also preformed exact tests to identify any deviations from Hardy Weinberg equilibrium with GENEALEX v 6.3 and tested for any evidence of linkage disequilibrium between all loci/population combinations with ARLEQUIN v 3.5.1.2 [39] P-values were adjusted for multiple comparisons using standard Bonferroni corrections [40]. Markov-chain parameters were 10 000 dememorization steps, 1 000 batches and 10 000 iterations per batch.

## Population Genetic Structure

We used a hierarchal analysis of molecular variance (AMOVA, n = 10 000 permutations) to determine the proportion of genetic variation that could be attributed to differences between sampling sites using GENEALEX v 6.3. Population differentiation ($F_{ST}$, [41]) was also measured across all loci with ARLEQUIN v 3.5.1.2 using sampling locations as population units. Significance values were based on 10 000 permutations. There is much debate as to whether the Stepwise Mutation Model (SMM) and associated $R_{ST}$ index is appropriate for population genetic studies using microsatellite loci [42]; therefore, we avoided using the $R_{ST}$ index. A Mantel test for Isolation by distance was conducted using linearized $F_{ST}$ ($F_{ST}/(1-F_{ST})$ and minimum oceanographic distance (km) between sites.

To visualise the genetic relationships between sampled populations, a Principle Coordinate Analysis (PCA) was conducted by GENEALEX 6.3 using the pairwise matrices of $F_{ST}$ and Nei's genetic distance ($D_s$).

Bayesian clustering analysis in STRUCTURE v 2.3.2 [43] was used to infer population structure using the admixture model with correlated allele frequencies among clusters and informed priors with a burn-in period of $10^4$ iterations and $10^5$ Markov Chain Monte Carlo (MCMC) repetitions and with K ranging from 1–7. Each individual in the data set is represented by a single vertical line, which is partitioned into K segments that represent that individual's estimated membership fraction in each of the K inferred clusters. The appropriate K value for the data set was determined by plotting the log probability (L(K)) and ΔK across multiple runs [44] as implemented in STRUCTURE HARVESTER [45]. Results from 10 runs were merged with CLUMPP [46] and visualized using DISTRUCT [47].

**Table 2.** Allele size variation (standardized allelic richness) and total number of alleles per locus (N) at nine microsatellite loci.

| Group A | SA | WAN | KG | KDS | RID | TCC | TOM |
|---|---|---|---|---|---|---|---|
| locus A7GC2 | 4 | 3 | 4 | 3 | 4 | 4 | 4 |
| locus ABNSW | 2 | 2 | 3 | 3 | 4 | 4 | 4 |
| locus BLX1H | 10 | 6 | 7 | 5 | 6 | 6 | 4 |
| locus BUDJ9 | 3 | 3 | 4 | 3 | 4 | 5 | 4 |
| locus BZ9GB | 3 | 3 | 2 | 2 | 3 | 2 | 2 |
| locus DK674 | 3 | 2 | 4 | 3 | 3 | 3 | 3 |
| locus DMYGB | 3 | 3 | 3 | 2 | 5 | 4 | 6 |
| locus GVO4K | 3 | 3 | 3 | 2 | 2 | 2 | 2 |
| locus GXZON | 5 | 2 | 4 | 4 | 4 | 5 | 3 |
| Mean | 4 | 3 | 4 | 3 | 4 | 4 | 4 |
| Total | 36 | 27 | 34 | 27 | 35 | 35 | 32 |
| UHe | 0.578 | 0.561 | 0.571 | 0.580 | 0.747 | 0.592 | 0.497 |
| **Group B** | **SA** | **WAN** | **KG** | **KDS** | **RID** | **TCC** | **TOM** |
| locus A7GC2 | 1 | 1 | 1 | 1 | 1 | 1 | 1 |
| locus ABNSW | 3 | 4 | 3 | 3 | 6 | 3 | 4 |
| locus BLX1H | 8 | 6 | 5 | 6 | 5 | 5 | 4 |
| locus BUDJ9 | 4 | 3 | 4 | 5 | 5 | 2 | 2 |
| locus BZ9GB | 4 | 3 | 4 | 5 | 4 | 3 | 5 |
| locus DK674 | 4 | 2 | 2 | 3 | 2 | 2 | 3 |
| locus DMYGB | 4 | 3 | 3 | 5 | 4 | 3 | 3 |
| locus GVO4K | 2 | 2 | 2 | 3 | 4 | 2 | 2 |
| locus GXZON | 7 | 7 | 6 | 8 | 8 | 4 | 8 |
| Mean | 4 | 3 | 3 | 4 | 4 | 3 | 4 |
| Total | 37 | 31 | 30 | 39 | 39 | 25 | 32 |
| UHe | 0.496 | 0.584 | 0.461 | 0.542 | 0.540 | 0.509 | 0.501 |

Genetic diversity indices including total number of alleles per population, mean number of alleles per loci, and Nei's unbiased Heterozygosity (UHe), are also shown.

Assignment tests were conducted using a Bayesian approach [48] in GENECLASS v 2.0 [49–50] and an analysis of first generation migrants was conducted using an exclusion threshold approach where individuals were excluded from their corresponding sampling site when probability of assignment to the reference population was less than 0.05 (Type I error; 50). This exclusion approach is likely to be more accurate than the "leave one out" methodology by Paetkau et al. [51] because it does not require that all true source populations be sampled, and in our case it is likely there are un-sampled populations. Excluded individuals were then reassigned to a source population when the probability of assignment was greater than 10%. When an excluded individual was re-assigned to more than one population ($P>0.10$) it was left unassigned. Those individuals that could not be re-assigned to any of the other populations were considered to have originated from a non-sampled location.

## Spatial Autocorrelation Analysis

To investigate fine-scale patterns of genetic relatedness, spatial autocorrelation tests were performed by GENEALEX 6.4.1 [52] to determine if the degree of genetic similarity between individual sponges was correlated with the geographic distance between them. This analysis was conducted using pairwise matrices of genetic distance and geographic distance across all loci among sponges from Sampela, where the specific location of each individual sponge was recorded along a 250 m transect (see above). The spatial autocorrelation coefficient of genetic distance (r) was calculated over 15 distance classes (up to 150 m) and illustrates the genetic similarity of individuals whose pairwise geographic distance falls within a specific distance class. The location where r crosses the x-intercept provides an estimate of positive spatial genetic structure [52], where genetic drift rather than gene flow is the primary force influencing spatial structure. For each distance class 95% confidence intervals were generated based on 10000 permutations and 10000 bootstrap replicates.

## Abundance and Size of Sponges

As we sampled all sponge specimens along a 250 m section of reef at each location, this provided us with an estimate of total population size at each location. At Karang Gurita and Sampela (considered to be one of the highest and lowest environmental quality sites, respectively) we also measured the approximate size of each barrel sponge. We generalised the shape of each sponge to a cylinder with a cone cut from the top to represent the spongocoel. We measured the maximum circumference, the height of the sponge, the diameter of the spongocoel and depth of the spongeocoel to calculate the approximate volume.

**Table 3.** Pairwise fixation index values $F_{ST}$ for group A (above diagonal) and group B (below diagonal).

| | SA | WAN | KG | KDS | RID | TCC | TOM |
|---|---|---|---|---|---|---|---|
| Sampela (SA) | | 0.038 | 0.029 | 0.037 | 0.051 | 0.032 | 0.049 |
| Wanci (WAN) | 0.044 | | 0.032 | 0.065 | 0.072 | 0.040 | 0.056 |
| Karang Gurita | 0.024 | 0.043 | | 0.072 | 0.066 | 0.036 | 0.070 |
| Kaledupa DS | 0.024 | 0.029 | 0.018 | | 0.049 | 0.066 | 0.101 |
| Ridge (RID) | 0.023 | 0.050 | 0.036 | 0.027 | | 0.051 | 0.088 |
| Tomea 1 (TCC) | 0.037 | 0.064 | 0.042 | 0.044 | 0.052 | | 0.026 |
| Tomea 2 (TOM) | 0.030 | 0.067 | 0.052 | 0.038 | 0.025 | 0.051 | |

Bold values indicate significance based on 10,000 permutations with a Bonferroni adjusted *P* value of 0.001.

## Results

### Preliminary Genetic Analyses

The PCA to examine the relationships between all of the samples across all seven sites showed clusters of samples, which were not consistent with their geographical location, and formed four distinct groups (Figure 2). AMOVA showed that 50% of the variation ($P<0.010$) could be attributed to among group differences. On re-examination of the photographs of each sampled sponge, those that fell within three of the groups could generally be distinguished based on their external morphology and we propose these likely represent different species (consistent with [19]). For the four group it was difficult to distinguish individuals based on morphology. There were no apparent differences between the geographical locations of the samples in any of the groups and all groups appeared to have the same distribution patterns. Only two of the groups had sufficient sample sizes to enable further analyses and the subsequent analyses were conducted separately for these two species (i.e. the sponges from the two other clusters were not included in any further analyses). Inclusion of the museum specimens within the PCA enabled us to determine whether any of the groups were consistent with the two described barrel sponges species, *Xestospongia testudinaria* and *X. bergquistia* (Figure 2). Unfortunately (presumably due to DNA degradation), it was only possible to amplify the microsatellites for two of the *X. testudinaria* samples and one of the *X. bergquistia* samples. The two *X. testudinaria* samples clustered with one of the two larger groups supporting that these samples are *X. testudinaria*, while the single *X. bergquistia* sample clustered with one of the smaller groups, suggesting the second large cluster from the Wakatobi may represent an undescribed species (hereafter referred to a *Xestospongia* sp.).

### Genetic Diversity

Allele discovery curves (Figure 3) showed that for most loci an asymptote was reached at low sample sizes ($<10$). We found no evidence of linkage disequilibrium between any pair of loci across all sample sites following standard Bonferroni corrections for either species. All populations exhibited a significant global heterozygote deficiency (Table 1) with inbreeding coefficients ($F_{IS}$) ranging from 0.131 to 0.431 for *X. testudinaria* and 0.054 to 0.364 for *Xestospongia* sp. We found deviations from HWE at the

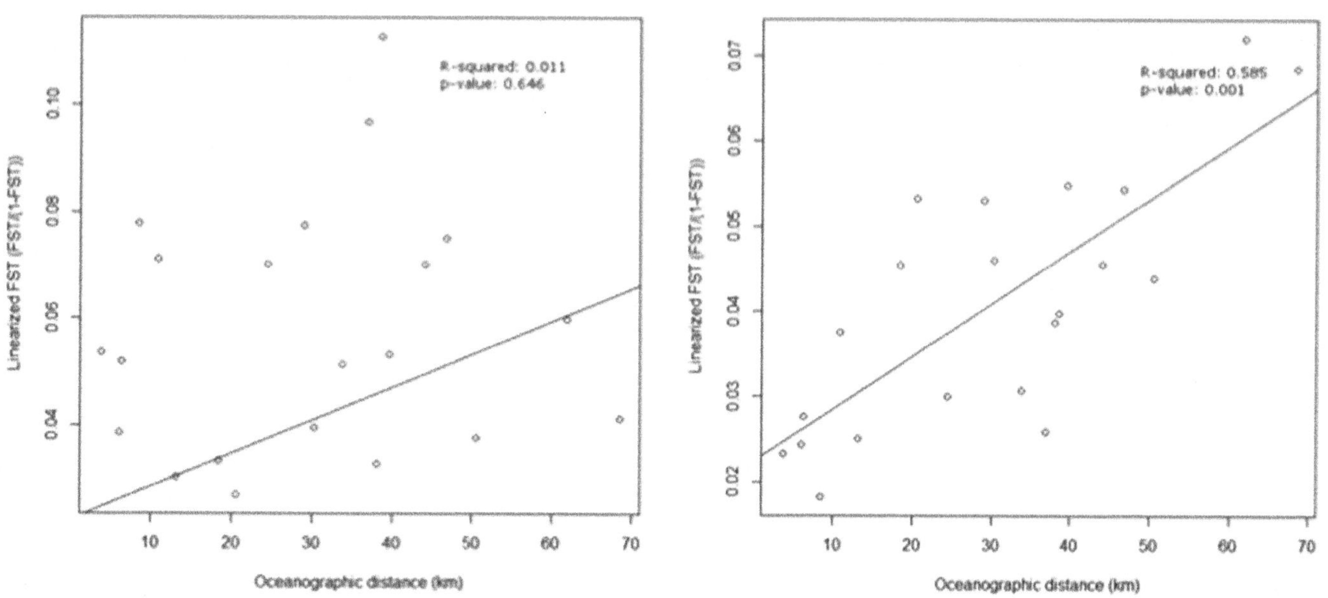


**Figure 4. Mantel test of Isolation by distance using linearized $F_{ST}$ with 95% confidence intervals for the two groups (Group – left, Group B – right).**

**Table 4.** Hierarchal analysis of molecular variance (AMOVA) was used to estimate levels of genetic differentiation between sampling sites (n = 7) for the two groups (P<0.01).

| | Source | df | SS | MS | Est. Var. | % |
|---|---|---|---|---|---|---|
| Group A | Among Populations | 6 | 92.240 | 15.373 | 0.543 | 7% |
| | Within Populations | 105 | 738.894 | 7.037 | 7.037 | 93% |
| | Total | 111 | 831.134 | | 7.580 | 100% |
| Group B | Among Populations | 6 | 63.647 | 10.608 | 0.280 | 5% |
| | Within Populations | 123 | 714.976 | 5.813 | 5.813 | 95% |
| | Total | 129 | 778.623 | | 6.093 | 100% |

locus level varied between sampling sites, although no population showed a significant deficiency at all loci (Table 2). The mean number of alleles per locus ranged from 3 to 4 for *X. testudinaria* and 2.77 to 4.33 for *Xestospongia* sp., while the mean number of alleles per population ranged from 3 to 4 for both species (Table 2). Locus A7GC2 was monomorphic in *Xestospongia* sp. Genetic diversity was lowest at Tomea (TOM) for *X. testudinaria* and Karang Gurita (KG) for *Xestospongia* sp. (Table 2). Comparisons of each genotype with each other genotype across all our samples revealed no evidence of clones within the sampled populations (i.e. no two sponges sampled had the same genotype).

## Population Genetic Structure

Significant $F_{ST}$ values were detected for all pairwise comparisons following standard Bonferroni corrections (Table 3). Isolation by distance plots (Figure 4) showed no relationship for *X. testudinaria* ($R^2 = 0.01$, $P = 0.65$), but a strong significant positive correlation for *Xestospongia* sp. ($R^2 = 0.68$, $P<0.001$). Our analysis revealed significant population structure for both species across the Wakatobi (Global $F_{ST}$ of 0.054, P = 0.010 for *X. testudinaria*; Global $F_{ST}$ of 0.039, P = 0.010 for *Xestospongia* sp.). Irrespective of species, AMOVA indicated only a small proportion (7 and 5% for *X. testudinaria* and *Xestospongia* sp., respectively; Table 4) of genetic variation could be attributed to differences between sampling sites.

PCA plots (Figure 5) show the genetic relationships between the sampled locations, and results were relatively consistent across the measures of differentiation ($D_s$ and $F_{ST}$). For *X. testudinaria*, the PCA showed evidence of four main clusters including the sites to the south (TOM and TCC), the north (WAN and KG), a central group (RID and KDS) and Sampela. For *Xestospongia* sp., there were no obvious population clusters consistent with isolation by distance structuring.

Differentiation patterns identified from PCA plots were supported by STRUCTURE. The optimal number of clusters according to Evanno et al. (2005) as implemented in STRUCTURE HARVESTER was determined to be K = 4 (Figure 6 and 7) for *X. testudinaria*: two central groups (SAM = 1; RID and KDS = 2), a northern group (KG and WAN) and a southern group (TOM and TCC). STRUCTURE HARVESTER identified K = 3 for *Xestospongia* sp., (SAM, RID and KDS = 2), a northern group (KG and WAN) and a southern group (TOM and TCC). However, this was not strongly supported (see Figure 6), which is consistent with a gradient and the isolation by distance patterns described above for this species.

## Assignment Testing and Self-recruitment

Our assignment testing identified high levels of self-recruitment for both species, although care is needed in the interpretation of the percentage of self-recruiting sponges for populations with small sample sizes (Table 5). For *X. testudinaria*, populations with sample sizes >10 generally had a low number of excluded individuals (< 12%) indicating high levels of self-recruitment. There was also evidence of sponges that could not be assigned to any of our sampled populations, with only 2–3% of sponges being assigned to an unsampled population.

## Spatial Autocorrelation Analyses

Our spatial autocorrelation analysis revealed small differences in the likely dispersal distances between the two groups at Sampela (Figure 8). Results for *X. testudinaria* indicated that genotypes were distributed randomly for the first 5 distance classes (20–100 metres), after which the *r* value showed a significant positive correlation ($p = 0.035$) among genotypes for sponges separated by distances of 120 metres. The correlation coefficient then dropped to become significantly negative ($p = 0.011$) at 140 metres. Spatial autocorrelation analysis of *Xestospongia* sp. indicated a significant positive correlation ($p = 0.008$) among genotypes of sponges within the first distance class (20 metres). Values then levelled of between the next two classes and dropped below the *x*-axis to become significantly negative ($p = 0.021$) among genotypes of sponges separated by distances greater than 80 metres. Values were again significantly negative ($p = 0.004$) at 160 metres. These results indicated that the neighbourhood size of this *X. testudinaria* was approximately 140 m and for *Xestospongia* sp. approximately 65 metres (based on the location where *r* crosses the *x*-axis; [53]).

## Abundance and Size of Sponges

Estimates of population size for the two species were only based on the densities of a single section of each reef (which we consider sites), but based on this we found the overall (across both species) highest abundance of sponges at Sampela and the lowest abundance at the Ridge (Table 1). However, there appeared to be no general pattern between abundance and reef quality based on our qualitative site assessments. The size distribution of sponges showed that the sponges at Sampela were generally much smaller than those at Karang Gurita (Figure 9) and the overall sponge biomass (for a 1250 $m^2$ section of reef) was lower at Sampela (337.60 $m^3$ compared with 581.26 $m^3$). At both these sites there appeared to be sponges of a range of size classes, suggesting recruitment has been relatively consistent. However, we did not find any small, very recent, recruits at ether site.

# Discussion

Despite connectivity being considered a key component in understanding population dynamics and in the development of

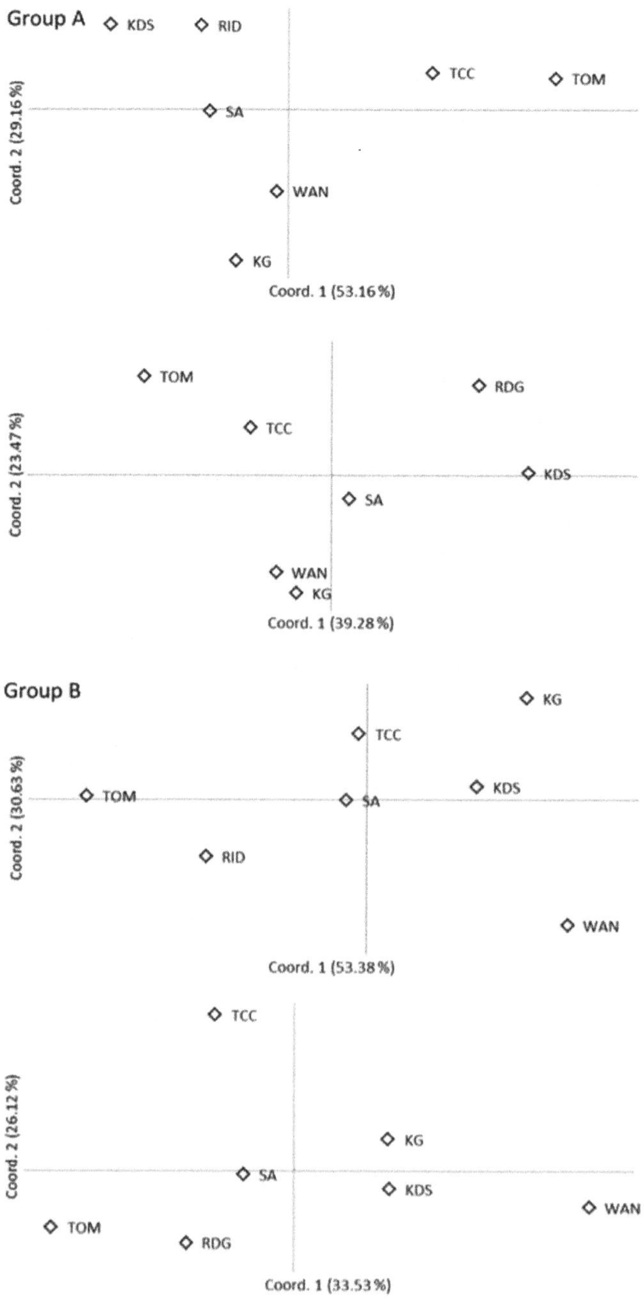


**Figure 5. Principle Coordinate Analysis (PCA) implemented in GENEALEX v 6.3 of pairwise (Nei's genetic distance (above) and $F_{ST}$ values (below) between populations of two groups (A and B) of *Xestospongia* samples collected in the Wakatobi Marine National Park.**

effective management plans, we still have a very poor understanding of the processes that link populations and the scales at which populations are connected. This is especially the case for non-coral reef invertebrates. Here we distinguish between multiple co-occurring species of Indo-Pacific barrel sponges based on microsatellite data, which is supported by gross morphological differences with one of the species being *X. testudinaria*, another *X. bergquistia* and the other two are likely undescribed species. Our results support the presence of a recently proposed *X. testudinaria* species complex in the Indo-Pacific [19]. We aimed to measure the levels of genetic differentiation and demographic connectivity between populations for the two most abundant species. Our genetic data showed strong levels of genetic differentiation between populations separated by as little as 2 km for both species. Furthermore, the results of assignment tests and spatial autocorrelation analyses suggest barrel sponges generally have small dispersal distances of several hundred meters and are likely to be reliant on self-recruitment. However, assignment tests did find some sponges that could not be assigned to any sampled populations and sponges that had come from more distant populations suggesting some potential for occasional longer-distance dispersal. We found different types of genetic structuring for the two species, with *X. testudinaria* following an island model [54] and *Xestospongia* sp. an isolation by distance model [55]. High levels of self-recruitment coupled with their small population census sizes and long-lived nature would appear to make barrel sponges susceptible to disturbance. However, barrel sponges were most abundant at one of the most heavily disturbed reefs suggesting some adaptive ability to deal with environmental disturbance. These results support a recent hypothesis that some sponge species may become more abundant on coral reefs in response to climate change and ocean acidification [16], as they may also benefit from other types of environmental degradation that results in declines in potential sponge predators and competition.

## Demographic Connectivity

There are currently very few estimates of demographic connectivity for most reefs organisms other than corals and fish, and even for these groups estimates are rare. While larval duration has been proposed as a predictor of larval dispersal potential, there is increasing evidence that populations may also be dependent on a high level of self-recruitment. For example, using DNA parentage data and marking via stable isotopes, Almany et al. [56] found that approximately 30% of juvenile clownfish (*Amphiprion polymnus*) settled within 100 m of their natal reefs, despite having a relatively short (compared to other marine species) 12-day larval duration. Estimating demographic connectivity is difficult for most marine species because of the problems in tracking microscopic larvae through the oceans. Parentage, assignment testing and spatial autocorrelation analyses are powerful tools for estimating demographic connectivity, but they are best suited for species with small census population sizes and limited dispersal capacity [57]. These features of barrel sponges make them suitable for such analyses, and our findings suggests these species have short dispersal distances and high levels of self-recruitment. However, there are some problems with using our findings for barrel sponges to infer connectivity amongst other sponge species within the region. Generally, the majority of sponges in this region comprise small (<10 $cm^2$) encrusting patches, whose densities can exceed hundreds of sponge patches per $m^2$ [58–59] so have a much larger census population size their population demography will be very different. However, given sponges generally have short-lived pelagic larvae (in the order of hours to days; Maldonado 2006), dispersal distances are also likely to be small, and it seems likely they will also be reliant on self-recruitment. Interestingly, at the Sampela site, the sponge *Lamellodysidea herbacea* was the most abundant in 2003 and has almost doubled in abundance over the last decade [17,58] (as coral has declined, 33). However, despite coral declines at other sites in the Wakatobi the same increases in *L. herbacea* have not been seen. Although environmental conditions at Sampela (heavy sedimentation and high turbidity) may favour this species, its rapid increase

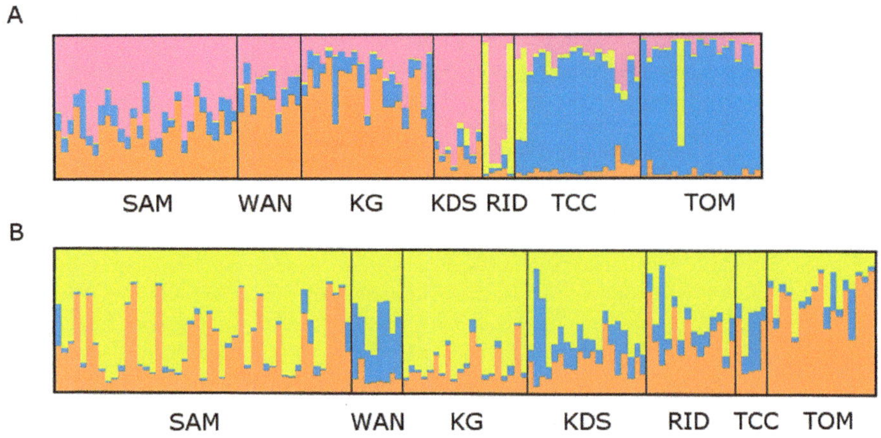


**Figure 6. Bayesian clustering analysis using STRUCTURE 2.3.2 for two groups of *Xestospongia* samples (group 1– upper chart, A: group 2 lower chart, B).** Sampling locations are along the x-axis and membership coefficient in each predefined cluster (K) is along the y-axis. K = 3 and K = 4 clustering scenarios according to STRUCTURE HARVESTER for group 1 and 2, respectively. Results were averaged across 10 runs with CLUMP and visualized with DISTRUCT.

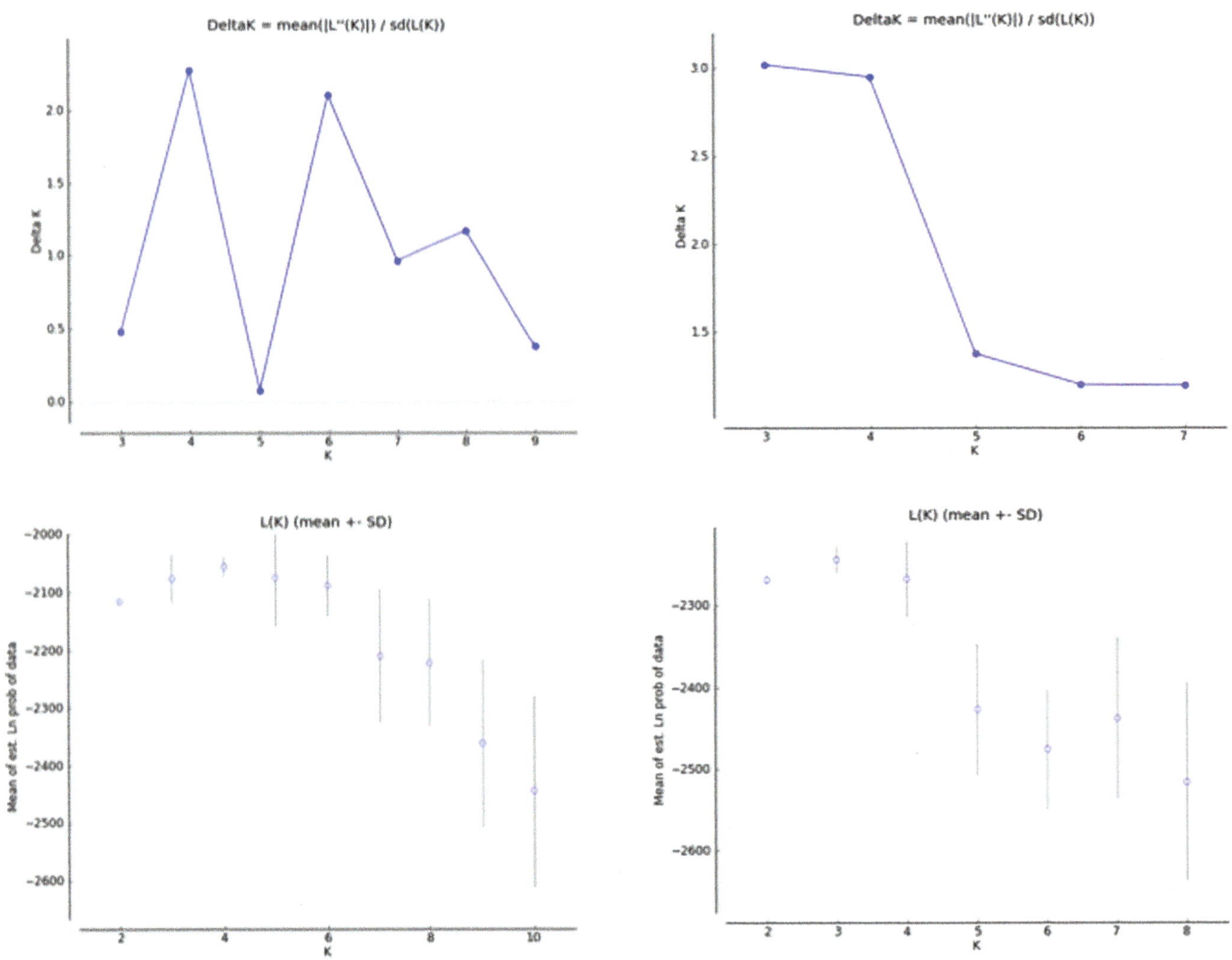


**Figure 7. Plots generated in STRUCTURE Harvester that show the mean log likelihood of the data [L(K)] and Evanno's delta K statistic for both sponge groups.**

**Table 5.** Assignment tests conducted in GeneClass v. 2.0.

| Population | # of Excluded Individuals | % of pop | Assigned Population | | | | | | | Not Assigned |
|---|---|---|---|---|---|---|---|---|---|---|
| | | | SAM | WAN | KG | KDS | RDG | TCC | TOM | |
| SAM (28) | 1 | 3% | – | 0 | 0 | 0 | 1 | 0 | 0 | 0 |
| WAN (10) | 1 | 10% | | – | | 0 | 1 | 0 | 0 | 0 |
| KG (12) | 1 | 5% | 0 | | – | 0 | 1 | 0 | 0 | 0 |
| KDS (8) | 3 | 38% | 0 | 0 | | – | 3 | 0 | 0 | 0 |
| RDG (5) | 2 | 40% | 0 | 0 | 0 | | – | 0 | 0 | 2 |
| TCC (19) | 2 | 10% | 0 | 0 | 0 | 0 | 2 | – | 0 | 0 |
| TOM (17) | 2 | 11% | 0 | 0 | 0 | 0 | 2 | 0 | – | 0 |
| Average | | 17% (+/−15) | | | | | | | | |
| **Population** | **# of Excluded Individuals** | **% of pop** | **Assigned Population** | | | | | | | **Not Assigned** |
| | | | **SAM** | **WAN** | **KG** | **KDS** | **RDG** | **TCC** | **TOM** | |
| SAM (46) | 4 | 9% | – | 1 | 0 | 2 | 0 | 0 | 0 | 1 |
| WAN (7) | 1 | 13% | 0 | – | 0 | 1 | | 0 | 0 | 0 |
| KG (20) | 3 | 15% | 0 | 1 | – | 0 | 2 | | 0 | 0 |
| KDS (19) | 2 | 11% | 0 | 0 | | – | 0 | 0 | 0 | 2 |
| RDG (14) | 1 | 7% | 0 | 0 | 0 | | – | 0 | 0 | 1 |
| TCC (5) | 1 | 20% | 0 | 0 | 0 | 0 | 1 | – | 0 | 0 |
| TOM (17) | 2 | 12% | 0 | 0 | 0 | 1 | 0 | 0 | – | 1 |
| Average | | 12%(+/−4) | | | | | | | | |

Detection of first generation migrants was done using an exclusion threshold approach where an individual was excluded from their corresponding sampling site when the probability of exclusion was >95%. Individuals that were excluded were assigned to another population when the probability of assignment was above a more conservative threshold of >0.10. Individuals that could not be re-assigned to any of the sampled populations with a p value > 0.10 were considered to have originated from a non-sampled location. Numbers in brackets in column 1 are the total sample sizes for each population.

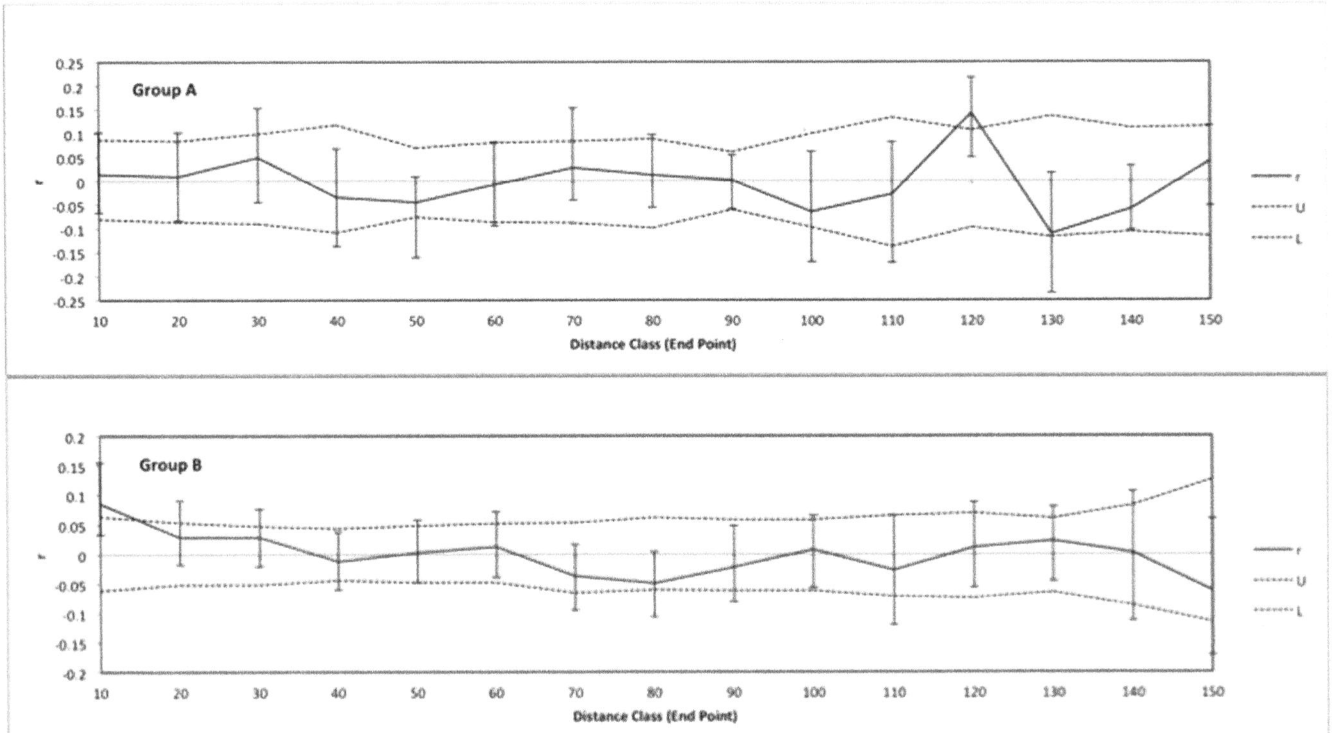


**Figure 8. Spatial autocorrelation analyses of the genetic correlation coefficient (r) as a function of distance for two genetically distinct groups of *Xestopongia* on the Wakatobi Marine National Park calculated with GenAlEx v6 (Peakall and Smouse 2006).** The bootstrapped 95% confidence error bars generated via 1000 bootstrap trials are shown.

in abundance suggests a local stock-recruitment relationship, which is likely to be supported by self-recruitment.

## Interpretation of Genetic Patterns

We found strong patterns of genetic differentiation between all sites for both species. *X. testudinaria* followed an island model, with three different groups consistent with their geographic position, while *Xestosponiga* sp. followed an isolation by distance model. Currently nothing is known of the oceanographic features of the Wakatobi or the location of any potential barriers to gene flow, so it is difficult for us to explain these patterns. However, given we found different patterns of genetic structuring for the two co-occurring barrel sponge species suggests different process are influencing gene flow between populations of these species. *X. testudinaria* had a larger estimated dispersal distance than *Xestospongia* sp. (120 m compared with 65 m), although these distances are still small compared to species with long-lived pelagic larvae, which might be in the order of 10 s to 1000 s km [60]. It is therefore possible that other ecological (e.g. longevity and fecundity) or physiological factors account for the differences between species. However, earlier reproductive studies [27] found little difference in the reproductive biology between *X. testudinaria* and *X. bergquistia*, although larval duration and behaviours have not yet been studied extensively; this might account for the differences we reported. We found high inbreeding coefficients for nearly all populations that we sampled for both groups, with an apparent excess of homozygotes, but no patterns between sites. While null alleles might account for this result, this seems unlikely as MICROCHECKER found no evidence for null alleles in the loci that we used (see Bell et al. [30] for discarded loci showing evidence of null alleles and selection). Given the small census population size, limited dispersal distances and consistency of this pattern across sites, it seems likely that inbreeding might account for these heterozygote deficiencies. This is also supported by individual loci showing few deviations from HWE.

Our results are consistent with a number of earlier genetic studies on sponges using allozymes and microsatellite markers. For example, Whalan et al. [19] found significant genetic differentiation between populations of *Haliclona* sp. over different spatial scales using allozymes, including differences between populations only several hundred meters apart. Currently, microsatellite markers have only be applied to two other sponge species, the Mediterranean species' *Crambe crambe* [18] and *Scopalina lophyropoda* [21]. Again these studies highlight high levels of genetic differentiation between populations separated at small (100 m) spatial scales. Furthermore, both these studies found high inbreeding coefficients within sponge populations and proposed asexual reproduction and inbreeding as possible explanations. While sponges are well known to reproduce through asexual propagation, we found no evidence for clones within our data and therefore propose inbreeding as the most likely cause of the high $F_{IS}$ values that we reported. A genetic analysis of the related barrel sponge *Xestospongia muta* in the Caribbean using the I3-M11 partition of the Cytochrome Oxidase (COI) gene found high levels of genetic differentiation between populations at spatial scales of several 100 km [20], nearly 1000 times the scale of our study. However, Given the conserved nature of the marker used by these authors they were unable to examine genetic differentiation at small-scale levels (100 m to kms). Interestingly, like our results for *X. testudinaria*, these authors also found no evidence of isolation by distance, and proposed that while limited larval dispersal may have led to differentiation among some of the populations, the

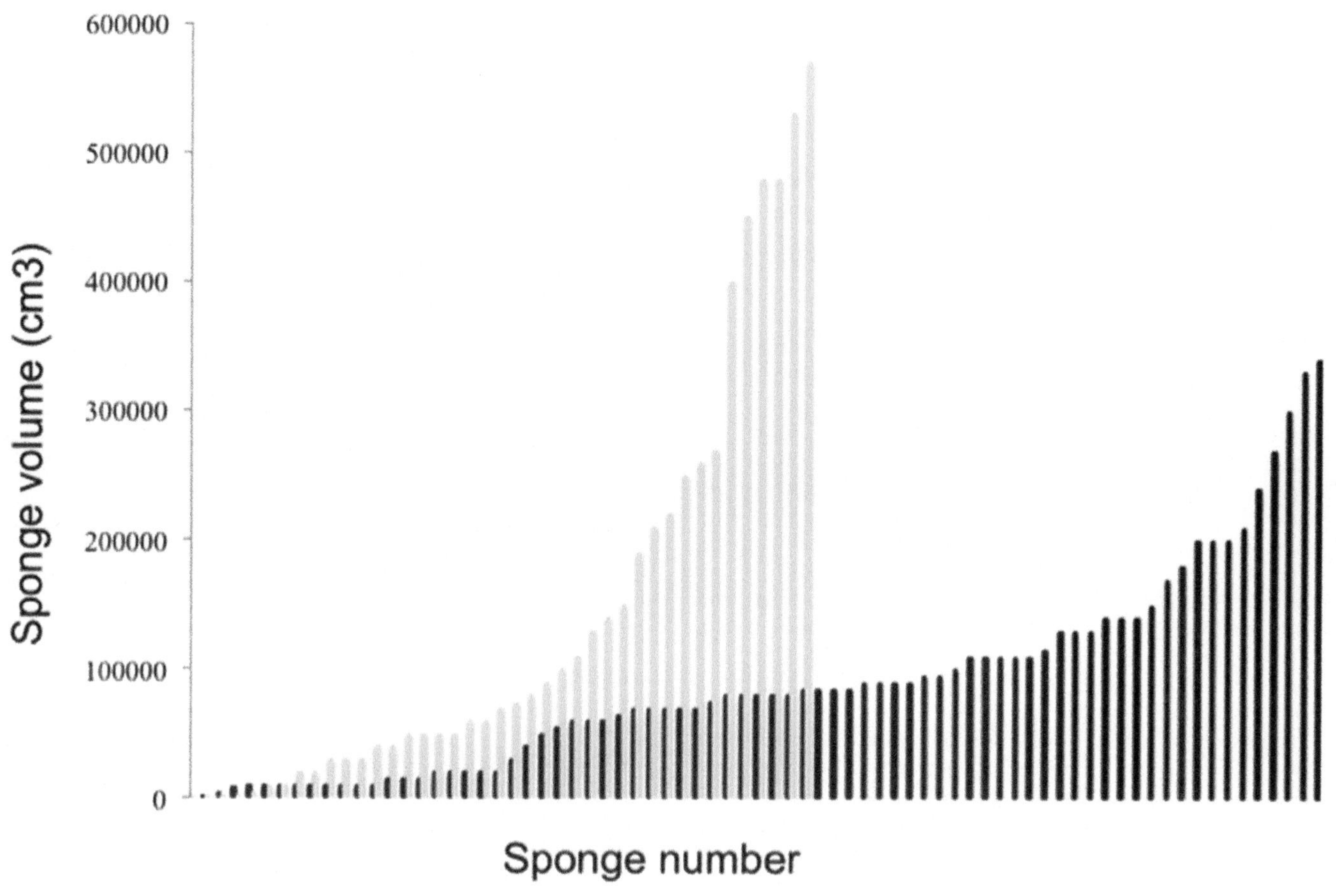


**Figure 9. Size distribution of all barrel sponges (all specimens) along a 250 m section of reef at two sites in the Wakatobi Marine National Park.** Grey lines are data from Karang Gurita (KG) and black lines for Sampela (SAM).

patterns of genetic structure appear to be most strongly related to patterns of ocean currents.

One of the primary limitations to our study is the small sample size for many of our populations. The identification of multiple *Xestospongia* species was unexpected during the initial collection of specimens for this study, but meant we had to split our data into two groups. Furthermore, some sponges were excluded from the analysis (n = 65) as they appeared to represent different species. As a result some care is needed in the interpretation of our data. However, the relatively low numbers of alleles for each of our loci based on an initial survey of 75 individuals [30] means it is unlikely there is a large amount of undetected variation within the populations we sampled; this is also supported by our allele discovery curves (Figure 3). Furthermore, our evidence for strong genetic differentiation between sites is supported by the autocorrelation analysis based on large single site sample size showing short dispersal distances giving confidence to our results.

It is interesting that we found evidence for possibly four *Xestospongia* sp. species, and this was consistent with differences in the external morphology of these specimens. The taxonomic relationships between these species will be the focus of future studies. Our study is not the first to report possible undescribed species within a *Xestospongia testudinaria* complex [19]. It is possible that one of the *Xestospongia* species that was not included in the genetic analysis does have a different local habitat distribution, as specimens were mainly collected from deeper water (21 m).

## Management and Conservation Implications

Barrel sponges play important functional roles through their ability to filter large volumes of water [12]. Our results demonstrate limited dispersal, high levels of self-recruitment and strong levels of genetic differentiation between populations of two *Xestospongia* species. Their small population size coupled with these features should make barrel sponges susceptible to environmental disturbance or there must be a large selective pressure for barrel sponges to tolerate a range of environmental conditions and be able to tolerate environmental perturbations. Sampela has experienced major declines in coral cover and fish in recent years [33], yet this site was where we found the highest numerical abundance of barrel sponges (although lower overall biomass). However, the size distribution of those sponges indicated they were generally smaller at Sampela compared to those at Karang Gurita. Size has been considered an indicator of age, based on growth models created for the Caribbean barrel sponge *Xestospongia muta* [22]. Although there are likely to be local and regional differences in growth rates and these are as yet are unqualified, extrapolation of these growth models from the Caribbean to sponges in the Wakatobi makes the largest sponges at Sampela to be around 50–60 years old and 250–300 years old at Karang Gurita. Interestingly, these ages at Sampela correspond to the settlement of a Bajo village very close to the Sampela reef that was only settled in the 1950s, and has been heavily exploited ever since. It is unclear if these events are related or whether size differences between sites are a functional of differential growth rates at the different sites, but this warrants further investigation and demonstrates the potential resilience of barrel sponges to environmental degradation and potential ability to recover following disturbance.

The development of management plans and any assessment of system resilience requires an understanding of connectivity patterns. However, determining the probability of inter-population connectivity needs to incorporate population dynamics and the factors that control larval longevity and survival. Such information is essential for managers, but collecting this information is time consuming and is generally impractical for all reef species. However, genetic data can be used to infer patterns of genetic and demographic connectivity, but it needs to be interpreted in the context of the species' ecology. Given that the degree of reef resilience can be altered through increased connectivity [3–4], without knowledge of dynamics, and population structure it is impossible to determine appropriate spatial scales of protection, conservation priority areas and likely outcomes of disturbance. Managers need to focus on enhancing system resilience either indirectly through the precautionary principle or directly through prioritising the protection of areas that are sources of potential recruits. Thus greater knowledge is required concerning the fundamental population biology of key reef organisms and in particular those species that play important functional roles.

In the case of barrel sponges, populations appear largely reliant on self-recruitment, which should have important implications for their management, which coupled with their long-lived nature and apparent low level of recruitment would appear to make them susceptible to physical disturbance or fatal disease outbreaks. Despite this, barrel sponges are abundant at sites considered to have experienced high levels of degradation in the past, suggesting some level of tolerance to disturbance. Furthermore, these sites also experience heavy levels of sedimentation, suggesting some level of adaptation to these conditions. While this might appear something of a paradox, the features of barrel sponge populations means they are likely to require management at relatively local-scales (several km) as population extinctions are unlikely to be replenished by distant larval sources.

## Conclusions

Here we report the first estimates of demographic connectivity in tropical sponges, and demonstrate barrel sponges in the Wakatobi have low dispersal distances and are likely to be heavily reliant on self-recruitment, with high levels of inbreeding. The features of these species' would make them susceptible to disturbance but this does not appear to be the case. This provides further evidence to support the hypothesis that sponges may be likely 'winners' in response to any further environmental degradation.

## Acknowledgments

We are grateful to John Hooper and Merrick Ekins from the Queensland Museum for providing museum specimens of *Xestospongia testudinaria* and *X. bergquistia*, and Amber Stubler for providing specimens of *X. muta*. We are also grateful to Jane Fromont for useful discussions. Logistical support for fieldwork was provided by Steve Oliver, Chris Majors and Operation Wallacea.

## Author Contributions

Conceived and designed the experiments: JJB. Performed the experiments: JJB DH LT. Analyzed the data: LT. Contributed reagents/materials/analysis tools: JJB AH JJ DS. Wrote the paper: JJB LT DS.

## References


1. Burke L, Reytar K, Spalding M, Perry A (2011) Reefs at Risk Revisited. World Resources Institute, Washington DC. 400 p.
2. Hoegh-Guldberg O, Mumby PJ, Hooten AJ, Steneck RS, Greenfield P, et al. (2007) Coral reefs under rapid climate change and ocean acidification. Science 318: 1737.
3. Hughes TP, Baird AH, Bellwood DR, Card M, Connolly SR, et al. (2003) Climate change, human impacts, and the resilience of coral reefs. Science 301: 929–933.
4. Hughes TP, Graham NA, Jackson JB, Mumby PJ, Steneck RS (2010) Rising to the challenge of sustaining coral reef resilience. TREE 25(11): 633–642.
5. Lowe WH, Allendorf FW (2010) What can genetics tell us about population connectivity? Mol Ecol 19: 3038–3051.
6. Allison GW, Gaines SD, Lubchenco J, Possingham HP (2003) Ensuring persistence of marine reserves: Catastrophes require adopting an insurance factor. Ecol App 13: 8–24.
7. Bell JJ, B Okamura (2005) Low genetic diversity in a marine nature reserve: re-evaluating diversity criteria in reserve design. Proc Royal Soc Biol Sci 272: 1067–1074.
8. Jones GP, Almany GR, Russ GR, Sale PF, Steneck RS, et al (2009) Larval retention and connectivity among populations of corals and reef fishes: history, advances and challenges. Coral Reefs 28: 307–325.
9. Saenz-Agudelo P, Jones GP, Thorrold SR, Planes S (2009) Estimating connectivity in marine populations: an empirical evaluation of assignment tests and parentage analysis under different gene flow scenarios. Mol Ecol 18: 1765–1776.
10. Underwood JN, Smith LD, Oppen MJV, Gilmour JP (2009). Ecologically relevant dispersal of corals on isolated reefs: implications for managing resilience. Ecol App 19: 18–29.
11. Horne JB, van Herwerden L, Abellana S, McIlwain JL (2013) Observations of migrant exchange and mixing in a coral reef fish metapopulation link scales of marine population connectivity. J Hered 104: 532–546.
12. Bell JJ (2008) The functional roles of marine sponges. Est Coast Shelf Sci 79: 341–353.
13. Wulff JL (2001) Assessing and monitoring coral reef sponges: Why and how? Bull Mar Sci 69: 831–846.
14. Webster NS, Taylor MW (2012) Marine sponges and their microbial symbionts: love and other relationships. Environ Micro 14: 335–346.
15. Knapp IS, Williams GJ, Carballo JL, Cruz-Barraza JA, Gardner JPA, et al. (2012) Restriction of sponges to an atoll lagoon as a result of reduced environmental. Mar Poll Bull 66: 209–220.
16. Bell JJ, Davy SK, Jones T, Taylor MW, Webster NS (2013a) Could some coral reefs become sponge reefs as our climate changes? Global Change Biology doi: 10.1111/gcb.12212.
17. Powell A (2013) The impacts of predation and habitat degradation on coral reef sponges. Victoria University of Wellington PhD thesis.
18. Duran S, Pascual M, Estoup A, Turon X (2004). Strong population structure in the marine sponge *Crambe crambe* (Poecilosclerida) as revealed by microsatellite markers. Mol Ecol 13: 511–522.
19. Whalan S, Johnson MS, Harvey E, Battershill C (2005) Mode of reproduction, recruitment, and genetic subdivision in the brooding sponge *Haliclona* sp. Mar Biol 146: 425–433.
20. López-Legentil S, Pawlik JR (2009) Genetic structure of the Caribbean giant barrel sponge *Xestospongia muta* using the I3-M11 partition of COI. Coral Reefs 28: 157–165.
21. Blanquer A, Uriz MJ, Caujapé-Castells J (2009) Small-scale spatial genetic structure in *Scopalina lophyropoda*, an encrusting sponge with philopatric larval dispersal and frequent fission and fusion events. Mar Ecol Prog Ser 380: 95–102.
22. McMurray SE, Blum JE, Pawlik JR (2008) Redwood of the reef: growth and age of the giant barrel sponge, *Xestospongia muta* in the Florida Keys. Mar Biol 155: 159–171.
23. McMurray SE, Henkel TP, Pawlik JR (2010) Demographics of increasing populations of the giant barrel sponge *Xestospongia muta* in the Florida Keys. Ecology 91: 560–570.
24. López-Legentil S, Song BK, McMurray SE, Pawlik JR (2008) Bleaching and stress in coral reef ecosystems: hsp70 expression by the giant barrel sponge *Xestospongia muta*. Mol Ecol 17: 1840–1849.
25. Fromont J (1991) Descriptions of species of the Petrosida (Porifera: Demospongiae) occuring in the tropical waters of the Great Barrier Reef - The Beagle, Records of the Northern Territory of Museum of Arts and Science 8(1): 73–96.
26. Fromont J, Kerr S, Kerr R, Riddle M, Murphy P (1994) Chemotaxonomic relationships within, and comparisons between, the orders Haplosclerida and Petrosida (Porifera: Demospongiae) using sterol complements. Biochemic Syst Ecol 22(7): 735–752.
27. Fromont J, Bergquist PR (1994) Reproductive biology of three sponge species of the genus *Xestospongia* (Porifera: Demospongiae: Petrosida) from the Great Barrier Reef. Coral Reefs 13(2): 119–126.

28. Maldonado M (2006) The ecology of the sponge larva. Can J Zool 84: 175–94.
29. Swierts T, Peijnenburg KT, de Leeuw C, Cleary DF, Hörnlein C, et al. (2013) Lock, Stock and Two Different Barrels: Comparing the Genetic Composition of Morphotypes of the Indo-Pacific Sponge *Xestospongia testudinaria*. PloS one 8(9): e74396.
30. Bell JJ, Smith DJ, Hannan D, Haris A, Thomas L (2013). Isolation and characterisation of twelve polymorphic microsatellite markers for *Xestospongia* spp. and their use for confirming species identity. Conservation Genetics Resources DOI 10.1007/s12686-013-0015-5.
31. Clifton J, Unsworth RFK (2010) Introduction to the Wakatobi National Park. Chapter 1 In: Clifton J, Unsworth RKF, Smith DJ (eds) Marine Conservation and Research in the Coral Triangle: The Wakatobi National Park?. Nova Publishers, New York, p 1–9.
32. Cullen LC (2010) Marine resource dependence and natural resource use patterns in a small Indo-Pacific island community: implications for management. Chapter 10 In: Clifton J, Unsworth RKF, Smith DJ (eds) Marine Research and Conservation in the Coral Triangle: the Wakatobi Marine National Park. Nova Publishers, New York, p 171–191.
33. McMellor S, Smith DJ (2010) Coral reefs of the Wakatobi: abundance and biodiversity Chapter 2 In: Clifton J, Unsworth RKF, Smith DJ (eds) Marine research and conservation in the Coral Triangle: the Wakatobi National Park. Nova Science Publishers, New York, p 11–26.
34. Holleley CE, Geerts PG (2009) Multiplex Manager 1.0: a cross-platform computer program that plans and optimizes multiplex PCR. BioTech 46: 511–517.
35. Nei M (1972) Genetic Distance between Populations. American Society of Naturalists 106: 283–292.
36. Goudet J (1995) FSTAT (Version 1.2): A Computer Program to Calculate F-Statistics. Heredity 86: 485–486.
37. Nei M (1987) Molecular Evolutionary Genetics (Columbia Univ. Press, New York). 512 p.
38. Peakall R, Smouse PE (2006) GenAlEx 6: genetic analysis in Excel. Population genetic software for teaching and research. Mol Ecol Notes 6: 288–295.
39. Excoffier L, Laval G, Schneider S (2005) Arlequin ver. 3.0: An integrated software package for population genetics data analysis. Evolutionary Biology Online 1: 47–50.
40. Rice WR (1989) Analyzing tables of statistical tests. Evolution 43: 223–235.
41. Weir BS, Cockerham CC (1984) Estimating F-statistics for the analysis of population structure. Evolution 1358–1370.
42. Meirmans PG, Hedrick PW (2011) Assessing population structure: F(ST) and related measures. Mol Ecol Res 11: 5–18.
43. Pritchard JK, Stephens M, Donnelly P (2000) Inference of population structure using multilocus genotype data. Genetics 155: 945–959.
44. Evanno G, Regnaut S, Goudet J (2005) Detecting the number of clusters of individuals using the software STRUCTURE: a simulation study. Mol Ecol 14: 2611–2620.
45. Earl DA (2012) STRUCTURE HARVESTER: a website and program for visualizing STRUCTURE output and implementing the Evanno method. Cons Gen Res 4: 359–361.
46. Jakobsson M, Rosenburg NA (2007) CLUMPP: a cluster matching and permutation program for dealing with label switching and multimodality in analysis of population structure. Bioinformatics 23: 1801–1806.
47. Rosenberg NA (2004) DISTRUCT: a program for the graphical display of population structure. Mol Ecol Notes 4: 137–138.
48. Rannala B, Mountain JL (1997) Detecting immigration by using multilocus genotypes. Proc Nat Acad Sci 94: 9197–9201.
49. Piry S, Alapetite A, Cornuet JM, Paetkau D, Baudouin L, et al. (2004) GeneClass2: A Software for Genetic Assignment and First-Generation Migrant Detection. Heredity 95: 536–539.
50. Berry O, Tocher MD, Sarre SD (2004) Can assignment tests measure dispersal? Mol Ecol 13: 551–561.
51. Paetkau D, Slade R, Burden M, Estoup A (2004) Direct, real-time estimation of migration rate using assignment methods: a simulation-based exploration of accuracy and power. Mol Ecol 13: 55–65.
52. Peakall R, Ruibal M, Lindenmayer DB (2003) Spatial autocorrelation analysis offers new insights into gene flow in the Australian bush rat, *Rattus fuscipes*. Evolution 57: 1182–1195.
53. Miller KJ, Ayre DJ (2008) Population structure is not a simple function of reproductive mode and larval type: insights from tropical corals. J Anim Ecol 77(4): 713–724.
54. Wright S (1931) Evolution in Mendelian populations. Genetics 16: 97–159.
55. Wright S (1943) Isolation by Distance. Genetics 28: 114–38.
56. Almany GR, Berumen ML, Thorrold SR, Planes S, Jones GP (2007) Local replenishment of coral reef fish populations in a marine reserve. Science 316: 742–744.
57. Selkoe KA, Toonen RJ (2006) Microsatellites for ecologists: a practical guide to using and evaluating microsatellite markers. Ecol Letters 9: 615–629.
58. Bell JJ, Smith DJ (2004) Ecology of sponge assemblages (Porifera) in the Wakatobi region, south-east Sulawesi, Indonesia: Richness and abundance. J Mar Biol Assoc UK 84: 581–591.
59. Bell JJ (2007) Contrasting patterns of species and functional composition of coral reef sponge assemblages. Mar Ecol Prog Ser 339: 73–81.
60. Kinlan BP, Gaines SD (2003) Propagule dispersal in marine and terrestrial environments: a community perspective. Ecology 84: 2007–2020.

# 5

# Different Relationships between Temporal Phylogenetic Turnover and Phylogenetic Similarity and in Two Forests Were Detected by a New Null Model

**Jian-Xiong Huang[1,2], Jian Zhang[3], Yong Shen[1,2], Ju-yu Lian[1], Hong-lin Cao[1], Wan-hui Ye[1], Lin-fang Wu[1], Yue Bin[1]***

**1** Key Laboratory of Vegetation Restoration and Management of Degraded Ecosystems, South China Botanical Garden, Chinese Academy of Sciences, Guangzhou, China, **2** University of Chinese Academy of Sciences, Beijing, China, **3** Department of Renewable Resources, University of Alberta, Edmonton, Alberta, Canada

## Abstract

***Background:*** Ecologists have been monitoring community dynamics with the purpose of understanding the rates and causes of community change. However, there is a lack of monitoring of community dynamics from the perspective of phylogeny.

***Methods/Principle Findings:*** We attempted to understand temporal phylogenetic turnover in a 50 ha tropical forest (Barro Colorado Island, BCI) and a 20 ha subtropical forest (Dinghushan in southern China, DHS). To obtain temporal phylogenetic turnover under random conditions, two null models were used. The first shuffled names of species that are widely used in community phylogenetic analyses. The second simulated demographic processes with careful consideration on the variation in dispersal ability among species and the variations in mortality both among species and among size classes. With the two models, we tested the relationships between temporal phylogenetic turnover and phylogenetic similarity at different spatial scales in the two forests. Results were more consistent with previous findings using the second null model suggesting that the second null model is more appropriate for our purposes. With the second null model, a significantly positive relationship was detected between phylogenetic turnover and phylogenetic similarity in BCI at a 10 m×10 m scale, potentially indicating phylogenetic density dependence. This relationship in DHS was significantly negative at three of five spatial scales. This could indicate abiotic filtering processes for community assembly. Using variation partitioning, we found phylogenetic similarity contributed to variation in temporal phylogenetic turnover in the DHS plot but not in BCI plot.

***Conclusions/Significance:*** The mechanisms for community assembly in BCI and DHS vary from phylogenetic perspective. Only the second null model detected this difference indicating the importance of choosing a proper null model.

**Editor:** Keping Ma, Institute of Botany, Chinese Academy of Sciences, China

**Funding:** This study was funded by the National Natural Science Foundation of China (31100312, 41371078), URL is http://www.nsfc.gov.cn/Portal0/default166.htm, Knowledge Innovation Project of The Chinese Academy of Sciences (KSCX2-EW-Z), Foreign Exchange Program National Founder (31011120470) and Chinese Forest Biodiversity Monitoring Network, URL is http://www.cfbiodiv.org. The funders had no role in study design, data collection and analysis, decision to publish, or preparation of the manuscript.

**Competing Interests:** The authors have declared that no competing interests exist.

* E-mail: binyue@scbg.ac.cn

## Introduction

A central challenge in community ecology is to understand community dynamics [1,2]. Community dynamics involve processes including birth, death, immigration and emigration that can be affected by various processes such as seed dispersal [2], habitat selection [3,4], competition [5], disturbance [6,7], or stochastic processes [2]. Ecologists have been documenting community dynamics, such as species-specific mortality rates, recruitment rates, population growth rates and the change of species composition, in many forest communities providing basic knowledge for biodiversity maintenance [8–10]. However, how species composition changes along a temporal axis (community dynamics) still remains poorly studied.

Community dynamics can results from both deterministic niche-based mechanisms [11] and stochastic mechanisms which consider only dispersal and ecological drift [2]. As species names do not convey critical information regarding the ecological and evolutionary similarity of species [12–15], exploring changes in taxonomic composition of species alone cannot distinguish deterministic mechanisms from stochastic processes [16]. A complete turnover in species composition of a community could result in a functionally analogous community, a directional change in the functional composition of a community, or changes in the community composition that are random with respect to species functions [14]. Thus, to make robust inferences, it is critical that the analysis of community turnover goes beyond traditional approaches of analyzing turnover in species composition by incorporating information pertaining to the ecological and evolutionary similarity of species [16].

With the increasing availability of phylogenetic data, computing power, and informatic tools, ecologists have witnessed a rapid expansion of studies that apply phylogenetic data and methods to community ecology (reviewed by Cavender-Bares et al. [17]). However, how phylogenetic similarity will influence phylogenetic temporal turnover is not well understood.

Two types of deterministic processes predict different phylogenetic temporal turnover when phylogenetic similarity is considered. Negative phylogenetic density dependence suggests that an individual surrounded with phylogenetically related neighbors will exhibit higher mortality than the reverse because they are likely susceptible to the same pathogens, fungi, etc [18,19]. For the same reason, the empty space produced by the death of an individual will likely be occupied by a phylogenetically unrelated species in the species pool. In other words, if co-occurring species are closely related in phylogeny, negative phylogenetic density dependence predicts a high phylogenetic temporal turnover (positive relationship between phylogenetic temporal turnover and phylogenetic similarity). Alternatively, abiotic filtering, also known as species sorting, predicts species with functional traits incompatible with their environment experience elevated rates of mortality [20] which results in phylogenetic clustering [21] if functional traits exhibit a phylogenetic signal (Close phylogenetic related species have similar functional traits) [22]. Thus, given the dominance of habitat filtering, a community with phylogenetically unrelated species is expected to experience higher phylogenetic temporal turnover, i.e. a negative relationship between temporal phylogenetic turnover and phylogenetic similarity of co-occurring species.

To quantify temporal phylogenetic turnover, a few metrics have been developed [23]. However, the raw level of these metrics does not allow for making inferences regarding the cause for the observed phylogenetic turnover. Swenson et al. [16] proposed to randomize species name across a potential species pool as a null model for measuring the temporal phylogenetic turnover expected by chance. This null model has merits such as maintaining the observed species abundance, species richness, species occupancy rates, and species turnover through time. When applied to phylogenetic turnover through time, however, it implicitly assumes that all species have identical dispersal capability, immigrant history and survival rate across all life stages. These assumptions seem unrealistic to a natural plant community. For example, both dispersal ability and survival rate have a wide variation among coexisting species [24,25]; additionally, species identity, survival rate also seems to be influenced by size [26,27]. Appropriately considering these differences among species in the null model can improve inference for the cause of phylogenetic turnover through time.

In this study, we investigated phylogenetic turnover through time in two long-term forest dynamic plots using Swenson et al.'s null model [16] and a proposed new null model. In the new null model careful consideration of the difference in dispersal ability among species, the differences in survival rate both among species and life stages was given. How phylogenetic similarity of coexisting species affected phylogenetic turnover through time was tested. To better understand their relationship, the significance of phylogenetic signal for a set of important functional traits was tested.

## Materials and Methods

### Study sites

In this study, data was collected from two long-term forest plots: a 50-hectare tropical lowland moist forest plot in the Barro Colorado Island (BCI), Panama (9°10′N, 79°51′W) [28], and a 20-hectare subtropical forest plot in the Dinghushan (DHS) National Natural Reserve in southern China (23°19′N, 112°30′E). In each plot, all free standing woody stems with ≥1 cm diameter at breast height (DBH) or 1.3 m from the ground [29] were tagged, mapped, measured for DBH, identified to species level, and recorded for survival status during re-census.

To detect temporal changes in species composition, the most recent censuses years (2005 and 2010) were selected for both sites. In BCI, a total of 208,371 living individuals were recorded in 2005, including 297 species belonging to 182 genera and 59 families. In 2010, the number of individuals decreased to 207,200, with 30,414 deaths and 29,243recruits, including 7 lost species and 3 recruited species. Eight traits for the species in BCI were available, including maximum height, leaf area, leaf percentage carbon, leaf percentage nitrogen, leaf percentage phosphorus, seed mass and wood density. Phylogenetic signal for these eight functional traits had been tested in previous studies [16]. In DHS, the number of living individuals was 71,462 in 2005, including 195 species belonging to 114 genera and 54 families. In 2010, it decreased to 60,072 individuals, with 18,072 deaths and 6,601 recruits, including 20 lost species and 3 recruited species. Seven traits for 137 species in DHS were collected. These traits included leaf area, specific leaf area, leaf thickness, leaf dry mass content, leaf chlorophyll, petiole dry weight/fresh weight and wood density. These traits were chosen as they represent fundamental axes of ecological strategy in trees [16].

### Phylogenetic tree construction

Although molecular phylogenies are currently available for most of species in these two plots [30,31], 16 (8.2%) and 27 (8.8%) rare species were not available for DHS and BCI plots, respectively. Rare species have low abundances and are sensitive to changes in habitat conditions and micro-successive stages, and thus are worth considering in our analysis. Therefore, phylogenies constructed by Phylomatic were used so that all species in these two censuses for both of plots were considered. A Phylomatic phylogeny was constructed using Phylomatic mega-tree by inputting all the species in the species list from the censuses into the plant phylogeny database [32,33]. Phylomatic uses the Angiosperm Phylogeny Group Classification as the backbone for mega-trees. Branch lengths of phylogenetic trees were adjusted using the BLADJ algorithm [21] with known molecular and fossil dates [34]. These dates are rough estimates, but they make the quality of phylogenetic analyses more robust than nodal distances [12].

### Quantifying phylogenetic turnover

In this study Rao's quadratic Entropy (RaoD) was used to quantify temporal phylogenetic turnover through time [35]. RaoD is widely used in quantifying beta-diversity because it measures differences of species relative abundance and species differences [35–38]. When RaoD is applied to quantify temporal phylogenetic turnover, RaoD is the expected phylogenetic difference of a quadrat between two census times:

$$\text{RaoD} = \sum_{i}^{S_A} \sum_{j}^{S_B} p_i \times d_{ij} \times p_j, \qquad (1)$$

where $S_A$ and $S_B$ are the numbers of species in a quadrat at time A and B, $p_i$ and $p_j$ are the relative abundances of species $i$ and $j$ in that quadrat, $d_{ij}$ is phylogenetic distance between species $i$ and $j$.

Two null models were used to obtain the RaoD expected by chance. The first null model (NM-I) we used was to randomize species names across all species types in a phylogeny, as in Swenson et al. (2012) [16]. The second null model (NM-II) simulated the recruitment process when considering the variation in dispersal ability among species and simulated the mortality process when considering the variations in mortality both among species and among size classes within a species. To simulate the recruitment process, species were first classified into either tree or shrub. According to advice from experts in DHS plot, trees and shrubs with DBH≥15 and 5 cm in the first census were considered mature and potential parents for the recruits, respectively (Zhongliang Huang, personal communication). The nearest conspecific mature individual within 50 m of this recruit was considered its parent. The distance between a recruit and its parent was recorded as *Dp*. A potential parent was then randomly selected as a recruit from all of conspecific mature individuals in the plot. The coordinates for the location of this recruit was randomly generated from a circle, for which the position of the selected parent is its center and *Dp* is its radius, If a recruit has no possible parents within 50 m, its position was randomly generated within the bounds of the plot using uniform distribution [39]. To simulate mortality, we first grouped all individuals for each species in the first census into size classes by DBH of 1–2, 2–3, ... 19–20, 20–25, 25–35, 35–45, 45–55 and ≥55 cm and calculated the survival rates for each of size class of each species. We then randomly assigned mortality to individuals in each size class for each species based on these species-specific and size-specific survival probabilities.

Randomized data sets were obtained for the second census of the BCI and the DHS plots using these two null models. For each plot, 499 such data sets were generated and calculated a RaoD for each of them. Then, the observed RaoD (RaoD.obz) for a quadrat was standardized to a standardized effect size (ses.RaoD) by the 499 null RaoDs (RaoD.null)as:

$$ses.RaoD = \frac{RaoD.obz - mean(RaoD.null)}{sd(RaoD.null)}, \quad (2)$$

where mean(RaoD.null) and sd(RaoD.null) are the mean and standard deviation of RaoD.null generated by the 499 simulated distributions. Thus, ses.RaoD can explain to what extent the observed temporal turnover is different from what expected by chance.

### Test of phylogenetic signal

The phylogenetic signal was examined among those species across the phylogeny using the statistic *K* which is the tendency of related species to resemble one another or the amount of phylogenetic signal [22]. *K* is close to zero when there is no phylogenetic signal for a functional trait. *K* is equal to one under a Brownian motion model, when traits change by small random amounts and at a constant rate through time. *K* is greater than one when close relatives are more similar than expected under Brownian motion evolution, indicating a strong phylogenetic signal for a functional trait. To evaluate the significance of the phylogenetic signal, a null expectation of *K* under no phylogenetic signal was generated by randomly shuffling the tips of the phylogeny 1000 times, and a probability that the observed *K* is higher than randomization is calculated. The probability generated indicates statistical significance of phylogenetic signal of a functional trait across a phylogeny.

### Statistical analysis

Each plot was divided into non-overlapping square quadrats with side lengths of 10, 20, 30, 40 and 50 m. The index ses.RaoD was used to quantify the temporal phylogenetic turnover against that expected by chance. Net phylogenetic relatedness index (NRI) was used to represent phylogenetic similarity of species in a quadrat [21]. NRI is the negative standardize effect size of mean phylogenetic distance, calculated as

$$NRI = -1 \times ses.MPD = -1 \times \frac{MPD_{sample} - mean(MPD_{rndsample})}{sd(MPD_{rndsample})}, \quad (3)$$

where $MPD_{sample}$ is observed MPD (mean of all pairwise phylogenetic distances between co-occurring taxa) for the quadrate, and mean ($MPD_{rndsample}$) and sd ($MPD_{rndsample}$) are mean and standard deviation of MPDs from 499 simulations. In each simulation, an independent swap algorithm randomization process was performed to generate the random assemblages [40]. This randomization process maintains species occurrence frequency and sample species richness as well.

A simultaneous autoregressive error model ($SAR_{err}$) with 1 grid-cell lag-distance was used to fit the relationship between ses.RaoD and *NRI*. Spatial autocorrelation is a frequent phenomenon in ecological data and can affect estimates of model coefficients and inference from statistical models and $SAR_{err}$ model is recommended when dealing with spatially autocorrelated species distribution data [41].

Spatial variations of ses.RaoD among quadrats at each of the five spatial scales explained by NRI and spatial structure represented by principal coordinates of neighbor matrices (PCNM) eigenfunctions were analyzed by variation partitioning [42,43]. Partitioning was based on adjusted $R^2$ statistic ($R^2_a$) in multiple linear regressions [44]. PCNM eigenfunctions describe all spatial scales that can be accommodated in the five quadrat sizes. They were obtained by principal coordinate analysis (PCoA) of a truncated geographic distance matrix among the sampling subplots. In the present study, all distances larger than those between the centers of diagonally adjacent subplots were replaced by four times their values before PCoA. A forward selection with permutation tests at 5% significance level based on $R^2$ increases at each step was used to remove those PCNM eigenfunctions that were not significantly correlated with the detrended ses.RaoD. Before variation partitioning, ses.RaoD was detrended using third-degree orthogonal polynomials, and the residuals were retained for variation partitioning [45].

All statistical analyses were performed in the software R 2.15.2 [46]. NRI was calculated using "picante" package [47]. The SAR analyses were computed using "spdep" package [48], PNCM eigenfunctions and variation partitioning were computed using the "vegan" package [49], and forward selection was computed using "packfor" package.

## Results

In DHS, five of seven traits exhibit significant phylogenetic signals ($P<0.05$) (Table 1). In BCI, *K* values were all less than 0.1 (close to zero) for all of the tested eight functional traits [16], indicating that phylogenetic signal was little to null for the eight functional traits for tree species in the BCI plot. The results suggest that using phylogenetic distance as a surrogate for difference of functional traits is proper for species in DHS plot but not reliable for species in BCI plot.

For both null models, spatial autocorrelation of residuals of models at five scales in the two plots were tested using Moran's I based on 999 random permutations [50]. Except at 50 m scale in DHS using NM-II, significant spatial autocorrelation was detected when simple linear regression was used but disappeared when $SAR_{err}$ was used (see Table S1) indicating that ses.RaoD was significantly spatially autocorrelated and that the spatial autocorrelation in ses.RaoD was successfully removed using $SAR_{err}$.

**Table 1.** Test of phylogenetic signal in seven functional traits for species in the DHS plot.

| Functional Trait | *K* | p-value |
|---|---|---|
| Leaf Area | 0.134 | 0.956 |
| Specific Leaf Area | 0.481 | 0.002 |
| Leaf thickness | 0.402 | 0.06 |
| Leaf dry mass content | 0.488 | 0.001 |
| Leaf chlorophyll | 0.434 | 0.004 |
| Petiole dry weight/fresh weight | 0.448 | 0.013 |
| Wood density | 0.569 | 0.001 |

The *K* statistic proposed by Blomberg *et al.* (2003) indicates the amount of phylogenetic signal, and p-value indicates statistical significance of phylogenetic signal of a functional trait across a phylogeny, to which the observed phylogenetic signal is different from null model.

When NM-I was used to calculate ses.RaoD, results of $SAR_{err}$ show that NRI was negatively correlated with phylogenetic turnover (evaluated by ses.RaoD) at all of five spatial scales in both of plots, suggesting that phylogenetically closely related species formed more temporally stable communities (Figure 1). When NM-II was used, there were markedly different results for the BCI and the DHS plots. In the BCI plot, the correlation between NRI and ses.RaoD was positive except at the largest scale but this relationship was significant only at the finest scale (Figure 1A). In contrast to the BCI plot, in DHS plot, NRI was negatively correlated with ses.RaoD at all scales, significant at the scales of 20, 40 and 50 m but not at scales of 10 m and 30 m (Figure 1B). The absolute values of regression coefficients increased with increasing scales except for 30 m. This trend was not observed with NM-I.

With NM-I, in the BCI plot, the proportion of variation of ses.RaoD explained by NRI totally in terms of [a+b] decreases from 41.7% at the 10 m scale to 6.8% at the 50 m scale (Figure 2A). In the DHS plot, the proportion of variance explained by NRI ranged from 0.6% to 18.7% (Figure 2C). No obvious trend was observed for the variation explained by NRI as scale increased, in contrast to the BCI plot.

NM-II showed a minimal contribution of NRI ([a+b]) to the variance of ses.RaoD in the BCI plot. PCNMs account for 25.1% to 49.6% of the variance at the five scales indicated by [c] (Figure 2B). In the DHS plot, the variation of ses.RaoD explained by NRI in terms of [a+b] increases from approximately 0% at 10 m scale to 17.7% at the 50 m scale (Figure 2D). In addition, for both plots, the proportion of variance explained by PCNMs increases at first and then decreases as scale increases. More details can be found in Table S2.

Comparison of residuals using variation partitioning analysis showed that the proportion of unexplained variations using NM-I were higher than using NM-II for eight of ten cases in the two plots, except for 30 and 50 m scales in DHS plot where the proportion of unexplained variations using NM-I was slightly lower than using NM-II (Figure 3).

## Discussion

To obtain the expected phylogenetic turnover by chance, a widely used method is to shuffle name of species across all tips of a phylogeny (NM-I) [12,16,51]. With NM-I, negative correlation between ses.RaoD and NRI were found in both BCI and DHS plots, suggesting that processes affecting phylogenetic turnover were generally similar in these two plots. This force could be abiotic filtering if species in the two plots represent niche conservatism. The importance of abiotic filtering is supported by previous studies in DHS where habitat conditions affect the distribution of 83% of species [52] and habitat factors explained

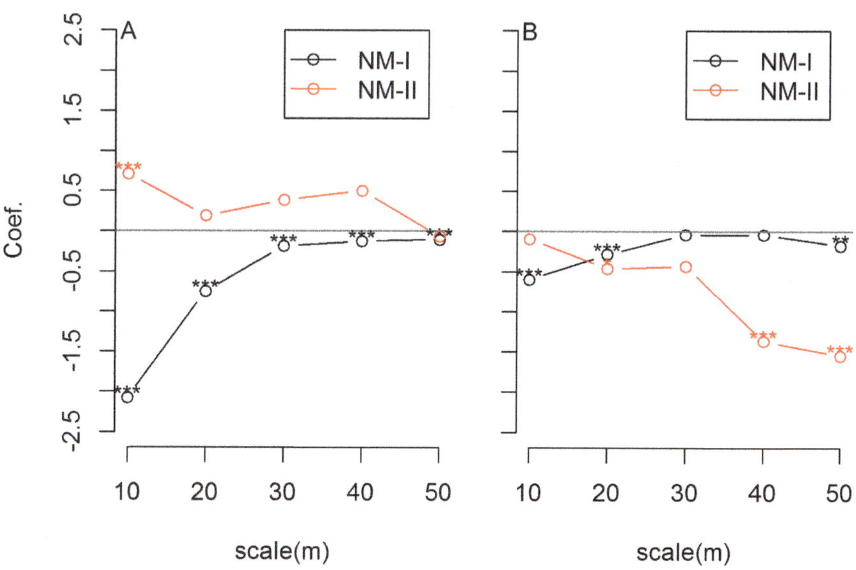


**Figure 1. The relationship between phylogenetic similarity and phylogenetic turnover along five spatial scales for BCI and DHS.** The relationship between phylogenetic similarity and phylogenetic turnover based on spatial simultaneous autoregressive error model across scales for the BCI (A) and DHS (B) plots. Y-axis is regression coefficient of ses.RaoD against NRI. The independent variable is phylogenetic diversity index NRI and the dependent variable is phylogenetic turnover ses.RaoD, which is the standardized effect size of RaoD of a null model. Null Model-I (NM-I) shuffles species names across all species in each plot, while Null Model-II (NM-II) randomizes the location of mortality according to species- and size-specific survival probability and the location of recruitment according to species-specific dispersal ability. The black line was from NM-I, and the red line from NM-II. '**' indicates p value<0.001 and >0.0001. '***' indicates p value<0.0001.

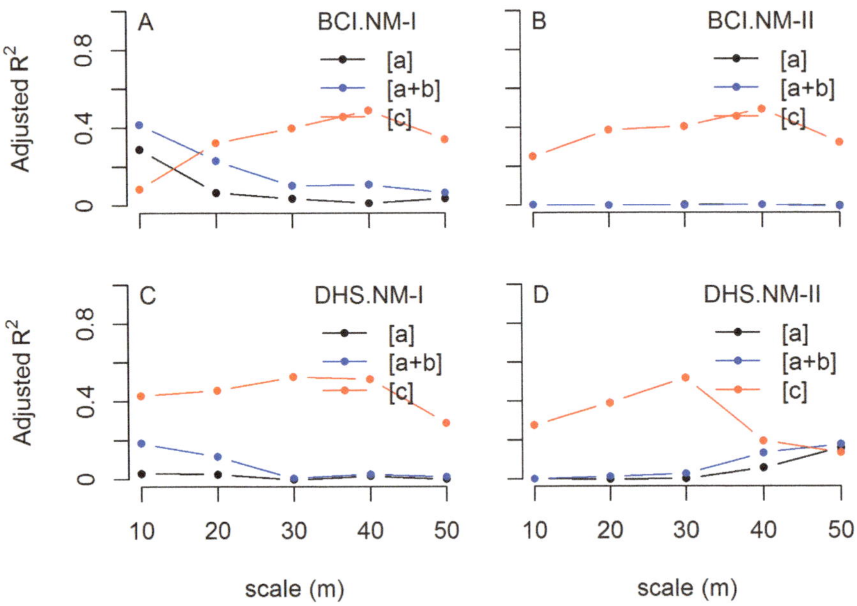


**Figure 2. Results of variation partitioning along five spatial scales for BCI and DHS.** The variation partitioning results for ses.RaoD using two Null models (NM-I and NM-II) for DHS and BCI plots along five spatial scales. [a] : the non-spatially structured part of variation explained by NRI, [a+b] : the total variation explained by NRI, [c] : the spatially structured variation not explained by NRI (the variance uniquely explained by PCNM eigenvectors). A) BCI plot using NM-I, B) BCI plot using NM-II, C) DHS plot using the NM-I, and D) DHS plot using the NM-II. Note that [a] and [a+b] are almost equal and are indistinguishable in panel A.

23.7% at the 10 m scale and 54.1% at the 100 m scale of the variation in species distribution [53]. However, in BCI, the main force for community structure may differ according to previous studies. Harms et al. [4] found that in the BCI plot, among 181 species, only 28 species were significantly associated with the most extensive habitat in the BCI plot, indicating a limited role of habitat condition for community structure. Hubbell [2] found that relative species abundance in the BCI plot could be fitted by neutral processes extremely well. In a comparative study, habitat factors were less important to species distribution in BCI than in DHS [54]. Taken together, the traditional null model generated results consistent with previous analyses in DHS plot but not with previous findings for the BCI plot. We speculated that the assumption of equal dispersal ability and survival rate of this model has reduced the ability to detect deterministic effect and led to this discrepancy.

To improve efficiency, the positions of the recruits were randomized based on distance to the nearest conspecific adult so that differences in dispersal ability and seedling survival rates can be considered to some degree. The occurrence of mortality was also randomized based on species-specific and size-specific survival rates in NM-II (the observed mortality of species at all size classes in BCI and DHS plots between 2005 and 2010 can be found in table S3 and table S4, respectively). Using NM-II, the direction of the correlation between NRI and ses.RaoD did not changed in DHS. In BCI, this correlation was significantly positive at the finest scale, positive but not significant at the three intermediate scales, and negative but not significant at the largest scale. A previous study found that phylogenetic similarity decreases with size at local spatial scale (10–15 m) in BCI [13], suggesting a larger phylogenetic turnover for phylogenetically close communities. This is consistent with the positive correlation between NRI and

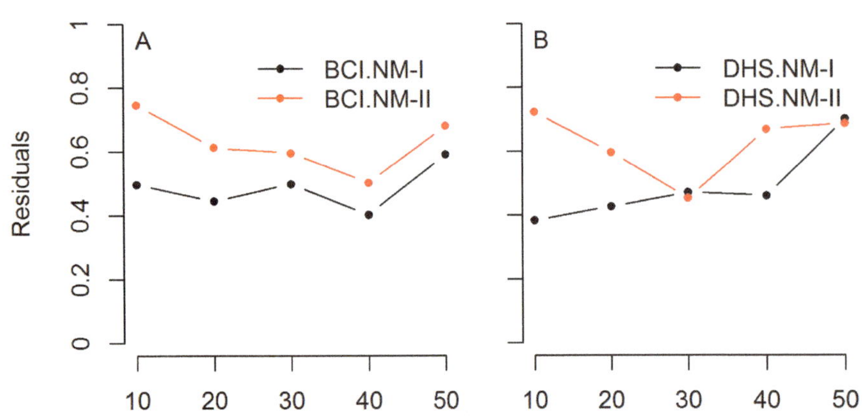


**Figure 3. Comparing of residuals in variation partitioning between two null models.** Comparing of residuals in variation partitioning between two Null models (NM-I and NM-II) along five spatial scales. A) BCI plot, B) DHS plot.

ses.RaoD found using NM-II. Therefore, using NM-II, the results for both the BCI and DHS plots were in agreement with previous studies.

As spatial scale increases, according to the well-established species-area relationship [55], species composition of a sampled community becomes more complete and community dynamics are likely to become more self-regulated [56,57]. Thus, the proportion of variance explained by NRI was expected to increase with spatial scale. This expectation was supported using NM-II but not the traditional null model. Additionally, the unexplained fraction of phylogenetic distance using NM-I was higher than NM-II for most of cases. Therefore, we believe that NM-II is more appropriate than the null model traditionally used in related studies.

According to the analysis based on NM-II, phylogenetic similarity positively influences temporal turnover only at the finest scale in BCI plot. In other words, a quadrat in the BCI plot with distantly related species is more phylogenetically stable than with phylogenetically closely related species between the census periods of 2005 to 2010. As mentioned above, this positive correlation is supported by previous study in BCI [13]. This can be explained by phylogenetic density dependent effect. When this effect dominates, individuals are more likely to be eliminated from a quadrat (or to introduce a turnover) when they are surrounded by closely related neighbors. However, the partition results showed that the proportion of variance explained by NRI was almost zero for all of five scales in BCI plot, suggesting that the significant positive correlation between ses.RaoD and NRI at 10 m scale detected using $SAR_{err}$ is weak.

In the DHS plot, there was a significant negative relationship between phylogenetic similarity and phylogenetic turnover at three of the five scales examined. This suggests that phylogenetically related communities in the DHS plot are more stable than phylogenetically unrelated communities, in contrast to the BCI plot. Considering that there is significant phylogenetic signal for most of tested functional traits, abiotic filtering is a reasonable explanation for this pattern. For given a habitat condition, species with certain traits that survive under that habitat condition should be selected by the habitat while other species that are not suitable are excluded. Therefore, different from phylogenetic density dependent effect, abiotic filtering accumulates species that are functionally similar, or phylogenetically close if functional traits show phylogenetic signal. Therefore, abiotic filtering expects communities with low NRI to have large phylogenetic turnover because a large proportion of the species in these communities are to be excluded by the habitat condition, leaving space for recruits that are favored by the habitat condition. Thus, the negative correlation between NRI and ses.RaoD indicates the importance of abiotic filtering, consistent with previous studies conducted in the same plot [52,53]. In the DHS plot, the proportion of variance in ses.RaoD explained by NRI increases from about 0% to 17.1% as the scale increased, which suggests that abiotic filtering have played an increasingly important role in the plot as scale increased.

In both plots, the proportion of variance of ses.RaoD explained by PCNMs is much higher than by NRI, suggesting the importance of spatial autocorrelation for phylogenetic turnover. In this study, spatial autocorrelation of temporal turnover consists of spatial autocorrelation of the occurrence of recruitment and mortality. On the one hand, because of dispersal limitation, neighbor quadrats are more likely to share similar species composition in the seed pool than quadrats that are spatially separated [58], which leads to spatial autocorrelation for the phylogenetic structure of recruits. On the other hand, limited dispersal and patchy habitat condition together results in autocorrelated species distribution [4,59,60]. This autocorrelation in species distribution can turn into autocorrelated phylogenetic structure of dead individuals because mortality rates also differ among species (see [61]). The strong autocorrelation in phylogenetic turnover indicates the importance of seed dispersal and survival probability and possibly other small-scale processes.

Although results using NM-II were more reasonable than using NM-I in this study, there were still limitations in the analysis. For example, when simulating the recruitment process, diameters of 15 cm and 5 cm were arbitrarily used to define mature individuals for the tree and shrub species respectively. About 60% recruits (4,683 of 6,679 in DHS plot and 19,472 of 36,525 in BCI plot) did not have a possible parent plant within 50 m. When simulating the mortality process, we took into account only the variation in mortality among size classes and among species. Other factors like habitat conditions that influence mortality risk were not considered. We believe the result can be improved by a more realistic procedure in future studies.

In conclusion, we found large spatial autocorrelation in phylogenetic turnover in both plots, indicating the importance of spatially structured processes in both of these two plots. Phylogenetic relatedness positively and negatively associated with phylogenetic turnover in BCI and DHS, respectively, implying different conservative strategies for species at these two sites.

## Supporting Information

**Table S1** Spatial autocorrelation of residuals of models at five scales in the two plots were tested using Moran's I based on 999 random permutations. P value less than 0.01 indicates significant spatial autocorrelation is detected using Moran's I.

**Table S2** Results of variation partitioning for ses.RaoD by NRI and PCNMs. NM-I, null model I, that considers deaths, recruits and the unvaried individuals separately. NM-II, null model II, that shuffles species names across all species in each of the plots.

**Table S3** Mortality of species for DBH size classes in BCI plot.

**Table S4** Mortality of species for DBH size classes in DHS plot.

## Author Contributions

Conceived and designed the experiments: JH. Performed the experiments: JH. Analyzed the data: JH. Contributed reagents/materials/analysis tools: JL HC WY LW YS. Wrote the paper: JH JZ YB.

## References


1. Chase JM, Leibold MA (2003) Ecological niches: linking classical and contemporary approaches: University of Chicago Press.
2. Hubbell S (2001) The Unified Neutral Theory of Biodiversity and Biogeography. Princeton University Press.
3. John R, Dalling JW, Harms KE, Yavitt JB, Stallard RF, et al. (2007) Soil nutrients influence spatial distributions of tropical tree species. Proceedings of the National Academy of Sciences of the United States of America 104: 864–869.
4. Harms KE, Condit R, Hubbell SP, Foster RB (2001) Habitat associations of trees and shrubs in a 50-ha neotropical forest plot. Journal of Ecology 89: 947–959.
5. Tilman D (1994) Competition and Biodiversity in Spatially Structured Habitats. Ecology 75: 2–16.

6. Pickett STA, White PS (1985) The ecology of natural disturbance and patch dynamics: Academic press.
7. Sousa WP (1984) The Role of Disturbance in Natural Communities. Annual Review of Ecology and Systematics 15: 353–391.
8. Condit R, Ashton PS, Manokaran N, LaFrankie JV, Hubbell SP, et al. (1999) Dynamics of the forest communities at Pasoh and Barro Colorado: comparing two 50–ha plots. Philosophical Transactions of the Royal Society of London Series B: Biological Sciences 354: 1739–1748.
9. Condit R, Hubbell SP, Foster RB (1992) Short-term dynamics of a neotropical forest. Bioscience 42: 822–828.
10. Shen Y, Santiago LS, Ma L, Lin G-J, Lian J-Y, et al. (2013) Forest dynamics of a subtropical monsoon forest in Dinghushan, China: recruitment, mortality and the pace of community change. Journal of Tropical Ecology 29: 131–145.
11. Levine JM, HilleRisLambers J (2009) The importance of niches for the maintenance of species diversity. Nature 461: 254–U130.
12. Webb CO (2000) Exploring the phylogenetic structure of ecological communities: An example for rain forest trees. American Naturalist 156: 145–155.
13. Swenson NG, Enquist BJ, Thompson J, Zimmerman JK (2007) The influence of spatial and size scale on phylogenetic relatedness in tropical forest communities. Ecology 88: 1770–1780.
14. Swenson NG, Anglada-Cordero P, Barone JA (2011) Deterministic tropical tree community turnover: evidence from patterns of functional beta diversity along an elevational gradient. Proceedings of the Royal Society B-Biological Sciences 278: 877–884.
15. Nipperess DA, Faith DP, Barton K (2010) Resemblance in phylogenetic diversity among ecological assemblages. Journal of Vegetation Science 21: 809–820.
16. Swenson NG, Stegen JC, Davies SJ, Erickson DL, Forero-Montana J, et al. (2012) Temporal turnover in the composition of tropical tree communities: functional determinism and phylogenetic stochasticity. Ecology 93: 490–499.
17. Cavender-Bares J, Kozak KH, Fine PVA, Kembel SW (2009) The merging of community ecology and phylogenetic biology. Ecology Letters 12: 693–715.
18. Webb CO, Gilbert GS, Donoghue MJ (2006) Phylodiversity-dependent seedling mortality, size structure, and disease in a bornean rain forest. Ecology 87: S123–S131.
19. Paine CET, Norden N, Chave J, Forget P-M, Fortunel C, et al. (2012) Phylogenetic density dependence and environmental filtering predict seedling mortality in a tropical forest. Ecology Letters 15: 34–41.
20. Harper JL (1977) Population biology of plants.
21. Webb CO, Ackerly DD, McPeek MA, Donoghue MJ (2002) Phylogenies and community ecology. Annual Review of Ecology and Systematics: 475–505.
22. Blomberg SP, Garland T, Ives AR (2003) Testing for phylogenetic signal in comparative data: Behavioral traits are more labile. Evolution 57: 717–745.
23. Swenson NG (2011) Phylogenetic Beta Diversity Metrics, Trait Evolution and Inferring the Functional Beta Diversity of Communities. PLoS One 6.
24. Jakobsson A, Eriksson O (2003) Trade-offs between dispersal and competitive ability: a comparative study of wind-dispersed Asteraceae forbs. Evolutionary Ecology 17: 233–246.
25. Bloor JMG, Grubb PJ (2003) Growth and mortality in high and low light: trends among 15 shade-tolerant tropical rain forest tree species. Journal of Ecology 91: 77–85.
26. Cook RE (1980) Germination and size-dependent mortality in Viola blanda. Oecologia 47: 115–117.
27. Harper J, White J (1974) The demography of plants. Annual review of ecology and systematics 5: 419–463.
28. Hubbell S, Foster R (1984) Diversity of canopy trees in a neotropical forest and implications for conservation. SPEC PUBL BR ECOL SOC 1984.
29. Condit R, Hubbell S (1998) Tropical forest census plots: Springer Austin.
30. Kress WJ, Erickson DL, Jones FA, Swenson NG, Perez R, et al. (2009) Plant DNA barcodes and a community phylogeny of a tropical forest dynamics plot in Panama. Proceedings of the National Academy of Sciences of the United States of America 106: 18621–18626.
31. Pei NC, Lian JY, Erickson DL, Swenson NG, Kress WJ, et al. (2011) Exploring Tree-Habitat Associations in a Chinese Subtropical Forest Plot Using a Molecular Phylogeny Generated from DNA Barcode Loci. PLoS One 6.
32. Webb C, Ackerly D, Kembel S (2008) Phylocom: software for the analysis of community phylogenetic structure and trait evolution. Version 4.2. Computer program and documentation distributed by the authors, website http://wwwphylodiversitynet/phylocom.
33. Webb CO, Donoghue MJ (2005) Phylomatic: tree assembly for applied phylogenetics. Molecular Ecology Notes 5: 181–183.
34. Wikstrom N, Savolainen V, Chase MW (2001) Evolution of the angiosperms: calibrating the family tree. Proceedings of the Royal Society B-Biological Sciences 268: 2211–2220.
35. Rao CR (1982) Diversity and Dissimilarity Coefficients - a Unified Approach. Theoretical Population Biology 21: 24–43.
36. Pavoine S, Ollier S, Pontier D (2005) Measuring diversity from dissimilarities with Rao's quadratic entropy: Are any dissimilarities suitable? Theoretical Population Biology 67: 231–239.
37. Ricotta C, Marignani M (2007) Computing β-diversity with Rao's quadratic entropy: a change of perspective. Diversity and Distributions 13: 237–241.
38. Ricotta C, Szeidl L (2006) Towards a unifying approach to diversity measures: Bridging the gap between the Shannon entropy and Rao's quadratic index. Theoretical Population Biology 70: 237–243.
39. Clark CJ, Poulsen JR, Bolker BM, Connor EF, Parker VT (2005) Comparative seed shadows of bird-, monkey-, and wind-dispersed trees. Ecology 86: 2684–2694.
40. Gotelli NJ, Entsminger GL (2001) Swap and fill algorithms in null model analysis: rethinking the knight's tour. Oecologia 129: 281–291.
41. Kissling WD, Carl G (2008) Spatial autocorrelation and the selection of simultaneous autoregressive models. Global Ecology and Biogeography 17: 59–71.
42. Legendre P, Borcard D, Peres-Neto PR (2005) Analyzing beta diversity: Partitioning the spatial variation of community composition data. Ecological Monographs 75: 435–450.
43. Legendre P, Mi XC, Ren HB, Ma KP, Yu MJ, et al. (2009) Partitioning beta diversity in a subtropical broad-leaved forest of China. Ecology 90: 663–674.
44. Dray S, Legendre P, Peres-Neto PR (2006) Spatial modelling: a comprehensive framework for principal coordinate analysis of neighbour matrices (PCNM). Ecological Modelling 196: 483–493.
45. Borcard D, Legendre P, Avois-Jacquet C, Tuomisto H (2004) Dissecting the spatial structure of ecological data at multiple scales. Ecology 85: 1826–1832.
46. R Core Team (2012) R: A Language and Environment for Statistical Computing. ISBN 3-900051-07-0, URL http://www.R-project.org/.
47. Kembel SW, Cowan PD, Helmus MR, Cornwell WK, Morlon H, et al. (2010) Picante: R tools for integrating phylogenies and ecology. Bioinformatics 26: 1463–1464.
48. Bivand R, Anselin L, Berke O, Bernat A, Carvalho M, et al. (2011) spdep: Spatial dependence: weighting schemes, statistics and models. R package version 0.5-31, URL http://CRAN.R-project.org/package=spdep.
49. Oksanen J, Blanchet FG, Kindt R, Legendre P, Minchin PR, et al. (2012) vegan: Community Ecology Package.
50. Cliff AD, Ord JK (1981) Spatial processes: models & applications: Pion London.
51. Mi X, Swenson NG, Valencia R, Kress WJ, Erickson DL, et al. (2012) The contribution of rare species to community phylogenetic diversity across a global network of forest plots. The American Naturalist 180: E17–E30.
52. Wang Z, Ye W, Cao H, Huang Z, Lian J, et al. (2009) Species-topography association in a species-rich subtropical forest of China. Basic and Applied Ecology 10: 648–655.
53. Lin G, Stralberg D, Gong G, Huang Z, Ye W, et al. (2013) Separating the Effects of Environment and Space on Tree Species Distribution: From Population to Community. PLoS One 8: e56171.
54. De Cáceres M, Legendre P, Valencia R, Cao M, Chang LW, et al. (2012) The variation of tree beta diversity across a global network of forest plots. Global Ecology and Biogeography 21: 1191–1202.
55. Connor EF, McCoy ED (2001) Species–area relationships. Encyclopedia of biodiversity 5: 397–411.
56. Hopkins B (1957) The concept of minimal area. The Journal of Ecology: 441–449.
57. Watt AS (1947) Pattern and process in the plant community. Journal of ecology 35: 1–22.
58. Dalling JW, Muller-Landau HC, Wright SJ, Hubbell SP (2002) Role of dispersal in the recruitment limitation of neotropical pioneer species. Journal of Ecology 90: 714–727.
59. Condit R, Ashton PS, Baker P, Bunyavejchewin S, Gunatilleke S, et al. (2000) Spatial Patterns in the Distribution of Tropical Tree Species. Science 288: 1414–1418.
60. Li L, Huang Z, Ye W, Cao H, Wei S, et al. (2009) Spatial distributions of tree species in a subtropical forest of China. Oikos 118: 495–502.
61. Condit R, Hubbell SP, Foster RB (1995) Mortality rates of 205 neotropical tree and shrub species and the impact of a severe drought. Ecological monographs 65: 419–439.

# 6

# Geospatial Analysis on the Distributions of Tobacco Smoking and Alcohol Drinking in India

**Sze Hang Fu[1], Prabhat Jha[1,2*], Prakash C. Gupta[3], Rajesh Kumar[4], Rajesh Dikshit[5], Dhirendra Sinha[6]**

**1** Centre for Global Health Research, Li Ka Shing Knowledge Institute, St. Michael's Hospital, University of Toronto, Toronto, Ontario, Canada, **2** Dalla Lana School of Public Health, University of Toronto, Toronto, Ontario, Canada, **3** Healis-Sekhsaria Institute of Public Health, Navi Mumbai, India, **4** School of Public Health, Post Graduate Institute of Medical Education and Research, Chandigarh, India, **5** Department of Epidemiology, Tata Memorial Hospital, Mumbai, India, **6** South East Asia Regional Office, World Health Organization, New Delhi, India

## Abstract

***Background:*** Tobacco smoking and binge alcohol drinking are two of the leading risk factors for premature mortality worldwide. In India, studies have examined the geographic distributions of tobacco smoking and alcohol drinking only at the state-level; sub-state variations and the spatial association between the two consumptions are poorly understood.

***Methodology:*** We used data from the Special Fertility and Mortality Survey conducted in 1998 to examine the geographic distributions of tobacco smoking and alcohol drinking at the district and postal code levels. We used kriging interpolation to generate smoking and drinking distributions at the postal code level. We also examined spatial autocorrelations and identified spatial clusters of high and low prevalence of smoking and drinking. Finally, we used bivariate analyses to examine the spatial correlations between smoking and drinking, and between cigarette and bidi smoking.

***Results:*** There was a high prevalence of any smoking in the central and northeastern states, and a high prevalence of any drinking in Himachal Pradesh, Arunachal Pradesh, and eastern Madhya Pradesh. Spatial clusters of early smoking (started smoking before age 20) were identified in the central states. Cigarette and bidi smoking showed distinctly different geographic patterns, with high levels of cigarette smoking in the northeastern states and high levels of bidi smoking in the central states. The geographic pattern of bidi smoking was similar to early smoking. Cigarette smoking was spatially associated with any drinking. Smoking prevalences in 1998 were correlated with prevalences in 2004 at the district level and 2010 at the state level.

***Conclusion:*** These results along with earlier evidence on the complementarities between tobacco smoking and alcohol drinking suggest that local public health action on smoking might also help to reduce alcohol consumption, and vice versa. Surveys that properly represent tobacco and alcohol consumptions at the district level are recommended.

**Editor:** Natalie Walker, The National Institute for Health Innovation, New Zealand

**Funding:** The authors have no support or funding to report.

**Competing Interests:** The authors have declared that no competing interests exist.

* Email: prabhat.jha@utoronto.ca

## Introduction

The World Health Organization has identified tobacco smoking and binge alcohol drinking as two of the leading risk factors for premature mortality worldwide [1,2]. In India, 14% per cent of the population above age 15 smoke tobacco (24% for men and 3% for women) [3]. Smoking is already responsible for about 1 in 5 deaths for men and 1 in 20 deaths for women at ages 30–69 [4]. India has among the lowest alcohol use per capita in the world, but reports growth in alcohol sales [5,6]. More importantly, male alcohol drinking in India is characterized by problematic drinking (e.g. binge-drinking), which has negative health and social impacts on the consumers and their families [5,7].

Most studies examined tobacco smoking and alcohol drinking behaviours separately among the Indian population. Few studies which examined the concurrent use of tobacco and alcohol found that smokers have a higher likelihood to drink alcohol than non-smokers [8], and vice versa [9,10].

Understanding the geographic distributions of tobacco smoking and alcohol drinking, along with the determinants contributing to their use, is essential for developing targeted tobacco and alcohol control policies. Previous studies have examined the distributions of tobacco smoking [11–13] and alcohol drinking [14] in India at the state level separately. State level analysis can potentially mask the sub-state spatial heterogeneity, especially in the bigger states (e.g. 17 Indian states have populations greater than 25 million [15]). Implementation of tobacco and alcohol control policies is the responsibility of the state government [6,16], thus studying the distributions of tobacco smoking and alcohol drinking as well as their spatial association at the sub-state level is desirable.

Advances in geographic information system (GIS) technology enable analysis of public health phenomena by taking "space" into account [17]. GIS is an integrated set of tools that allow both the analytical manipulation and the visual presentation of spatial data [17,18]. Epidemiologic studies have employed GIS methods for disease mapping and cluster detection [19–21], and although these types of research often focus on diseases, GIS methods can also study risk factors for diseases.

Here, we utilized GIS methods to analyze the distributions of tobacco smoking and alcohol drinking in India. Our objective was to identify sub-state regions with significantly high and low prevalence of smoking and drinking. We also explored whether smoking and drinking were spatially correlated.

## Methods

### Dataset

The Special Fertility and Mortality Survey (SFMS) was a one-time study conducted in February 1998 [22]. The SFMS covered over one million nationally-representative households, based on the Registrar General of India's Sample Registration System (SRS) [22,23]. The SRS aims to provide reliable estimates of fertility and mortality at the state and national levels for rural and urban areas separately [24]. The SFMS interviewed household heads to provide various health indicators for the household residents, including their tobacco smoking and alcohol drinking behaviours. They were asked "does 'household member' smoke/drink alcohol", "age at which 'household member' started smoking/drinking", "what does he/she smoke: cigarette, bidi", and "how frequently 'household member' drinks alcohol: daily, 3–4 times a week, once a fortnight, once a month, less frequently". These definitions are similar to the National Family Health Survey (NFHS) 1998–1999 ("Does 'household member' smoke tobacco/drink alcohol?") [25] and the Global Adult Tobacco Survey (GATS) 2009–2010 ("Do you currently smoke tobacco on a daily basis, less than daily, or not at all?") [3]. The GATS survey interviewed respondents directly while the earlier surveys relied on proxy responses, usually from the head of the household.

We focused on males between 30 and 69 years old, as this age group comprises the majority of smokers (80%) and drinkers (83%), similar to the results from other surveys [11,26,27]. Among Indian females, only about 3% smoke regularly [3,11] and 1% drink regularly [27,28], thus females were excluded from the analysis. We analyzed the following smoking and drinking variables: any smoking, early smoking, cigarette smoking, bidi smoking, and any alcohol drinking. Excluding the missing records or those reporting initiation of smoking before age 10 years, the mean age of initiating any smoking, cigarette smoking, and bidi smoking was 21 (SD 4.8), 23 (SD 4.7), and 20 (SD 4.6), respectively. Therefore, we defined early smoking as those who initiated smoking before age 20. The mean age of initiating any alcohol drinking was 24 (SD 5.4).

The SFMS covered 445 out of the 467 districts from the 1991 Census. Three island districts were excluded. The geographic locations of the SRS units were further geocoded to the associated Indian postal codes, allowing analyses at the district and postal code levels.

### Geospatial analyses

Three GIS methods were used to analyse the geographic patterns of tobacco smoking and alcohol drinking and identify areas with high and low prevalence. 1) ordinary kriging interpolation was used to describe the prevalence at the postal code level, 2) Moran's I was used to measure the overall spatial autocorrelation at the district level, 3) and the local indicators of spatial association (LISA) statistic was used to identify spatial clusters at the district level. These methods are described next, and detailed explanations can be found in Appendix S1.

Kriging is a technique that generates an estimated interpolation surface from a set of data points by incorporating the inference from the spatial structure of data points [29,30]. First, the study population at the SRS units was aggregated to the Indian postal code locations. Crude prevalences (i.e. without age standardization) were calculated for the smoking and drinking variables at the postal code locations. The kriging procedure used the calculated crude prevalence to create an interpolation surface for the entire country. Kriging was implemented using ArcGIS software (ESRI, 2011. Environmental Systems Research Institute, http://www.esri.com/software/arcgis/). Results from kriging interpolation allowed us to validate the calculated prevalence at the district level using the SFMS data, which was designed for state level analysis [24].

Moran's I is an index that measures the "global" spatial autocorrelation of values over the entire dataset [29]. An index score derived from the calculation indicates the strength of spatial autocorrelation of the values, with a score of 0.3 or more and −0.3 or less indicating relatively strong positive and strong negative autocorrelations, respectively [29,31]. While Moran's I determines the overall spatial autocorrelation of a variable, it does not provide a measure for spatial clusters. Hence, we used the LISA statistics to identify spatial clusters of the smoking and drinking variables. The LISA statistic gives an indication of the extent of significant spatial clustering of similar or dissimilar values around a spatial feature [32]. Four types of spatial associations can be derived from this statistic, with high-high (HH) and low-low (LL) types for spatial

**Table 1.** Prevalence of smoking and drinking among men age 30–69 years.

| | District level | | | Postal code level | | |
|---|---|---|---|---|---|---|
| | # of observations | Population (n) | Prevalence (%) | # of observations | Population (n) | Prevalence (%) |
| **Study population** | – | 1,031,625 | 100 | – | 959,411 | 100 |
| Any smokers | 422,353 | 1,031,625 | 40.9 | 394,719 | 959,411 | 41.1 |
| Cigarette smokers† | 66,244 | 837,433† | 7.9 | 60,908 | 783,426† | 7.8 |
| Bidi smokers† | 245,274 | 837,433† | 29.3 | 231,164 | 783,426† | 29.5 |
| Early smokers (before age 20)‡ | 170,162 | 419,419‡ | 40.6 | 158,835 | 391,980‡ | 40.5 |
| Any drinkers | 192,515 | 1,031,625 | 18.7 | 179,392 | 959,411 | 18.7 |

†Six states (Haryana, Bihar, Sikkim, Meghalaya, Mizoram, and Andhra Pradesh) were excluded from the analysis for smoking type as responses were not provided for majority of the smokers (>20%) from these states.

‡2934 and 2739 records from the smoking population at district and postal code level, respectively, were excluded due to invalid information for the resident's age at which he started smoking.

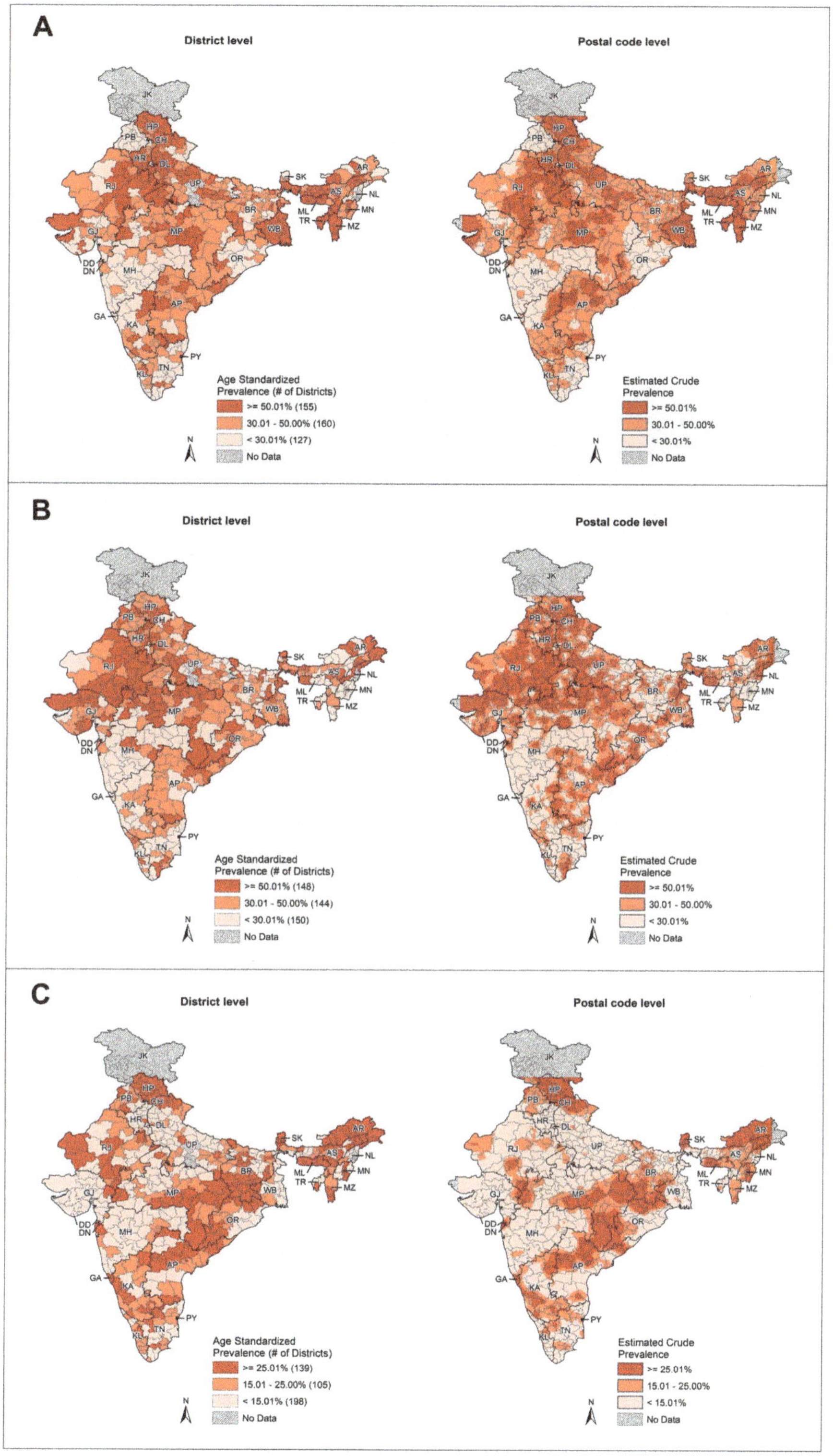


Notes on prevalence calculations:
Prevalence at the district level was age standardized using Census of India 2001 national population;
prevalence at the postal code level was estimated crude prevalence from kriging interpolation

**Figure 1. Smoking and drinking prevalence at the district and postal code levels.** Prevalence at district level was age standardized using Census of India 2001 national population; prevalence at postal code level was estimated crude prevalence from kriging interpolation. State abbreviations: AP - Andhra Pradesh, AR - Arunachal Pradesh, AS - Assam, BR - Bihar, CH - Chandigarh, DD - Daman and Diu, DL - Delhi, DN - Dadra & Nagar Haveli, GA - Goa, GJ - Gujarat, HP - Himachal Pradesh, HR - Haryana, JK - Jammu & Kashmir, KA - Karnataka, KL - Kerala, MG - Meghalaya, MH - Maharashtra, MN - Manipur, MP - Madhya Pradesh, MZ - Mizoram, NL - Nagaland, OR - Orissa, PB - Punjab, PD - Pondicherry, RJ - Rajasthan, SK - Sikkim, TN - Tamil Nadu, TR - Tripura, UP - Uttar Pradesh, WB - West Bengal. **A.** Any smoking prevalence at district and postal code levels. **B.** Early smoking (before age 20) prevalence at district and postal code levels. **C.** Any drinking prevalence at district and postal code levels.

**Table 2.** Univariate global Moran's I for smoking and drinking variables.

| Variable | Moran's I |
|---|---|
| Any smoking | 0.4204** |
| Cigarette smoking | 0.5849** |
| Bidi smoking | 0.4620** |
| Early smoking (before age 20) | 0.2798** |
| Any drinking | 0.3707** |

**p<0.01.

clustering of similar values, and high-low (HL) and low-high (LH) types for spatial clustering of dissimilar values, that is, spatial outliers [31]. For example, the high-high type indicates that a district with high prevalence is surrounded by neighbouring districts also with high prevalence; while low-high type indicates that a district with low prevalence is surrounded by neighbouring districts with high prevalence. We used the first order queen spatial weight for identifying neighbours for the spatial features. The first order queen weight defines neighbours as districts that share either a common border or a vertex with a given district. Preliminary analysis showed that the first order queen weight produced superior results to higher order weights (data not shown).

Two types of Moran's I and LISA analyses were performed: univariate and bivariate. Using the district-level age standardized prevalence (Appendix S1), we performed univariate Moran's I and LISA to examine the spatial autocorrelation and spatial clusters, respectively, for each smoking and drinking variable. We used bivariate Moran's I and LISA to examine the spatial associations between smoking and drinking, as well as between cigarette and bid smoking.

A permutation approach was used to assess the statistical significance of the Moran's I and LISA results ($p<0.05$). The permutation approach produced a reference distribution for significance testing by randomly shuffling the actual data values over space a given number of times [33]. Moran's I and LISA statistics were calculated using GeoDa software [34]. Results from both kriging interpolation and LISA analysis were visualized using ArcGIS software.

## Results

Among men aged 30–69 years in 1998, 40.9% were smokers and 18.7% were drinkers; among smokers, 40.6% were early smokers (before age 20) (Table 1). For smoking types, six states were excluded due to greater than 20% missing data on smoking types–Haryana, Bihar, Sikkim, Meghalaya, Mizoram, and Andhra Pradesh. In the remaining states, 39.7% of men aged 30–69 years were smokers, among which 7.9% were cigarette smokers and 29.3% were bidi smokers, while 2.5% of the records had missing data for smoking type. At the postal code level, about 7% of the study population was excluded as they could not be aggregated to postal code locations. Nevertheless, the prevalence for both smoking and drinking status at the two geographic levels remained similar.

**Table 3.** Bivariate global Moran's I for smoking versus drinking, and cigarette smoking versus bidi smoking.

| Variable | Moran's I |
|---|---|
| Any smoking *vs.* any drinking | 0.0308 |
| Cigarette smoking *vs.* any drinking | 0.1190** |
| Bidi smoking *vs.* any drinking | −0.0525 |
| Cigarette smoking *vs.* Bidi smoking | −0.1115* |

*p<0.05;
**p<0.01.

The geographic distributions for both smoking and drinking were similar between the district and postal code levels in India (Figure 1A–C, Figure S1D–E). Kriging interpolation identified the sub-district level patterns that were concealed in the district level maps. It also helped to provide estimations for districts not covered by the SFMS, such as those in Uttar Pradesh. The prevalence of smoking was high in the central, northern (except Punjab), and the northeastern states, while the prevalence was lower in Punjab, Maharashtra and Orissa (Figure 1A). Some neighbouring states showed contrasting results (e.g. high prevalence in Madhya Pradesh vs. low prevalence in Maharashtra). Prevalence of early smoking (before age 20) was high in the central and northern states (Figure 1B). Although Punjab had low smoking prevalence, it had high prevalence of early smoking (before age 20). Cigarette and bidi smoking showed distinctly different geographic patterns: high cigarette smoking prevalence was found in the southern and northeastern states (Figure S1D); while high bidi smoking prevalence was found in the central and northern (except Punjab) states (Figure S1E). West Bengal had high prevalence for both cigarette and bidi smoking. The geographic pattern of bidi smoking was similar to early smoking (before age 20). High prevalence of any drinking was located in Himachal Pradesh, parts of the northeastern states, and the central eastern part of the country (Figure 1C).

In Moran's I analysis, all smoking and drinking variables were statistically significant ($p<0.01$) and had relatively strong autocorrelations (index scores close to or more than 0.3) (Table 2). Bivariate Moran's I results showed that cigarette smoking was positively correlated to any drinking (index score = 0.119, $p<0.01$) and negatively correlated to bidi smoking (index score = −0.112, $p<0.05$) (Table 3). However, any smoking did not show any significant spatial correlation with any drinking (index score = 0.031, $p = 0.14$) or drinking more than 3 days a week (data not shown).

The univariate LISA results highlighted districts with significantly high prevalence as high-high clusters and low prevalence as low-low clusters (Figure 2A–C, Figure S2D–E). Spatial outliers, specifically high-low and low-high types, also existed in the maps. Compared to the high-high and low-low clusters, fewer districts belonged to the high-low and low-high types, and these outliers did not often neighbour other outliers in clusters. For early smoking (before age 20) and cigarette smoking, outliers were found near the boundary shared by Rajasthan, Uttar Pradesh, and Madhya Pradesh, and they were surrounded by spatial clusters in the adjacent regions (Figure 2B, Figure S2D).

The results from bivariate LISA analysis showed that compared to univariate LISA, sizes of high-high and low-low clusters were smaller and the number of high-low and low-high outliers was higher (Figure 3A–B, Figure S3C–D). This indicates that the bivariate spatial correlations between smoking and drinking were not as strong as those found in the univariate analysis. The bivariate LISA results were supported by the bivariate Moran's I results as most of the bivariate pairs had weak and non-significant results (Table 3). Nonetheless, it can be observed that for smoking

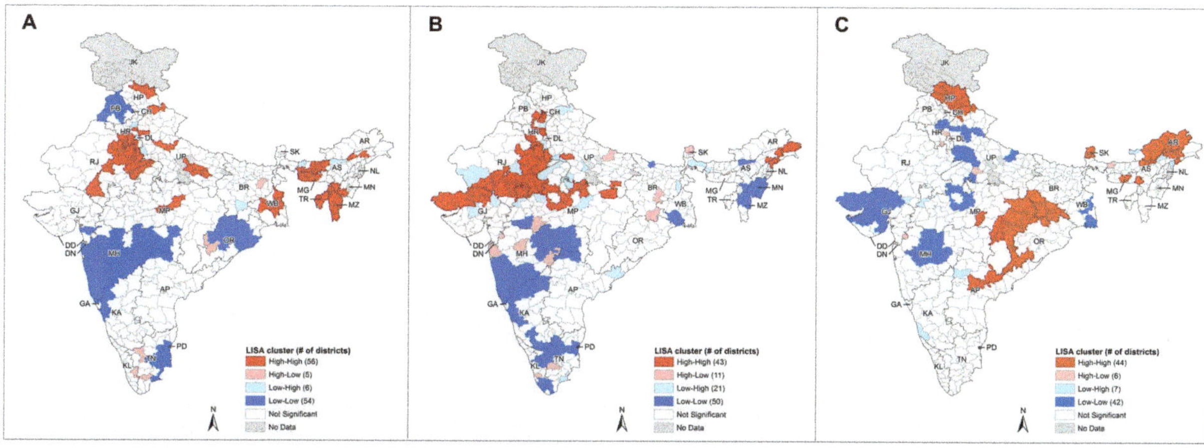


**Figure 2. LISA cluster maps for smoking and drinking prevalence at the district level.** State abbreviations: AP - Andhra Pradesh, AR - Arunachal Pradesh, AS - Assam, BR - Bihar, CH - Chandigarh, DD - Daman and Diu, DL - Delhi, DN - Dadra & Nagar Haveli, GA - Goa, GJ - Gujarat, HP - Himachal Pradesh, HR - Haryana, JK - Jammu & Kashmir, KA - Karnataka, KL - Kerala, MG - Meghalaya, MH - Maharashtra, MN - Manipur, MP - Madhya Pradesh, MZ - Mizoram, NL - Nagaland, OR - Orissa, PB - Punjab, PD - Pondicherry, RJ - Rajasthan, SK - Sikkim, TN - Tamil Nadu, TR - Tripura, UP - Uttar Pradesh, WB - West Bengal. **A.** LISA cluster map for any smoking prevalence. **B.** LISA cluster map for early smoking (before age 20) prevalence. **C.** LISA cluster map for any drinking prevalence.

versus drinking, high-high clusters exist in Himachal Pradesh, and low-low clusters exist in Gujarat and Maharashtra (Figure 3A). Cigarette smoking had the strongest spatial correlation with drinking (Figure S3C). For cigarette smoking versus bidi smoking, high-high clusters were found in West Bengal, and low-low clusters were found in Punjab, Maharashtra, and Orissa (Figure 3B). The low-high outliers found in the central states clearly showed that low prevalence of cigarette smoking coexisted with high prevalence of bidi smoking, and vice versa for the states of Kerala and Tamil Nadu. Overall, there was high prevalence for both cigarette and bidi smoking in West Bengal, and low prevalence for both smoking and drinking in Gujarat and Maharashtra (Table 4).

We compared our results on smoking prevalences with the living respondents in the Million Death Study (MDS) [23] and adults interviewed in the Global Adult Tobacco Survey (GATS) 2009–2010 [3]. We found correlations of smoking prevalences between SFMS in 1998 and MDS in 2004 at the district level (any smoking: $R^2 = 0.58$, $p < 0.01$; cigarette smoking: $R^2 = 0.52$, $p < 0.01$; bidi smoking: $R^2 = 0.56$, $p < 0.01$), and high correlations between SFMS in 1998 and GATS in 2010 at the state level (any

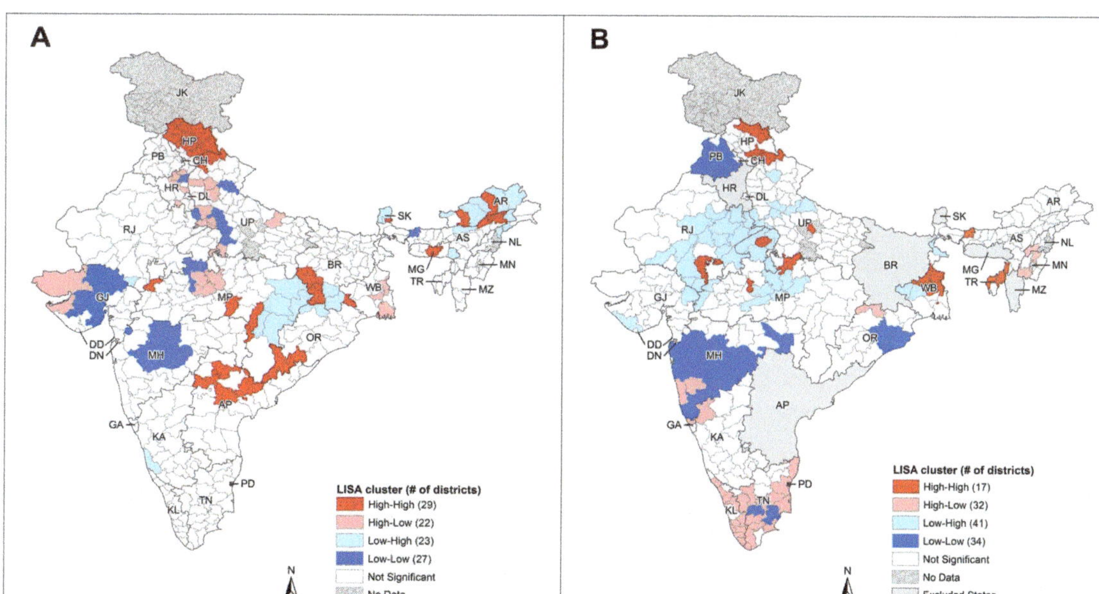


**Figure 3. Bivariate LISA cluster maps.** State abbreviations: AP - Andhra Pradesh, AR - Arunachal Pradesh, AS - Assam, BR - Bihar, CH - Chandigarh, DD - Daman and Diu, DL - Delhi, DN - Dadra & Nagar Haveli, GA - Goa, GJ - Gujarat, HP - Himachal Pradesh, HR - Haryana, JK - Jammu & Kashmir, KA - Karnataka, KL - Kerala, MG - Meghalaya, MH - Maharashtra, MN - Manipur, MP - Madhya Pradesh, MZ - Mizoram, NL - Nagaland, OR - Orissa, PB - Punjab, PD - Pondicherry, RJ - Rajasthan, SK - Sikkim, TN - Tamil Nadu, TR - Tripura, UP - Uttar Pradesh, WB - West Bengal. **A.** Bivariate LISA cluster map for any smoking prevalence versus any drinking prevalence. **B.** Bivariate LISA cluster map for cigarette smoking prevalence versus bidi smoking prevalence.

**Table 4.** Broad geographic patterns by regions across India for smoking and drinking variables.

| Region | Variable (State name if applicable) | |
|---|---|---|
| | **with high prevalence** | **with low prevalence** |
| North | Any drinking (Himsachal Pradesh) | Any smoking (Punjab); Bidi smoking (Punjab) |
| Northeast | Any smoking; Cigarette smoking; Any drinking (Arunachal Pradesh, Sikkim) | n/a |
| East | Any smoking (West Bengal); Bidi smoking (West Bengal); Cigarette smoking (West Bengal) | Any smoking (Orissa); Bidi smoking (Orissa) |
| Central | Any smoking; Bidi smoking; Early smoking (before age 20) | Cigarette smoking |
| West | n/a | Any smoking (Maharashtra); Cigarette smoking (Gujarat, Maharashtra); Bidi smoking (Maharashtra); Early smoking (before age 20) (Maharashtra); Any drinking (Gujarat, Maharashtra) |
| South | Cigarette smoking (Kerala) | Any smoking (Tamil Nadu); Bidi smoking (Tamil Nadu) |

smoking: $R^2 = 0.67$, $p<0.01$; cigarette smoking: $R^2 = 0.64$, $p<0.01$; bidi smoking: $R^2 = 0.70$, $p<0.01$).

## Discussion

Our results confirmed that smoking prevalence is significantly higher in the northeastern and central states, while it is significantly lower in Punjab, Maharashtra, and Orissa. These state level results of smoking are comparable to the findings from previous research studies [11,12] and GATS [3]. Punjab had significantly lower smoking prevalence as the majority of its population practices Sikhism, which prohibits tobacco use [11,12]. The high prevalence in the northeastern states is consistent with the GATS report [3]. Moreover, high prevalence has been reported among students in this region [35].

There was a high prevalence of any drinking in Himachal Pradesh, south Bihar, eastern Madhya Pradesh (presently Chhattisgarh state), and in the northeastern states (Sikkim and Arunachal Pradesh in particular). Many of these states have high proportion of tribal population [36] and liquor drinking is common among the tribal persons [37], as well as a low proportion of Muslim population [38], in whom the reported alcohol drinking prevalence is low [14,39]. The low prevalence in Gujarat and Maharashtra could be attributed to the alcohol prohibition in these states, with Gujarat under complete prohibition (theoretically) on the production and sale of alcohol [7]. Surprisingly, we found that Tamil Nadu had low drinking prevalence even though other research has found high drinking prevalence [40,41].

Our results showed the sub-state variations of smoking and drinking prevalence and the coexistence of high-high and low-low clusters with outliers within a state. For example, in Madhya Pradesh low-low clusters for cigarette smoking were found in the western and eastern part of the state, with high-low outliers found in the central part of the state. These spatial heterogeneities of tobacco and alcohol use at the sub-state levels were also found in previous studies [12,14].

Cigarette smoking was high in the northeastern and the southern states, while bidi smoking was high in the central states and in West Bengal. These two smoking types are characterized by different users in India. Cigarette smoking is common among the educated and people with skilled occupations, while bidi smoking is common among the rural population, the urban poor, and the less educated [9,42]. The relatively inexpensive bidis are more affordable and attractive to young people [43]. Our results show that the geographic patterns of bidi smoking closely resemble those of early smoking (before age 20). Control policies on bidi smoking thus may reduce smoking uptake among young people.

Since 1998, data has shown an increase in sales and the affordability of tobacco [44], despite the implementation of several tobacco control policies [45,46]. Limited data suggested an increase in smoking prevalence, in particular among younger people [44]. Evidence also suggests that cigarettes are displacing bidis among younger males [47]. Meanwhile, increases in alcohol sales and use have been observed [5,48], along with the more relaxed restrictions on consumption in many states [49]. For example, alcohol prohibition is slowly being liberalized and consumption is increasing in Maharashtra. The increase in popularity of both cigarette smoking and drinking required our attention as we found an association between the two. Previous studies also found complementarities between tobacco smoking and alcohol drinking among Indians [8–10]. Smoking cessation may enhance alcohol abstinence [50] and certain types of smoking cessation medication can help reduce the frequency of alcohol consumption [51]. Thus, local public health action on smoking might help reduce alcohol consumption, and vice versa. Tobacco and alcohol control is currently implemented at the state level [6,16], and thus reliable monitoring of changes in consumption patterns and complementarities between cigarette smoking and alcohol drinking at the district level is required to enable effective control.

This study has several limitations. First, the SFMS was aimed to provide information representative at the state level [24], similar to the NFHS 1998–1999 [25] and GATS 2009–2010 [3]. But its large sample size yielded over 1 million individuals for this study, which provide sufficient number for stable district level analysis. Results from the kriging interpolations also confirmed the geographic patterns at the district level. Second, the SFMS was conducted over a decade ago and the current level of smoking and drinking among Indians may now differ. However, smoking prevalences in 1998 were correlated with prevalence in 2004 at the district level (SFMS *vs.* MDS) and with prevalences in 2010 at the state level (SFMS *vs.* GATS). Notwithstanding the different survey designs between SFMS and GATS (i.e. the use of household informants in SFMS and selected individuals in GATS), conclusions derived from our results at sub-state level are still likely to apply to the current Indian population. Third, the use of household informants may lead to under-reporting of smoking and drinking behaviours as they may not be aware of these consumption behaviours of the household residents (such as number of days alcohol was used), especially among younger males. Under-reporting could also be caused by the social desirability bias. For example, in states where prohibitions on tobacco and alcohol use were in effect, respondents may be less likely to report the smoking and drinking behaviours of their household residents. Fourth, our results are nationally representative and applicable to Indian males aged 30–69, which comprises the majority of smokers in India [11,26]. The exclusion of women was reasonable as smoking and drinking prevalences among women remained low in the last decade [3,11,27,28]. Similarly, the exclusion of six states (with 19% of study population) from the analysis of cigarette and bidi geographic patterns should not greatly alter the national results for the remaining 23 states. Finally, although the prevalence for the various smoking and drinking variables were not markedly different between the district level and postal code level study populations, it is possible that the 7% of records excluded in the postal code level were spatially clustered. Therefore, the kriging interpolations only provide the approximate distributions of smoking and drinking at the sub-district level.

## Conclusion

India has over 100 million current tobacco smokers and it accounts for approximately one-fifth of the world's tobacco-related deaths [52,53]. There is an urgent need of ongoing research to determine the current level of smoking in the population. Our study informs Indian tobacco control policy by filling the research gap of tobacco use and the use of cigarette and bidi smoking at the sub-state level in India. We found that cigarette smoking dominated the northeastern and southern states and bidi smoking dominated the central states, with early smoking (before age 20) closely resembling the geographic patterns of bidi smoking. We also identified high-burden areas of drinking and its spatial association with cigarette smoking. Finally, the sub-state variations of smoking and drinking require the attention of state policy-makers. In order to control tobacco and alcohol use more effectively, state governments should implement the existing national acts of tobacco and alcohol control at the district level. Surveys that properly represent their consumption at the district level are recommended. Reductions in smoking may simultaneously reduce alcohol consumption, and vice versa.

## Supporting Information

**Figure S1 Cigarette smoking and bidi smoking prevalence at the district and postal code levels.** Prevalence at district level was age standardized using Census of India 2001 national population; prevalence at postal code level was estimated crude prevalence from kriging interpolation. State abbreviations: AP - Andhra Pradesh, AR - Arunachal Pradesh, AS - Assam, BR - Bihar, CH - Chandigarh, DD - Daman and Diu, DL - Delhi, DN - Dadra & Nagar Haveli, GA - Goa, GJ - Gujarat, HP - Himachal Pradesh, HR - Haryana, JK - Jammu & Kashmir, KA - Karnataka, KL - Kerala, MG - Meghalaya, MH - Maharashtra, MN - Manipur, MP - Madhya Pradesh, MZ - Mizoram, NL - Nagaland, OR - Orissa, PB - Punjab, PD - Pondicherry, RJ - Rajasthan, SK - Sikkim, TN - Tamil Nadu, TR - Tripura, UP - Uttar Pradesh, WB - West Bengal. **D.** Cigarette smoking prevalence at district and postal code levels. **E.** Bidi smoking prevalence at district and postal code levels.

**Figure S2 LISA cluster maps for cigarette smoking and bidi smoking prevalence at the district level.** State abbreviations: AP - Andhra Pradesh, AR - Arunachal Pradesh, AS - Assam, BR - Bihar, CH - Chandigarh, DD - Daman and Diu, DL - Delhi, DN - Dadra & Nagar Haveli, GA - Goa, GJ - Gujarat, HP - Himachal Pradesh, HR - Haryana, JK - Jammu & Kashmir, KA - Karnataka, KL - Kerala, MG - Meghalaya, MH - Maharashtra, MN - Manipur, MP - Madhya Pradesh, MZ - Mizoram, NL - Nagaland, OR - Orissa, PB - Punjab, PD - Pondicherry, RJ - Rajasthan, SK - Sikkim, TN - Tamil Nadu, TR - Tripura, UP - Uttar Pradesh, WB - West Bengal. **D.** LISA cluster map for cigarette smoking prevalence. **E.** LISA cluster map for bidi smoking prevalence.

**Figure S3 Bivariate LISA cluster maps.** State abbreviations: AP - Andhra Pradesh, AR - Arunachal Pradesh, AS - Assam, BR - Bihar, CH - Chandigarh, DD - Daman and Diu, DL - Delhi, DN - Dadra & Nagar Haveli, GA - Goa, GJ - Gujarat, HP - Himachal Pradesh, HR - Haryana, JK - Jammu & Kashmir, KA - Karnataka, KL - Kerala, MG - Meghalaya, MH - Maharashtra, MN - Manipur, MP - Madhya Pradesh, MZ - Mizoram, NL - Nagaland, OR - Orissa, PB - Punjab, PD - Pondicherry, RJ - Rajasthan, SK - Sikkim, TN - Tamil Nadu, TR - Tripura, UP - Uttar Pradesh, WB - West Bengal. **C.** Bivariate LISA cluster map for cigarette smoking prevalence versus any drinking prevalence. **D.** Bivariate LISA cluster map for bidi smoking prevalence versus any drinking prevalence.

**Appendix S1 Explanation on age standardization, kriging method, Moran's I and LISA statistics.**

## Acknowledgments

The authors thank Chinthanie Ramasundarahettige for her support in data preparation, suggestions on the methodology, and comments on the manuscript, and Samir Gambhir and Peter Rodriguez for their support in GIS analysis. We also thank Marvin Hsiao for his valuable comments on the manuscript.

## Author Contributions

Conceived and designed the experiments: SHF PJ. Performed the experiments: SHF. Analyzed the data: SHF PJ PG RK. Contributed reagents/materials/analysis tools: SHF PJ. Wrote the paper: SHF PJ PG RK. Reviewed the manuscript: SHF PJ PG RK RD DS.

## References

1. Mathers C, Stevens G, Mascarenhas M (2009) Global Health Risks: Mortality and burden of disease attributable to selected major risks. Geneva: WHO.
2. Jha P, Nugent R, Verguet S, Bloom D, Hum R (2012) Chronic disease prevention and control. Challenge Paper for the Copenhagen Consensus 2012.
3. International Institute of Population Sciences (IIPS) (2010) Global Adult Tobacco Survey: India 2009–2010. New Delhi: Ministry of Health and Family Welfare, Government of India.
4. Jha P, Jacob B, Gajalakshmi V, Gupta PC, Dhingra N, et al. (2008) A nationally representative case-control study of smoking and death in India. N Engl J Med 358: 1137–1147.
5. Prasad R (2009) Alcohol use on the rise in India. Lancet 373: 17–18.
6. Benegal V (2005) India: alcohol and public health. Addiction 100: 1051–1056.
7. Mahal A (2000) What works in alcohol policy? Evidence from rural India. Economic Political Weekly 35: 3959–3968.
8. Mohan D, Chopra A, Sethi H (2002) The co-occurrence of tobacco & alcohol in general population of metropolis Delhi. Indian J Med Res 116: 150–154.

9. Narayan KM, Chadha SL, Hanson RL, Tandon R, Shekhawat S, et al. (1996) Prevalence and patterns of smoking in Delhi: cross sectional study. BMJ 312: 1576–1579.
10. Gupta PC, Maulik PK, Pednekar MS, Saxena S (2005) Concurrent alcohol and tobacco use among a middle-aged and elderly population in Mumbai. Natl Med J India 18: 88–91.
11. Rani M, Bonu S, Jha P, Nguyen SN, Jamjoum L (2003) Tobacco use in India: prevalence and predictors of smoking and chewing in a national cross sectional household survey. Tob Control 12: e4.
12. Subramanian SV, Nandy S, Kelly M, Gordon D, Davey Smith G (2004) Patterns and distribution of tobacco consumption in India: cross sectional multilevel evidence from the 1998–9 national family health survey. BMJ 328: 801–806.
13. John RM (2005) Tobacco consumption patterns and its health implications in India. Health Policy 71: 213–222.
14. Subramanian SV, Nandy S, Irving M, Gordon D, Davey Smith G (2005) Role of socioeconomic markers and state prohibition policy in predicting alcohol consumption among men and women in India: a multilevel statistical analysis. Bull World Health Organ 83: 829–836.
15. Census of India (2011) Primary Census Abstract Data Tables. New Delhi: Registrar General of India, Government of India.
16. Mehrotra R, Mehrotra V, Jandoo T (2010) Tobacco control legislation in India: past and present. Indian J Cancer 47 Suppl 1: 75–80.
17. Ricketts TC (2003) Geographic information systems and public health. Annu Rev Public Health 24: 1–6.
18. Beale L, Abellan JJ, Hodgson S, Jarup L (2008) Methodologic issues and approaches to spatial epidemiology. Environ Health Perspect 116: 1105–1110.
19. Singh A, Pathak PK, Chauhan RK, Pan W (2011) Infant and child mortality in India in the last two decades: a geospatial analysis. PLoS One 6: e26856.
20. Kandwal R, Augustijn EW, Stein A, Miscione G, Garg PK, et al. (2010) Geospatial analysis of HIV-Related social stigma: a study of tested females across mandals of Andhra Pradesh in India. Int J Health Geogr 9: 18.
21. Srivastava A, Nagpal BN, Joshi PL, Paliwal JC, Dash AP (2009) Identification of malaria hot spots for focused intervention in tribal state of India: a GIS based approach. Int J Health Geogr 8: 30.
22. Registrar General of India (2005) Special Fertility & Mortality Survey, A Report of 1.1 Million Indian Households. New Delhi: Registrar General.
23. Jha P, Gajalakshmi V, Gupta PC, Kumar R, Mony P, et al. (2006) Prospective study of one million deaths in India: rationale, design, and validation results. PLoS Med 3: e18.
24. Registrar General of India (1996) Sample Registration System Statistical Report 1996. New Delhi: Registrar General of India.
25. International Institute of Population Sciences (IIPS) (2000) National family health survey–II (1998–1999). Bombay: International Institute of Population Studies.
26. Registrar General of India (2004) Sample Registration System (SRS): Baseline Survey Report -2004. New Delhi: Office of the Registrar General of India.
27. Neufeld KJ, Peters DH, Rani M, Bonu S, Brooner RK (2005) Regular use of alcohol and tobacco in India and its association with age, gender, and poverty. Drug Alcohol Depend 77: 283–291.
28. International Institute for Population Sciences (IIPS) and Macro International (2007) National Family Health Survey (NFHS-3), 2005–06: India: Volume I. Mumbai: IIPS.
29. O'Sullivan D, Unwin D (2010) Geographic Information Analysis. Hoboken, New Jersey: John Wiley & Sons, Inc.
30. Oliver MA, Webster M (1990) Kriging: a method of interpolation for GIS. International Journal of Geographical Information Systems 4: 313–332.
31. Pfeiffer DU, Robinson TP, Stevenson M, Stevens KB, Rogers DJ, et al. (2008) Spatial Analysis in Epidemiology: Oxford University Press.
32. Anselin L (1995) Local Indicators of Spatial Association - Lisa. Geographical Analysis 27: 93–115.
33. Anselin L (2005) Exploring Spatial Data with GeoDa: A Workbook. Spatial Analysis Laboratory, Department of Geography University of Illinois and Center for Spatially Integrated Social Science, IL, USA.
34. Anselin L, Syabri I, Kho Y (2006) GeoDa: An Introduction to Spatial Data Analysis. Geographical Analysis 38: 5–22.
35. Sinha DN, Gupta PC, P G (2007) Tobacco use among students and school personnel in India. Asian Pac J Cancer Prev 8: 417–421.
36. Census of India (2001) Census Reference Tables, A – Series: General Population Tables. New Delhi: Registrar General of India, Government of India.
37. Joshi RA (2010) Role of NGOs In Removing Tribal Problems. In: Kanmony JC, editor. Dalits and Tribes of India. New Delhi: Krishan Mittal.
38. Census of India (2001) Census Reference Tables, C – Series: Social and Cultural Tables. New Delhi: Registrar General of India, Government of India.
39. Gupta PC, Saxena S, Pednekar MS, Maulik PK (2003) Alcohol consumption among middle-aged and elderly men: a community study from western India. Alcohol Alcohol 38: 327–331.
40. Chakravarthy C (1990) Community workers estimate of drinking and alcohol-related problems in rural areas. Indian Journal of Psychological Medicine 13: 49–56.
41. John A, Barman A, Bal D, Chandy G, Samuel J, et al. (2009) Hazardous alcohol use in rural southern India: Nature, prevalence and risk factors. The National Medical Journal of India 22: 123–125.
42. Rahman M, Fukui T (2000) Bidi smoking and health. Public Health 114: 123–127.
43. Gupta PC, Asma S (2008) Bidi Smoking and Public Health. New Delhi: Ministry of Health and Family Services, Government of India.
44. John RM, Rao RK, Rao MG, Moore J, Deshpande RS, et al. (2010) The Economics of Tobacco and Tobacco Taxation in India. Paris: International Union Against Tuberculosis and Lung Disease.
45. Kaur J, Jain DC (2011) Tobacco control policies in India: implementation and challenges. Indian J Public Health 55: 220–227.
46. Government of India (2000) Cable Television Networks (Amendment) Act.
47. Jha P., Guindon E., Joseph R A., Nandi A., John R M., et al. (2011) A Rational Taxation System of Bidis and Cigarettes to Reduce Smoking Deaths in India. Economic & Political Weekly 46: 44–51.
48. Jacob KS (2010) Alcohol and public health policies in India. The National Medical Journal of India 23: 224–225.
49. Das SK, Balakrishnan V, Vasudevan DM (2006) Alcohol: Its health and social impact in India. The National Medical Journal of India 19: 94–99.
50. Kalman D, Kim S, DiGirolamo G, Smelson D, Ziedonis D (2010) Addressing tobacco use disorder in smokers in early remission from alcohol dependence: the case for integrating smoking cessation services in substance use disorder treatment programs. Clin Psychol Rev 30: 12–24.
51. McKee SA, Young-Wolff KC, Harrison EL, Cummings KM, Borland R, et al. (2013) Longitudinal associations between smoking cessation medications and alcohol consumption among smokers in the International Tobacco Control Four Country survey. Alcohol Clin Exp Res 37: 804–810.
52. Tiwari R, Deb P, Debbarma A, Chaudhuri R, Chakraborty A, et al. (2006) Tobacco use and cardiovascular disease: a knowledge, attitude and practice study in rural Kerala. Indian J Med Sci 60: 271–276.
53. Jha P, Peto R (2014) Global effects of smoking, of quitting, and of taxing tobacco. N Engl J Med 370: 60–68.

7

# Fine-Scale Distribution and Spatial Variability of Benthic Invertebrate Larvae in an Open Coastal Embayment in Nova Scotia, Canada

**Rémi M. Daigle[1]*, Anna Metaxas[1], Brad deYoung[2]**

**1** Department of Oceanography, Dalhousie University, Halifax, Nova Scotia, Canada, **2** Department of Physics and Physical Oceanography, Memorial University, St. John's, Newfoundland, Canada

## Abstract

This study quantified the fine- scale (0.5 km) of variability in the horizontal distributions of benthic invertebrate larvae and related this variability to that in physical and biological variables, such as density, temperature, salinity, fluorescence and current velocity. Larvae were sampled in contiguous 500-m transects along two perpendicular 10-km transects with a 200-µm plankton ring net (0.75-m diameter) in St. George's Bay, Nova Scotia, Canada, in Aug 2009. Temperature, conductivity, pressure and fluorescence were measured with a CTD cast at each station, and currents were measured with an Acoustic Doppler Current Profiler moored at the intersection of the 2 transects. Gastropod, bivalve and, to a lesser extent, bryozoan larvae had very similar spatial distributions, but the distribution of decapod larvae had a different pattern. These findings suggest that taxonomic groups with functionally similar larvae have similar dispersive properties such as distribution and spatial variability, while the opposite is true for groups with functionally dissimilar larvae. The spatial variability in larval distributions was anisotropic and matched the temporal/spatial variability in the current velocity. We postulate that in a system with no strong oceanographic features, the scale of spatially coherent physical forcing (*e.g.* tidal periodicity) can regulate the formation or maintenance of larval patches; however, swimming ability may modulate it.

**Editor:** Michael Hart, Simon Fraser University, Canada

**Funding:** Funding for this study was provided by NSERC (Natural Sciences and Engineering Research Council of Canada) grants to A.M. and B.deY.: Discovery (A.M.: 222932, B.deY.: 200291) and the Strategic Network (290606) - Canadian Healthy Oceans Network (CHONe). Funding to R.D. was through a NSERC CGS M and PGS D, as well as Graduate Scholarships and President Awards from Dalhousie University. The funders had no role in study design, data collection and analysis, decision to publish, or preparation of the manuscript.

**Competing Interests:** The authors have declared that no competing interests exist.

* Email: daigleremi@gmail.com

## Introduction

The dynamics and persistence of populations of marine benthic invertebrates are affected by connectivity, which is in turn regulated by dispersal during the planktonic larval phase [1,2]. For example, spatial and temporal variability in larval supply and settlement influences the patterns of recruitment of new individuals to adult populations [3,4]. The scale of temporal variability in larval abundance can range from decadal climatic variations [5] to diel or tidal vertical migrations [6]. Spatially, variability in larval abundance can result from the variation in water properties at regional scales [7] to small scale features of the water column that may aggregate and transport invertebrate larvae, such as up- or downwelling [8,9], internal tidal bores [10], or frontal systems and Langmuir cells [11]. Many different physical processes can influence the spatial or temporal pattern of larval dispersal and connectivity. At large spatial or temporal scales (km or y), the variability in larval abundance leads to spatially dependant variability in recruitment, and consequently, spatially dependant species assemblages. At smaller scales (m or h), physical processes aggregate larvae and alter larval transport, ultimately affecting larval dispersal. Additionally, patchiness or spatial heterogeneity generated by variability in larval abundance, affects the perceived patterns in the distribution of planktonic larvae, and may contribute to the detection of erroneous patterns through spatial aliasing or masking of the true pattern by high sampling variability [11].

Quantifying larval dispersal and connectivity is difficult since larvae, while numerous, are distributed patchily and are generally difficult to sample accurately [1,2]. Traditional net sampling provides estimates of abundance at one point in time and space, but does not provide information on areas of larval release, or potential settlement. Geochemical tracers and genetic markers have been used to quantify realized dispersal, but bio-physical modelling remains the only method currently used to predict trajectories of larval dispersal for all species. There are examples where the dispersal of relatively large, short-lived tunicate larvae has been visually tracked [12], but this technique is not practical for very small or long lived larvae. Most biophysical models are based on either numerical circulation models or advection-diffusion models used to quantify the effects of the physical properties of the ocean on larval dispersal [13]. The validity of the bio-physical model is limited by the model's ability to resolve relevant physical features of the water column with which larvae could interact behaviourally, such as up- or downwelling, as well as by the incorporation of this behaviour into the "bio" component of the bio-physical model. Studying patterns of larval distribution and their mechanisms of formation at different scales [14,15] enable development of valuable parameterization of the biological components of these models.

Biological components typically incorporated in models, such as longer pelagic larval duration (PLD), mortality, and vertical migration, can have a significant impact on larval dispersal [6,16–19]. Larvae with longer PLDs may be advected further, and be associated with longer dispersal distances. However, observations of the spread of invasive species, genetic studies of population structure, direct observations, and experimental estimates of dispersal distances show that this is not always true [19,20]. Advection does depend on larval duration, but is also influenced by larval behaviour, larval characteristics (*e.g.* swimming ability) and the local ocean current regime. Species with short PLDs can be associated with a surprisingly homogeneous population structure and species with longer PLDs can be retained locally [19,20]. Some of this variability may be explained by the biases related to the quantification methods; however, there is certainly variability in the relationship between PLD and dispersal distance that can be attributed to larval behaviour and interactions with oceanographic features. An example of such a behavioural interaction is selective tidal stream transport [6,21,22], where larvae of certain species migrate vertically with the same

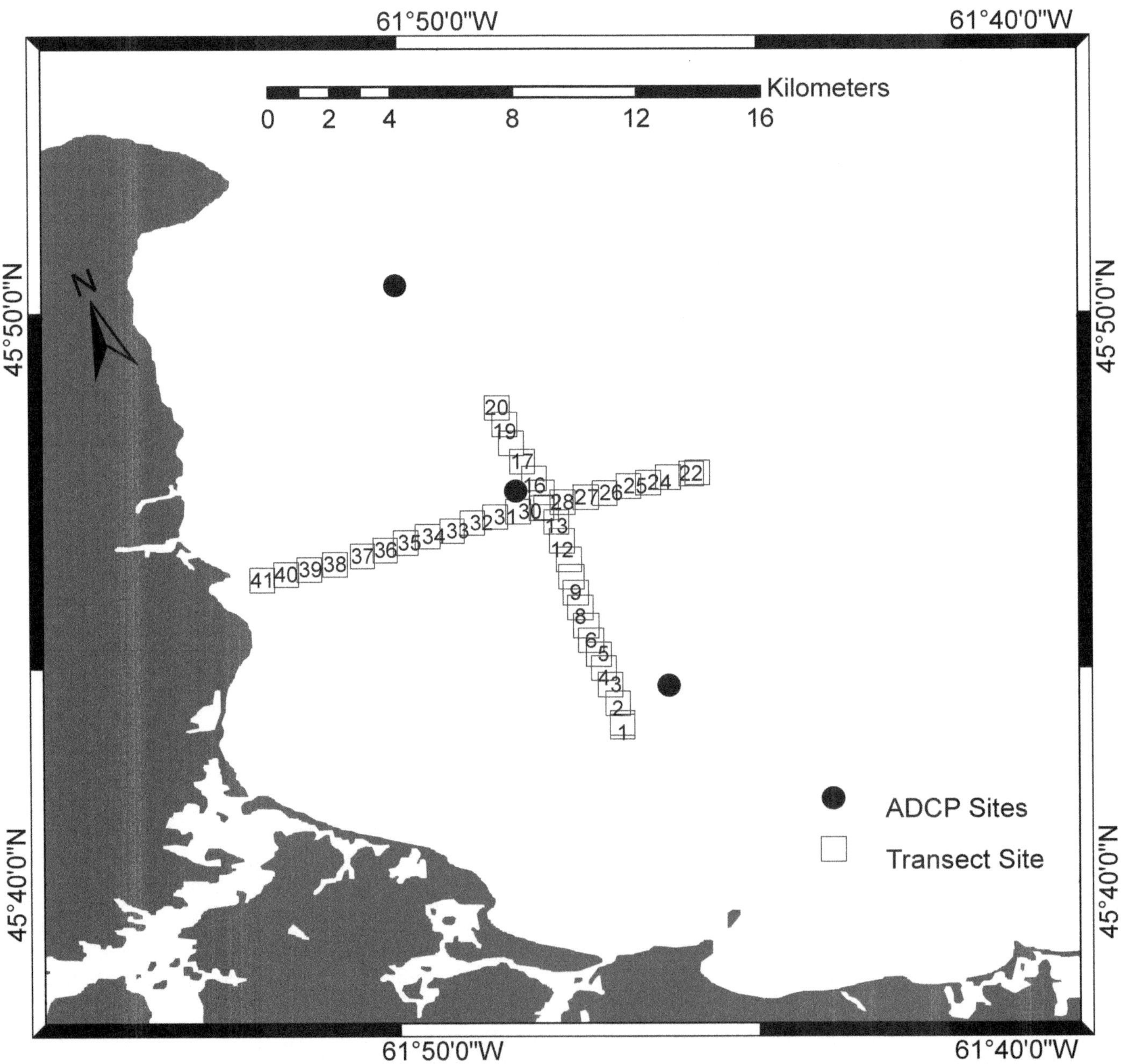


**Figure 1. Map of St. George's Bay, Nova Scotia, Canada indicating sites along 2 perpendicular ~10-km transects which were sampled continuously (one larval sample every ~500 m for 5 min) at 3 m depth, with a 200-μm plankton ring net (0.75 m diameter).** The entire water column was profiled with a CTD between larval samples. The locations of the ADCP moorings are also shown.

**Table 1.** Mean, minimum and maximum abundance (individuals $m^{-3}$) of larval taxonomic groups, during plankton sampling in St. George's Bay, Nova Scotia, Canada, in Aug 2009.

| **A) Abundance** | | | | | | | |
|---|---|---|---|---|---|---|---|
| **Bryozoans** | | **Bivalves** | | **Gastropods** | | **Decapods** | |
| Mean | 4.80 | | 20.6 | | 26.0 | | 0.30 |
| Minimum | 0.32 | | 0.46 | | 1.92 | | 0.04 |
| Maximum | 10.6 | | 114 | | 119 | | 0.87 |
| **B) Composition** | | | | | | | |
| **Species** | **%** | **Species** | **%** | **Species** | **%** | **Species** | **%** |
| *Electra pilosa* | 99.9 | *Mytilus* spp. | 34.6 | *Margarites* spp. | 49 | *Cancer irroratus* | 71.7 |
| *Membranipora membranacea* | 0.1 | *Modiolus modiolus* | 14.6 | *Astyris lunata* | 32.5 | *Crangon septemspinosa* | 16.4 |
| | | *Anomia simplex* | 5.4 | *Diaphana minuta* | 6.2 | *Neopanopeus sayi* | 8 |
| | | other | 45.4 | *Crepidula spp.* | 4.6 | *Carcinus maenas* | 3.9 |
| | | | | *Arrhoges occidentalis* | 4.1 | | |
| | | | | *Bittiolum alternatum* | 3.1 | | |
| | | | | other | 0.5 | | |

Proportional species composition for each group is also shown.

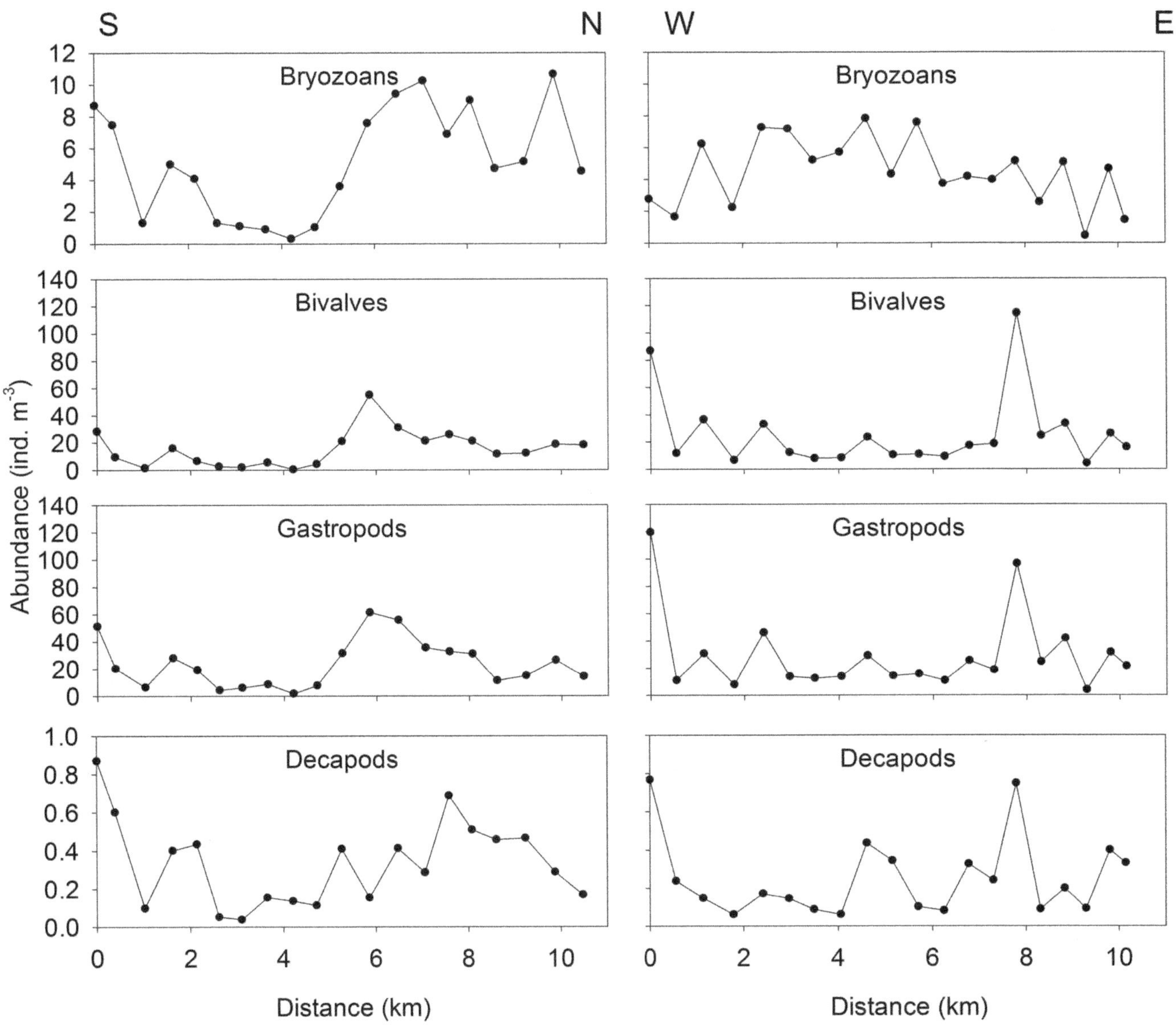


**Figure 2. Larval abundance from North to South and West to East transects in St. George's Bay, Nova Scotia, Canada, which were sampled continuously (one sample every ~500 m for 5 minutes) at 3 m depth with a 200-μm plankton ring net (0.75 m diameter).**

periodicity as the tidal cycle, towards the surface at flood tide and towards the sea floor at ebb tide, taking advantage of the differential transport to reach the head of the estuary.

By comparing the spatial scale in larval patchiness with that of the physical variables of the water column (temperature, salinity etc.), we can determine the relative contribution of different physical features (upwelling, internal tidal bore etc.) in generating patterns in larval abundance. For example, a similarity between the scale of larval patch size and the variability in temperature, but not of salinity, implies that temperature is more important in determining the larval pattern. Both a correlation in space between larval abundance and physical variables of the water column, and a similarity in the scale of spatial variability for biological and physical variables can be used as indicators of the physical features that regulate larval abundance [23]. In this study, we determined the patch size through spatial autocorrelation in larval abundance of different taxonomic groups (bryozoans, bivalves, gastropods, decapods) and in the physical variables of the water column (temperature, salinity, fluorescence, density, depth of the fluorescence maximum and depth of the mixed layer), and identified relevant physical or biological processes that may influence larval distributions at fine spatial scales (0.5 km).

## Bay-scale larval distribution

In a concurrent study [24], we investigated mechanisms of pattern formation in the larval distributions of benthic invertebrates by relating the spatial and temporal variability in the larval distributions to that of physical and biological variables at scales of 10 s km's. At bay-wide scales, gastropod, bivalve and, to a lesser extent, bryozoan larvae had very similar spatial distributions, but the distribution of decapod larvae followed a different pattern. These results suggested that taxonomic groups that have functionally (i.e. swimming ability) similar larvae (e.g. bivalves and gastropods) also show similar dispersion properties (distribution and spatial variability), while the opposite is true for groups with functionally dissimilar larvae (e.g. bivalves and decapods).

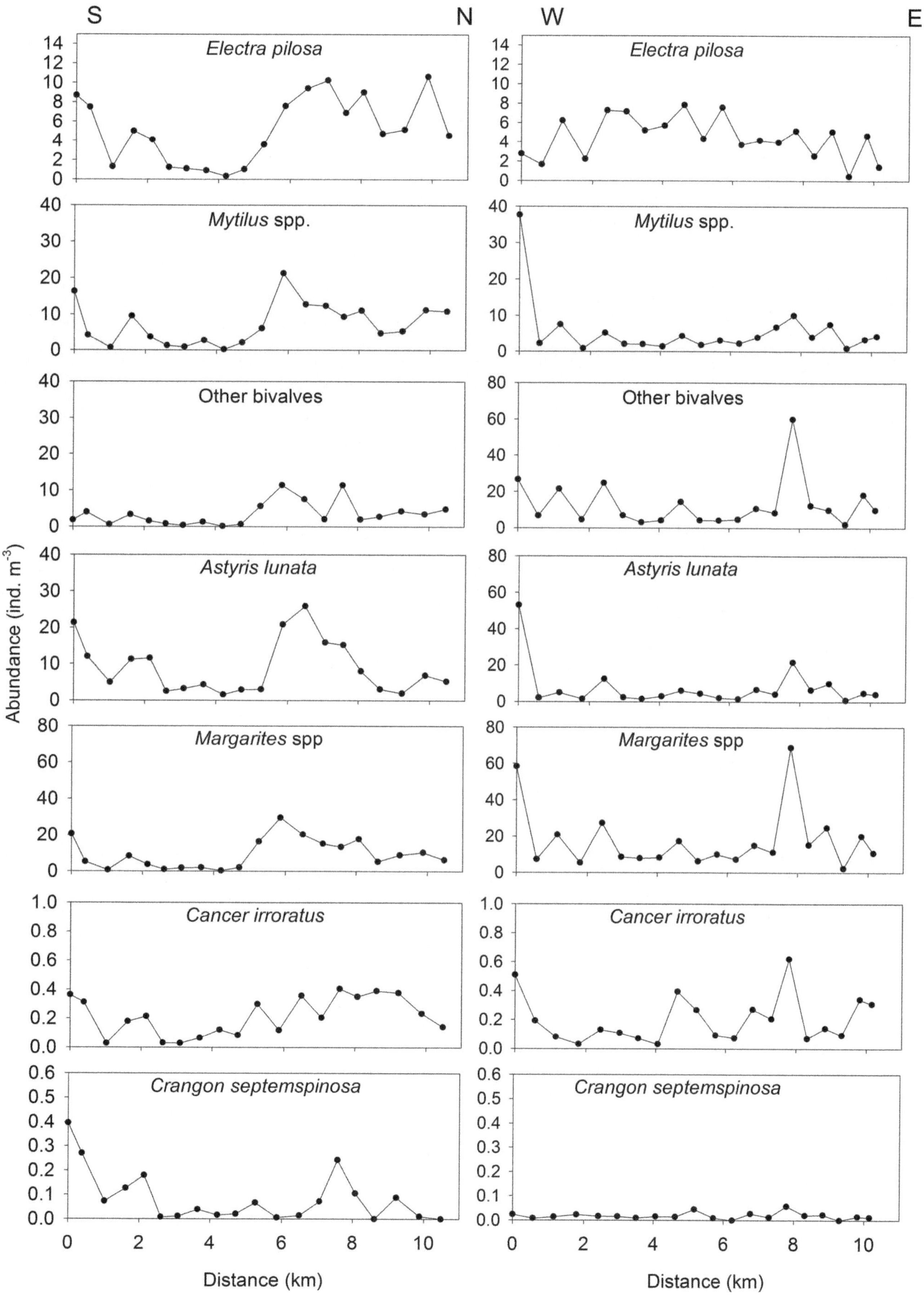
S
N
W
E
Electra pilosa
Mytilus spp.
Other bivalves
Astyris lunata
Margarites spp
Cancer irroratus
Crangon septemspinosa
Abundance (ind. m-3)
Distance (km)

**Figure 3. Larval abundance from North to South and West to East transects in St. George's Bay, Nova Scotia, Canada, which were sampled continuously (one sample every ~500 m for 5 minutes) at 3 m depth with a 200-μm plankton ring net (0.75 m diameter).**

Patterns in observed horizontal distribution at the broad scale were determined during the larval phase, and that the primary mechanism for pattern formation was larval interaction with physical oceanographic structures (e.g. stratification and tidal currents).

## Physical Characteristics of St. George's Bay

St. George's Bay is an open coastal embayment, ~45 by 45 km, generally shallow, with a mean depth of about 20 m and a maximum depth of 35–40 m. There is little freshwater runoff into the Bay, and the primary forcing of the circulation is from the tides and wind, both locally and indirectly from the neighbouring Gulf of St. Lawrence. The tidal currents in the Bay are generally weak mixed diurnal to semidiurnal, with a tidal range of about 1.5 m (Canadian Hydrographic Service, water.charts.gc.ca/twl-mne/index-eng.asp). The dominant tidal constituent, the semi-diurnal $M_2$ velocity has an amplitude near the mouth of ~0.10 m $s^{-1}$. Other tidal constituents are much smaller.

There is relatively weak horizontal spatial structure in the density field, and essentially no geostrophic circulation given the weak lateral stratification. The surface mixed-layer in summer is formed by solar heating and wind-forcing and deepens over the summer period, with very weak gradients in the top 10 m. However, there is a strong, often nearly linear, gradient in temperature, salinity, and density from the base of the surface mixed-layer (10 m) to at least 25 m depth (Fig. S1). Consequently, there is a strong density gradient and a clear boundary between the well-mixed surface waters (0 to 10 m) and the stratified near-bottom water from 10 to at least 25 m below the surface. St. George's Bay was described in greater detail in a previous study [24].

# Methods

## Field sampling

Larval abundance and biological and physical variables were sampled in St. George's Bay, Nova Scotia, on 15 Aug 2009. No specific permissions were required for these locations/activities. The field study did not involve endangered or protected species. The sampling area was from 45.713 to 45.804°N and from 61.882 to 61.759°W, specific site coordinates can be found in the Data S1. Two 10-km transects (Fig. 1) were sampled in contiguous 500-m transects for 5 min each at 3 m depth with a 200-μm plankton ring net (0.75-m diameter). For each tow, we repositioned the boat to ensure that the coordinates of the start position of the tow matched the end position of the previous tow. The net was towed at ~1.7 m $s^{-1}$ for 5 min and the volume of filtered water was quantified using a flowmeter. Temperature, conductivity, pressure and fluorescence were measured through the entire water column by doing vertical casts with a Seabird 25 Conductivity-Temperature-Depth (CTD) recorder, and an attached SCUFA fluorometer. CTD casts were done between plankton tows and at both ends of the transects. Each transect consisted of 20 plankton samples to quantify larval abundance, and 21 CTD casts. The CTD recorded at 1 Hz, but the data from 1–5 m depth (which was always in the surface mixed layer at all sites) were averaged to provide a representative estimate of the physical variables at 3 m. The depth of the mixed layer was calculated by determining the shallowest depth at which the density differences between consecutive measurements exceeded 2 standard deviations of all density differences between consecutive measurements of the entire density profile between 1–5 m. Each transect took 6 h to complete and was timed so that high tide (which coincided with high noon) would occur between transects. By ensuring that transects were sampled during a single phase of the tidal cycle (and light cycle), any effect of tidal or diel vertical migration would be of the scale of the entire transect (10 km). The first transect was sampled from South to North (low to high tide) and the second transect was sampled from East to West (high to low tide). Using a net of this mesh size may under-estimate abundance of small larvae (<200 μm). However, it is a necessary compromise in this multi-species study to allow capture of a wide range of larval types at sufficient numbers (e.g. very abundant but small gastropods to larger but rare decapods). All plankton samples were preserved in

**Table 2.** Pearson correlation examining the relationship in abundance for pairs of taxa: bryozoans (Bz), bivalves (Bv), gastropods (Gp) and decapods (Dp).

| **A)** | **Bz** | **Bv** | **Gp** | **Dp** |
|---|---|---|---|---|
| Bz | 1 | **0.646** | **0.684** | **0.429** |
| Bv | **<0.0001** | 1 | **0.943** | **0.567** |
| Gp | **<0.0001** | **<0.0001** | 1 | **0.654** |
| Dp | **0.0057** | **0.0001** | **<0.0001** | 1 |
| **B)** | **Bz** | **Bv** | **Gp** | **Dp** |
| Bz | 1 | 0.231 | **0.317** | **0.402** |
| Bv | 0.1514 | 1 | **0.929** | **0.556** |
| Gp | **0.0464** | **<0.0001** | 1 | **0.649** |
| Dp | **0.0102** | **0.0002** | **<0.0001** | 1 |

Relationships were assessed for both the logarithm (base 10) of the abundances variables (A) and the residuals of the abundances from the large scale spatial regression (B). The upper half of the matrix indicates the correlation coefficients for $\log_{10}(x+1)$ where x is larval abundance, lower half indicates the p-value. Statistically significant correlations are indicated in bold.

**Table 3.** Pearson correlation examining the relationship in abundance for pairs of species: *Electra pilosa* (Bz1), *Mytilus* spp. (Bv1), Other bivalves (Bv2), *Astyris lunata* (Gp1), *Margarites* spp. (Gp2), *Cancer irroratus* (Dp1), *Crangon septemspinosa* (Dp2).

| A) | Bz1 | Bv1 | Bv2 | Gp1 | Gp2 | Dp1 | Dp2 |
|---|---|---|---|---|---|---|---|
| Bz1 | 1 | **0.631** | **0.419** | **0.491** | **0.668** | **0.405** | 0.302 |
| Bv1 | **<0.0001** | 1 | **0.506** | **0.811** | **0.779** | **0.602** | 0.279 |
| Bv2 | **0.0072** | **0.0009** | 1 | **0.435** | **0.853** | **0.480** | −0.142 |
| Gp1 | **0.0013** | **<0.0001** | **0.0051** | 1 | **0.636** | **0.557** | **0.432** |
| Gp2 | **<0.0001** | **<0.0001** | **<0.0001** | **<0.0001** | 1 | **0.592** | 0.073 |
| Dp1 | **0.0094** | **<0.0001** | **0.0017** | **0.0002** | **0.0001** | 1 | **0.365** |
| Dp2 | 0.0583 | 0.0807 | 0.3811 | **0.0053** | 0.6548 | **0.0205** | 1 |
| **B)** | **Bz1** | **Bv1** | **Bv2** | **Gp1** | **Gp2** | **Dp1** | **Dp2** |
| Bz1 | 1 | 0.300 | 0.105 | 0.240 | 0.257 | **0.359** | **0.391** |
| Bv1 | 0.0602 | 1 | **0.341** | **0.905** | **0.692** | **0.479** | 0.311 |
| Bv2 | 0.5187 | **0.0314** | 1 | **0.445** | **0.881** | **0.523** | −0.084 |
| Gp1 | 0.1350 | **<0.0001** | **0.0040** | 1 | **0.731** | **0.480** | **0.372** |
| Gp2 | 0.1093 | **<0.0001** | **<0.0001** | **<0.0001** | 1 | **0.609** | 0.065 |
| Dp1 | **0.0230** | **0.0018** | **0.0005** | **0.0017** | **<0.0001** | 1 | **0.315** |
| Dp2 | **0.0126** | 0.0506 | 0.6069 | **0.0181** | 0.6883 | **0.0477** | 1 |

Relationships were assessed for both the logarithm (base 10) of the abundances (A) and the residuals of the abundances from the large scale spatial regression (B). The upper half of the matrix indicates the correlation coefficients for $\log_{10}(x+1)$ where x is larval abundance, lower half indicates the p-value. Statistically significant correlations are indicated in bold.

**Table 4.** Pearson correlation coefficients examining the relationship among physical variables of the water column and abundance of A) taxonomic groups [bryozoans (Bz), bivalves (Bv), gastropods (Gp) and decapods (Dp)] and B) species [Electra pilosa (Bz1), Mytilus spp. (Bv1), Other bivalves (Bv2), Astyris lunata (Gp1), Margarites spp. (Gp2), Cancer irroratus (Dp1), Crangon septemspinosa (Dp2)] from 15 Aug 2009 for sampling depths of 3 m (n = 20).

**Physical variables and logarithm (base 10) of abundance**

| A) | Temperature | Salinity | Fluorescence | Z ($F_{max}$). | Mixed Layer |
|---|---|---|---|---|---|
| Bz | −0.267 | **0.637** | **−0.662** | 0.096 | **−0.445** |
| | (0.2553) | **(0.0025)** | **(0.0015)** | (0.6866) | **(0.0491)** |
| Bv | −0.194 | **0.479** | **−0.593** | 0.237 | −0.281 |
| | (0.4116) | **(0.0327)** | **(0.0058)** | (0.3134) | (0.2294) |
| Gp | −0.053 | 0.381 | −0.431 | 0.125 | −0.120 |
| | (0.8234) | (0.0971) | (0.0580) | (0.5993) | (0.6153) |
| Dp | 0.160 | 0.274 | −0.414 | −0.267 | −0.076 |
| | (0.5009) | (0.2429) | (0.0695) | (0.2555) | (0.7489) |
| **B)** | **Temperature** | **Salinity** | **Fluorescence** | **Z ($F_{max}$)** | **Mixed Layer** |
| Bz1 | −0.266 | **0.637** | **−0.662** | 0.095 | −0.444 |
| | (0.2574) | **(0.0025)** | **(0.0015)** | (0.6890) | (0.0496) |
| Bv1 | −0.224 | **0.508** | **−0.600** | 0.186 | −0.310 |
| | (0.3434) | **(0.0223)** | **(0.0052)** | (0.4320) | (0.1841) |
| Bv2 | −0.300 | **0.491** | **−0.532** | 0.235 | −0.300 |
| | (0.1987) | **(0.0280)** | **(0.0158)** | (0.3192) | (0.1995) |
| Gp1 | 0.110 | 0.557 | −0.223 | −0.041 | 0.098 |
| | (0.6456) | (0.2740) | (0.3448) | (0.8643) | (0.6795) |
| Gp2 | −0.119 | 0.427 | **−0.573** | 0.268 | −0.245 |
| | (0.6178) | (0.0602) | **(0.0082)** | (0.2537) | (0.2974) |
| Dp1 | −0.056 | **0.461** | **−0.651** | 0.032 | −0.352 |
| | (0.8140) | **(0.0409)** | **(0.0019)** | (0.8927) | (0.1275) |
| Dp2 | 0.429 | −0.093 | 0.022 | **−0.615** | 0.318 |
| | (0.0594) | (0.6952) | (0.9281) | **(0.0039)** | (0.1722) |

Relationships were assessed for the logarithm (base 10) of the abundances and the physical variables. The value in brackets indicates the p-value. Statistically significant correlations are indicated in bold.

95% ethanol and larvae were identified and enumerated under a Nikon SMZ 1500, as described in Lloyd et al. [25]. Samples were split into subsamples using a Folsom plankton splitter. Samples from 8 of the stations were split to 1/64 of the original volume and all subsamples were processed. Based on those samples, we determined that at least 20 individuals of each species were required to obtain an estimate of abundance that was within 5% of the true sample abundance. The remainder of the samples were split to 1/128–1/1 to ensure that ≥20 individuals of the most abundant species (*Margarites* spp., *Astyris lunata*, *Mytilus* spp., *Electra pilosa* and *Cancer irroratus*) were enumerated. Individual larval species have also been combined into broader taxonomic groupings to generalize the interpretation of results.

Three 600 kHz Teledyne RDI Workhorse Sentinel Acoustic Doppler Current Profilers (ADCP) were deployed on the seafloor, sampling the full water column in 1-m depth bins every 20 min from11 July to 22 Aug 2009 (Fig. 1). We only included the horizontal velocities in the bin centered at 3 m depth in our analyses. The reliability of the current velocities at 3 m was assessed by plotting the variance in current velocities for each 1-m depth bin for the entire water column; the variance in the 2.5–3.5 m bin was not greater than that in the mixed layer below it (~4–12 m). Cross-correlation and coherence-squared analyses were applied to the current data collected by the 3 ADCP in Aug 2009. For the spatial/temporal analyses, the data record was truncated to 4 days (12–15 Aug) to reflect recent hydrodynamic conditions that may have affected larval distributions at the time scale of our study. The mean horizontal speed over the week preceding the study was weak, 28.6 mm $s^{-1}$ (0.1 km $h^{-1}$), corresponding to a passive particle being transported roughly10 km (scale of larval sampling) over 4 days.

## Data analyses

We examined the relationships among abundances of different larval groups or species, as well as the relationship between the abundance of each larval group or species and each physical variable (temperature, salinity, fluorescence, density, depth of the fluorescence maximum and depth of the mixed layer) using Pearson correlations. The logarithm (base 10) of larval abundance data was used for all statistical tests because it improved the normality of count data [26].

Since this study was designed to examine small scale patterns specifically, we removed large scale trends in larval abundance by linearly regressing larval abundance through space and using the

**Table 5.** Pearson correlation coefficients examining the relationship among physical variables of the water column and abundance of A) taxonomic groups [bryozoans (Bz), bivalves (Bv), gastropods (Gp) and decapods (Dp)] and B) species [Electra pilosa (Bz1), Mytilus spp. (Bv1), Other bivalves (Bv2), Astyris lunata (Gp1), Margarites spp. (Gp2), Cancer irroratus (Dp1), Crangon septemspinosa (Dp2)] from 15 Aug 2009 for sampling depths of 3 m (n = 20).

**Residuals of physical variables and residuals of abundance**

| A) | Temperature | Salinity | Fluorescence | Z ($F_{max}$) | Mixed Layer |
|---|---|---|---|---|---|
| Bz | −0.151 | **0.601** | **−0.708** | −0.139 | −0.331 |
| | (0.5253) | **(0.0051)** | **(0.0005)** | (0.5596) | (0.1541) |
| Bv | 0.254 | 0.273 | **−0.510** | 0.202 | 0.010 |
| | (0.2794) | (0.2238) | **(0.0217)** | (0.3920) | (0.9680) |
| Gp | 0.209 | 0.360 | **−0.581** | 0.110 | −0.030 |
| | (0.3768) | (0.1189) | **(0.0073)** | (0.6454) | (0.8996) |
| Dp | −0.243 | 0.331 | **−0.645** | −0.270 | −0.235 |
| | (0.3021) | (0.1546) | **(0.0022)** | (0.2498) | (0.3196) |
| **B)** | **Temperature** | **Salinity** | **Fluorescence** | **Z ($F_{max}$)** | **Mixed Layer** |
| Bz1 | −0.151 | **0.601** | **−0.708** | −0.140 | −0.329 |
| | (0.5259) | **(0.0051)** | **(0.0005)** | (0.5547) | (0.1569) |
| Bv1 | 0.135 | 0.328 | **−0.559** | 0.009 | −0.071 |
| | (0.5698) | (0.1586) | **(0.0104)** | (0.9710) | (0.7677) |
| Bv2 | 0.101 | 0.257 | −0.430 | 0.202 | 0.033 |
| | (0.6709) | (0.2741) | (0.0586) | (0.3925) | (0.8911) |
| Gp1 | 0.203 | **0.520** | **−0.705** | −0.009 | −0.075 |
| | (0.3904) | **(0.0189)** | **(0.0005)** | (0.9701) | (0.7547) |
| Gp2 | 0.269 | 0.277 | **−0.528** | 0.178 | 0.015 |
| | (0.2519) | (0.2378) | **(0.0166)** | (0.4625) | (0.9505) |
| Dp1 | −0.192 | 0.327 | **−0.608** | −0.108 | −0.302 |
| | (0.4165) | (0.1599) | **(0.0044)** | (0.6489) | (0.1952) |
| Dp2 | −0.306 | 0.319 | **−0.627** | **−0.488** | −0.090 |
| | (0.1893) | (0.1708) | **(0.0031)** | **(0.0290)** | (0.7061) |

Relationships were assessed for the residuals of the abundances and residuals of the physical variables from the large scale spatial regression. The value in brackets indicates the p-value. Statistically significant correlations are indicated in bold.

residuals for the spatial analyses. The smaller scale variability was still within the residuals, but the large scale variability (possibly related to vertical migration, sampling design, etc.) was removed. To quantify spatial and temporal autocorrelation, the residuals of larval abundances (to remove large scale variables) and the physical variables were analysed using Moran's I calculated as:

$$I(d)=\frac{\frac{1}{W}\sum_{h=1}^{n}\sum_{i=1}^{n}w_{hi}(y_h-\bar{y})(y_i-\bar{y})}{\frac{1}{n}\sum_{h-1}^{n}(y_h-\bar{y})^2} for\, h\neq i \quad (1)$$

In this equation, $y_h$ and $y_i$ are the residual values of the variable of interest (abundance or physical parameter) at sites $h$ and $i$, $d$ is the distance class, $w_{hi}$ are the weights which take the value of 1 when sites $h$ and $i$ are at distance $d$ or are equal to 0 otherwise, and $W$ is the sum of those weights [27]. Moran's I is a spatial autocorrelation statistic that takes on a null value when no autocorrelation is detected and varies from 1 to −1, indicating positive and negative spatial autocorrelation, respectively. Spatial lags were calculated for 12 bins with equal numbers of site pairs in each bin (n = 15–16 and 17–18, for larval abundances and physical variables, respectively) to allow bins to have the same statistical power. For the velocity data from the ADCP, temporal lags were calculated for 15 bins (n = 2755) to maintain a resolution similar to that of the spatial autocorrelation. To allow comparisons between the two, the temporal autocorrelation was converted to spatial autocorrelation by multiplying the time lag with the mean speed measured by the ADCP from 12–15 Aug (32.9 and 22.5 mm $s^{-1}$, in the S-N and W-E directions, respectively). All autocorrelation analyses were done in SAM v4.0 [28].

For consistency and comparability, we chose to analyse all spatial and temporal patterns using Moran's I. This metric was chosen because all data series except current velocity are relatively short, making conventional spectral analysis difficult (very coarse resolution). Additionally, with Moran's I, the width of the distance or time bins can be customized to allow the use of the same number of variable pairs for each data point.

Patterns in the differences in physical variables among sites were explored using Nonmetric Multidimensional Scaling (nMDS). The nMDS plot was created using the "metaMDS" function in the "vegan" package [29] in R 2.14.1 [R Core 30]. The nMDS plot was then rotated in order for the first dimension (NMDS1) to be

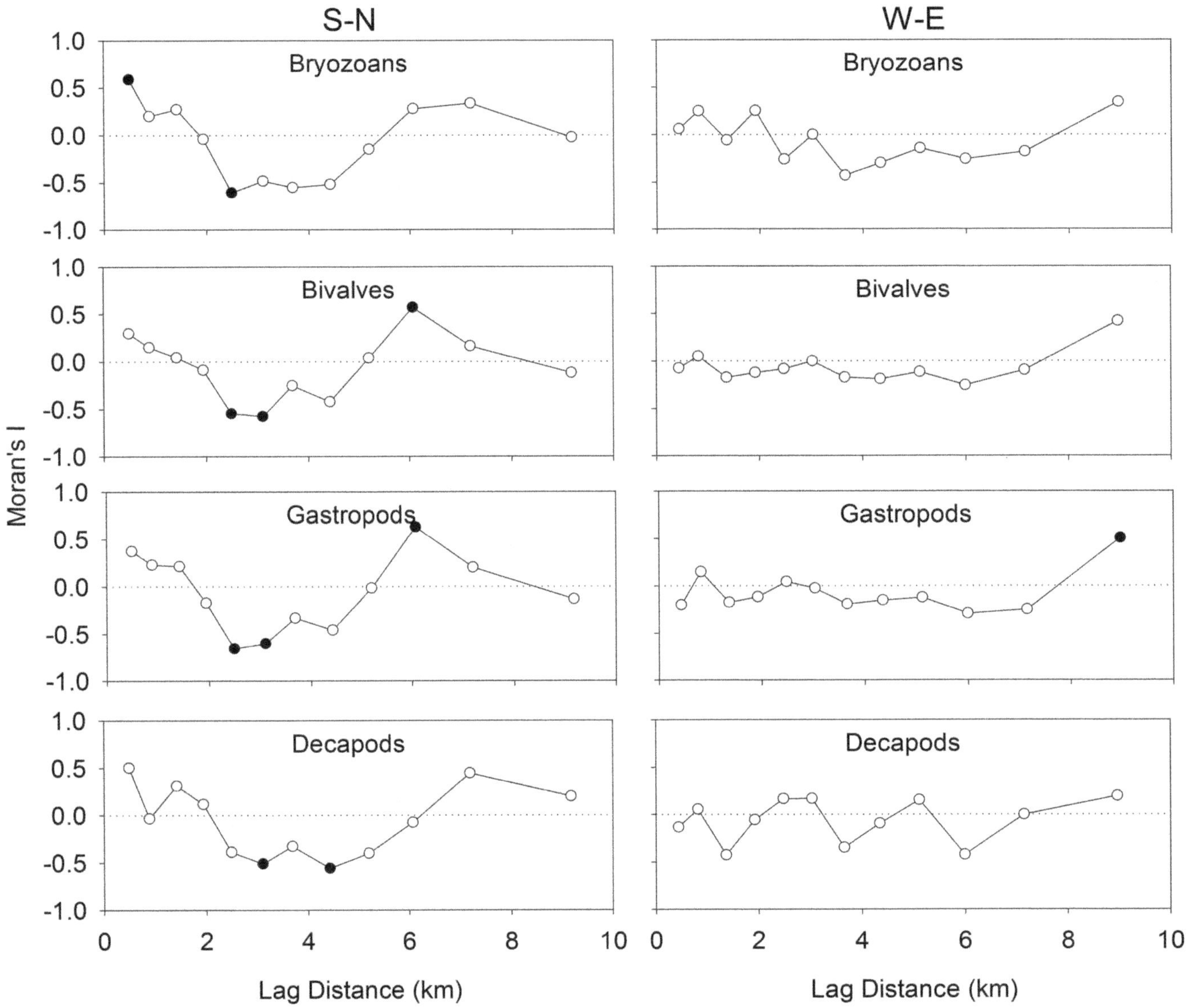


**Figure 4. Spatial analysis of the abundance for each of 4 larval groups (*Electra pilosa, Mytilus* spp., Other bivalves, *Astyris lunata, Margarites* spp., *Cancer irroratus, Crangon septemspinosa*) using Moran's I.** Positive and negative values of Moran's I indicate positive and negative spatial autocorrelation, respectively. Filled circles indicate significant spatial autocorrelation at that scale.

parallel to the abundance of each larval group, using "MDSrotate". This function rotates the nMDS plot so that the dispersion of points of the abundance of a particular larval group is highest along NMDS1. This rotation allowed the visualization of the relationship among sites that were physically similar (a single water mass) or dissimilar (different water masses), and larval abundance.

## Results

### Larval distributions

Collected larvae were categorized as bryozoans, bivalves, gastropods or decapods (Tab. 1). The bryozoans consisted almost entirely of *Electra pilosa*, whereas the bivalves were largely unidentifiable to species, but the most abundant identifiable genus was *Mytilus* spp. The gastropods consisted of *Margarites* spp. and *Astyris lunata*, and the decapods were mostly *Cancer irroratus*. In the S-N transect, the most striking feature in larval distribution of taxonomic groups was a peak in abundance at 5.8 km for both the bivalves and gastropods, and at the same location, an increase in abundance of bryozoan larvae (Fig. 2). At the species level, the patterns were similar except that the peak at 5.8 km was not as pronounced for the 'other bivalves' (Fig. 3). In the W-E transect, peaks in the abundance of bivalves, gastropods and decapods occurred at 0 km (closest to shore) and 7.8 km (Fig. 2). At the species level, the pattern was similar, but the peak at 7.8 km was less pronounced for *Mytilus* sp. and *A. lunata*, and there are no large peaks in the distribution of *Crangon septemspinosa* (Fig. 3).

The abundances of bivalves and gastropods were highly correlated with one another, while the correlation between the abundance of bryozoans and decapods was much weaker (Tab. 2). Intermediate correlation coefficients were obtained for the remaining pairs of taxonomic groups (Tab. 2). The abundances of bivalves and gastropods were also highly correlated at the species level, except for *A. lunata* (Tab. 3). Additionally, the abundance of *C. irroratus* was significantly correlated with that of all other species whereas that of *C. septemspinosa* was only

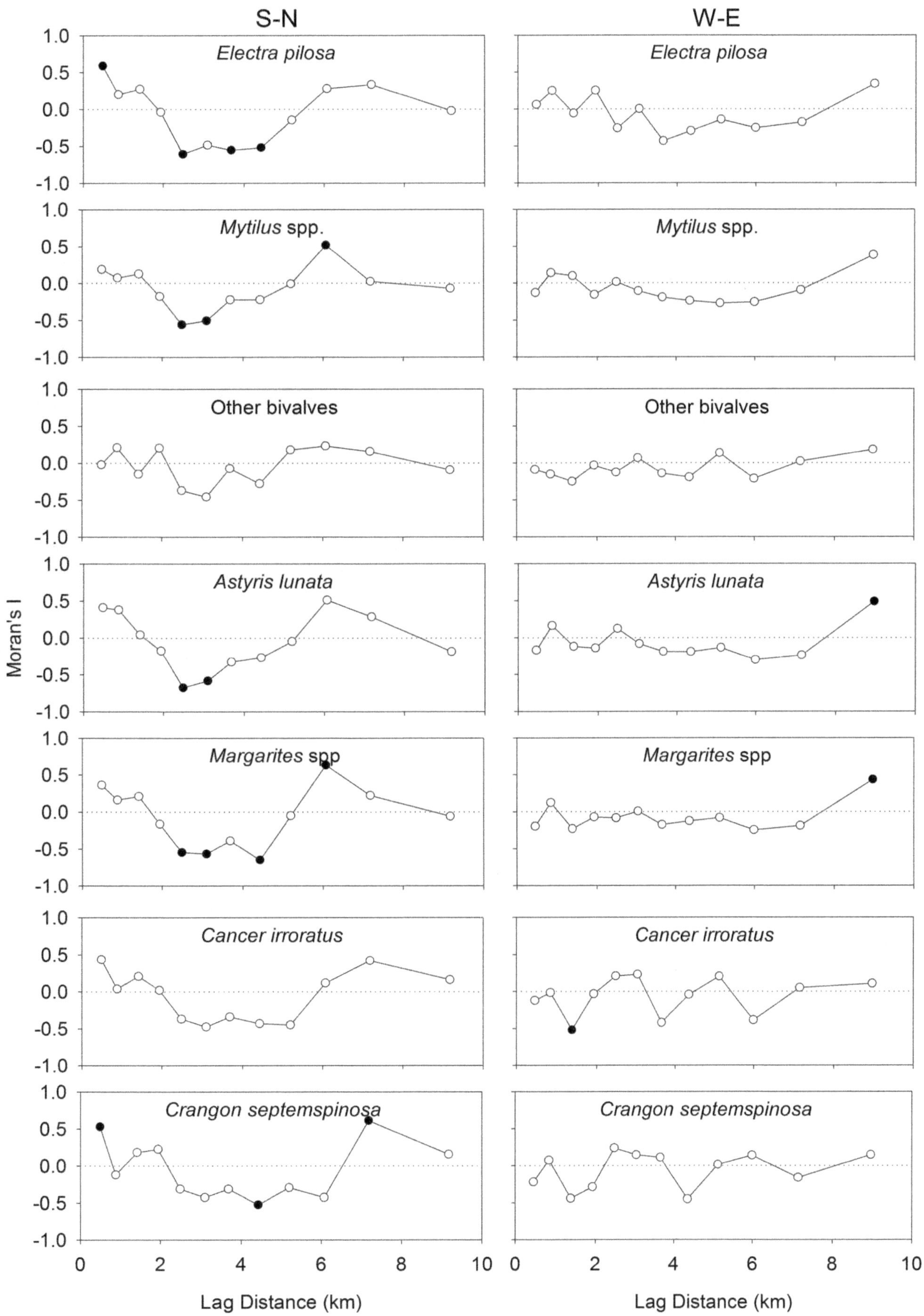
S-N
W-E
Electra pilosa
Mytilus spp.
Other bivalves
Astyris lunata
Margarites spp
Cancer irroratus
Crangon septemspinosa
Moran's I
Lag Distance (km)
1.0
0.5
0.0
-0.5
-1.0
0
2
4
6
8
10

**Figure 5. Spatial analysis of the abundance for each of 7 larval species (gastropods, bivalves, bryozoans, and decapods) using Moran's I.** Positive and negative values of Moran's I indicate positive and negative spatial autocorrelation, respectively. Filled circles indicate significant spatial autocorrelation at that scale.

correlated with *E. pilosa* and *A. lunata*. In general, there were more correlations between the physical variables and the abundance of bryozoans, gastropods and bivalves than for decapods (Tab. 4). Interestingly, all species were negatively correlated with fluorescence (in most cases significantly) and positively correlated with salinity (not always significantly) except *C. septemspinosa,* which was significantly negatively correlated with the depth of the fluorescence maximum. Bryozoan and bivalve abundance was significantly correlated with both fluorescence and salinity. Overall, fluorescence and salinity had the highest number of significant correlations with the abundance of all groups and species (Tab. 4).

When large scale spatial gradients were removed from the abundances of larvae and from the physical variables through spatial regression, the patterns were similar, but the relationship with fluorescence was stronger (Tab. 5). The abundance of *E. pilosa* and *A. lunata* was significantly correlated with fluorescence, and that for all other species abundance was also significantly negatively correlated with fluorescence except 'other bivalves' (Tab. 5). Lower fluorescence corresponded to a high larval abundance at the local scale (<10 km).

On the S-N transect, significant positive spatial autocorrelation of larval abundance was observed at the smallest observable scale of 0.5 km for bryozoans, and at 7.2 km for the gastropods and bivalves (Fig. 4). Significant negative autocorrelation of abundance was observed at 2.5 km for bryozoans, 2.5 and 3.0 km for bivalves and gastropods, and 3.0 and 4.3 km for decapods. In the W-E transect, no statistically significant spatial autocorrelation of larval abundance was observed for any taxonomic group, except gastropods at 9.2 km. Unsurprisingly, a similar pattern is present at the species level (Fig. 5). In the S-N transect, all species except 'other bivalves' and *C. irroratus* had significant negative spatial autocorrelation between 2.5 and 4.3 km and positive spatial autocorrelation between 6 and 7.2 km.

## Distribution of physical variables in the transects

In the S-N transect, there was a large increase in salinity (front) at ~6 km, with a corresponding decrease in fluorescence (Fig. 6). In the W-E transect, the most identifiable feature was the low temperature and salinity combined with high fluorescence, density and depth of fluorescence maximum at 0 km, which was closest to shore. Significant positive spatial autocorrelation was observed in temperature, density and fluorescence at the smallest observable scale of 0.4 km on both transects, and in the depth of the fluorescence maxima and mixed layer in the S-N transect (Fig. 7). Additionally, in the S-N transect, 3 out of 6 of the variables were significantly negatively autocorrelated at 3.2, 3.9 and 5.5 km. In the W-E transect, more than one of the physical variables were significantly negatively autocorrelated at 3.2, 5.3 and 6.3 km. Salinity was significantly positively autocorrelated at 5.5 km in the S-N transect, and at 5.3 km in the W-E transect.

The cross-correlation in current velocity among the ADCP moorings was just under 0.6 for moorings separated by 3–4 km, and 0.2–0.3 for moorings separated by 5–6 km. Only at tidal periods was coherence squared greater than the significance threshold of 0.3 (just below 0.6 for neighbouring moorings). The current velocity data showed significant temporal autocorrelation at most time lags because of the large sample size (Fig. 8). An important difference in the pattern of autocorrelation between directions is that the peak at 8.5 h is positive in the S-N direction; therefore, the first negative peak in autocorrelation is at 33 h in the S-N direction and 8.5 h in the W-E direction. After being converted to distance by multiplying by the mean speed, these time lags correspond to 3.9 and 0.7 km, in the S-N and W-E direction, respectively. Spectral analysis (Daigle, unpublished data) also indicated that the dominant peak in the periodogram for the S-N transect is at a lower frequency than in W-E transect. This suggests that the oscillations along the S-N transect occur over a longer time period, translating into larger spatial scales, than in the W-E direction.

## Biophysical interactions

The dispersion of sites classified by their physical variables in the rotated nMDS plots was relatively parallel to NMDS1 for bryozoans, bivalves and gastropods compared to NMDS1 for the decapods (Fig. 9). The loadings for the physical variables were oriented perpendicular to the gradient of larval abundance for the decapods. For the other taxonomic groups, the loadings of the physical variables varied in parallel to the larval gradient. Fluorescence and the depth of the mixed layer had the largest influence on the separation of the sites indicated by their letter symbols being furthest from the origin.

# Discussion

In this study, as on the scale of the entire bay [24], the distributions of gastropods and bivalves were most similar to one another, and those of bryozoans and decapods were most different from one another. Gastropods and bivalves are similar morphologically (*i.e.* calcareous shells), have similar swimming ability, and both groups display vertical migration [25,31]. If different types of larvae behave similarly or have similar early life history strategies, then they will likely aggregate in the same types of oceanographic features [10,32,33]. Conversely, contrasting life strategies result in very different dispersal properties, and consequently, different larval distributions. Bryozoans are among the slowest swimmers [31,34], and decapods are among the fastest of the invertebrate larvae [35,36]. It is well established that larval behaviour can be species- and even stage-specific, which can lead to differences in patterns of horizontal distribution [18,22,25,37]. In this study, species specific behaviour may be observed in *C. septemspinosa*. However, the similarity in larval distribution among most species is suggestive of common mechanisms of pattern formation.

The consistent relationship between bryozoan abundance and salinity suggests that bryozoans were being advected like a passive tracer such as salinity. We found that the abundance of bryozoans was negatively correlated with fluorescence although others have found that the peak in the vertical bryozoan abundance was associated with the fluorescence maximum [25]. Similarly, for all species, lower fluorescence corresponded to a high larval abundance at the local scale (<10 km). Given the relatively small horizontal differences in fluorescence compared to vertical ones, the association with lower fluorescence likely reflects an association with a particular water mass with low fluorescence rather than with food availability. Additionally, we found that the abundance of bryozoans, gastropods and bivalves had a stronger relationship than the decapods with the physical variables in multivariate space. While the abundance of *C. irroratus* had generally weaker

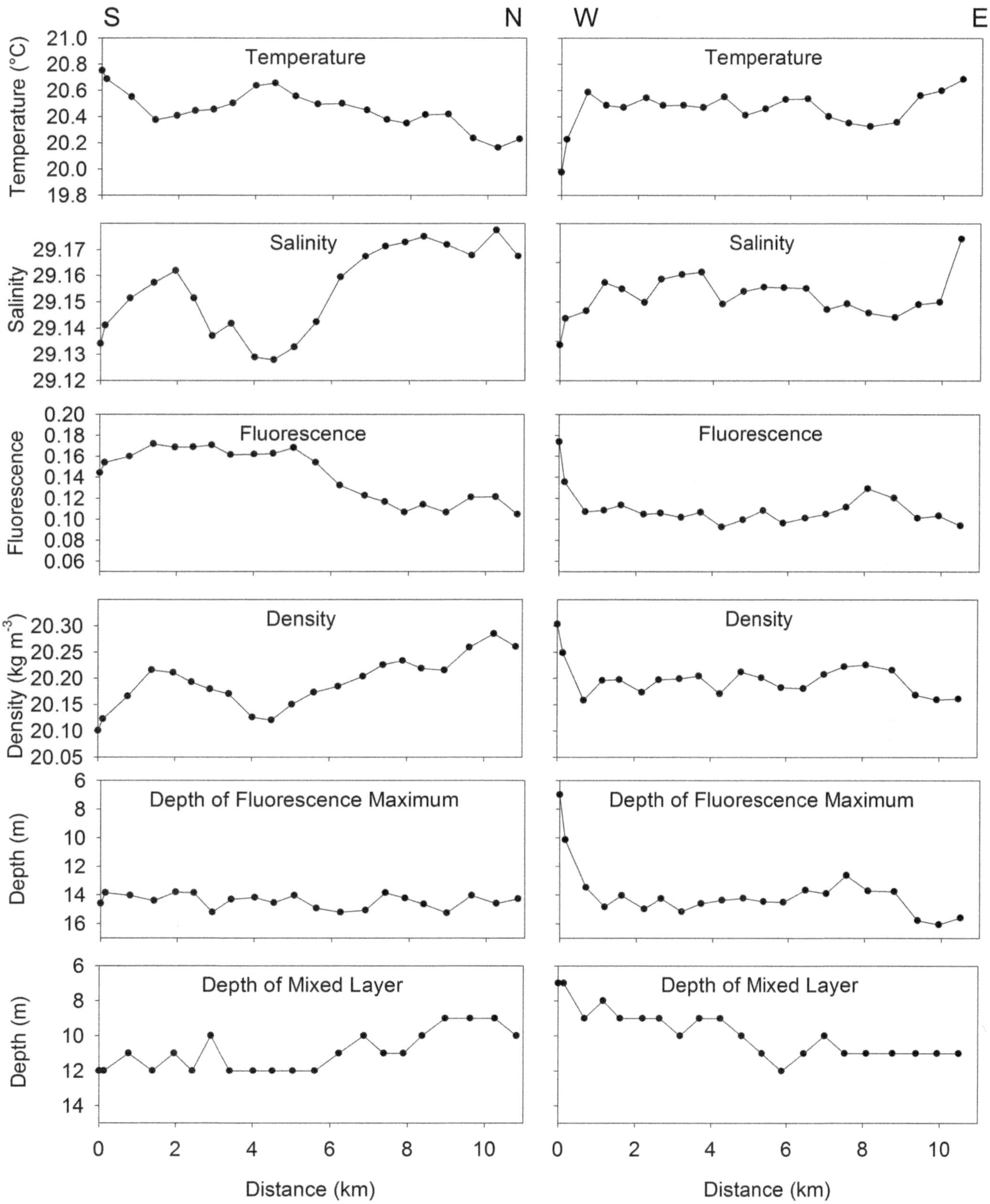


**Figure 6. Physical variables in the water column of North to South and East to West transects in St. George's Bay, Nova Scotia, Canada which were sampled once every ~500 m.** Data represents the average for the water column from 1–5 m depth.

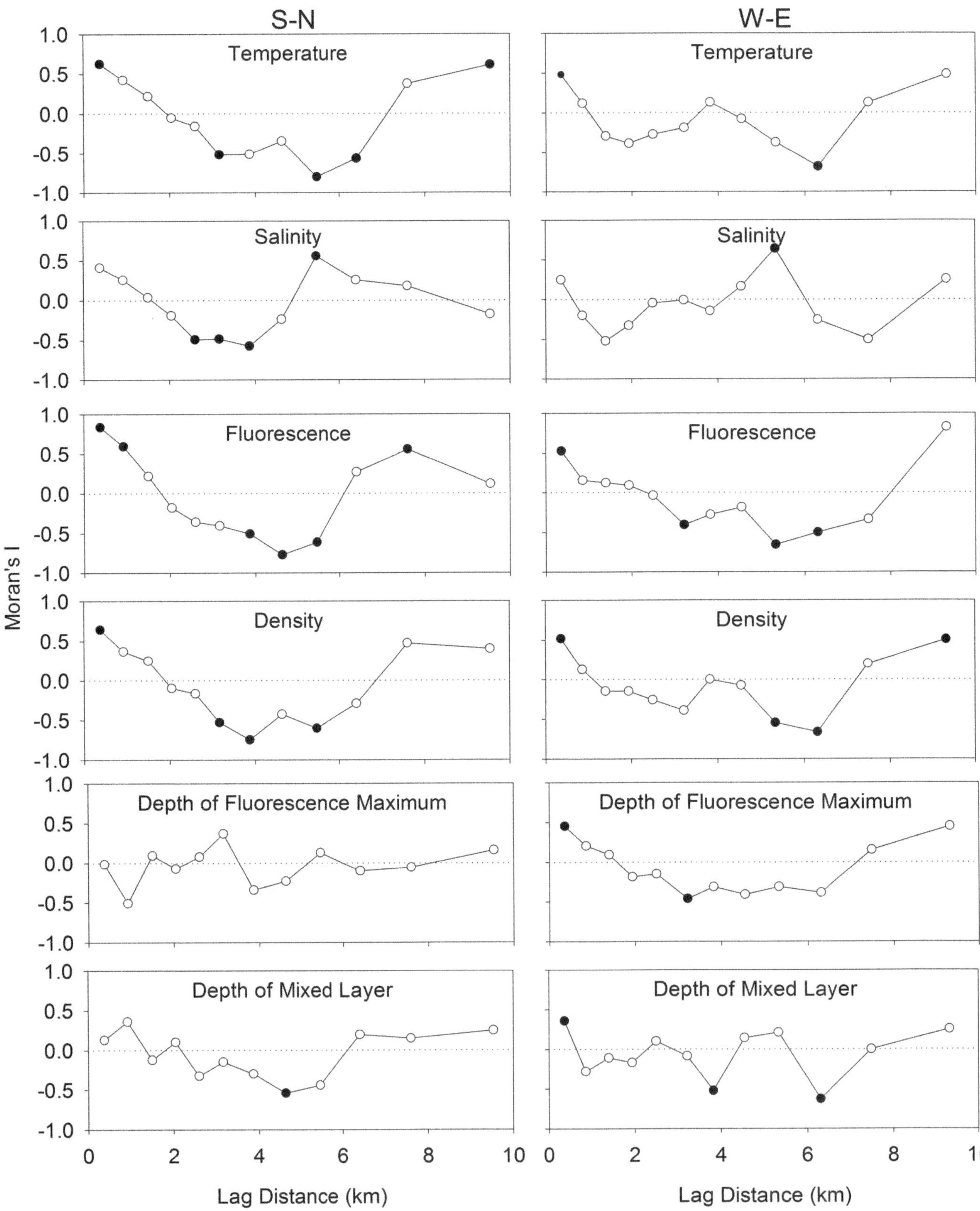


**Figure 7. Spatial analyses of the physical variables using Moran's I.** Positive and negative values of Moran's I indicate positive and negative spatial autocorrelation, respectively. Filled circles indicate significant spatial autocorrelation at that scale.

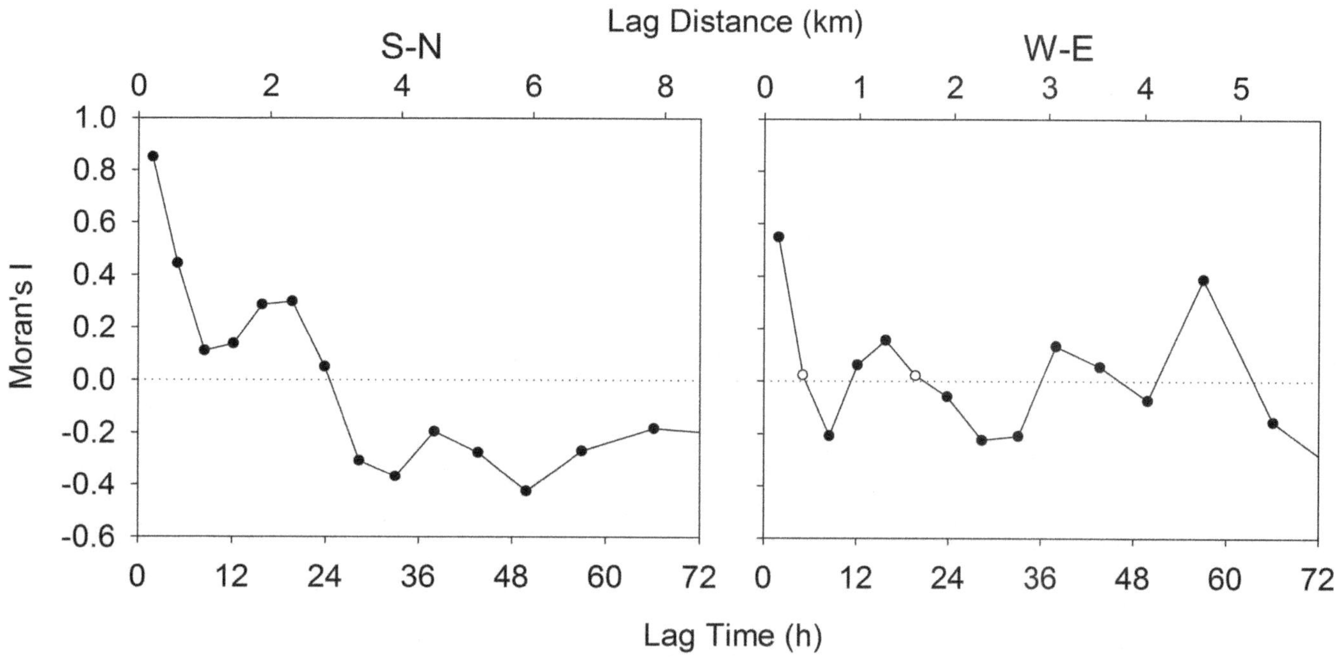


**Figure 8. Temporal analyses of horizontal current velocities using Moran's I.** Positive values of Moran's I indicate positive spatial autocorrelation and vice versa. Filled circles indicate significant spatial autocorrelation at that scale. Equivalent lag distances were calculated by multiplying the time lag by the mean speed (see Methods).

relationship with physical variables than that of bryozoan, bivalve and gastropod species, the pattern differed for *C. septemspinosa*. The negative relationship between the abundance of *C. septemspinosa* and the depth of the fluorescence maximum combined with the relatively weak relationship with the other physical variables, suggests that: either 1) larvae aggregated in areas where the fluorescence maximum is relatively shallower (*i.e.* available to them during the deep phase of their vertical migration); or 2) larval mortality was lower in these regions. However, the horizontal distribution for this species suggests that the larvae are largely retained in the southern portion of the bay.

For both the larvae (all groups, most species) and all the physical properties of the water column, except the fluorescence maximum, there was significant negative autocorrelation at distance lags between 2.5 and 5 km in the S-N direction. The first peak in negative autocorrelation of current velocity at time lags of 28 or 33 h corresponded to spatial lags of 3.3 to 3.9 km. While the peaks in autocorrelation of current velocity didn't quite correspond with the M2 tidal cycle period (12.42 h), the offset in periodicity likely reflects an interaction between the tidal cycle and the diel cycle (24 h) of wind driven circulation. Given the scale of spatial autocorrelation in the W-E direction, it is not surprising that our transect sampling did not reveal patterns similar to the N-S direction. In the W-E direction, the first peak (at 8.5 h) in negative autocorrelation of the ADCP current velocities was as large as all other peaks. This indicates that the data were dominated by the signal that corresponds to that peak. Since this peak corresponds to a distance lag of 1 km, any spatial autocorrelation along the W-E transect would not have been properly resolved by the larval sampling at a frequency of 500 m. The method used to convert temporal to spatial autocorrelation in current velocity (multiplying the time lag with the mean speed), has similar assumptions to a progressive vector method which converts Eulerian observations into Lagrangian data. In both cases, it is assumed that the current is spatially uniform and only varies in time. This method is justified since there was stronger autocorrelation in the current velocities among the ADCPs that were closer together (3–4 km), indicating some spatial uniformity (coherence) of current velocities at this scale.

While it is well known that larval distribution can be affected by salinity, temperature, density or food availability [38–40], the observed variance in these variables in our study is biologically irrelevant. For example, a change in temperature of 0.075°C over a 500-m distance (which was the average thermal gradient) is unlikely to elicit a larval behavioural response. Such a gradient also should not have any direct consequences for growth or mortality. It is also possible that larval patches can result from predator-prey diffusive instabilities, where predator and prey disperse at different rates (possibly due to swimming ability) and the resulting predator-prey interactions can cause spatial heterogeneity in larval abundance [41]. In this case, larvae could be either the predator or the prey. However, diffusive instabilities are thought to occur at larger scales (10–100 km), and those instabilities do not explain the relationships among physical variables of the water column and the larval distributions.

We propose that the observed relationships between the abundance of larval groups and physical variables are the result of the non-uniform mixing of water masses. Given the weak mean currents ($\sim 0.02$ m $s^{-1}$) it is clear that the residence time for larvae in St. George's Bay can be quite long, on the order of days to tens of days, O(40 km/0.02 m $s^{-1} \approx 20$ days), a result supported by numerical simulations (B. deYoung, unpublished data). This residence time is also very similar to the estimate of 15 d based on current meter observations in 1974–1975 (Petrie and Drinkwater, 1978). The relatively low cross-correlation in velocity among moorings suggests that large scale, coherent flow features in St. George's Bay are relatively weak. The persistent tide ($M_2$ tide is dominant, with a period of 12.42 hours) and the weak mean circulation generate an ideal dispersive environment. The long residence time and the weak spatial coherence in the velocity field

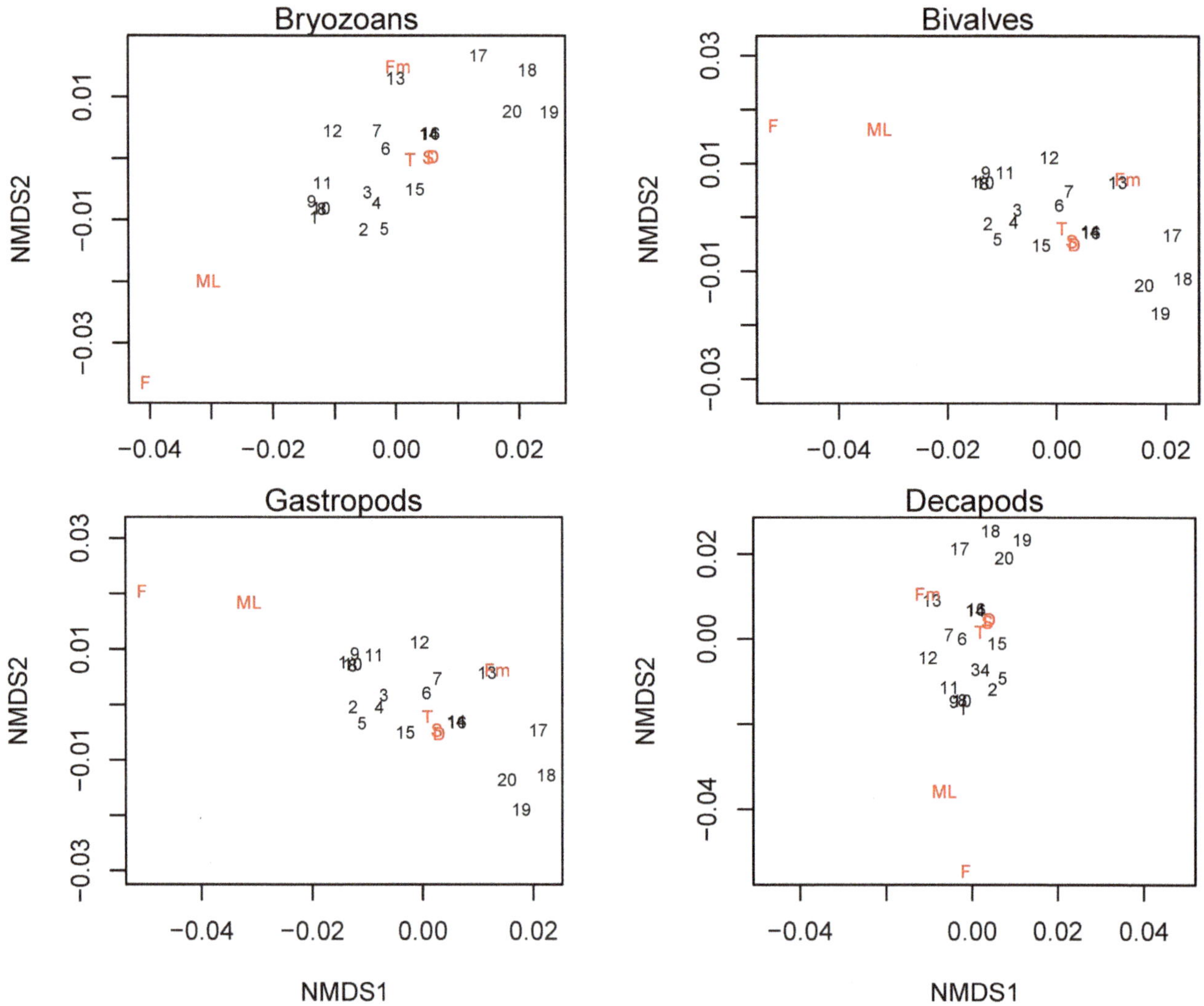


**Figure 9. Ordination from nonmetric multidimensional scaling of the sampling sites based on physical variables.** The ordinations were rotated so that NMDS1 is parallel to an external variable (abundances of bryozoans, bivalves, gastropods and decapods). Larval abundance increases from left to right. Numbers indicate the location of the site in nMDS ordination space and the letters represent the physical variables of the water column [temperature (T), salinity (S), fluorescence (F), density (D), depth of the fluorescence maximum (Fm) and depth of the mixed layer (ML)]. The letter's distance from the origin represents the weight of that factor in the ordination space and the values for the physical variables increase from the origin to the letters indicating that variable. Stress = 0.0486.

at long periods and large spatial scales, combined with the regular tidal dispersion, twice daily at short spatial scales (km's) will lead to significant dispersion in larval patterns [42]. By definition, flow structures with low coherence act to enhance mixing within a water-mass and coherent flow structures act to transport a water mass. We argue that the significant spatial coherence at tidal periods reduced advective diffusion (diffusion of larvae by means of advection as opposed to random diffusion, which is relatively constant) at that scale, and the low coherence at smaller scales enhances advective diffusion; spatial structures at those smaller scales will be erased while structure at the scale of the tidal period will be preserved.

We suggest that the scale of cyclical periodicity of coherent structures in current velocity (*i.e.* tidal period in this case) affects not only the physical properties of the water column, but also larval distributions. This effect of cyclical periodicity in current velocities should be particularly noticeable in the autocorrelation patterns of the poor swimmers (i.e. quasi-passive tracers), since their distribution is most related to the physical properties of the water column. Indeed, the peak in negative autocorrelation is lower for bryozoans than for decapods. The difference in patterns between the two transects suggests that the scale of cyclical periodicity in current velocities may play a directionally dependent role in the scaling of larval patchiness. Consequently, to properly resolve larval patchiness, biophysical models used to predict larval dispersal should at the very least resolve to the scale of the tidal excursion or the scale of the wind driven circulation, whichever is smallest.

Overall, we found significant negative autocorrelation at a similar spatial scale in all larval groups (most species) and all the physical properties of the water column, yet relatively few correlations between larval distributions and physical properties of the water column. This is not surprising since matching scales of negative spatial autocorrelation does not imply linear correlation between those variables. For example, a larval aggregation at an oceanographic front [32,43] would be manifested as a larval patch centered on the boundary of two "physical" patches. If this scenario is repeated over space, we would detect consistent scales of spatial autocorrelation, yet no correlation between larval distributions and physical properties of the water column.

Our observations provide a snapshot of the conditions in the bay on that particular day, since sampling was conducted over a single 12-h period, and may not be representative of average conditions. However, observations from three different sampling periods at the scale of the entire bay also suggested that distributions of gastropods and bivalves were most similar to, and those of bryozoans and decapods were most different from one another [24]. Additionally, the negative relationship between fluorescence and larval abundance was remarkably consistent across taxa. We contend that the most parsimonious explanation for the similarity in scales of the spatial coherence in the tidal period, the larval distributions and the physical variables is that they are related due to the reduced advective diffusion at that scale and/or the advective-diffusive history of the water masses. In either case, the scale of the coherent structure (tides) is important in determining the patterns in larval distribution. Furthermore, we suggest that the scale of coherent structure in current velocities may be important in forming larval distribution patterns in other systems, which similarly to ours are not dominated by strong oceanographic features (e.g. estuarine plumes, upwelling events, or markedly different water masses). While our sampling design may limit our observations to depths of 3 m, it is reasonable to apply our suggested mechanism for scaling of larval patches to coherent structure in current velocities to other depths; the caveat would be that the spatial scale of the patches should scale to the local scale of coherent structures, which may not necessarily be ~3 km.

The effects of the spatial distribution of the adult habitat and the influence of large scale hydrographic conditions (i.e. estuarine plume, eddies) on larval distributions [44–46] or small scale aggregating processes [10,32] have been identified previously. However, these effects influence larval distributions at very large scales (>100 km), in locations with very different water masses, or at scales below the resolution of our study. Our study shows that small scale larval patchiness (<10 km) can exist even in areas of relatively homogenous physical properties. It is well established that the distribution of phytoplankton and holoplankton are affected by a variety of processes with multiple and overlapping scales [41,47,48], but the relative importance of all these processes on meroplankton is poorly known at present [10,49,50]. These studies have shown that larvae are not passive particles randomly drifting in the ocean. Their active interactions with the water column can affect their horizontal distribution at multiple scales from small-scale aggregations at fronts, to medium scale patches such as those observed, to large scale oceanographic features [1,2,22]; ultimately these larval behaviours affect larval dispersal.

Unlike holoplankton, the larvae of benthic invertebrates settle on the seafloor before taking their adult form, and the patchiness of larval distributions can affect the distributions of adults [49,51,52]. This effect can result from direct settlement of "patches", but also indirectly by changing larval dispersal pathways through aggregations at fronts, advection along with water masses or vertical migration. We have shown that larval groups with similar life strategies (*e.g.* swimming ability) have similar distributions and that the weaker swimming larvae have a demonstrably tighter relationship with the physical properties of the water column, suggesting that they have dispersal properties similar to that of a passive tracer. Therefore, incorporating larval behaviour in biophysical models of larval transport would be more important in larvae with strong swimming abilities, whereas it may not be as important for weak swimmers. Additionally, we have shown that the scale of larval patchiness corresponds with that of both the physical variables of the water column and the tidal/diel periodicity in the current velocities. Factors such as larval swimming ability and the scale of coherent structures (*e.g.* tidal period), which determine larval distribution and patch size, will affect larval dispersal and ultimately the recruitment of benthic adults.

## Supporting Information

**Figure S1 Depth profiles of physical variables in the water column of N-S and W-E transects in St. George's Bay, Nova Scotia, Canada, were sampled every ~500 m.** White areas in the W-E transect represent the seafloor.

**Data S1 Larval abundance (ind. m$^3$) from North to South and West to East transects in St. George's Bay, Nova Scotia, Canada, which were sampled continuously (one sample every ~500 m for 5 minutes) at 3 m depth with a 200-µm plankton ring net (0.75 m diameter).** Physical variables in the water column were sampled once every ~500 m. Physical variables data represents the average for the water column from 1–5 m depth.

## Acknowledgments

We thank Michelle Lloyd, Ryan Stanley, Stacey Henderson, Marielle Lesperance, Jessie Short, Donny Ross, and Jack Foley for help with field work. We also thank Keith Thompson, Eric Oliver, Jean-Pierre Auclair and Kim Davies for assistance with Matlab.

## Author Contributions

Conceived and designed the experiments: RMD AM. Performed the experiments: RMD AM BdY. Analyzed the data: RMD AM BdY. Contributed reagents/materials/analysis tools: AM BdY. Contributed to the writing of the manuscript: RMD AM BdY. Processed the samples: RMD.

## References

1. Levin LA (2006) Recent progress in understanding larval dispersal: new directions and digressions. Integr Comp Biol 46: 282–297. Available: http://www.ncbi.nlm.nih.gov/pubmed/21672742. Accessed 2012 July 13.
2. Cowen RK, Sponaugle S (2009) Larval Dispersal and Marine Population Connectivity. Ann Rev Mar Sci 1: 443–466. Available: http://www.annualreviews.org/doi/abs/10.1146/annurev.marine.010908.163757. Accessed 2012 July 13.
3. Underwood AJ, Fairweather PG (1989) Supply-side ecology and benthic marine assemblages. Trends Ecol Evol 4: 16–20. Available: http://www.sciencedirect.com/science/article/pii/0169534789900086. Accessed 2012 September 14.
4. Shanks AL, Brink LA (2005) Upwelling, downwelling, and cross-shelf transport of bivalve larvae: test of a hypothesis. Mar Ecol Prog Ser 302: 1–12.
5. Menge BA, Chan F, Nielsen KJ, Di Lorenzo E, Lubchenco J (2009) Climatic variation alters supply-side ecology: impact of climate patterns on phytoplankton and mussel recruitment. Ecol Monogr 79: 379–395. Available: http://www.esajournals.org/doi/abs/10.1890/08-2086.1. Accessed 2012 August 30.
6. Tamaki A, Mandal S, Agata Y, Aoki I, Suzuki T, et al. (2010) Complex vertical migration of larvae of the ghost shrimp, Nihonotrypaea harmandi, in inner shelf waters of western Kyushu, Japan. Estuar Coast Shelf Sci 86: 125–136. Available: http://linkinghub.elsevier.com/retrieve/pii/S0272771409005071. Accessed 2012 September 12.
7. Jillett JB (1976) Zooplankton associations off otago peninsula, south-eastern New Zealand, related to different water masses. New Zeal J Mar Freshw Res 10: 543–557. Available: http://www.tandfonline.com/doi/abs/10.1080/00288330.1976.9515639. Accessed 2012 August 10.
8. DiBacco C, Fuchs HL, Pineda J, Helfrich K (2011) Swimming behavior and velocities of barnacle cyprids in a downwelling flume. Mar Ecol Prog Ser 433:

131–148. Available: http://www.int-res.com/prepress/m09186.html. Accessed 2012 August 30.
9. Poulin E, Palma AT, Leiva G, Narvaez D, Pacheco R, et al. (2002) Avoiding offshore transport of competent larvae during upwelling events: The case of the gastropod Concholepas concholepas in Central Chile. Limnol Oceanogr 47: 1248–1255.
10. Pineda J (1991) Predictable upwelling and the shoreward transport of planktonic larvae by internal tidal bores. Science (80-) 253: 548. Available: http://www.sciencemag.org/content/253/5019/548.short. Accessed 2012 August 15.
11. Omori M, Hamner WM (1982) Patchy distribution of zooplankton: behavior, population assessment and sampling problems. Mar Biol 72: 193–200. Available: http://www.springerlink.com/index/Q384517P486J2335.pdf. Accessed 2012 August 30.
12. Olson RR (1985) The consequences of short-distance larval dispersal in a sessile marine invertebrate. Ecology 66: 30–39.
13. Metaxas A, Saunders MI (2009) Quantifying the "bio-" components in biophysical models of larval transport in marine benthic invertebrates: advances and pitfalls. Biol Bull 216: 257–272. Available: http://www.ncbi.nlm.nih.gov/pubmed/19556593.
14. Daigle RM, Metaxas A (2012) Modeling of the larval response of green sea urchins to thermal stratification using a random walk approach. J Exp Mar Bio Ecol 438: 14–23. Available: http://linkinghub.elsevier.com/retrieve/pii/S0022098112003346. Accessed 2012 October 31.
15. Butler MJ I, Paris C, Goldstein J, Matsuda H, Cowen R (2011) Behavior constrains the dispersal of long-lived spiny lobster larvae. Mar Ecol Prog Ser 422: 223–237. Available: http://www.int-res.com/abstracts/meps/v422/p223-237/. Accessed 2014 July 14.
16. Fujimura AG, Reniers AJHM, Paris CB, Shanks AL, MacMahan JH, et al. (2014) Numerical simulations of larval transport into a rip-channeled surf zone. Limnol Oceanogr 59: 1434–1447. Available: http://www.aslo.org/lo/toc/vol_59/issue_4/1434.html. Accessed 2014 July 16.
17. Higgins FA, Bates AE, Lamare MD (2012) Heat tolerance, behavioural temperature selection and temperature-dependent respiration in larval Octopus huttoni. J Therm Biol 37: 83–88. Available: http://linkinghub.elsevier.com/retrieve/pii/S0306456511001598. Accessed 2012 July 13.
18. Tapia FJ, Pineda J (2007) Stage-specific distribution of barnacle larvae in nearshore waters: potential for limited dispersal and high mortality rates. Mar Ecol Prog Ser 342: 177–190. Available: https://darchive.mblwhoilibrary.org/handle/1912/4521. Accessed 2012 September 13.
19. Weersing K, Toonen R (2009) Population genetics, larval dispersal, and connectivity in marine systems. Mar Ecol Prog Ser 393: 1–12. Available: http://www.int-res.com/abstracts/meps/v393/p1-12/. Accessed 2013 March 3.
20. Shanks AL (2009) Pelagic larval duration and dispersal distance revisited. Biol Bull 216: 373–385. Available: http://www.ncbi.nlm.nih.gov/pubmed/19556601.
21. Forward RB Jr, Tankersley RA, Welch JM (2003) Selective tidal-stream transport of the blue crab Callinectes sapidus: an overview. Bull Mar Sci 72: 347–365. Available: http://www.ingentaconnect.com/content/umrsmas/bullmar/2003/00000072/00000002/art00008. Accessed 2013 January 21.
22. DiBacco C, Sutton D, McConnico L (2001) Vertical migration behavior and horizontal distribution of brachyuran larvae in a low-inflow estuary: implications for bay-ocean exchange. Mar Ecol Prog Ser 217: 191–206. Available: http://www.int-res.com/abstracts/meps/v217/p191-206/. Accessed 2012 August 10.
23. Sokal RR (1978) Spatial autocorrelation in biology 2. Some biological implications and four applications of evolutionary and ecological interest. Biol J Linn Soc 10: 199–228.
24. Daigle RM, Metaxas A, de Young B (2014) Bay-scale patterns in the distribution, aggregation and spatial variability of larvae of benthic invertebrates. Mar Ecol Prog Ser 503: 139–156. Available: http://www.int-res.com/abstracts/meps/v503/p139-156/. Accessed 2014 April 29.
25. Lloyd MJ, Metaxas A, DeYoung B (2012) Patterns in vertical distribution and their potential effects on transport of larval benthic invertebrates in a shallow embayment. Mar Ecol Prog Ser 469: 37–52. Available: http://www.int-res.com/prepress/m09983.html. Accessed 2012 September 18.
26. Zar JH (1999) Biostatistical analysis. Upper Saddle River, NJ, Prentice Hall.
27. Legendre P, Legendre L (1998) Numerical ecology. 2nd Englis. Amsterdam: Elsevier. Available: http://books.google.com/books?hl = en&lr = &id = 4hwfTDGe5rgC&oi = fnd&pg = PP2&dq = Numerical +Ecology&ots = -1z_vXYFbx&sig = WMCYT6gySX54YrTcSPOoq0A1_tQ. Accessed 2012 August 20.
28. Rangel TF, Diniz-Filho JAF, Bini LM (2010) SAM: a comprehensive application for Spatial Analysis in Macroecology. Ecography (Cop) 33: 46–50. Available: http://doi.wiley.com/10.1111/j.1600-0587.2009.06299.x. Accessed 2012 July 13.
29. Oksanen J, Kindt FGBR, Legendre P, Minchin PR, O'Hara RB, et al. (2012) Community Ecology Package.
30. Team RC (2012) R: A language and environment for statistical computing. R Found Stat Comput. Available: http://www.r-project.org/.
31. Chia F-S, Buckland-Nicks J, Young CM (1984) Locomotion of marine invertebrate larvae: a review. Can J Zool 62: 1205–1222. Available: http://www.nrcresearchpress.com/doi/abs/10.1139/z84-176.
32. Shanks AL, Largier JL, Brink L, Brubaker J (2000) Demonstration of the onshore transport of larval invertebrates by the shoreward movement of an upwelling front. Limnol Oceanogr. Available: http://www.jstor.org/stable/10.2307/2670803. Accessed 2012 August 10.
33. Folt CL, Burns CW (1999) Biological drivers of zooplankton patchiness. Trends Ecol Evol 14: 300–305. Available: http://www.sciencedirect.com/science/article/pii/S016953479901616X. Accessed 2012 August 10.
34. Ryland JS (1977) Taxes and tropisms of bryozoans. In: Woollacott RM, Zimmer RL, editors. Biology of Bryozoans. New York: Academic Press. pp. 441–436.
35. Young CM (1995) Behavior and locomotion during the dispersal phase of larval life. In: McEdward L, editor. Ecology of Marine Invertebrate Larvae. CRC Press, Vol. 6. pp. 249–277.
36. Shanks AL (1995) Orientated swimming by megalopae of several eastern North Pacific crab species and its potential role in their onshore migration. J Exp Mar Bio Ecol 186: 1–16. Available: http://linkinghub.elsevier.com/retrieve/pii/0022098194001443.
37. Miller SH, Morgan SG (2013) Interspecific differences in depth preference: regulation of larval transport in an upwelling system. Mar Ecol Prog Ser 476: 301–306. Available: http://www.int-res.com/abstracts/meps/v476/p301-306/. Accessed 2014 May 25.
38. Daigle RM, Metaxas A (2011) Vertical distribution of marine invertebrate larvae in response to thermal stratification in the laboratory. J Exp Mar Bio Ecol 409: 89–98. Available: http://linkinghub.elsevier.com/retrieve/pii/S0022098111003789. Accessed 2012 July 13.
39. Tremblay MJ, Sinclair MM (1990) Sea scallop larvae Placopecten magellanicus on Georges Bank: Vertical distribution in relation to water column stratification and food. Mar Ecol Prog Ser 61: 1–15. Available: http://www.csa.com/partners/viewrecord.php?requester = gs&collection = ENV&recid = 2646540. Accessed 2012 September 12.
40. Gallager SM, Manuel JL, Manning DA, O'Dor RK (1996) Ontogenetic changes in the vertical distribution of giant scallop larvae, Placopecten magellanicus, in 9-m deep mesocosms as a function of light, food, and temperature stratification. Mar Biol 124: 679–692. Available: http://www.springerlink.com/index/10.1007/BF00351049.
41. Steele JH (1976) Patchiness. In: Cushing DH, Walsh JJ, editors. The Ecology of the Seas. Philadelphia: W. B. Saunders Co.
42. Okubo A, Levin SA (2001) Diffusion and ecological problems: modern perspectives. 2nd ed. Springer.
43. Epstein A, Beardsley R (2001) Flow-induced aggregation of plankton at a front: a 2-D Eulerian model study. Deep Sea Res Part II Top Stud Oceanogr 48: 395–418. Available: http://www.sciencedirect.com/science/article/pii/S0967064500000862. Accessed 2013 October 27.
44. Vázquez E, Ameneiro J, Putzeys S, Gordo C, Sangrà P (2007) Distribution of meroplankton communities in the Bransfield Strait, Antarctica. Mar Ecol Prog Ser 338: 119–129.
45. Mackas DL (1984) Spatial Autocorrelation of Plankton Community Composition in a Continental Shelf Ecosystem. Limnol Oceanogr 29: 451–471.
46. Shanks AL, Largier JL, Brink LA, Brubaker J, Hooff R (2002) Observations on the distribution of meroplankton during a downwelling event and associated intrusion of the Chesapeake Bay estuarine plume. J Plankton Res 24: 391–416. Available: http://plankt.oxfordjournals.org/content/24/4/391.short. Accessed 2012 September 13.
47. Okubo A (1977) Horizontal dispersion and critical scales for phytoplankton patches. In: Steele JH, editor. Spatial Pattern in Plankton Communities. New York: Plenum Press. pp. 21–42.
48. Martin AP (2003) Phytoplankton patchiness: the role of lateral stirring and mixing. Prog Oceanogr 57: 125–174. Available: http://linkinghub.elsevier.com/retrieve/pii/S0079661103000855. Accessed 2012 July 16.
49. Fuchs HL, Neubert MG, Mullineaux LS (2007) Effects of turbulence-mediated larval behavior on larval supply and settlement in tidal currents. Limnol Oceanogr 52: 1156–1165. Available: http://www.aslo.org/lo/toc/vol_52/issue_3/1156.html.
50. Manuel JL, O'Dor RK (1997) Vertical migration for horizontal transport while avoiding predators: I. A tidal/diel model. J Plankton Res 19: 1929–1947. Available: http://plankt.oxfordjournals.org/cgi/doi/10.1093/plankt/19.12.1929.
51. Hughes TP, Baird AH, Dinsdale EA, Moltschaniwskyj NA, Pratchett MS, et al. (2000) Supply-side ecology works both ways: the link between benthic adults, fecundity, and larval recruits. Ecology 81: 2241–2249. Available: http://www.esajournals.org/doi/abs/10.1890/0012-9658(2000)081[2241:SSEWBW]2.0.CO;2. Accessed 2012 August 10.
52. Grosberg RK, Levitan DR (1992) For adults only? Supply-side ecology and the history of larval biology. Trends Ecol Evol 7: 130–133. Available: http://www.ncbi.nlm.nih.gov/pubmed/21235979.

# 8

# Soil Nutrient Content Influences the Abundance of Soil Microbes but Not Plant Biomass at the Small-Scale

**Kadri Koorem[1*¤], Antonio Gazol[1], Maarja Öpik[1], Mari Moora[1], Ülle Saks[1], Annika Uibopuu[1], Virve Sõber[2], Martin Zobel[1]**

**1** Department of Botany, Institute of Ecology and Earth Sciences, University of Tartu, Tartu, Estonia, **2** Department of Zoology, Institute of Ecology and Earth Sciences, University of Tartu, Tartu, Estonia

## Abstract

Small-scale heterogeneity of abiotic and biotic factors is expected to play a crucial role in species coexistence. It is known that plants are able to concentrate their root biomass into areas with high nutrient content and also acquire nutrients via symbiotic microorganisms such as arbuscular mycorrhizal (AM) fungi. At the same time, little is known about the small-scale distribution of soil nutrients, microbes and plant biomass occurring in the same area. We examined small-scale temporal and spatial variation as well as covariation of soil nutrients, microbial biomass (using soil fatty acid biomarker content) and above- and belowground biomass of herbaceous plants in a natural herb-rich boreonemoral spruce forest. The abundance of AM fungi and bacteria decreased during the plant growing season while soil nutrient content rather increased. The abundance of all microbes studied also varied in space and was affected by soil nutrient content. In particular, the abundance of AM fungi was negatively related to soil phosphorus and positively influenced by soil nitrogen content. Neither shoot nor root biomass of herbaceous plants showed any significant relationship with variation in soil nutrient content or the abundance of soil microbes. Our study suggests that plants can compensate for low soil phosphorus concentration via interactions with soil microbes, most probably due to a more efficient symbiosis with AM fungi. This compensation results in relatively constant plant biomass despite variation in soil phosphorous content and in the abundance of AM fungi. Hence, it is crucial to consider both soil nutrient content and the abundance of soil microbes when exploring the mechanisms driving vegetation patterns.

**Editor:** Kurt O. Reinhart, USDA-ARS, United States of America

**Funding:** This research was funded by grants from the Estonian Science Foundation (9050, 9157), targeted financing (IUT 20-28), European Union's Seventh Framework Programme for research, technological development and demonstration under grant agreement no GA-2010-267243 – PLANT FELLOWS, ERMOS programme grant 14 (co-funded by Marie Curie Actions), and the European Regional Development Fund (Centre of Excellence FIBIR). The funders had no role in study design, data collection and analysis, decision to publish, or preparation of the manuscript.

* E-mail: kadri.koorem@ut.ee

¤ Current address: Department of Terrestrial Ecology, Netherlands Institute of Ecology, NIOO-KNAW, Wageningen, The Netherlands

## Introduction

Resource heterogeneity in space and time is a ubiquitous phenomenon in natural ecosystems and plays a fundamental role in determining plant community dynamics [1,2]. Small-scale resource heterogeneity can be especially important in determining plant productivity since plants can concentrate their root biomass in areas of high nutrient content [3–5]. However, plant growth can also be strongly influenced by soil microorganisms [6]. Symbiotic microorganisms, such as mycorrhizal fungi, contribute to plant nutrient uptake, thereby affecting plant growth and biomass allocation [7]. Similarly, bacteria play an important role in increasing nutrient availability for plants [6]. Despite increasing recognition of the roles played by soil nutrients and especially soil microbes in determining vegetation patterns, information about how soil nutrients, microbes and plant biomass covary in natural ecosystems remains limited.

The spatial distribution of soil microbes is heterogeneous at various scales [8,9] and can be closely related to soil fertility [7,10–12]. Less is known about how temporal variation in soil nutrient content [1] influences soil microbial communities, though the abundance of fungi and bacteria has been shown to display significant temporal variation, with maximum abundance in spring and autumn [13]. Indeed, the abundance of soil microbes is expected to be high in spring and autumn when nutrient availability is high and decline as the plant growth season proceeds and nutrient availability subsequently decreases [14]. However more evidence is needed to better understand these dynamics and the mechanisms driving them.

Studies have found significant small-scale patchiness in the abundance in AM fungi, ectomycorrhizal fungi, other fungi and bacteria [15–17] but knowledge about the simultaneous dynamics of different microbial groups remains limited. Both positive and negative associations in the abundance of co-occuring microbes have been reported: in saprotrophic fungi and bacteria [18] and AM fungi and bacteria [19,20]. Positive interactions may be the result of one group providing substrate for the other [19], while negative interactions may indicate competition for nutrients [20]. A recent study demonstrated that high N:P ratio in soil can provide an advantage, in the form of plant assimilated carbon, to

AM fungi, in competition with non-mycorrhizal fungi that also inhabit plant roots [21].

Mycorrhizal fungi contribute to plant nutrient acquisition from the soil [7] and their abundance has been found to decrease with increasing nutrient availability [10,11,22]. Hence, mycorrhizal fungi and other soil microbes are expected to play an important role in determining plant productivity in ecosystems with low soil nutrients [6]. However, it is unknown whether the effect of soil microbes on plant productivity remains dependent on soil nutrient content at small scales. A recent experimental study did not find any change in the effect of AM fungi on plant productivity in conditions where local soil fertility varied [23], but that study did not consider the effect of microbes besides AM fungi.

We simultaneously measured the temporal and spatial variation of soil nutrient content, soil microbial biomass and plant biomass in a natural herb-rich boreonemoral spruce forest. We focused on a scale that corresponds to the approximate size of adult plant individuals in the understory (i.e. small scale, 15 $cm^2$). The study site is characterized by heterogeneous light and soil conditions, patchy understory vegetation [24], and remarkably taxon rich AM fungal communities in the roots of herbaceous species [25,26]. We ask:

i) Does soil nutrient content and microbial abundance vary during the first half of the growing season when plant growth is rapid?

ii) Are there negative associations in the small-scale spatial distribution of soil microbial groups?

iii) Is the spatial distribution of soil microbes and plant biomass related to soil nutrient content?

## Materials and Methods

### Study site and sampling design

The study was carried out in a boreonemoral forest near Koeru, Estonia (58°58′N; 26°03′E). The climate of the study site is transitional between maritime and continental. The mean annual precipitation in the area is 665 mm and the mean annual air temperature is 5.0 °C (−5.4 °C in January and 16.6 °C in July on average; Estonian Meteorological and Hydrological Institute, 2013). The study site is a 130 ha patch of *Hepatica* type forest on a calcareous cambisol (cf. [24]). The soil of the area is relatively uniform in terms of pH, ranging from 5.2–5.7, but exhibits some variation in nutrient content [27,28]. The tree layer is dominated by *Picea abies* (L.) Karst. with individuals of *Fraxinus excelsior* L. and *Acer platanoides* L. also present. The shrub layer is dominated by *Corylus avellana* L. Common species in the herb layer are *Oxalis acetosella* L., *Hepatica nobilis* Schreb., *Paris quadrifolia* L. and *Viola mirabilis* L. The forest can be classified as relatively undisturbed old-growth forest, but clear cutting of approximately 1–2 ha patches has taken place repeatedly in parts of the study site. Detailed descriptions of the area are provided by Moora *et al.* [24] and Zobel *et al.* [27]. This study complies with the laws of Estonia in which no permits are required to carry out research on public land, unless specified otherwise in legislation. No regulation applies to the Koeru site and no protected species were sampled during the study.

Three 105×105 cm plots (A, B, C) were positioned at 5 m intervals in an area of homogenous forest in terms of vegetation cover and light conditions as estimated by eye. All plots were divided into 49 (15×15 cm) quadrats. During preliminary measurements conducted in summer 2007, five plant individuals of the five most common understory species (*Oxalis acetosella, Hepatica nobilis, Fragaria vesca* L., *Galeobdolon luteum* Huds., *Viola mirabilis*) were excavated to a depth of 15 cm, and the root systems of those plants measured with a ruler to determine the size of an average adult plant root system. Soil cores studied during the preliminary study harboured most of the roots in the upper 10 cm of the soil and the average size of plant roots systems was 15×15 cm (data not presented). The size of quadrats used in the main experiment therefore corresponds to the average size of the root system of an adult plant (the average of the species listed above), and is expected to harbour the majority of the biomass of herbaceous plant roots and microbes present in the rhizosphere. Plot A was used to address temporal variation, and plots A, B and C were used to address spatial variation in soil nutrient content and soil microbial activity at small scales. Plot A was sampled twice, at the beginning (26.05.2008) and in the middle (21.07.2008) of the growing season. The small spatial scale of the study meant that it was not possible to address temporal variation in the abundance of soil nutrients and microbes over a longer time period, as additional sampling would have necessitated considerable disturbance. Soil samples (50 g and 10 g) were collected from the soil surface (depth 10 cm) at both sampling times for soil chemical, and fatty acid analysis respectively. Soil samples collected at beginning of the plant growing season (in plot A only) were excavated with a soil corer (2 cm diameter and 10 cm deep) at the corner of each quadrat ($n = 49$) in order to minimize disturbance to the plots. The hole was filled with soil taken from nearby, to minimize sampling effects on soil conditions. Collected soil was mixed (to avoid any effect of vertical structure on measured parameters) and subjected to soil chemical and fatty acid analysis (see below). Soil samples collected in the middle of the growing season (July) were collected during destructive harvesting (see below). One core was taken from each quadrat; in plot A, the locations where the first samples were collected were avoided. The soil corer used during destructive harvesting had a diameter of 15 cm and a depth of 10 cm. All soil cores were transported to the laboratory where, after the removal of soil samples (50 and 10 grams as before) from all of the soil cores ($n = 147$), the shoots and roots of herbaceous plant species were separated carefully by hand from the samples from plots B and C ($n = 98$). Plant biomass was dried to a constant weight at 70 °C, and weighed. As the vegetation in plot A was disturbed by the sampling at the beginning of the growing season, biomass was not collected from that plot.

### Soil chemical and fatty acid analysis

Soil samples for chemical analyses were kept at −80°C prior to analysis, in order to prevent the loss of soil nutrients. Analysis of soil nutrient content followed International Organization for Standardization (ISO) protocols. Specifically, determination of soil ammonium ($NH_4$) and nitrate ($NO_3$) content (mg/kg) was based on ISO 14256–2: 2005. Organic carbon (C) and total nitrogen (N) content (%) were determined based on ISO 10694: 1995 and ISO 13878: 1998, respectively. As soil $NH_4$, $NO_3$ and C content were highly correlated with total N content, only the latter was used in further analysis. The Mehlich III procedure was used to determine soil phosphorus (P) and potassium (K) content (mg/kg): P was extracted by reaction with acetic acid and fluoride compounds; exchangeable K was extracted by the action of ammonium nitrate and nitric acid [29]. Chemical analyses were performed at the laboratory of the Agricultural Research Centre in Saku, Harjumaa, Estonia.

Ester-linked fatty acid (ELFA) content in soil samples was used as a proxy for the biomass of major microbial groups: AM fungi, other fungi and bacteria. Fatty acid profiling is widely used to measure microbial biomass and coarse microbial community composition [10,30–35]. The abundance of soil microbes can also

**Table 1.** Summary statistics of the Linear Mixed-Effect Models analysis fitted to study the influence of sampling time (May or July) on soil nutrient content and ester-linked fatty acid (ELFA) biomarkers of arbuscular mycorrhizal (AM) fungi, other fungi and bacteria measured from plot A.

| Variable | Num DF | Den DF | F | P |
|---|---|---|---|---|
| Nitrogen (%) | 1 | 45 | 3.43 | 0.07 |
| Phosphorus (mg/kg) | 1 | 45 | 0.15 | 0.70 |
| Potassium (mg/kg) | 1 | 45 | 5.20 | 0.03 |
| AM fungi (ELFA 16:1ω5c µg g−1 soil) | 1 | 45 | 4.87 | 0.03 |
| Other fungi (ELFA 18:2ω6,9 µg g−1 soil) | 1 | 45 | 1.16 | 0.29 |
| AM fungi: other fungi | 1 | 45 | 0.79 | 0.38 |
| Bacteria (ELFA µg g−1 soil) | 1 | 45 | 6.53 | 0.01 |

The degrees of freedom of the numerator (Num DF) and denominator (Den DF), F statistic (F) and associated probability (P) for the sampling time (i.e. fixed factor) are presented.

be estimated by quantifying their DNA by qPCR [35] but the advantage of fatty acid profiling is that it enables the biomass of different microbial groups to be determined simultaneously. Three grams of fresh soil (randomly taken from collected 10 g) was lyophilized immediately after separation and stored at room temperature. Phospholipids were converted to ester-linked fatty acids and extracted from the soil following the protocol of Schutter & Dick [36], which methylates only ester-linked and not free fatty acids. Briefly, 15 ml of 0.2 M KOH in methanol was added to 2 g of dry soil, mixed and incubated with shaking at 200 rpm at 37°C for 1 h. The sample was neutralized with 3 ml of 1 M acetic acid. Fatty acid methyl esters were extracted by adding 10 ml of hexane, the sample was mixed by vortexing and left to settle into two layers. The upper hexane layer was transferred to a clean tube and evaporated under a vacuum in a Jouan RC1022 Vacuum Concentrator. Fatty acids were re-suspended in 100 µl hexane and transferred to a GC vial for analysis using gas chromatography.

The ester-linked fatty acid peak areas were normalized against the internal standard and expressed as µg per g of dry soil. Fatty acid nomenclature follows Ranneklev & Bååth [37]. Ester-linked fatty acid 16:1ω5c was used as a biomarker for AM fungi. This marker is widely used as an indicator for AM fungi [30,32], though the different fatty acid fractions of cells may be of different indicative value for AM fungi. Namely, NLFA (neutral lipid fatty acids, storage lipids) 16:1ω5c occurs only in AM fungi, while PLFA (phospholipid fatty acids, structural lipids in cell membranes) 16:1ω5c can also occur in some Gram-negative bacteria [38,39]. NLFA is therefore proposed as a more suitable indicator of AM fungi in samples with potentially high-bacterial content, such as soils [39,40]. However, ELFA 16:1ω5c has been shown to correlate very well with AM fungal inoculation in an experiment containing both AM fungi and microbes from the AM fungal inocula, where ELFA 16:1ω5c was absent in microbial wash-treated, but AM fungal-free soils [41]. Therefore, ELFA measurements used in this paper represent whole-cell fatty acids, i.e. they contain both the PLFA and NLFA fraction. Fatty acid 18:2ω6,9 was used as a biomarker for fungi other than AM fungi, as it is a dominant fatty acid in most fungi [31] but is only found at very low levels in AM fungi [30]. We calculated the ratio between the levels of fatty acids 16:1ω5c and 18:2ω6,9 (hereafter AM fungi: other fungi ratio) to estimate the relative abundance of AM fungi in the soil fungal community. Fatty acids i15:0, a15:0, 15:0, i16:0, 16:1ω9c, br17:0, 10Me16:0, i17:0, a17:0, 17:0, cy17:0, br18:0, 10Me17:0, 18:1ω7c, 10Me18:0, cy19:0 ω 9, cy19:0 ω 7 were used as bacterial biomarkers [31,42].

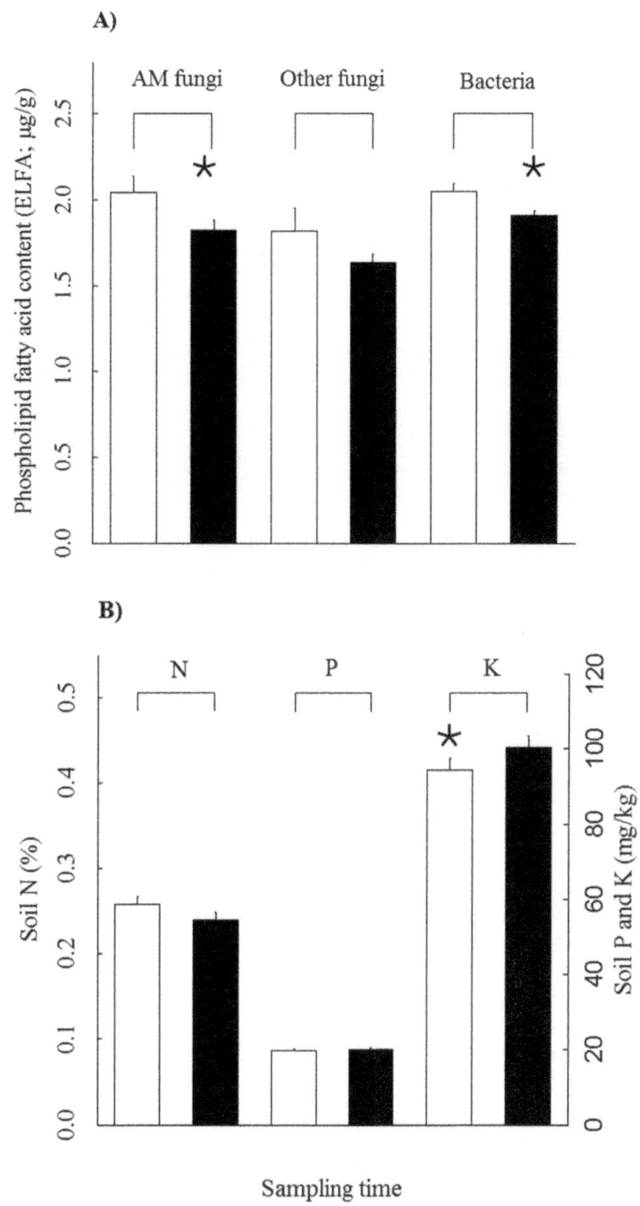


**Figure 1. The abundance of soil microbes (A) and nutrients (B) in May (open bars) and July (closed bars).** The abundance of arbuscular mycorrhizal (AM) fungi, other fungi and bacteria was estimated as the content of respective ester-linked fatty acid (y-axis on graph A, see Methods for details). Soil N concentration was measured in % (left y-axis of graph B) whereas P and K content were measured as mg/kg (right y-axis of graph B). Significant differences (Linear Mixed-Effect Models; $p<0.05$) in the measured parameters over time are marked with *.

## Statistical analyses

Linear Mixed-Effect Models (LMM) [43] were used to address temporal differences in soil nutrient content (N, P, K) and soil microbial biomass (fatty acid biomarkers for AM fungi, other fungi, AM fungi: other fungi ratio, bacteria). The abundance of other fungi and bacteria as well as the AM fungi: other fungi ratio

**Table 2.** Results of the modified t-Test for correlation performed to study the relationship between the abundance of arbuscular mycorrhizal (AM) fungi, other fungi and bacteria in the three plots (A, B, C).

| | | Other fungi | Bacteria |
|---|---|---|---|
| AM fungi | Plot A | 0.44** | 0.53** |
| | Plot B | 0.54** | 0.36* |
| | Plot C | 0.81** | 0.56** |
| Other fungi | Plot A | – | 0.41** |
| | Plot B | – | 0.54** |
| | Plot C | – | 0.55** |

For each plot, the Pearson's moment correlation coefficient is presented, accompanied by its significance corrected by the presence of spatial autocorrelation. * $P<0.05$; ** $P<0.01$.

was log transformed prior to analysis to meet the assumptions of parametric tests. For each variable, a model using time (sampling time: May or July) as a fixed factor and quadrat identity as a random factor were fitted. The homogeneity assumption for model residuals was evaluated graphically [44]. When any sign of heterogeneity was detected in the residuals, it was tested if keeping the variance fixed between the two time periods resulted in homogeneity for the residuals (i.e. including a fixed variance structure in the model; [44]). Models were fitted using Restricted Maximum Likelihood (REML) in order to compare models with different variance structures [44].

Moran's I was used to determine the presence of spatial autocorrelation in the distribution of soil microbes. Moran's I calculates the degree of correlation between observations as a function of the spatial distance separating them [45]. A correlogram is a graphical representation of the Moran's I statistic against distance classes, and provides information about the spatial autocorrelation of a variable and the size of patches. We constructed a spatial correlogram for each group of soil microbes (AM fungi, other fungi and bacteria) in each plot (A, B and C). Moran's I was evaluated at different distance classes (0 to 50 cm) and its significance was assessed by performing 9999 unrestricted permutations. Since the quadrats were distributed regularly, all correlograms were constructed using distance classes with increments of 0.075 m up to a maximum distance of 0.75 m. The first distance class (0.000–0.075) was discarded since it did not contain any pairs of quadrats.

The Modified t-Test for correlation [46] was used to study pairwise relationships between the abundance of soil microbes: AM fungi, other fungi and bacteria in each plot (A, B and C). This method evaluates the correlation between two spatial patterns using the Pearson's product-moment correlation, but corrects the significance of the correlation for the presence of spatial autocorrelation.

Generalized least square models (GLS) [43] were used to determine which factors best explain variations in soil microbial abundance (in plots A, B, C) and both above- and belowground biomass of herbaceous plants (in plots B, C). The set of predictor variables were different depending on the response variable considered. Specifically, in the models of soil microbial abundance (AM fungi, other fungi, bacteria and AM fungi: other fungi ratio), the influence of plot identity (A, B and C) and soil nutrients (N, P, K) were considered. In the models of above- and belowground biomass of herbaceous plants, the influence of plot identity, soil nutrients and microbial abundance were considered. In order to select the most parsimonious model for each variable, the protocol proposed by Zuur *et al.* [44] was followed. First, a model which included the full set of covariates as potential predictors was fitted for each variable. After that, the homogeneity of the residuals of the model was evaluated graphically. In addition, due to the nested structure of the sampling design, within plot spatial autocorrelation of residuals was studied. When spatial autocorrelation was present, the inclusion of correlation structures of different form were evaluated [47]. Finally, models were fitted with Maximum Likelihood (ML) and the best set of predictor variables was selected using Akaike's Information Criterion (AIC). If several models showed similar AIC values (difference lower than two), the most parsimonious one, i.e. the one with fewer fixed components, was chosen [44]. The final model was then refitted using REML to obtain estimates of factor effects and their significance.

LMM, GLS and modified t-Tests and correlograms were performed in the R environment [48], using the packages nlme [49], SpatialPack [50] and ncf [51]. Maps of the spatial patterns of AM fungal abundance and soil P content were created using ArcGIS 9.3 software (ESRI Redlands, NY, USA). For each variable, the "geostatistical analyst" extension of ArcGIS 9.3 was used to generate semivariograms of density and create kriging maps.

## Results

### Temporal variation

After the removal of outliers, data from 46 quadrats from plot A was subjected to analysis of temporal variation in soil nutrient content and soil microbial abundance. The linear mixed-effect models (LMMs) revealed significant temporal variation in the abundance of arbuscular mycorrhizal (AM) fungi, bacteria and soil K content, but not in the other variables (Table 1, Figure 1).

**Table 3.** Results of the generalized least square models performed to study the influence of environmental conditions on soil microbes.

| | Plot B | Plot C | Nitrogen | Phosphorus | Potassium | AIC weight | $R^2$ |
|---|---|---|---|---|---|---|---|
| AM fungi | 0.206* | 1.051** | 2.874* | −0.074** | – | 0.36 | 0.57 |
| Other fungi | 0.014 | 0.627** | – | – | 0.003* | 0.71 | 0.37 |
| AM: other fungi | 0.151** | 0.166** | – | -0.024** | – | 0.23 | 0.23 |
| Bacteria | 1.612** | 3.874** | – | – | 0.017* | 0.36 | 0.36 |

For each soil microbial group, the coefficient associated with the explanatory variable is presented. In addition, the AIC weight of the regression model (i.e. its importance as compared to other models containing a different subset of explanatory variables) and the $R^2$ are shown. * $P<0.05$; ** $P<0.01$.

**Table 4.** The mean, standard deviation (SD) and the range of the soil nutrient content, ester-linked fatty acid (ELFA) biomarkers of AM fungi, other fungi, bacteria and vegetation characteristics measured at 1 $m^2$ scale.

| Variable | Plot A | | Plot B | | Plot C | |
|---|---|---|---|---|---|---|
| | Mean (±SD) | Range | Mean (±SD) | Range | Mean (±SD) | range |
| Nitrogen (%) | $0.24 \pm 0.07^a$ | 0.15–0.43 | $0.27 \pm 0.05^b$ | 0.15–0.41 | $0.24 \pm 0.03^a$ | 0.18–0.30 |
| Phosphorus (mg/kg) | $20.11 \pm 3.02^a$ | 13.00–28.00 | $20.63 \pm 2.63^a$ | 15.00–30.00 | $14.16 \pm 1.64^b$ | 12.00–18.00 |
| Potassium (mg/kg) | $100.43 \pm 21.07^a$ | 57.00–159.00 | $92.80 \pm 22.58^a$ | 54.00–159.00 | $68.59 \pm 10.56^b$ | 47.00–93.00 |
| AM fungi (ELFA 16:1ω5c μg g-1 soil) | $1.83 \pm 0.42^a$ | 1.16–2.95 | $2.09 \pm 0.52^a$ | 1.22–3.07 | $3.32 \pm 0.81^b$ | 1.56–4.98 |
| Other fungi (ELFA 18:2ω6,9 μg g-1 soil) | $1.65 \pm 0.30^a$ | 1.02–2.36 | $1.64 \pm 0.35^a$ | 1.02–2.67 | $2.18 \pm 0.39^b$ | 1.49–2.88 |
| AM fungi: other fungi | $1.13 \pm 0.24^a$ | 0.59–1.67 | $1.29 \pm 0.24^b$ | 0.68–1.72 | $1.52 \pm 0.22^c$ | 1.04–1.93 |
| Bacteria (ELFA μg g-1 soil) | $19.14 \pm 1.61^a$ | 16.08–22.38 | $20.58 \pm 1.75^b$ | 16.20–25.50 | $22.54 \pm 2.30^c$ | 15.64–26.85 |
| Root biomass of woody species (g) | – | – | 10.30±6.76 | 2.30–37.00 | 8.92±5.53 | 1.80–28.10 |
| Shoot biomass of herbaceous species (g) | – | – | 0.81±0.42 | 0.16–2.06 | 0.79±0.36 | 0.04–1.53 |
| Root biomass of herbaceous species (g) | – | – | 0.57±0.63 | 0.03–3.07 | 0.64±0.53 | 0.03–1.88 |

Due to destructive sampling, vegetation characteristics were only measured in plots B and C (see explanation in Methods). Different letters (when present) mark a significant difference among means according to Tukey HSD test ($p<0.05$).

Means of the measured variables are given in Table S1. The abundance of AM fungal and bacterial fatty acid biomarkers in the soil was higher in May than in July. Soil K content was lower at the beginning of the growing season. The model fitted to explain temporal variation in the abundance of AM fungi included a fixed variance structure as it substantially improved the model fit.

## Spatial variation

After the outliers were removed 47, 46 and 44 quadrats from plots A, B and C respectively were subjected to analysis to test the spatial variation of soil nutrient content and the abundance of soil microbes. Limited evidence was found for the presence of spatial autocorrelation in the abundance of soil microbes studied in each plot, since only the abundance of bacteria showed significant autocorrelation (Figure S1). The abundance of AM fungi and soil bacteria tended to show positive spatial autocorrelation up to a distance of 20 cm, but this trend was significant only in the case of bacteria for plots A and C (Figure S1). The Modified t-Test for correlation revealed significant correlations between the abundance of AM fungi and other fungi, AM fungi and bacteria and between bacteria and other fungi (Table 2).

Generalized least square models revealed that plot identity was an important predictor of the abundance of AM fungi (Table 3); the mean abundance was significantly lower in plots A and B compared to plot C (Table 4). In addition, the abundance of AM fungi was significantly positively related to the content of soil N and negatively affected by soil P content (Table 3, Figure 2). Plot identity also played a significant role in predicting the abundance of other fungi (Table 3); the mean abundance was considerably higher in plot C compared to plots A and B (Table 4). In addition, the abundance of other fungi was significantly positively influenced by soil K (Table 3). The AM fungi: other fungi ratio was also predicted by plot identity (Table 3); the mean AM fungi: other fungi ratio differed significantly between all of the plots, being lowest in plot A and highest in plot C (Table 4). In addition, the AM fungi: other fungi ratio was negatively associated with soil P content (Table 3). The abundance of bacteria was the only variable that was best explained by a model where a spatial correlation structure was included, indicating small-scale autocorrelation in bacterial biomass. The abundance of bacteria was explained by plot identity (Table 3), as the mean differed significantly between all plots, being lowest in plot A and highest in plot C (Table 4). In addition, the abundance of bacteria was significantly positively influenced by soil K content (Table 3).

Of all the variables considered, the AM fungi: other fungi ratio was the only one with a significant and positive effect on the shoot biomass of herbaceous plants ($F = 5.29$; $P = 0.02$). However, the mean shoot biomass of herbaceous plants did not differ between plots (Table 4). None of the studied variables explained the root biomass of herbaceous plants, which also did not vary significantly between plots (Table 4).

## Discussion

The small-scale abundance of soil microbes varied through time in the studied boreonemoral herb-rich spruce forest. We recorded a lower abundance of soil microbes in the middle compared to the beginning of the growing season, whereas no decrease in soil nutrient content occurred. Contrary to our expectations, the small-scale spatial abundance of different soil microbes, including bacteria, arbuscular mycorrhizal (AM) fungi and other fungi, was positively related, suggesting little or no importance of resource competition among these groups in our study system. However, the abundance of soil microbes was significantly influenced by soil nutrient content. In particular, the abundance of AM fungi was strongly negatively related to soil P content. Plant biomass on the other hand showed little variation. Based on these results, we suggest that plants potentially compensate for low soil P content with alternative resource acquisition mechanisms, for example via symbiosis with AM fungi, and therefore achieve equal biomass production despite variations in soil nutrient content and the abundance of soil microbes. Soil microbes (especially plant

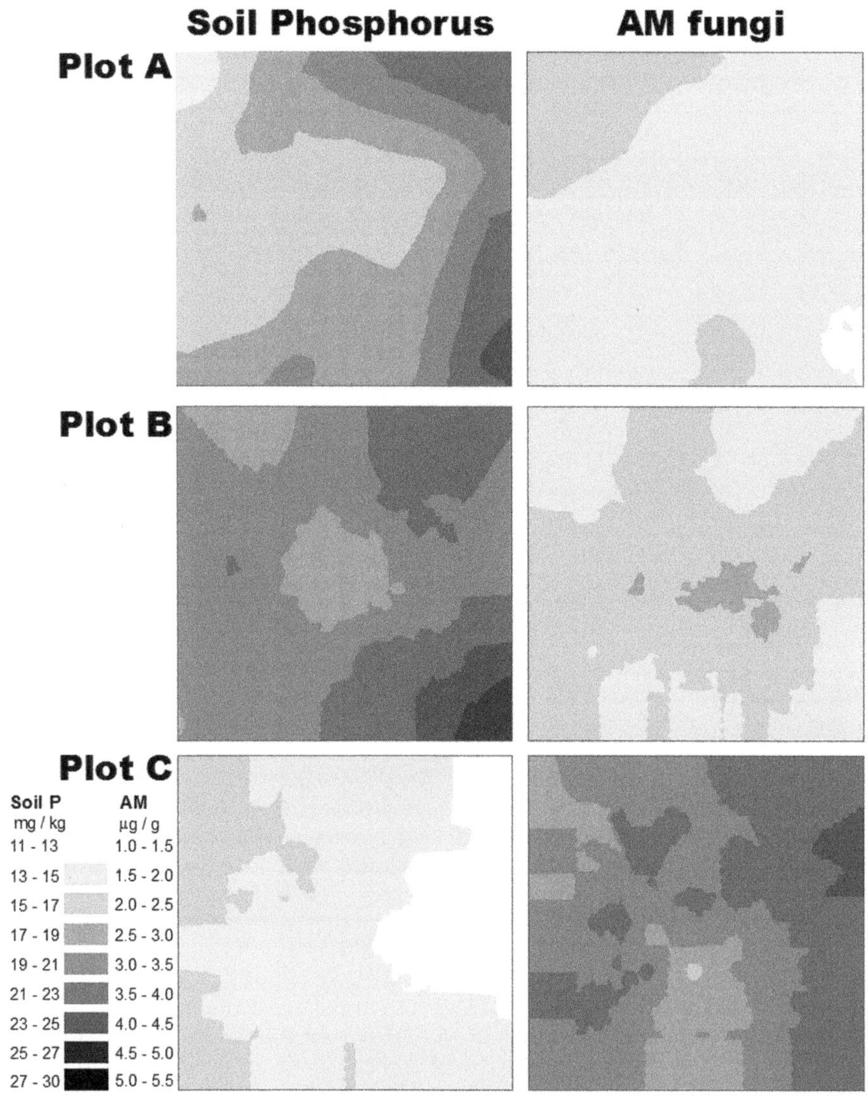


**Figure 2. Spatial pattern of soil phosphorus content and abundance of arbuscular mycorrhizal fungi (ester-linked fatty acid 16:1ω5c in soil) in plots A, B and C.** Maps were created by kriging, using data from soil samples collected from 15×15 cm quadrats in each 105×105 cm plot.

symbionts such as AM fungi) thus play a fundamental role in the relationship between soil nutrients and plant productivity.

The abundance of soil microbes (especially AM fungi and bacteria) was lower in the middle of the plant growing season, which matches the findings of previous studies that have addressed the temporal variation of soil microbes at larger and smaller spatial scales [13,52]. Active plant growth during the first half of the growing season did not bring about a decrease in soil nutrient content. Thus, the importance of soil nutrients in determining temporal changes in the abundance of soil microbes seem to be minor [52]. Some previous studies have reported a contrasting pattern of AM fungal abundance during the plant growing season: an initial increase with a peak in the summer, followed by a decrease in the autumn [53,54]. A potential explanation for this inconsistent variation of AM fungal abundance in soil can be the particular weather conditions of the sampling year, as soil AM fungal hyphal density is considerably affected by temperature and precipitation [54,55]. Thus, data from long-term measurements with multiple sampling times during the growing season is needed in order to draw solid conclusions about the dynamics of soil microbes.

We found a strong positive association between the abundance of bacteria, AM fungi and other fungi, suggesting a limited role of resource competition among these microbial groups in a natural ecosystem. Several previous experimental studies have reported negative associations between the abundance of ectomycorrhizal fungi, AM fungi and bacteria [20,56], as well as between the abundance of ectomycorrhizal and saprotrophic fungi [57], and AM fungi and other fungi [21,58]. On the other hand, studies conducted in natural conditions have demonstrated similar biomass dynamics of fungi other than AM fungi and bacteria during both the growing season [13] and over decades [59]. These contrasting results highlight the importance of studies addressing the pattern of soil microbes in natural environments. The strong positive relationship between the abundance of bacteria, AM fungi and other fungi found in this study suggests that the use of slightly different resources by soil microbes promotes their coexistence. Indeed, AM fungi, which are expected not to be highly saprotrophic themselves, have been shown to promote the activity of saprotrophic microbes [60]. In addition, fungal hyphae and spores can function as substrates for bacteria [19,61], and carbon released from dead bacteria can promote the growth of saprotrophic fungi [32]. Also, living roots excrete carbon compounds [62], which can be consumed by saprotrophic soil microorganisms [13]. In addition, AM fungi as well as ectomycorrhizal fungi have been shown to increase their growth in the presence of root exudates [63,64].

The abundance of all groups of soil microorganisms tested depended on the sampling location (plot identity), which confirms the small-scale spatial heterogeneity of soil microbes previously demonstrated [12]. In addition, soil nutrients played an important role in determining the abundance of soil microorganisms. The negative effect of soil P content on the abundance of AM fungi has also been reported in earlier field fertilization experiments in grasslands and forests [65,66]. As AM fungi can improve the acquisition of soil nutrients, especially P, by plants [7,67], they can play an important role when soil nutrient content is low. When soil nutrient levels increase, the cost to a plant of supporting AM fungi may outweigh the benefits, and can result in a decrease of AM fungal colonization of plant roots [7]. However, soil AM fungal fatty acid content has also been reported to be positively or insignificantly related to soil P content [68,69]. We recorded a positive effect of soil N on the abundance of AM fungi, a pattern that has also been registered at a large scale [35] and can be explained by the high N demand of AM fungi [70]. In addition, AM fungal communities have been shown to be highly influenced by soil pH [71]. However, Dumbrell *et al.* [71] focused on sites with high variation in soil pH (ranging from 4.5–8), whereas our study area is characterised by fairly uniform pH values [27,28]. Therefore, the role of soil nutrients in determining the abundance of AM fungi in current study site could be expected to be more pronounced, while the effect of small-scale variation in soil pH remains to be fully considered in future studies.

Saprotrophic and ectomycorrhizal fungi as well as bacteria were positively influenced by soil K content. A recent study demonstrated an increase in bacterial biomass, and growth of extramatrical mycelia of ectomycorrhizal fungi in response to K addition in peatland soils with severe K deficiency [72]. Our results suggest that K also has an important role to play in determining the abundance of soil microbes in soils where K content is not extremely low. On the other hand, the observed positive relationship can also reflect increased weathering and K influx resulting from active growth of bacteria and saprotrophic fungi [73,74].

Plants have been shown to concentrate their biomass into patches with high nutrient content [3–5], but in the current study, conducted in an old and stable ecosystem, we found no direct effect of nutrients on plant biomass. One explanation for the absence of the expected pattern is that plants are continuously foraging, while the pattern of biomass measured in studies such as this one reflects dynamics that have operated over numerous years. Greenhouse experiments have shown that nutrient poor patches contain higher plant biomass when next to nutrient rich patches [4,75] but those experiments were short-term. In the longer term, there will be natural turnover of nutrient rich patches (both temporal and spatial), and active foraging can therefore result in relatively uniform plant biomass. Secondly, the natural variation in nutrient content could be too low to generate distinct patterns. Indeed, understory plant biomass in this study area has not shown changes in response to induced variation in soil nutrient content either [23]. Thirdly, the effect of soil nutrients can be mitigated by the influence of soil microbes such as AM fungi. AM fungi seemed to be the only group of soil microbes to affect understory plant growth, as the relative increase of AM fungi compared to other fungi resulted in increased plant shoot biomass. As AM fungi can greatly improve plant P acquisition [7,67], the presence of AM fungi can be more important for plant growth when soil P content is low and less beneficial for plants when soil nutrient content is high. In accordance with this, a decrease of AM fungal hyphal density in soil and AM colonization in plant roots has been reported in response to increased P availability [22,76]; whereas the nutrient content of plant tissues and plant biomass did not exhibit any change [22,68]. Our study, based on data from a natural environment, demonstrates stable plant biomass in conditions of low abundance of AM fungi and high P content and *vice versa*. We support the suggestion of Beauregard *et al.* [68] that plants can compensate for low available nutrient concentration with more efficient symbiosis with soil microbes. Hence, it is crucial to consider both the soil nutrient content and the abundance of soil microbes when exploring the mechanisms driving vegetation patterns.

## Supporting Information

**Figure S1** Spatial correlograms of microbes. Spatial correlograms of arbuscular mycorrhizal fungi (a–c), other fungi (d–f), and bacteria (g–i) in plot A (a, d, g), B (b, e, h) and C (c, f, i). Distances with significant spatial autocorrelation are marked with solid circles.

**Table S1** Temporal variation in the abundance of soil nutrients and microbes. Summary statistics of the soil nutrient content and ester-linked fatty acid (ELFA) biomarkers of arbuscular mycorrhizal (AM) fungi, other fungi and bacteria measured from plot A at the beginning (May) and in the middle (July) of the growing season. The mean, standard deviation (SD) and the range are presented for each variable.

## Acknowledgments

We thank Sandra Caul at The James Hutton Institute, Dundee, UK, for help with fatty acid analyses. We are grateful to Jodi Price and John Davison for their comments on earlier drafts of this manuscript.

## Author Contributions

Conceived and designed the experiments: MZ MM MÖ ÜS AU KK. Performed the experiments: KK MÖ MM ÜS AU VS MZ. Analyzed the data: AG MZ MÖ MM KK. Contributed reagents/materials/analysis tools: AG MM MZ MÖ. Wrote the paper: KK AG MÖ MM ÜS AU VS MZ.

## References

1. Farley RA, Fitter AH (1999) Temporal and spatial variation in soil resources in a deciduous woodland. J. Ecol. 87: 688–696.
2. Pausas JG, Carreras J, Ferre A, Font X (2003) Coarse-scale plant species richness in relation to environmental heterogeneity. J. Veg. Sci. 14: 661–668.
3. Hutchings MJ, Wijesinghe DK (1997) Patchy habitats, division of labour and growth dividends in clonal plants. Trends Ecol. Evol. 12: 390–394.
4. Day KJ, Hutchings MJ, John EA (2003) The effects of spatial pattern of nutrient supply on yield, structure and mortality in plant populations. J. Ecol. 91: 541–553.
5. Hodge A (2004) The plastic plant: root responses to heterogeneous supplies of nutrients. New Phytol. 162: 9–24.

6. van der Heijden MGA, Bardgett RD, van Straalen NM (2008) The unseen majority: soil microbes as drivers of plant diversity and productivity in terrestrial ecosystems. Ecol. Lett. 11: 296–310.
7. Smith SE, Read DJ (2008) Mycorrhizal Symbiosis. Third edition. Academic Press, Amsterdam
8. Ettema CH, Wardle DA (2002) Spatial soil ecology. Trends Ecol. Evol. 17: 177–183.
9. Baldrian P, Vetrovsky T (2012) Scaling down the analysis of environmental processes: monitoring enzyme activity in natural substrates on a millimeter resolution scale. Appl. Environ. Microbiol. 78: 3473–3475.
10. Olsson PA, Bååth E, Jakobsen I (1997) Phosphorus effects on the mycelium and storage structures of an arbuscular mycorrhizal fungus as studied in the soil and roots by analysis of fatty acid signatures. Appl. Environ. Microbiol. 63: 3531–3538.
11. Nilsson LO, Bååth E, Falkengren-Grerup U, Wallander H (2007) Growth of ectomycorrhizal mycelia and composition of soil microbial communities in oak forest soils along a nitrogen deposition gradient. Oecologia 153: 375–384.
12. Franklin RB, Mills AL (2009) Importance of spatially structured environmental heterogeneity in controlling microbial community composition at small spatial scales in an agricultural field. Soil Biol. Biochem. 41: 1833–1840.
13. Kaiser C, Koranda M, Kitzler B, Fuchslueger L, Schnecker J, et al. (2010) Belowground carbon allocation by trees drives seasonal patterns of extracellular enzyme activities by altering microbial community composition in a beech forest soil. New Phytol. 187: 843–858.
14. Paul EA, Clark FE (1996) Soil Microbiology and Biochemistry. Second edition. Academic Press, San Diego.
15. Klironomos JN, Rillig MC, Allen MF (1999) Designing belowground field experiments with the help of semi-variance and power analyses. Appl. Soil Ecol. 12: 227–238.
16. Mummey DL, Rillig MC (2008) Spatial characterization of arbuscular mycorrhizal fungal molecular diversity at the submetre scale in a temperate grassland. FEMS Microbiol. Ecol. 64: 260–270.
17. Baldrian P, Merhautova V, Cajthaml T, Petrankova M, Snajdr J (2010) Small-scale distribution of extracellular enzymes, fungal, and bacterial biomass in *Quercus petraea* forest topsoil. Biol. Fertil. Soils 46: 717–726.
18. Rousk J, Bååth E (2011) Growth of saprotrophic fungi and bacteria in soil. FEMS Microbiol. Ecol. 78: 17–30.
19. Lecomte J, St-Arnaud M, Hijri M (2011) Isolation and identification of soil bacteria growing at the expense of arbuscular mycorrhizal fungi. FEMS Microbiol. Lett. 317: 43–51.
20. Leigh J, Fitter AH, Hodge A (2011) Growth and symbiotic effectiveness of an arbuscular mycorrhizal fungus in organic matter in competition with soil bacteria. FEMS Microbiol. Ecol. 76: 428–438.
21. Chagnon P-L, Bradley RL (2013) Evidence that soil nutrient stoichiometry controls the competitive abilities of arbuscular mycorrhizal *vs.* root-borne non-mycorrhizal fungi. Fungal Ecol. 6: 557–660.
22. McNaughton SJ, Oesterheld M (1990) Extramatrical mycorrhizal abundance and grass nutrition in a tropical grazing ecosystem, the Serengeti National Park, Tanzania. Oikos 59: 92–96.
23. Koorem K, Saks Ü, Sõber V, Uibopuu A, Öpik M, et al. (2012) Effects of arbuscular mycorrhiza on community composition and seedling recruitment in temperate forest understory. Basic Appl. Ecol. 13: 663–672.
24. Moora M, Daniell TJ, Kalle H, Liira J, Püssa K, et al. (2007) Spatial pattern and species richness of boreonemoral forest understorey and its determinants – a comparison of differently managed forests. For. Ecol. Manage. 250: 64–70.
25. Öpik M, Metsis M, Daniell TJ, Zobel M, Moora M (2009) Large-scale parallel 454 sequencing reveals host ecological group specificity of arbuscular mycorrhizal fungi in a boreonemoral forest. New Phytol. 184: 424–437.
26. Davison J, Öpik M, Daniell TJ, Moora M, Zobel M (2011) Arbuscular mycorrhizal fungal communities in plant roots are not random assemblages. FEMS Microbiol. Ecol. 78: 103–115.
27. Zobel M, Kalamees R, Püssa K, Roosaluste E, Moora M (2007) Soil seed bank and vegetation in mixed coniferous forest stands with different disturbance regimes. For. Ecol. Manage. 250: 71–76.
28. Koorem K, Moora M (2010) Positive association between understory species richness and a dominant shrub species (*Corylus avellana*) in a boreonemoral spruce forest. For. Ecol. Manage. 260: 1407–1413.
29. Sen Tran T, Simard RR (1993) Mehlich III-extractable elements. In: Carter MR, editor. Soil sampling and methods of analysis. pp. 43–48.
30. Graham JH, Hodge NC, Morton JB (1995) Fatty acid methyl ester profiles for characterization of Glomalean fungi and their endomycorrhizae. Appl. Environ. Microbiol. 61: 58–64.
31. Frostegård A, Bååth E (1996) The use of phospholipid fatty acid analysis to estimate bacterial and fungal biomass in soil. Biol. Fertil. Soils 22: 59–65.
32. Hinojosa MB, Carreira JA, Garcia-Ruiz R, Dick RP (2005) Microbial response to heavy metal-polluted soils: community analysis from phospholipid-linked fatty acids and ester-linked fatty acids extracts. J. Environ. Qual. 34: 1789–1800.
33. De Deyn GB, Quirk H, Bardgett RD (2011) Plant species richness, identity and productivity differentially influence key groups of microbes in grassland soils of contrasting fertility. Biol. Lett. 23: 75–78.
34. Schnoor TK, Martensson LM, Olsson PA (2011) Soil disturbance alters plant community composition and decreases mycorrhizal carbon allocation in a sandy grassland. Oecologia 167: 809–819.
35. Birkhofer K, Schöning I, Alt F, Herold N, Klarner B, et al. (2012) General relationships between abiotic soil properties and soil biota across spatial scales and different land-use types. PLoS ONE 7: e43292.
36. Schutter ME, Dick RP (2000) Comparison of fatty acid methyl ester (FAME) methods for characterizing microbial communities. Soil Sci. Soc. Am. J. 64: 1659–1668.
37. Ranneklev SB, Bååth E (2003) Use of phospholipid fatty acids to detect previous self-heating events in stored peat. Appl. Environ. Microbiol. 69: 3532–3539.
38. Nichols P, Stulp BK, Jones JG, White DC (1986) Comparison of fatty acid content and DNA homology of the filamentous gliding bacteria *Vitreoscilla*, *Flexibacter*, *Filibacter*. Arch. Microbiol. 146: 1–6.
39. Olsson PA (1999) Signature fatty acids provide tools for determination of the distribution and interactions of mycorrhizal fungi in soil. FEMS Microbiol. Ecol. 29: 303–310.
40. Frostegård A, Tunlid A, Bååth E (2011) Use and misuse of PLFA measurements in soils. Soil Biol. Biochem. 43: 1621–1625.
41. Walder F, Niemann H, Lehmann MF, Boller T, Wiemken A, et al. (2013) Tracking the carbon source of arbuscular mycorrhizal fungi colonizing C3 and C4 plants using carbon isotope ratios (delta 13C). Soil Biol. Biochem. 58: 341–344.
42. Zelles L (1999) Fatty acid patterns of phospholipids and lipopolysaccharides in the characterisation of microbial communities in soil: a review. Biol. Fertil. Soils 29: 111–129.
43. Pinheiro JC, Bates DM (2000) Mixed-Effect Models in S and S-PLUS. Springer.
44. Zuur AF, Ieno EN, Walker NJ, Saveliev AA, Smith GM (2009) Mixed effects models and extensions in ecology with R. Springer, New York.
45. Legendre P, Legendre L (1998) Numerical Ecology. Second English edition. Elsevir Science, Amsterdam.
46. Dutilleul P, Clifford P, Richardson S, Hemon D. (1993) Modifying t test for assessing the correlation between two spatial processes. Biometrics 49: 305–314.
47. Dormann CF, McPherson JM, Araujo MB, Bivand R, Bolliger J, et al. (2007) Methods to account for spatial autocorrelation in the analysis of species distributional data: a review. Ecography 30: 609–628.
48. R Development Core Team (2011) R: A language and environment for statistical computing. R Foundation for Statistical Computing, Vienna, Austria. ISBN 3-900051-07-0, URL http://www.R-project.org/ Accessed 2013 February 27
49. Pinheiro J, Bates D, DebRoy S, Sarkar D, the R Development Core Team (2011) Nlme: linear and nonlinear mixed effects models. R package version 3: 1–98.
50. Osorio F, Vallejos R, Cuevas F (2012) SpatialPack: package for analysis of spatial data.
51. Bjornstad ON (2012) Ncf: spatial nonparametric covariance functions. R package version 1.1–4.
52. Hamel C, Hanson K, Selles F, Cruz AF, Lemke R, et al. (2006) Seasonal and long-term resource-related variations in soil microbial communities in wheat-based rotations of the Canadian prairie. Soil Biol. Biochem. 38: 2104–2116.
53. Kabir Z, O'Halloran IP, Fyles JW, Hamel C (1998) Dynamics of the mycorrhizal symbiosis of corn (*Zea mays* L.): effects of host physiology, tillage practice and fertilization on spatial distribution of extra-radical mycorrhizal hyphae in the field. Agric. Ecosyst. Environ. 68: 151–163.
54. Staddon PL, Thompson K, Jakobsen I, Grime JP, Askew AP, et al. (2003) Mycorrhizal fungal abundance is affected by long-term climatic manipulations in the field. Glob. Chang. Biol. 9: 186–194.
55. Hu Y, Rillig MC, Xiang D, Hao Z, Chen B (2013) Changes of AM fungal abundance along environmental gradients in the arid and semi-arid grasslands of Northen China. PLoS ONE 8: e57593.
56. Olsson PA, Chalot M, Bååth E, Finlay RD, Söderström B (1996) Ectomycorrhizal mycelia reduce bacterial activity in a sandy soil. FEMS Microbiol. Ecol. 21: 77–86.
57. Pereira E, Coelho V, Tavares RM, Lino-Neto T, Baptista P (2012) Effect of competitive interactions between ectomycorrhizal and saprotrophic fungi on *Castanea sativa* performance. Mycorrhiza 22: 41–49.
58. Olsson PA, Francis R, Read DJ, Söderström B (1998) Growth of arbuscular mycorrhizal mycelium in calcareous dune sand and its interaction with other soil microorganisms as estimated by measurement of specific fatty acids. Plant Soil 201: 9–16.
59. Campbell CD, Cameron CM, Bastias BA, Chen CR, Cairney JWG (2008) Long term repeated burning in a wet sclerophyll forest reduces fungal and bacterial biomass and responses to carbon substrates. Soil Biol. Biochem. 40: 2246–2252.
60. Cheng L, Booker FL, Tu C, Burkey KO, Zhou L, et al. (2012) Arbuscular mycorrhizal fungi increase organic carbon decomposition under elevated $CO_2$. Science 337: 1084–1087.
61. Roesti D, Ineichen K, Braissant O, Redecker D, Wiemken A, et al. (2005) Bacteria associated with spores of the arbuscular mycorrhizal fungi *Glomus geosporum* and *Glomus constrictum*. Appl. Environ. Microbiol. 71: 6673–6679.
62. Jones DL, Hodge A, Kuzyakov Y (2004) Plant and mycorrhizal regulation of rhizodeposition. New Phytol. 163: 459–480.
63. Buee M, Rossignol M, Jauneau A, Ranjeva R, Becard G (2000) The pre-symbiotic growth of arbuscular mycorrhizal fungi is induced by a branching factor partially purified from plant root exudates. Mol. Plant Microbe Interact. 13: 693–698.

64. Kikuchi K, Matsushita N, Suzuki K, Hogetsu T (2007) Flavonoids induce germination of basidiospores of the ectomycorrhizal fungus *Suillus bovinus*. Mycorrhiza 17: 563–570.
65. Groffman PM, Fisk MC (2011) Phosphate additions have no effect on microbial biomass and activity in a northern hardwood forest. Soil Biol. Biochem. 43: 2441–2449.
66. Liu YJ, Shi GX, Mao L, Cheng G, Jiang SJ, et al. (2012) Direct and indirect influences of 8 yr of nitrogen and phosphorus fertilization on Glomeromycota in an alpine meadow ecosystem. New Phytol. 194: 523–535.
67. Smith SE, Smith FA, Jakobsen I (2003) Mycorrhizal fungi can dominate phosphate supply to plants irrespective of growth responses. Plant Physiol. 133:16–20.
68. Beauregard MS, Hamel C, Nayyar A, St-Arnaud M (2010) Long-term phosphorus fertilization impacts soil fungal and bacterial diversity but not AM fungal community in alfalfa. Microb. Ecol. 59: 379–389.
69. Thoms C, Gattinger A, Jacob M, Thomas FM, Gleixner G (2010) Direct and indirect effects of tree diversity drive soil microbial diversity in temperate deciduous forest. Soil Biol. Biochem. 42: 1558–1565.
70. Hodge A, Fitter AH (2010) Substantial nitrogen acquisition by arbuscular mycorrhizal fungi from organic material has implications for N cycling. Proc. Natl. Acad. Sci. U.S.A. 107: 13754–13759.
71. Dumbrell AJ, Nelson M, Helgason T, Dytham C, Fitter AH (2010) Relative roles of niche and neutral processes in structuring a soil microbial community. ISME J. 4: 337–345.
72. Potila H, Wallander H, Sarjala T (2009) Growth of ectomycorrhizal fungi in drained peatland forests with variable P and K availability. Plant Soil 316: 139–150.
73. Hoffland E, Giesler R, Jongmans AG, van Breemen N (2003) Feldspar tunneling by fungi along natural productivity gradients. Ecosystems 6: 739–746.
74. Calvaruso C, Turpault MP, Frey-Klett P (2006) Root-associated bacteria contribute to mineral weathering and to mineral nutrition in trees: a budgeting analysis. Appl. Environ. Microbiol. 72: 1258–1266.
75. Gazol A, Tamme R, Price JN, Hiiesalu I, Laanisto L, et al. (2013) A negative heterogeneity–diversity relationship found in experimental grassland communities. Oecologia 173: 545–555.
76. Nayyar A, Hamel C, Forge T, Selles F, Jefferson PG, et al. (2008) Arbuscular mycorrhizal fungi and nematodes are involved in negative feedback on a dual culture of alfalfa and Russian wildrye. Appl. Soil Ecol. 40: 30–60.

9

# Repeatability of Feather Mite Prevalence and Intensity in Passerine Birds

Javier Diaz-Real[1,2], David Serrano[1], Javier Pérez-Tris[3,4], Sofía Fernández-González[3], Ana Bermejo[4], Juan A. Calleja[4,5], Javier De la Puente[4], Diana De Palacio[4], José L. Martínez[4], Rubén Moreno-Opo[4], Carlos Ponce[4,6], Óscar Frías[6], José L. Tella[1], Anders P. Møller[7], Jordi Figuerola[1], Péter L. Pap[8,9], István Kovács[10], Csongor I. Vágási[8], Leandro Meléndez[11], Guillermo Blanco[6], Eduardo Aguilera[1], Juan Carlos Senar[12], Ismael Galván[1], Francisco Atiénzar[13], Emilio Barba[13], José L. Cantó[14], Verónica Cortés[13], Juan S. Monrós[13], Rubén Piculo[13], Matthias Vögeli[15], Antoni Borràs[12], Carlos Navarro[16], Alexandre Mestre[17], Roger Jovani[1]*

**1** Estación Biológica de Doñana (CSIC), Sevilla, Spain, **2** Departamento de Ecoloxía e Bioloxía Animal. Universidade de Vigo, Campus As Lagoas Marconsende, Vigo, Pontevedra, Spain, **3** Departamento de Zoología y Antropología Física. Universidad Complutense de Madrid, Madrid, Spain, **4** Grupo Ornitológico SEO-Monticola. Unidad de Zoología. Universidad Autónoma de Madrid, Madrid, Spain, **5** Departamento Biología Animal, Vegetal y Ecología, Universidad Autónoma de Barcelona, Bellaterra, Barcelona, Spain, **6** Department of Evolutionary Ecology, Museo Nacional de Ciencias Naturales (CSIC), Jose Gutiérrez Abascal, Madrid, Spain, **7** Laboratoire d'Ecologie, Systématique et Evolution, CNRS UMR 8079, Université Paris-Sud 11, Bâtiment 362, Orsay, France, **8** Evolutionary Ecology Group, Hungarian Department of Biology and Ecology, Babeş-Bolyai University, Cluj Napoca, Romania, **9** MTA-DE "Lendület" Behavioural Ecology Research Group, Department of Evolutionary Zoology and Human Biology, University of Debrecen, Debrecen, Hungary, **10** 'Milvus Group' Bird and Nature Protection Association, Târgu Mureş, Romania, **11** Unidad Mixta de Investigacion en Biodiversidad. Instituto Cantábrico de Biodiversidad (CSIC-Universidad de Oviedo), Oviedo, Spain, **12** Unidad Asociada de Ecología Evolutiva y del Comportamiento, Museo de Ciencias Naturales de Barcelona (CSIC), Barcelona, Spain, **13** Cavanilles Institute of Biodiversity and Evolutionary Ecology, University of Valencia, Paterna, Spain, **14** Parque Natural del Carrascal de la Font Roja, Alcoi, Spain, **15** Federal Office for the Environment FOEN, Species, Ecosystems, Landscape Division, Bern, Switzerland, **16** Profesor Gonzalo Sanchez Vazquez, Sevilla, Spain, **17** Department of Microbiology and Ecology, University of Valencia, Burjassot, Spain

## Abstract

Understanding why host species differ so much in symbiont loads and how this depends on ecological host and symbiont traits is a major issue in the ecology of symbiosis. A first step in this inquiry is to know whether observed differences among host species are species-specific traits or more related with host-symbiont environmental conditions. Here we analysed the repeatability (R) of the intensity and the prevalence of feather mites to partition within- and among-host species variance components. We compiled the largest dataset so far available: 119 Paleartic passerine bird species, 75,944 individual birds, ca. 1.8 million mites, seven countries, 23 study years. Several analyses and approaches were made to estimate R and adjusted repeatability ($R_{adj}$) after controlling for potential confounding factors (breeding period, weather, habitat, spatial autocorrelation and researcher identity). The prevalence of feather mites was moderately repeatable ($R = 0.26–0.53$; $R_{adj} = 0.32–0.57$); smaller values were found for intensity ($R = 0.19–0.30$; $R_{adj} = 0.18–0.30$). These moderate repeatabilities show that prevalence and intensity of feather mites differ among species, but also that the high variation within species leads to considerable overlap among bird species. Differences in the prevalence and intensity of feather mites within bird species were small among habitats, suggesting that local factors are playing a secondary role. However, effects of local climatic conditions were partially observed for intensity.

**Editor:** Csaba Moskát, Hungarian Academy of Sciences, Hungary

**Funding:** Funding was provided by the Spanish Ministerio de Economía y Competitividad (Ramón y Cajal research contract RYC-2009-03967 to RJ, CGL2012-38262 research project to JCS, CGL2010-15734/BOS to JPT, CGL2009-11445 to JF, CGL2011-24466 to RJ), Babeş-Bolyai Univeristy (post-doc grant GTC 34062/2013 to CIV), Junta de Andalucía (Excellence Projects RNM1274 and RNM4014 to MV, DS and JLT, RNM-7038 to JF), Romanian Ministry of Education and Research (research grant PN II. RU TE 291/2010 to PLP and CIV) and San Martín de la Vega town hall and SEO/BirdLife (Las Minas Ringing Station, Madrid, Spain). The funders had no role in study design, data collection and analysis, decision to publish, or preparation of the manuscript.

**Competing Interests:** The authors have declared that no competing interests exist.

* Email: jovani@ebd.csic.es

## Introduction

Why some organisms are abundant while others are rare and how large populations can grow are major questions in ecology. Symbiosis is the most abundant life style in nature [1], although the above questions are still poorly understood in the host-symbiont context. Host-symbiont systems represent an interesting study subject because two complementary approaches exist to explain commonness and rarity: either from the point of view of the host (i.e. why some host species harbour more symbionts than others) or the symbiont (i.e. why some symbiont species are more abundant than others). Under these two approaches much has been done to understand how species-specific and environmental factors affect symbiont abundance. Typically, comparative studies from the host's point of view have attempted to understand which species-specific traits of hosts (e.g. body size, food resources for symbionts) shape the abundance of symbionts (e.g. [2,3]). Host-focused intraspecific studies have shown how host individual features (e.g. age, sex, body condition) and environmental variables (e.g. weather, abundance of symbiont vectors) shape the abundance of symbionts among host individuals or populations within host species (e.g. [4–7]). The same approach has been used from the symbionts' point of view [8,9]. Note that in case of this approximation the symbiont's environment, especially for ecto-symbionts, is not only the host itself but the host's habitat as well.

Overall, there is an unresolved conflict between studies showing a huge (environmentally-governed) variability of symbionts within host species (i.e. between individuals or populations), and studies showing that there are species-specific features that can explain either why some host species have a higher abundance of symbionts than others, or why some symbionts are more abundant than others irrespective of the inhabited host. Surprisingly, little has been done to solve this apparent contradiction between the two approaches, by analysing the relative importance of species-specific vs. local environmental variables to understand why some symbiont species are common while others are rare in host-symbiont systems [10].

As far as we know, few studies have analysed (from the symbiont species or community point of view) symbiont abundance and prevalence repeatabilities when symbionts are occurring on different host species. Previous findings showed that the abundance and the intensity of infection are more repeatable than the prevalence [10–16]. Only two previous studies (to our knowledge) have focused on the repeatability of abundance, prevalence, and intensity of infection or richness of symbionts among host species (from the host point of view). They found that abundance and the prevalence of parasites were repeatable among host species, while the repeatability of prevalence was weaker than for abundance [17,18]. Moreover, a few comparative tests also showed that the studied variable was repeatable at the host species level before using that variable in comparative analyses (e.g. [15,19]). Therefore, current evidence shows that, while environmental factors shape host-symbiont interactions [20], there are species-specific traits of both hosts and symbionts that consistently shape the outcome of the interaction under environmental stochasticity [7]. Thus, this ultimately leads to higher similarity in symbiont population attributes within host species than between host species.

Here we investigate astigmata feather mites that live on the surface of the wing feathers of birds and are the commonest avian ectosymbionts [21–23]. The nature of the biological relationship between feather mites and birds is still poorly understood, and empirical studies show a puzzling scenario: some studies have shown that feather mite abundance correlates positively with bird's body condition [24,25], while others have found no significant correlation suggesting commensalism [26] or negative correlations and experimental evidence suggesting parasitism [27–29]. A recent correlative study analysing a large dataset from 83 species has shown a largely positive relationship with host condition though with a small effect size, which suggests a commensal interaction [30] agreeing with previous experimental results [31]. Another recent correlative study found that uropygial secretions and feather mites reduce hatching failure in birds by reducing bacteria loads in eggshells, showing a mutualist relationship [32]. Overall, this suggests a multifaceted complex of interactions resulting in conditional outcomes for both host and symbiont [33].

In this system, there is also an apparent conflict between intraspecific and interspecific comparative studies. The former show important differences between individuals (e.g. by age, sex, body size, body condition, [24–26]) and populations (e.g. by migratory status of the population, [34]), or along environmental gradients (e.g. salt concentration in the air, [6]). Multispecies comparisons have shown how mean abundance of feather mites correlate with other species-specific traits of birds such as migratory status, sociality, body size, plumage coloration or size of the uropygial gland [2,3,27,35,36]. However, it has never been tested whether feather mite intensity and prevalence could be considered a species-specific trait in birds. If abiotic environmental variables (but not bird species) were the main determinants of differences among bird species, previous studies would be (at least partially) indicating a hidden correlation between species features and the habitats where these bird species live.

To do so, we analysed for the first time repeatability of intensity and prevalence of feather mites on the flight feathers of birds using the largest dataset to date, which comprised 119 species of passerine birds occurring in distant localities from seven North African, European and Asian countries in the Northern hemisphere. We tested for the repeatability of intensity and prevalence of feather mites because they have different meanings and ecological implications. We also controlled for potentially confounding biological and methodological factors that could systematically bias our repeatability estimates.

## Material and Methods

### Ethics statement

The study was conducted under the current laws of the different countries where it was done. All sampling was conducted with permission from the local government at the sampling site and under the appropriate permits when required (refer to Information S1 for the license numbers). No endangered species were involved. All birds were studied with non-invasive methods and released at sampling locality some minutes after capture.

### Dataset

Data were obtained from "FeatherMites", a collaborative dataset on feather mite occurrence with data gathered by 89 researchers between 1989 and 2012 from seven countries (Fig. 1). See Information S1 for a summary of the dataset (dataset available under request).Birds were mostly captured using mist nests and kept individually in cloth bags until they were banded, inspected for feather mites and released. Feather mite occurrence was assessed exposing the wing against the sunlight and inspected from the dorsal and ventral surface of each primary, secondary and tertial feathers of one wing (the number of feather mites on both wings of a bird are highly repeatable; [37,38]). Moreover, for a subsample of birds with at least one feather mite on the wing, the total number of feather mites in one wing was also counted. In 853 individuals (<0.02% of the whole dataset) of Greenfinch (*Carduelis chloris*) the total number of feather mites was not counted, but it was estimated by linear regression from data on the number of primary feathers with more than five mites ($R^2 = 0.48$), or from counts of the number of mites on primaries only ($R^2 = 0.87$). In all analyses, to avoid pseudo-replication, only the first observation of each individual bird was used in the analyses.

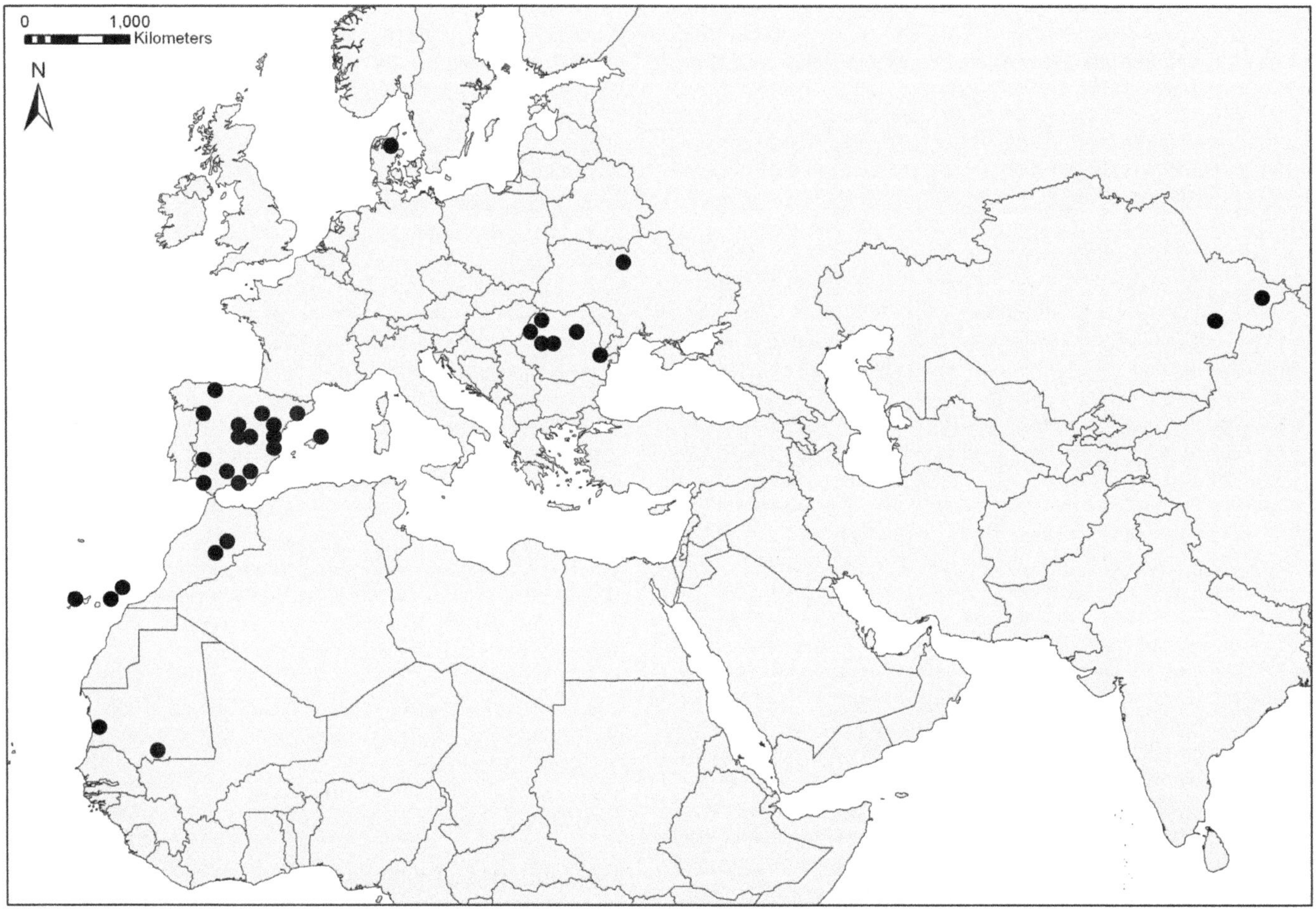


**Figure 1. Study area and spatial distribution of data.** Close sampling points are summarised by black circles. Countries (sampled bird individuals): Mauritania (85), Morocco (105), Spain (70,321), Denmark (2,394), Romania (1,827), Ukraine (1,175) and Kazakhstan (37).

The dataset was analysed separately for the prevalence and the intensity of feather mites. For the analysis of prevalence, we used data on the occurrence of feather mites (whether the individual bird had one or more feather mites on the wing, 1, or not, 0). For the analysis of the intensity of feather mites, we used data of the number of feather mites on the wing of birds with at least one mite. Only bird species with at least 25 individuals were used to avoid model convergence issues during prevalence analyses. After applying these constraints the final sample size for the prevalence dataset comprised 75,944 individuals from 97 species of passerine birds (from 57 genera and 27 families) inspected for feather mites. The intensity database contained 27,457 individuals from 119 species belonging to 62 genera and 28 families. We analysed the distribution of variance in intensity and prevalence across taxonomic levels in a nested analysis of variance for species, genus and family [36,39]. Species accounted for the highest proportion of total variance (see Results) so repeatability analyses were centred to the species level.

## Statistical analyses

**Repeatability.** Repeatability (the intra-class correlation coefficient) is the proportion of variation that can be attributed to between-group differences [40]; in our case study, between bird species. Here we used this statistic to test whether prevalence and intensity of feather mites could be considered bird species-specific traits; i.e. traits that are more variable between than within bird species.

Repeatability of feather mite prevalence was calculated as $R_{logitM} = \sigma_\alpha^2/(\sigma_\alpha^2+\sigma_\varepsilon^2)$, where $\sigma_\alpha^2$ is the between-groups variance (e.g. species) and $\sigma_\varepsilon^2$ is the residual variance (i.e. between individuals within species). Residual variance was calculated as $\sigma_\varepsilon^2 = \omega \times (\pi^2/3)$, where $\omega$ is the multiplicative overdispersion parameter, and $\pi^2/3$ is the distribution-specific variance for the logit model. Because prevalence is a proportion variable, these variances were obtained by fitting Generalized Linear Mixed-effects Models (GLMMs) with binomial distribution of errors and logit link function in which bird species identity was included as a random effect. In general, penalized-quasi likelihood (PQL) estimation was used with multiplicative overdispersion [40] except in two analyses, where additive overdispersion was used due to convergence issues. In any case, R values from additive and multiplicative models for binary data have been found to be similar in the presence of overdispersion [40]. We used parametric bootstrapping to obtain 95% confidence intervals (CI; nboot = 1000, npermut = 1000).

Feather mite intensity data were $\log_{10}$-transformed and R was calculated as $R = \sigma_\alpha^2/(\sigma_\alpha^2+\sigma_\varepsilon^2)$. Linear mixed-effects models (LMEs with normal error distribution and identity link function) with bird species identity as a random variable were used to retrieve $\sigma_\alpha^2$ and $\sigma_\varepsilon^2$. Restricted maximum likelihood (REML) was used for parameter estimation because of its flexibility to control for confounding factors ([40]; see below). Parametric bootstrapping was used, as detailed above, to obtain 95% CI.

**Adjusted repeatability.** Adjusted repeatability ($R_{adj}$) is the repeatability after statistically controlling for confounding effects [40]. Confounding effects are biological and methodological factors that can potentially systematically bias the intensity or prevalence of feather mites, thus artificially either increasing or decreasing repeatability estimates. Thus, $R_{adj}$ is closer to real repeatability after known biases are taken into account. The following confounding factors were considered:

*Observer.* — Some researchers could produce consistently higher feather mite counts or prevalence estimates than others. This, for instance, could lead to higher within-species variance when clumping data from different observers, thus reducing R estimates. Alternatively, species sampled only by one or few observers could misleadingly increase assessed feather mite load differences among species (i.e. increases $\sigma^2_\alpha$); then leading to R overestimation. To correct for this potential observer bias we included observer identity (N = 89) as a random factor ("observer") in statistical analyses.

*Breeding period.* — Feather mites live permanently on birds, and the main dispersal mode is thought to be from bird to bird when birds are in contact [23]. Thus, the main transmission moment seems to be in the nest from parents to offspring [23]; but also between related and unrelated conspecifics in communal roosts [24,25]. Therefore, it is not surprising that feather mite intensity and prevalence vary during the year, with birds having the lowest intensities of feather mites during the breeding season when the number of feather mites per bird dilutes, and feather mite populations in each individual have not started to recover [3,24,35,41,42]. We thus included the categorical variable "breeding period" with two levels, breeding and non-breeding period, as a fixed effect in statistical analyses. The year was divided in two equal periods of six months starting at egg laying based on breeding phenology for each species [43].

*Local climatic conditions.* — As ectosymbionts are ectotherms, feather mites are not only exposed to factors directly governed by the bird host, but also to abiotic environmental factors such as precipitation and temperature. Accordingly, feather mites are known to move within their hosts to meet their favourable conditions [24,44]. In our study, we analysed the possible differences between localities in terms of environmental factors by two different approaches: weather description and habitat classification. For weather description, we used six different climatic variables: annual mean temperature, mean temperature of the warmest quarter of the year, mean temperature of the coldest quarter, and the same 3 variables for precipitation. Data were obtained from BIOCLIM (http://www.worldclim.org) for each locality. All six variables were highly correlated, and a principal component analysis (PCA) with varimax rotation was used to summarise weather information. The PCA was carried out using "prcomp" function from the "stats" package in R [45] with default settings. The first axis of the PCA (PC1) accounted for 84% of the variance, positioning localities along an axis from dry and warm to wet and cold. Thus, we included the continuous variable "PC1" as a fixed effect in the statistical models.

*Spatial autocorrelation.* — A spatial term of the form $x+y+x^2+xy+y^2+x^3+x^2y+xy^2+y^3$ [46] was included to control for spatial autocorrelation as fixed effect in statistical analyses, where $x$ and $y$ are longitude and latitude coordinates of the sampling sites, respectively. Prior to the analyses, coordinates were centred on their respective means to reduce collinearity with higher order terms [45] and standardized to unit variance.

*Habitat.* — Different habitats could provide different environmental conditions for feather mites thus shaping their populations. The habitat of each study locality was classified as Atlantic forest, crops, Mediterranean and continental forest, river forest, steppe, subalpine meadow, or wetland and included as fixed effect in statistical models.

**Complementary analyses.** To test for consistency in our results, complementary analyses were performed. R and $R_{adj}$ were also obtained separately for the datasets from researchers with at least 25 species sampled at least in two different habitats: five researchers with an N range = 1,256–3,217 birds sampled for intensity and seven researchers with N range = 2,024–3,454 for prevalence. Moreover, given that the habitat consistently entered as confounding factor in the models, we explored whether species had a consistent prevalence and intensity independent of the habitat where they were captured. To do so, we analysed a subsample of ten well-sampled resident bird species (N>25 individuals of each species in each habitat, N = 28,340 birds for prevalence and N = 7,136 for intensity) captured in three habitats (wetland, Mediterranean and continental forest, and river forest).

The rptR package [47] for software R [45] was used to calculate R and its 95% CI for intensity and prevalence. It was also used to calculate $R_{adj}$ of feather mite intensity and its 95% CI. However, binomial errors are not implemented in this package for $R_{adj}$ of prevalence (not elsewhere for $R_{adj}$, H. Schielzeth pers. comm.). Thus, we used SAS 9.2 software (SAS Institute, Cary, NC, USA) for estimating $R_{adj}$ from $\sigma^2_\alpha$ and $\sigma^2_\varepsilon$ estimates retrieved with SOLUTION statement, and $\omega$ with the RANDOM statement (indicating "_residual_") in the GLIMMIX procedure. Significance of $R_{adj}$ estimates for prevalence was calculated comparing twice the difference in log-likelihoods of models with and without the random effect, against the $\chi^2$ distribution with one degree of freedom [40]. We used backward stepwise GLMMs and LMEs starting from the saturated model to assess the effect of each of the explanatory variables on prevalence and intensity (variables with $p>0.05$ were excluded). Thus, only statistical significant variables were used in $R_{adj}$ estimations.

## Results

### Prevalence

Mite prevalence was very variable among species (Fig. 2), ranging from 0% in *Sylvia hortensis* up to 100% in *Rhodospiza obsoleta* (see Table S1 for detailed results for each species). Analysing the whole dataset we found a moderately high repeatability of prevalence among species (R = 0.494). This result, as well as all other repeatability estimates reported in this study was statistically significant ($p<0.001$). Prevalence was higher during the non-breeding season, differed among habitats, and showed spatial autocorrelation (Table 1). After controlling for these factors we found an $R_{adj} = 0.409$, showing that while several variables explained part of the among-species variation, differences in prevalence of feather mites were consistent at the species level (Fig. 3).

**Different observers.** The same analyses were repeated separately for datasets from the researchers contributing the highest sample sizes. Similar results were found as when analysing the whole dataset (R range: 0.325–0.531; $R_{adj}$ range: 0.367–0.571; see Fig. 3 for the confounding effects retained in the analyses of each subsample).

**Different habitats.** Well-sampled resident bird species were quite consistent in their prevalence of mites independently of the habitat where they were sampled (Fig. 4). Repeatability calculated

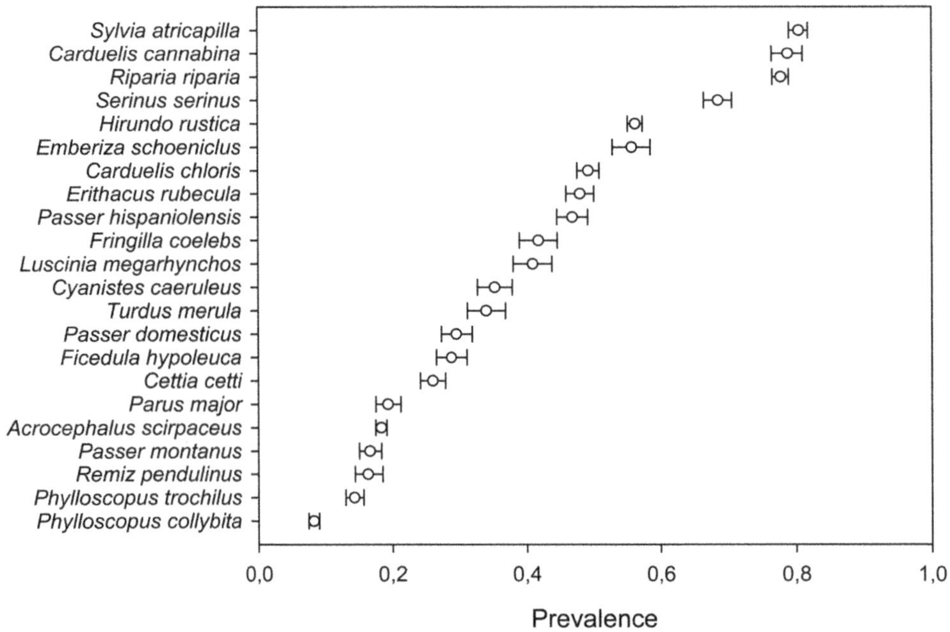


**Figure 2. Prevalence of feather mites of birds.** Feather mite prevalence (proportion of birds with feather mites) in bird species with data for more than 1,000 individual birds with corresponding 95% confidence interval estimates.

from this subsample led to a lower repeatability estimate ($R = 0.255$, $R_{adj} = 0.324$; see Fig. 4 for retained confounding variables).

**Taxonomic relationships.** To explore the distribution of variance in intensity and prevalence across taxonomic levels we ran a GLMM including a nested random effect of species, genus,

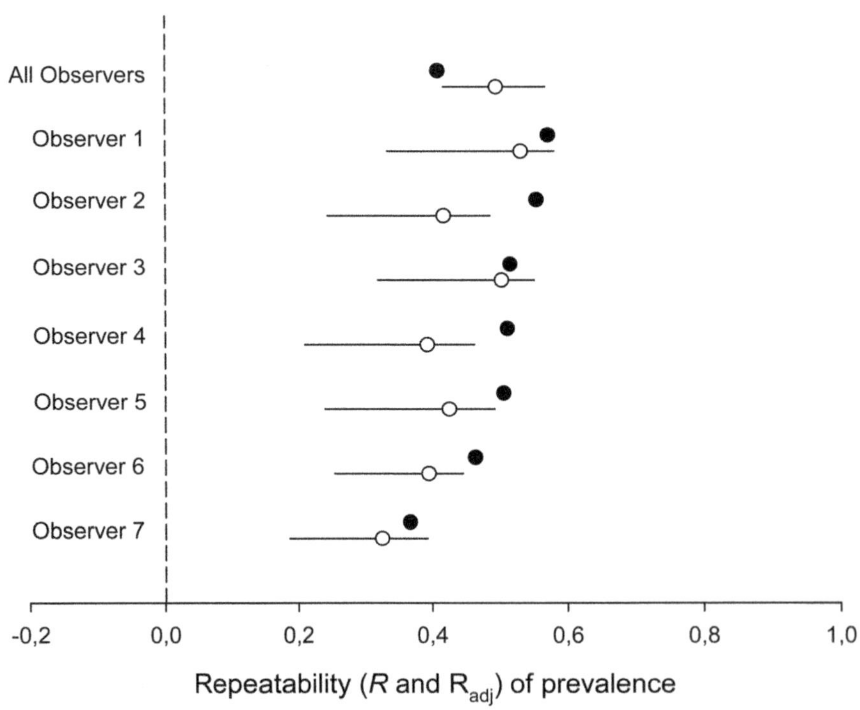


**Figure 3. Species repeatability and adjusted repeatability of feather mite prevalence.** Species repeatability (R; white circles) and adjusted repeatability ($R_{adj}$; black circles) for feather mite prevalence with corresponding 95% confidence interval estimates for different intensity data subsets (either the entire dataset, "All observers", or for data from seven different researchers separately). The following confounding effects were retained in the models: "breeding" was retained for observers 1–7; "habitat" was also retained for observers 2, 5 and 6; the "PC1" was also added for observers 1 and 4; finally, six variables of the spatial autocorrelation term were retained for observer 4. For "All Observers" the final model included the fixed effects shown in Table 1. All R and $R_{adj}$ estimates were statistically significant at $\alpha = 0.001$. 95% CI could not be calculated for $R_{adj}$ (see Methods).

**Table 1.** Parameter estimates for GLMM of the dataset of prevalence of feather mites on birds.

| Factor | Description | Estimate | SE | df | t | F | P |
|---|---|---|---|---|---|---|---|
| Intercept | | 0.681 | 0.527 | 88 | 1.29 | | 0.1996 |
| Breeding period | | | | 1, 75747 | | 1088.46 | <.0001 |
| | Non breeding | 0.916 | 0.028 | 75747 | 32.99 | | <.0001 |
| | Breeding | 0 | . | . | . | | . |
| Habitat | | | | 6, 7532 | | 18.20 | <.0001 |
| | Atlantic forest | −1.583 | 0.514 | 75747 | −3.08 | | 0.0021 |
| | Crops | −1.809 | 0.497 | 75747 | −3.64 | | 0.0003 |
| | Steppe | −1.450 | 0.511 | 75747 | −2.84 | | 0.0045 |
| | Wetland | −1.785 | 0.499 | 75747 | −3.58 | | 0.0003 |
| | M. and C. forest | −1.972 | 0.499 | 75747 | −3.95 | | <.0001 |
| | River forest | −1.585 | 0.499 | 75747 | −3.18 | | 0.0015 |
| | Sub. meadow | 0 | . | . | . | | . |
| Autocorrelation | | | | | | | |
| | $y$ | 0.416 | 0.052 | 75747 | 8.06 | | <.0001 |
| | $xy$ | 0.658 | 0.055 | 75747 | 11.88 | | <.0001 |
| | $y^2$ | −0.217 | 0.033 | 75747 | −6.50 | | <.0001 |
| | $x^3$ | 0.008 | 0.001 | 75747 | 6.45 | | <.0001 |
| | $x^2y$ | −0.136 | 0.015 | 75747 | −8.93 | | <.0001 |

For continuous variables, only statistically significant variables are shown.

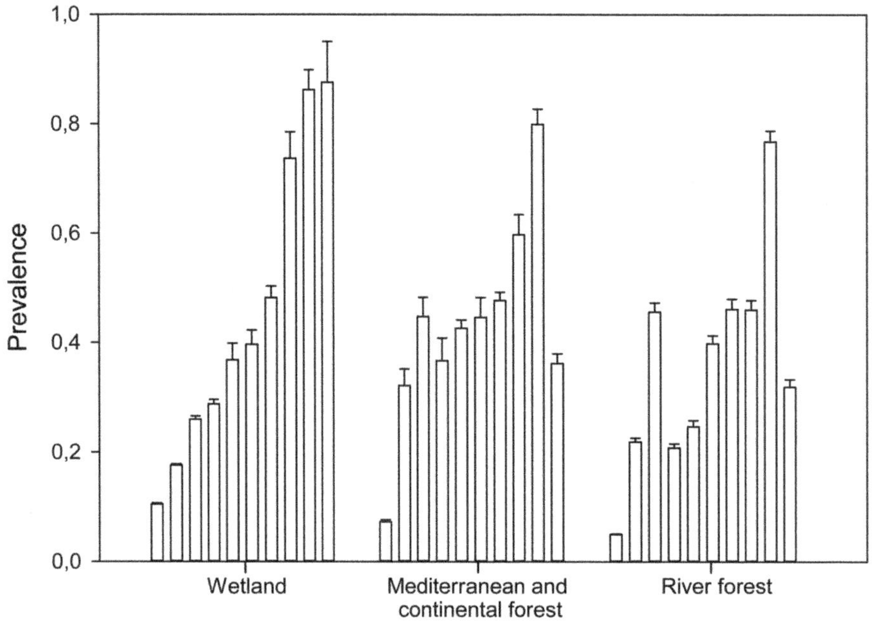


**Figure 4. Feather mite prevalence (proportion of birds with feather mites) for ten species of well-sampled resident passerines in three habitats.** Species are ordered from left to right within each habitat according to their prevalences in wetlands as follows: *Phylloscopus collybita, Acrocephalus scirpaceus, Carduelis chloris, Cettia cetti, Cyanistes caeruleus, Luscinia megarhynchos, Erithacus rubecula, Serinus serinus, Sylvia atricapilla* and *Fringilla coelebs*. "Breeding" and "habitat" variables were retained as fixed factors in the GLMM, while "observer" and "species" were included as random factors.

and family for the whole dataset. Species accounted for 49.7% of total variance on feather mite prevalence, while less was explained by family (21.9%) and very little by genus (0.02%).

## Intensity

Mite intensity varied among individuals by four orders of magnitude (range = 1–10,000 mites), and by two orders of magnitude between species: median range = from 1, e.g. *Oriolus oriolus* and *Petronia petronia*, to 486 mites in *Acrocephalus melanopogon* (Fig. 5; see Table S2 for detailed results for each species). We found a moderate low repeatability for intensity ($R = 0.253$). Small R could be due to either small between-group (i.e. species) variance or large within-group residual variance (i.e. between individuals of a species). Provided that the between-species variance was quite large, the relatively small R shows that within species variation was also considerable (Fig. 5). Intensity was higher during the non-breeding season and differed among habitats and latitude (Table 2). Moreover, a significant effect of climatic conditions as reflected by PC1 of the PCA analysis was observed, showing that drier and warmer localities held higher feather mite intensities. Controlling for all these confounding variables, the adjusted repeatability was similar to the R estimate ($R_{adj} = 0.208$, Fig. 6).

**Different observers.** Similar results were found when analysing data separately for five researchers with the greater contribution to the dataset (R range: 0.190–0.304, $R_{adj}$ range: 0.184–0.303; see Fig.6 for the confounding effects retained in the analysis of each subsample).

**Different habitats.** Again, we analysed a subset of ten well-sampled resident species in three different habitats. Species were consistent in their intensity independently of the habitat where they were found (Fig. 7), leading to $R = 0.266$ and $R_{adj} = 0.226$ (see Fig. 7 for retained confounding variables).

**Taxonomic relationships.** Following the scheme for prevalence, we estimated the proportion of total variance among taxonomical levels for feather mite intensity. Species accounted for 20.3% of the variance, while very little was explained by family (0.05%) and nothing by genus (0%).

## Discussion

Despite considerable progress in our understanding of the relationship between host characteristics and feather mite burdens, there was no information on within vs. among bird species variation in prevalence and intensity of mites. Filling this gap is important for knowing at which level (intraspecific or interspecific) future research efforts could be more productive. Feather mite population sizes were highly variable within bird species, leading to considerable overlap between species (e.g. Fig. 5). However, prevalence showed a moderate ($R = 0.494$) and intensity a lower repeatability ($R = 0.208$). Interestingly, while several factors potentially biased repeatability estimates, adjusted repeatabilities showed similar results as unadjusted repeatabilities (prevalence: $R_{adj} = 0.409$; intensity: $R_{adj} = 0.253$). Overall, thus, bird species consistently differed in their prevalence and intensity of feather mites, while there was a high within-species variance and many species showed similar (average) values. In this regard, differences in prevalence were consistent among species, but high within-species variation was also evident suggesting that analyses at both levels are required. For intensity, however, evidence was much stronger for within-species than for among-species differences. Hence, species-specific approaches could probably be more fruitful for determining causes underlying variation, at least from the host's point of view. Repeatabilities estimated for a sample of researchers were similar, although different variables were retained in each analysis for each observer (likely because researchers differed in species and habitats sampled).

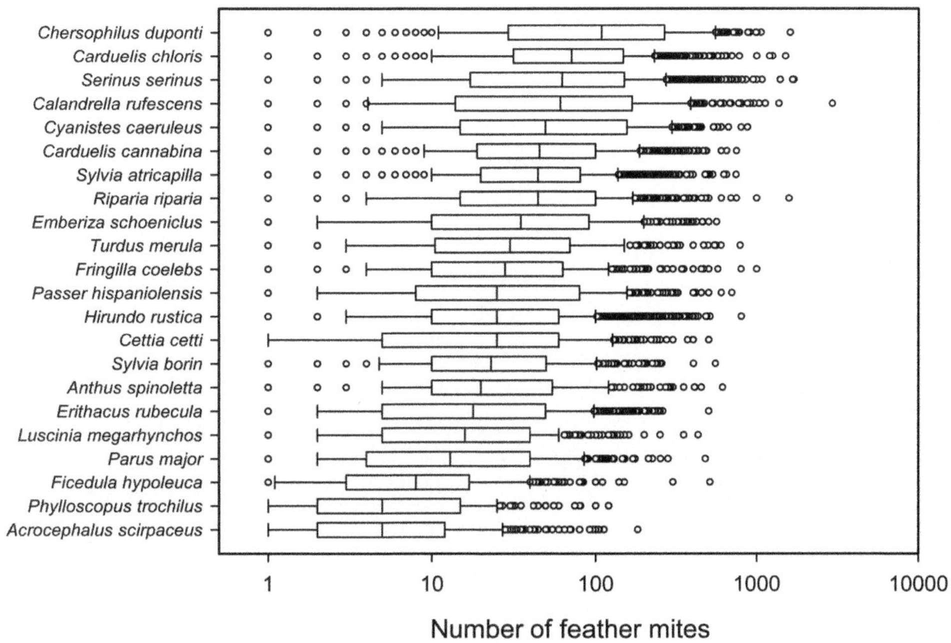


**Figure 5. Feather mite intensity of birds.** Box-plot of feather mite intensity (on $\log_{10}$ axis) for species with more than 300 records. Species are ordered according to their median intensity.

Different reasons could be behind the higher repeatability found for prevalence than for intensity. Prevalence could be more linked to feather mite transmission capacity between habitat islands (bird individuals) of the same island type (host species). For instance, some bird (e.g. breeding sociality) or mite-related species-specific traits (e.g. transmission propensity) could drive the differences in feather mite prevalence between birds. On the other hand, feather mite population size of successfully colonised islands (i.e. intensity)

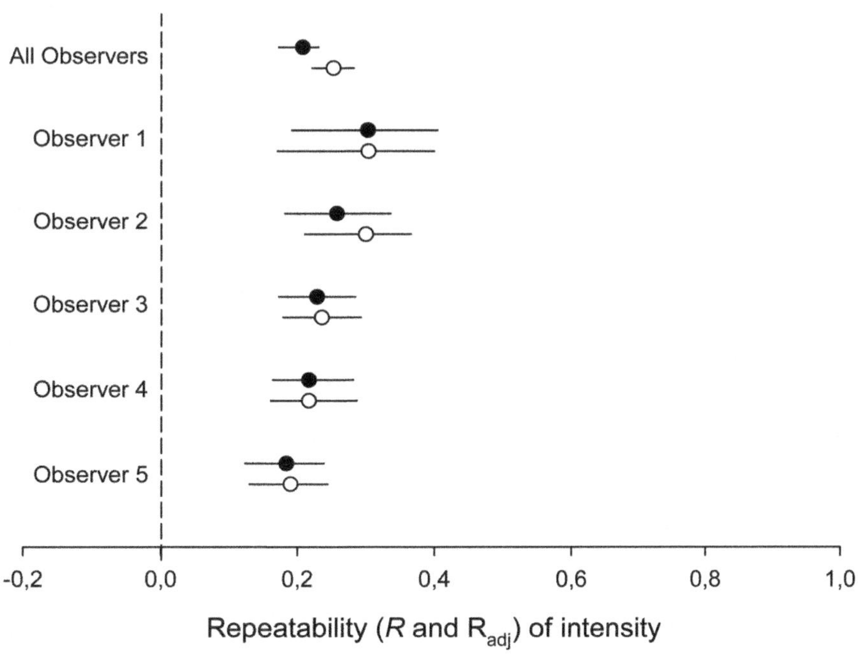


**Figure 6. Species repeatability and adjusted repeatability of feather mite intensity.** Species repeatability (R; white circles), and adjusted repeatability ($R_{adj}$; black circles) for feather mite intensity with corresponding 95% confidence interval estimates for different intensity data subsets (either the entire dataset, "All observers", or for data from five different researchers separately). Results are shown separately for five different observers (see Methods), and also for the entire dataset ("All observers"). Adjusted repeatability estimates for observers 1, 3, 4 and 5 were obtained using "breeding" as a fixed effect. For observer 2 "habitat" was retained as a fixed effect. For observers 1 and 2, two and one variables of the autocorrelation term were also added as fixed effects, respectively (see Methods).

**Table 2.** Parameter estimates for LME of the entire dataset of $log_{10}$-intensity of feather mites in birds.

| Factor | Description | Estimate | SE | df | t | F | P |
|---|---|---|---|---|---|---|---|
| Intercept | | 1.789 | 0.1269 | 83 | 14.09 | | <.0001 |
| Breeding period | | | | 1, 27246 | | 320.72 | <.0001 |
| | Non breeding | 0.184 | 0.010 | 27246 | 17.88 | | <.0001 |
| | Breeding | 0 | . | . | . | | . |
| Habitat | | | | 6, 27246 | | 9.86 | <.0001 |
| | M. and C. forest | −0.428 | 0.118 | 27246 | −3.62 | | 0.0003 |
| | Atlantic forest | −0.580 | 0.119 | 27246 | −4.89 | | <.0001 |
| | River forest | −0.515 | 0.119 | 27246 | −4.34 | | <.0001 |
| | Crops | −0.562 | 0.118 | 27246 | −4.75 | | <.0001 |
| | Steppe | −0.555 | 0.125 | 27246 | −4.43 | | <.0001 |
| | Wetland | −0.500 | 0.119 | 27246 | −4.20 | | <.0001 |
| | Sub. meadow | 0 | . | . | . | | . |
| | | | | | | | |
| Climatic conditions | PC1 | 0.001 | <0.001 | 27246 | −5.42 | 29.41 | <.0001 |
| Autocorrelation | | | | | | | |
| | y | 0.078 | 0.012 | 27246 | −6.37 | 40.54 | <.0001 |

For continuous variables, only statistically significant variables are shown.

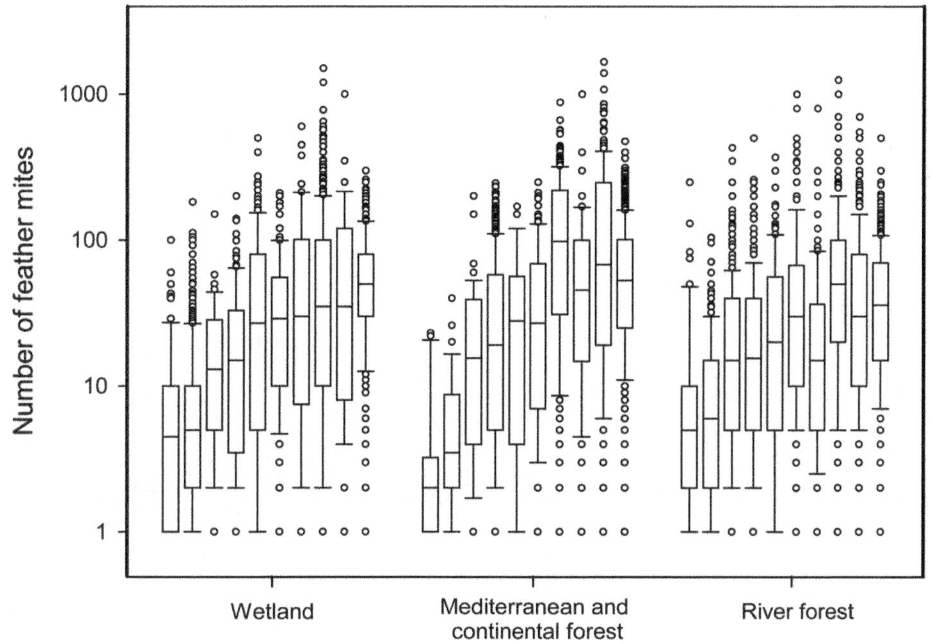


**Figure 7. Boxplot of feather mite intensity for ten different species of well-sampled resident passerines in three habitats.** Species are ordered from left to right within each habitat according to the intensities of feather mites in individuals captured in wetlands, as follows: *Phylloscopus collybita, Acrocephalus scirpaceus, Luscinia megarhynchos, Erithacus rubecula, Cettia cetti, Fringilla coelebs, Cyanistes caeruleus, Carduelis chloris, Serinus serinus* and *Sylvia atricapilla*. "Breeding", "habitat" and four variables from the spatial autocorrelation term were retained as fixed factors in LME, while "observer" and "species" were included as random factors.

could be more variable within host species because it is potentially affected by different attributes of the colonised island per se. For instance, mite population growth might depend on attributes that are related to the pool of exploitable resources such as bird body condition, [24,30] or the size of the uropygial gland [3,48,49]. Moreover, these mechanisms could be related to environmental variables other than those exerted by the host (e.g. weather conditions hosts do experience, [6,44,50]). Interestingly, the first principal component of local climatic conditions was retained only in the model for the intensity, but not for the prevalence of feather mites. This suggests that intensity of feather mite could be partially shaped by environmental factors, while prevalence could be more related with variables shaping the transmission of feather mites such as winter sociality [36]. Otherwise, it is also possible that the differences in repeatability for prevalence and intensity are also influenced by differences in measurement error for both parameters. Although we controlled for systematic bias in intensity and prevalence estimation among observers, we did not control for intra-observer measurement error, which is likely larger when counting the number of feather mites (intensity) than when simply recording whether a bird had or not feather mites (occurrence). In that case, intensity repeatability would be more affected by measurement error than prevalence, leading to lower repeatability estimates for intensity.

Differences among species could be due to variation at higher taxonomic levels. However, we found that species accounted for the highest proportion of the variance, while family and genus accounted for less variance. That shows that intensity and prevalence of feather mites only could be considered a species-specific trait, and not affected by patterns at higher taxonomic levels. Thus, future studies on passerines should be focused on variation in feather mite population in the host at the species level. Moreover, future work including non-passerine species will show whether the pattern found here holds for birds in general.

Ectosymbionts are potentially more affected by environmental conditions than endosymbionts. However, we found small differences in prevalence and intensity of feather mites in bird species among habitats (Figs. 4 and 7). Moreover habitat (for prevalence) and habitat and local climatic conditions (for intensity) played a secondary role compared to the breeding period (Tables 1 and 2, see below). However, climatic conditions seemed to affect the intensity more than habitat. This suggests that while feather mites are highly exposed to external environmental conditions (e.g. humidity, solar radiation), their demography seems more related to bird-mite interactions.

Intensity and prevalence of feather mites were lower during the breeding period, supporting the hypothesis that feather mites are mostly transmitted vertically from parents to offspring by direct physical contact [23,51]. Hence, our results encourage further studies of the reasons for seasonal changes in feather mite population size, and population dynamics before and after the breeding season. Moreover, we strongly suggest controlling for breeding season in future studies of mite intensity and prevalence as it was the most important variable retained in the statistical models.

Spatial autocorrelation was observed between sampling points in both prevalence and intensity analyses. Interestingly, the latitudinal component of the autocorrelation term was retained in intensity of feather mite models, showing that the number of feather mites was lower at northern latitudes. This pattern deserves further exploration in the future.

Future studies (along the lines initiated by Rózsa [2] and Galván et al. [3]) should continue searching for such species-specific traits of birds that could explain differences in populations of feather mites among bird species (e.g. wing area, uropygial gland size)

However, recent approaches to this problem [3] have found that such variables explain a relatively small amount of variance in differences among species, suggesting that complementary approaches are needed. Until now, all attempts (including the present study) have entirely focused on features of bird species. The next approaches will need to incorporate a feather mites' point of view in these analyses (e.g. [34]). A first step would be to test if repeatability of intensity and prevalence of feather mites vary among feather mite species (instead of among bird species), or if repeatability is higher when considering pairs of bird-feather mite species.

## Supporting Information

**Table S1 Intensity of feather mites (Min = Minimum, Max = Maximum, Med = Median) and sample size (N) for each bird species and country.**

**Table S2 Prevalence of feather mites (Prev = Prevalence; i.e. proportion of birds with at least one feather mite; see main text for details) and sample size (N) for each bird species and country.**

**Information S1 Extended ethical statement.**

## Acknowledgments

Logistical support and data on weather variables were provided by Isabel Afán and David Aragonés at the Remote Sensing and GIS Laboratory of Estación Biológica de Doñana, CSIC (LAST-EBD).

## Author Contributions

Conceived and designed the experiments: JD-R DS RJ. Performed the experiments: JD-R DS JP-T SF-G AB JAC JDLP DDP JLM RM-O CP OF JLT APM JF PLP IK CIV LM GB EA JCS IG FA EB JLC VC JSM RP MV AB CN AM RJ. Analyzed the data: JD-R DS RJ. Contributed reagents/materials/analysis tools: JD-R DS JP-T SF-G AB JAC JDLP DDP JLM RM-O CP OF JLT APM JF PLP IK CIV LM GB EA JCS IG FA EB JLC VC JSM RP MV AB CN AM RJ. Contributed to the writing of the manuscript: JD-R DS RJ.

## References


1. Lafferty KD, Dobson AP, Kuris AM (2006) Parasites dominate food web links. Proc Natl Acad Sci U S A 103: 11211–11216. Available: http://www.pubmedcentral.nih.gov/articlerender.fcgi?artid=1544067&tool=pmcentrez&rendertype=abstract.
2. Rózsa L (1997) Wing-feather mite (Acari: Proctophyllodidae) abundance correlates with body mass of passerine hosts: a comparative study. Can J Zool 1539: 1535–1539.
3. Galván I, Barba E, Piculo R, Cantó JL, Cortés V, et al. (2008) Feather mites and birds: an interaction mediated by uropygial gland size? J Evol Biol 21: 133–144. Available: http://www.ncbi.nlm.nih.gov/pubmed/18028353. Accessed 28 March 2012.
4. Senar JC, Copetel JL, Domenechi J, Walter GVON (1994) Prevalence of louse-flies Diptera, Hippoboscidae parasiting a cardueline finch and its effect on body condition. Ardea 82: 157–160.
5. Sol D, Jovani R, Torres J (2000) Geographical variation in blood parasites in feral pigeons: the role of vectors. Ecography (Cop) 23: 307–314. Available: http://www.blackwell-synergy.com/links/doi/10.1034/j.1600-0587.2000.d01-1639.x.
6. Dowling DK, Richardson DS, Blaakmeer K, Komdeur J (2001) Feather mite loads influenced by salt exposure, age and reproductive stage in the Seychelles Warbler *Acrocephalus sechellensis*. J Avian Biol 4: 364–369.
7. Szöllosi E, Cichoń M, Eens M, Hasselquist D, Kempenaers B, et al. (2011) Determinants of distribution and prevalence of avian malaria in blue tit populations across Europe: separating host and parasite effects. J Evol Biol 24: 2014–2024. Available: http://www.ncbi.nlm.nih.gov/pubmed/21726328. Accessed 13 January 2014.
8. Tripet F, Christe P, Møller AP (2002) The importance of host spatial distribution for parasite and speciation: a comparative study of bird specialization fleas (Siphonaptera: Ceratophyllidae). J Anim Ecol 71: 735–748.
9. Krasnov BR, Shenbrot GI, Khokhlova IS, Degen AA (2004) Flea species richness and parameters of host body, host geography and host "milieu." J Anim Ecol 73: 1121–1128.
10. Poulin R (2006) Variation in infection parameters among populations within parasite species: intrinsic properties versus local factors. Int J Parasitol 36: 877–885. Available: http://www.ncbi.nlm.nih.gov/pubmed/16620823. Accessed 13 July 2011.
11. Arneberg P, Skorping A, Read AF (1997) Is population density a species character? Comparative analyses of the nematode parasites of mammals. Oikos 80: 289–300.
12. Krasnov BR, Shenbrot GI, Khokhlova IS, Poulin R (2006) Is abundance a species attribute? An example with haematophagous ectoparasites. Oecologia 150: 132–140. Available: http://www.ncbi.nlm.nih.gov/pubmed/16896773. Accessed 2011 July 23.
13. Korallo-Vinarskaya NP, Krasnov BR, Vinarski MV, Shenbrot GI, Mouillot D, et al. (2007) Stability in abundance and niche breadth of gamasid mites across environmental conditions, parasite identity and host pools. Evol Ecol 23: 329–345. Available: http://www.springerlink.com/index/10.1007/s10682-007-9229-x. Accessed 2011 July 23.
14. Thieltges DW, Fredensborg BL, Poulin R (2009) Geographical variation in metacercarial infection levels in marine invertebrate hosts: parasite species character versus local factors. Mar Biol 156: 983–990. Available: http://www.springerlink.com/index/10.1007/s00227-009-1142-0. Accessed 2012 September 5.
15. Garamszegi LZ, Møller AP (2012) The interspecific relationship between prevalence of blood parasites and sexual traits in birds when considering recent methodological advancements. Behav Ecol Sociobiol 66: 107–119. Available: http://link.springer.com/10.1007/s00265-011-1259-2. Accessed 2013 November 20.
16. Møller AP (1991) Parasites, sexual ornaments and mate choice in the barn swallow *Hirundo rustica*. In: Loye JE, Zuk M, editors. Ecology, behavior, and evolution of bird-parasite interactions. Oxford University Press, Oxford. pp. 328–343.
17. Poulin R, Mouritsen K (2003) Large-scale determinants of trematode infections in intertidal gastropods. Mar Ecol Prog Ser 254: 187–198. Available: http://www.int-res.com/abstracts/meps/v254/p187-198/.
18. Krasnov BR, Stanko M, Morand S (2007) Host community structure and infestation by ixodid ticks: repeatability, dilution effect and ecological specialization. Oecologia 154: 185–194. Available: http://www.ncbi.nlm.nih.gov/pubmed/17684769. Accessed 2013 September 16.
19. Scheuerlein A, Ricklefs RE (2004) Prevalence of blood parasites in European passeriform birds. Proc Biol Sci 271: 1363–1370. Available: http://www.pubmedcentral.nih.gov/articlerender.fcgi?artid=1691737&tool=pmcentrez&rendertype=abstract. Accessed 2011 June 10.
20. Møller AP, Merino S, Soler JJ, Antonov A, Badás EP, et al. (2013) Assessing the effects of climate on host-parasite interactions: a comparative study of European birds and their parasites. PLoS One 8: e82886. Available: http://www.pubmedcentral.nih.gov/articlerender.fcgi?artid=3876993&tool=pmcentrez&rendertype=abstract. Accessed 2014 January 22.
21. Gaud J, Atyeo WT (1996) Feather mites of the world (Acarina, Astigmata). The supraespecific taxa. Part I. Ann Mus R l'Afrique Cent Sci Zool 277: 3–193.
22. Dabert J, Mironov SV (1999) Origin and evolution of feather mites (Astigmata). Exp Appl Acarol 23: 437–454.
23. Proctor HC (2003) Feather mites (Acari: Astigmata): ecology, behavior, and evolution. Annu Rev Entomol 48: 185–209. Available: http://www.ncbi.nlm.nih.gov/pubmed/12208818. Accessed 2012 March 13.
24. Blanco G, Tella JL, Potti J (1997) Feather mites on group-living Red-billed Choughs: a non-parasitic interaction? J Avian Biol 28: 197–206.
25. Jovani R, Blanco G (2000) Resemblance within flocks and individual differences in feather mite abundance on long-tailed tits, *Aegithalos caudatus* (L.). Ecoscience 7: 428–432.
26. Blanco G, Seoane J, De la Puente J (1999) Showiness, non-parasitic symbionts, and nutritional condition in a passerine bird. Ann Zool Fennici 362: 83–91.
27. Harper DGC (1999) Feather mites, pectoral muscle condition, wing length and plumage coloration of passerines. Anim Behav 58: 553–562.
28. Bridge E. (2003) Densities and distributions of commensal feather mites (*Zachvatkinia caspica*) among the primaries of Caspian Terns. Int J Acarol 29: 389–398.
29. Figuerola J, Domènech J, Senar JC (2003) Plumage colour is related to ectosymbiont load during moult in the serin, *Serinus serinus*: an experimental study. Anim Behav 65: 551–557. Available: http://linkinghub.elsevier.com/retrieve/pii/S0003347203920720. Accessed 2013 August 24.
30. Galván I, Aguilera E, Atiénzar F, Barba E, Blanco G, et al. (2012) Feather mites (Acari: Astigmata) and body condition of their avian hosts: a large correlative study. J Avian Biol 43: 273–279. Available: http://doi.wiley.com/10.1111/j.1600-048X.2012.05686.x. Accessed 2013 November 12.
31. Pap PL, Tökölyi J, Szép T (2005) Host-symbiont relationship and abundance of feather mites in relation to age and body condition of the barn swallow (*Hirundo rustica*): an experimental study. Can J Zool 83: 1059–1066.

32. Soler JJ, Peralta-Sánchez JM, Martín-Platero a M, Martín-Vivaldi M, Martínez-Bueno M, et al. (2012) The evolution of size of the uropygial gland: mutualistic feather mites and uropygial secretion reduce bacterial loads of eggshells and hatching failures of European birds. J Evol Biol 25: 1779–1791. Available: http://www.ncbi.nlm.nih.gov/pubmed/22805098. Accessed 2012 October 23.
33. Blanco G, Tella JL, Potti J, Baz A (2001) Feather mites on birds: costs of parasitism or conditional outcomes? J Avian Biol 32: 271–274.
34. Fernández-González S, De la Hera I, Pérez-Rodríguez A, Pérez-Tris J (2013) Divergent host phenotypes create opportunities and constraints on the distribution of two wing-dwelling feather mites. Oikos 122: 1227–1237. Available: http://doi.wiley.com/10.1111/j.1600-0706.2012.00241.x. Accessed 2013 November 11.
35. McClure E (1989) Occurrence of feather mites (Proctophyllodidae) among birds of ventura county lowlands, California. J F Ornithol 60: 431–450.
36. Figuerola J (2000) Ecological correlates of feather mite prevalence in passerines. J Avian Biol 31: 489–494. Available: http://www.blackwell-synergy.com/links/doi/10.1034/j.1600-048X.2000.310408.x.
37. Jovani R, Serrano D (2004) Fine-tuned distribution of feather mites (Astigmata) on the wing of birds: the case of blackcaps *Sylvia atricapilla*. J Avian Biol 35: 16–20.
38. Behnke J, McGregor P, Cameron J, Hartley I, Shepherd M, Gilbert F, Barnard C, Hurst J, Gray S, Wiles R (1999) Semi-quantitative assessment of wing feather mite (Acarina) infestations on passerine birds from Portugal. J Zool 248: 337–347.
39. Harvey PH, Pagel MD (1991) The Comparative Method in Evolutionary Biology. Oxford: Oxford University Press.
40. Nakagawa S, Schielzeth H (2010) Repeatability for Gaussian and non-Gaussian data: a practical guide for biologists. Biol Rev Camb Philos Soc 85: 935–956. Available: http://www.ncbi.nlm.nih.gov/pubmed/20569253. Accessed 2011 June 24.
41. Blanco G, Frías O (2001) Symbiotic feather mites synchronize dispersal and population growth with host sociality and migratory disposition. Ecography (Cop) 24: 113–120.
42. Pap PL, Vágási CI, Osváth G, Mureşan C, Barta Z (2010) Seasonality in the uropygial gland size and feather mite abundance in house sparrows *Passer domesticus*: natural covariation and an experiment. J Avian Biol 41: 653–661. Available: http://doi.wiley.com/10.1111/j.1600-048X.2010.05146.x. Accessed 2013 June 26.
43. Cramp S, Simmons KEL (1977) Handbook of the birds of Europe, the Middle East and North Africa: the birds of the Western Palaearctic. Oxford University Press.
44. Dubinin VB (1951) Feather mites (Analgesoidea). Part I. Introduction to their study. Fauna USSR 6: 1–363.
45. R Development Core Team (2012) R: A language and environment for statistical computing. R Foundation for Statistical Computing, Vienna, Austria. ISBN 3-900051-07-0, URL http://www.R-project.org/. Available: http://www.r-project.org/.
46. Legendre P, Legendre L (1998) Numerical Ecology. Elsevier Science, Amsterdam.
47. Schielzeth H, Nakagawa S (2011) rptR: Repeatability for Gaussian and non-Gaussian data. R package version 0.6.404/r42. http://R-Forge.R-project.org/projects/rptr/. Available: http://r-forge.r-project.org/projects/rptr/.
48. Galván I, Sanz JJ (2006) Feather mite abundance increases with uropygial gland size and plumage yellowness in Great Tits *Parus major*. Ibis (Lond 1859) 148: 687–697.
49. Møller AP, Erritzøe J, Rózsa L (2010) Ectoparasites, uropygial glands and hatching success in birds. Oecologia 163: 303–311. Available: http://www.ncbi.nlm.nih.gov/pubmed/20043177. Accessed 2013 December 20.
50. Wiles PR, Cameron J, Behnke JM, Hartley IR, Gilbert FS, et al. (2000) Season and ambient air temperature influence the distribution of mites (*Proctophyllodes stylifer*) across the wings of blue tits (*Parus caeruleus*). Can J Zool 78: 1397–1407.
51. OConnor BM (1994) Life-history modifications in astigmatid mites. Houck, M. A. (ed.). Mites: Ecological and Evolutionary Analyses of Life-history Patterns. Chap- man & Hall, New York. pp. 136–139.

# Fine-Scale Genetic Structure and Cryptic Associations Reveal Evidence of Kin-Based Sociality in the African Forest Elephant

**Stephanie G. Schuttler[1]*, Jessica A. Philbrick[1], Kathryn J. Jeffery[2,3,4], Lori S. Eggert[1]**

**1** Division of Biological Sciences, University of Missouri, Columbia, Missouri, United States of America, **2** Agence Nationale des Parcs Nationaux, Libreville, Gabon, **3** School of Natural Sciences, University of Stirling, Stirling, Scotland, **4** Institut de Recherche en Ecologie Tropicale, Libreville, Gabon

## Abstract

Spatial patterns of relatedness within animal populations are important in the evolution of mating and social systems, and have the potential to reveal information on species that are difficult to observe in the wild. This study examines the fine-scale genetic structure and connectivity of groups within African forest elephants, *Loxodonta cyclotis*, which are often difficult to observe due to forest habitat. We tested the hypothesis that genetic similarity will decline with increasing geographic distance, as we expect kin to be in closer proximity, using spatial autocorrelation analyses and Tau $K_r$ tests. Associations between individuals were investigated through a non-invasive genetic capture-recapture approach using network models, and were predicted to be more extensive than the small groups found in observational studies, similar to fission-fusion sociality found in African savanna (*Loxodonta africana*) and Asian (*Elephas maximus*) species. Dung samples were collected in Lopé National Park, Gabon in 2008 and 2010 and genotyped at 10 microsatellite loci, genetically sexed, and sequenced at the mitochondrial DNA control region. We conducted analyses on samples collected at three different temporal scales: a day, within six-day sampling sessions, and within each year. Spatial autocorrelation and Tau $K_r$ tests revealed genetic structure, but results were weak and inconsistent between sampling sessions. Positive spatial autocorrelation was found in distance classes of 0–5 km, and was strongest for the single day session. Despite weak genetic structure, individuals within groups were significantly more related to each other than to individuals between groups. Social networks revealed some components to have large, extensive groups of up to 22 individuals, and most groups were composed of individuals of the same matriline. Although fine-scale population genetic structure was weak, forest elephants are typically found in groups consisting of kin and based on matrilines, with some individuals having more associates than observed from group sizes alone.

**Editor:** Claire Wade, University of Sydney, Australia

**Funding:** The authors thank the USFWS African Elephant Conservation Fund (agreement #98210-8-G753 to LSE), National Science Foundation GK-12 Grant #1045322, and University of Missouri Life Sciences fellowship for financial support. The funders had no role in study design, data collection and analysis, decision to publish, or preparation of the manuscript.

**Competing Interests:** The authors have declared that no competing interests exist.

* E-mail: stephanie.schuttler@gmail.com

## Introduction

Spatial patterns of relatedness between individuals have important evolutionary consequences, as they influence the formation of mating and social systems within a species [1]. Individuals interact non-randomly, such that mate choice and sex-biased dispersal lead to fine-scale genetic structure. The resulting structure may create opportunities for kin selection [1,2], affect inbreeding or outbreeding rates [3], or influence local adaptations [4,5]. Therefore, understanding patterns of underlying genetic relationships offers insight on evolutionary processes within a species, as well as direct applications to conservation and management.

Mammals typically have male-mediated sex-biased dispersal while females are philopatric, creating the potential for matrilocal social groups [6]. One of the best-described mating systems is the breeding group, where one to a few males form permanent or semi-permanent associations with a female group. Groups have high co-ancestry within, but genetic differentiation between due to kin-based relationships within females and shared paternal genes [7]. As geographic distance between individuals increases, genetic structure deteriorates, and kin groups in close spatial proximity to one another will exhibit elevated genetic structure [8,9].

Species with less rigid or dynamic social systems can still have genetic structure, and even non-social species may exhibit isolation by distance. However, patterns in such species may not be as pronounced as those in species with polygynous breeding groups. For example, species with larger groups such as herd-living ungulates include more members and a larger proportion of the genetic diversity present in the population, weakening genetic patterns [10,11]. Fission-fusion societies may also have diluted effects, as group sizes change over time. African savanna elephants (*Loxodonta africana*) not only have fission-fusion sociality, but also differ from classic breeding groups because males are not associated with one matrilocal group, but search for mates across the population. Female groups change size over time and space, with some aggregations reaching hundreds of elephants [12,13]. Matrilocality creates coancestry within and genetic differentiation between core groups, but also cohorts of similar-aged paternal relatives across groups from male gene flow in their prime reproductive years [14].

Here we examine the fine-scale genetic structure of the African forest elephant (*Loxodonta cyclotis*), a species that is not well understood due to the difficulty of conducting studies in its remote and heavily forested habitat. Characteristics of fission-fusion sociality have been detected [15], but observed group sizes are much smaller. Groups appear to be nuclear families typically consisting of an adult female and her dependent calves, with males dispersing [16,17]. Dung samples collected together revealed relationships between adult females to consist of sisters or half-sisters, and juvenile offspring with occasional instances of more than one reproductive female per group [18]. Associated dung piles included individuals that were significantly more related to each other than to non-associated individuals, but only in one of two populations [18]. Observational studies also indicate social complexity, as some adult females have preferred associations, but individuals are not always found in the same groups or in groups of the same size [19,20].

The goals of our study were to examine fine-scale genetic structure in a forest elephant population, and investigate the genetic composition and connectivity within and between groups. First, we tested the hypothesis that forest elephants have genetic structure across small scales. We predicted genetic patterns at smaller geographic scales would reveal a decline in genetic similarity with increasing geographic distance. As declining genetic distance with increasing geographic distance could also reflect short-distance dispersal, we also examined associations between individuals using a genetic capture-recapture approach with dung samples. We tested the hypothesis that forest elephants associate with multiple related individuals, consistent with fission-fusion sociality. We predicted that samples collected in a group would be kin-based and include multiple reproductive females, but not always the same individuals. If forest elephants do not have fission-fusion sociality, groups would consist of small nuclear families and individuals would rarely associate with other groups. Therefore dung samples collected together would consist of an adult female and dependent offspring only. If groups reflect aggregations at preferred resources rather than social preferences, genetic composition of dung collected together would be random, and individuals will be largely unrelated. Consistent with results from other elephant species, we predicted that forest elephants would have additional associations outside their nuclear family, and that associations would be based on matrilines and kinship.

## Materials and Methods

### Study Area and Sample Collection

Field research was conducted in the Station d'Etudes des Gorilles et Chimpanzees (SEGC) study zone of Lopé National Park (LNP), Gabon (Figure 1) under permits AR0005/08 and AR0023/09 issued by the Centre National de la Recherche Scientifique et Technologique, the Gabonese government, and Agence Nationale des Parcs Nationaux. Much of LNP is dominated by mature forest, but the northeastern section is characterized by a variety of forest types and savannas [16]. The study zone (approximately 200 km$^2$) makes up 3% of the park (4,910 km$^2$), but has higher elephant densities with approximately 3.0 elephants/km$^2$ [21]. This study was conducted concurrently with an observational study where elephant individuals and groups were searched for throughout the SEGC study zone, individuals were identified, and where possible, dung was collected for genetic analyses from identified groups.

The SEGC study zone was divided into 1.0 km$^2$ sections (n = 196) according to Universal Transverse Mercator (UTM) gridlines. Sampling sessions were chosen to capture the range of seasonality. There are two dry seasons, June-August, and a less defined one from December-February. Wet seasons occur in October/November and March/April, although variation occurs between years. Our study included four, five-to-six-day sampling sessions (Table 1); session 1: September 23$^{rd}$–26$^{th}$, 2008 (end of dry season); session 2: October 20$^{th}$–24$^{th}$, 2008 (wet season); session 3: March 21$^{st}$–27$^{th}$, 2010 (beginning of wet season); and session 4: May 10$^{th}$–15$^{th}$, 2010 (end of short wet season, beginning of dry season). Rainfall data were collected daily at SEGC. Sampling sessions were conducted within short time frames to reduce the potential for individuals to move large distances and therefore to obtain dung sample locations that more accurately represent elephant positions in relation to one another.

Within each sampling session, between 47–61 1.0 km$^2$ sections were searched for fresh (≤24 hours) elephant dung. Sections were chosen randomly each day and teams searched simultaneously in different sections throughout the study zone. Once a section was sampled, the section was not re-sampled during the same session to capture the largest spatial extent. Some samples were collected opportunistically; for example if found on the road while driving to a section. In forested sections, teams searched on and around known elephant trails. In savanna sections, teams searched on elephant paths identified by freshly broken vegetation. If no recent signs were observed, the section was abandoned. Each team included at least one experienced Gabonese field assistant.

Elephant dung can remain visible for months, but changes in appearance and odor make it possible to discern fresh samples [22]. Fresh dung (≤24 hours) was characterized as having sheen, intact boli (unless crushed by elephants or disrupted by insects), and a strong odor (pers. obs). When a fresh sample was found, approximately 10 grams were collected in polypropylene tubes for genetic analyses and GPS coordinates were recorded. If dung piles were intact, up to three bolus circumferences were recorded and averaged to infer the age class of the individual [23]. Samples were boiled for 15 minutes to destroy pathogens and preserved with Queen's College buffer (20% DMSO, 100 mM Tris pH 7.5, 0.25 EDTA, saturated with NaCL; [24]). Samples were stored at room temperature in the dark until the end of each field season (approximately four months), and imported into the US under USDA permit #48529. In the US, samples were stored in the laboratory at 4°C until DNA was extracted, which began within weeks of arrival.

### Laboratory Methods

DNA was extracted using QIAamp Mini Stool Kits (QIAGEN) and modifications in Archie et al. [25] or following the Guanadine Thiocyanate method of Eggert et al. [26]. Extractions took place in a separate room designated for DNA extractions from non-invasive samples to reduce the possibility of contamination.

We genotyped samples using 12 polymorphic microsatellite loci: FH60R, FH94R, FH48R, FH19R, LA6R, LafMS02R [27], FH67, FH126, FH103R [28] FH129 (5′-3′ F-TGGCTAAAATGCCTAT-CACTCA, R-CCAGCTAAACTAAGTCTGCTCTTTT, [29]), LaT05, [25], and LaT13R [30]. We redesigned primers FH103R (5′-3′ F-GCTGCCACTTCCTACACCTT, R-CCTTTGCT-TTTCTAATGAGTCC) and LafMS02R (5′-3′ F-GTCTATCTC-CACCCCCTGCT, R-TGTCTGTTGTAAAANTCGCTTG) to shorten fragments. Polymerase chain reactions (PCRs) were performed in a PCR Workstation (Fisher Scientific) with ultraviolet light used between PCRs to decontaminate surfaces.

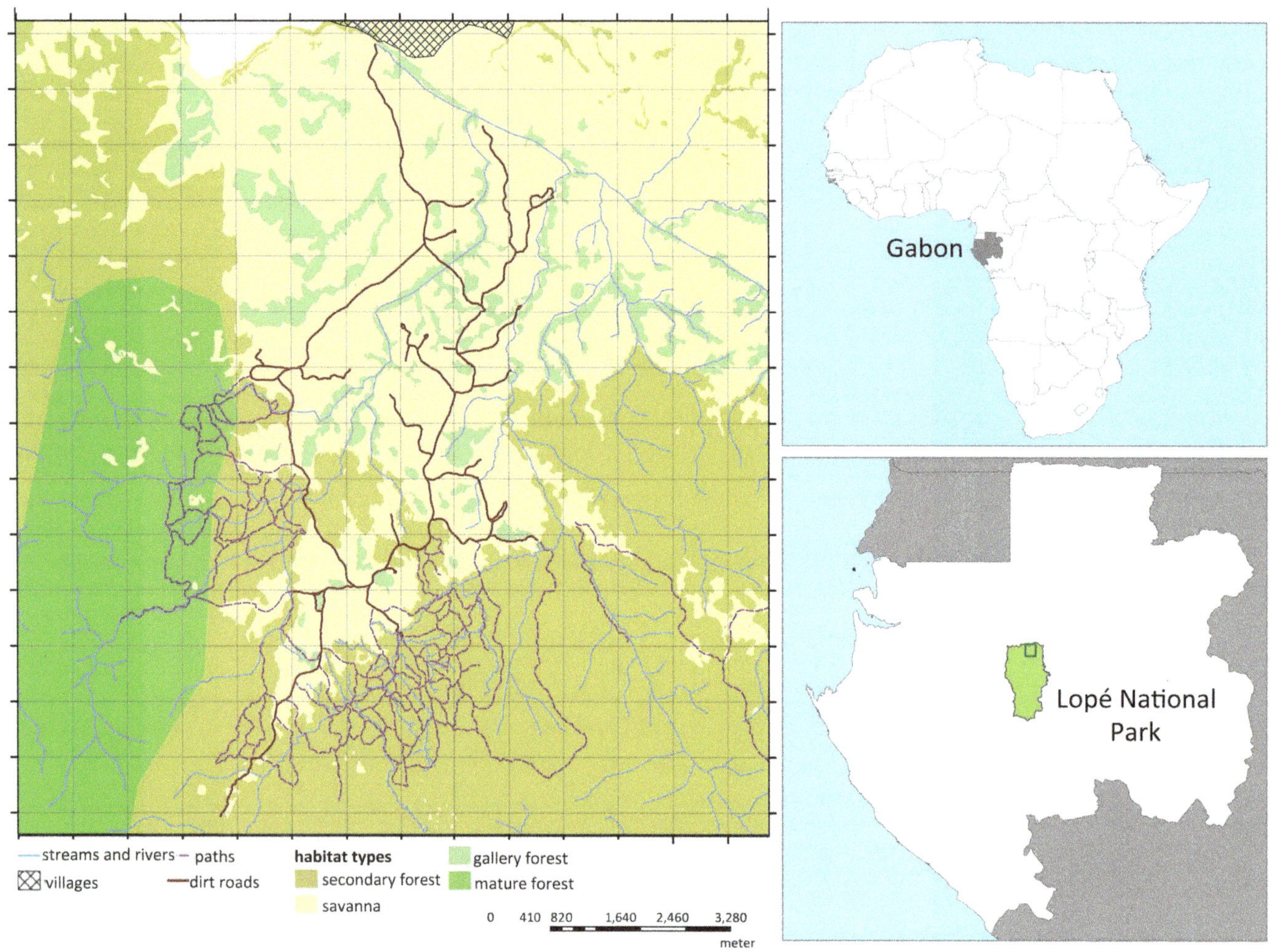


**Figure 1. Map showing the SEGC study zone within Lopé National Park, Gabon.**

Samples were amplified in single locus reactions or in four multiplex reactions. Single locus reactions contained 0.5 U AmpliTaq Gold Polymerase (Applied Biosystems), 1X PCR Gold Buffer (Applied Biosystems), 0.4 µM fluorescently labeled forward primer, 0.4 µM unlabeled reverse primer, 2 mM MgCl2, 0.2 mM each dNTP, 1.5 µl 20X BSA, and 3 µl DNA extract in 25 µL reactions. Thermocycling consisted of 95°C for 10 minutes, 45 cycles denaturation at 95°C for 1 minute, primer annealing at locus specific temperatures for 1 minute, and primer extension at 72°C for 1 minute, followed by an extension of 72°C for 10

**Table 1.** Summary of sample collection, rainfall, genotyped samples, age categories, and sexes of unique individuals.

| | | | | | unique females | | | unique males | | |
|---|---|---|---|---|---|---|---|---|---|---|
| **sampling session** | **dates** | **total monthly rainfall (mm)** | **samples collected** | **samples genotyped** | **adult** | **juvenile** | **unknown age** | **adult** | **juvenile** | **unknown age** |
| single day | 10/24/08 | - | 40 | 34 | 16 | 2 | 0 | 1 | 3 | 1 |
| session 1 | 9/23–9/26/08 | 134.8 | 48 | 39 | 18 | 3 | 3 | 3 | 4 | - |
| session 2 | 10/20–10/24/08 | 418.5 | 102 | 85 | 25 | 11 | 2 | 8 | 3 | 3 |
| session 3 | 3/21–3/27/10 | 157.9 | 63 | 37 | 14 | 3 | 5 | 1 | 4 | 4 |
| session 4 | 5/10–5/15/10 | 115.5 | 56 | 51 | 15 | 6 | 1 | 4 | 5 | 1 |
| year 2008 | 8/12–11/7/08 | 728.2 | 239 | 196 | 55 | 16 | 6 | 12 | 14 | 4 |
| year 2010 | 2/17–5/10/10 | 423.5 | 262 | 202 | 57 | 13 | 11 | 6 | 17 | 6 |

For the year analyses, samples were combined with those from a separate observational study (n = 88 from 2008, n = 142 from 2010).

minutes. Samples amplified at either 58°C or 60°C and were arranged into four multiplex reactions (Table S1). Loci LaT05 and LaT13R were included in all multiplexes because of larger fragment sizes and yielded the same genotypes at both temperatures. Loci LA6R and FH126 had different annealing temperatures from the single to multiplex reactions (54°C to 58°C, and 58°C to 60°C). These loci were tested on two positive controls and 16 samples to confirm that different temperatures resulted in the same genotypes. Multiplex reactions were performed in 8.0 μL volumes containing 4.0 μL Master multiplex mix (QIAGEN), 0.2 μM diluted primer mix, 0.8x BSA, and 1.0–2.5 μL DNA extract. Amplifications were performed with an initial cycle of 95°C for 15 minutes, followed by 40–45 cycles of denaturation at 94°C for 0.5 minutes, primer annealing at 58°C or 60°C for 1.5 minutes, primer extension at 72°C for 1 minute, and a final extension at 60°C for 30 minutes. Each reaction included a positive control to standardize allele scoring and a negative control to detect contamination. PCR products were visualized in 2% agarose gels containing Gel Star (Lonza) to verify amplification.

Fragment analysis was performed using an ABI 3730 DNA Analyzer with Liz 600 size standard (Applied Biosystems) and genotypes were scored in GeneMarker v1.6 (Soft Genetics LLC). Samples were amplified in PCR reactions separately and scored at each locus at least two to three times. Matching heterozygotes from the same sample were scored at least twice and matching homozygotes were scored at least three times to obtain a consensus genotype [31,32].

We calculated $PID_{sib}$, the power to differentiate between siblings [33], using a subset of 20 genotyped individuals in GenAlEx version 6.41 [34]. We chose the $PID_{sib}$ test over $PID_{random}$ test because it is more conservative and elephants may be found in groups of related individuals [35]. Based on the results, genotypes with at least six loci ($PID_{sib} = 0.002$) were included in the analyses. $PID_{sib}$ was recalculated once all samples were genotyped, and results did not change. Locus LafMS02R did not amplify reliably and was removed.

Genotyping error rate was calculated using 25 randomly selected samples in Reliotype [36]. We used default settings and 10,000 bootstrap replicates.

We used Dropout [37] to identify genotypes that differed at two or fewer loci, and those genotypes were checked manually. We considered genotypes to represent the same individual if they met the following criteria: (1) at least six matching loci (2) the other loci either did not amplify, or mismatches could be explained by allelic dropout, and (3) one mismatch was allowed if the alleles were difficult to score. Once samples were identified as the same individual, molecular sexes, mitochondrial DNA (mtDNA) haplotypes, and bolus circumferences were compared to ensure accordance. For matches that differed in bolus circumferences, field notes were reviewed to determine if differences were explained by field conditions.

Sex was determined following methods in Munshi-South et al. [38] or Ahlering et al. [39]. A subset of samples was tested for consistency between methods, and because band dropout (*failure to amplify the Y-chromosome*) may occur in bands in samples from males, three independent runs confirmed females, while males were confirmed twice.

We amplified a 627 bp fragment of the mtDNA control region for all individuals identified through unique genotypes using the primers MDL3 and MDL5 [40] or AFDL1, AFDL2, AFDL3, and AFDL4 [41]. Products were sequenced in both directions on an ABI 3730 DNA Analyzer (Applied Biosystems) using the Big Dye Terminator cycle sequencing chemistry. Sequences were aligned and edited in Sequencher 4.5 (Gene Codes Corporation). Haplotypes were identified by at least one base pair difference.

## Data Analysis

We used the average circumferences of three dung boli to establish age classes of juvenile (pre-reproductive) and adult (reproductive) individuals. Eggert et al. [23] considered average bolus circumferences greater than 32 cm to be adults, calibrated from samples based on the age distribution of savanna elephant. As reproductive age has been shown to vary across savanna elephant populations [42], we compared these estimates to samples from known reproductive and pre-reproductive individuals from an accompanying observational study in LNP and adjusted the criteria; samples ≥30 cm were considered adults, while ≤30 cm were juveniles. For individuals with multiple captures and sample averages above and below 30 cm, we averaged all measurements to determine age class. We also looked for evidence of damage (e.g. rain, insects) in field notes, in which case, we relied on measurements of undamaged samples to determine age class.

We used Micro-Checker version 2.2.3 [43] to test for null alleles, stuttering, and large allelic dropout. We used Genepop version 4.0 [44] to calculate observed and expected heterozygosities, allelic diversity for each locus, and to test for deviations from expected heterozygosity values under Hardy-Weinberg equilibrium and for linkage disequilibrium.

To test if genetic similarity declines with increasing geographic distance, we conducted spatial autocorrelation (SA) analyses for adult females using GenAlEx version 6.41 [34]. As samples collected over short time periods resulted in small sample sizes, we tested several scenarios: samples collected together on the same day (single day), each sampling session (week), and combined samples from each year (year) for correlations between genetic and geographic distance. We were only able to conduct one single day analysis due to sample size. For year analyses, samples collected during the spatial genetics sampling were combined with samples collected during the observational study, which were collected differently. In the observational study, circuits by vehicle were conducted almost daily for four-month periods after sunrise and before sunset to search for forest elephant groups and individuals. Circuits covered all roads within the study area (Figure 1), and directions and starting locations were altered to reduce bias in sampling areas. Elephants were observed in savanna habitats, and areas were searched for dung samples after individuals left. Although samples were not searched for randomly in sections, samples were collected throughout the study zone.

Spatial autocorrelation analyses are based on a single location per individual. For repeat captures of individuals within a sampling session, we calculated a midpoint between two locations, or created centroids using minimum convex polygons with Hawth's tools v3.2 (www.spatialecology.com/htools/) in ArcGIS 9.2 (ESRI, Redlands, CA) with more than two locations. Each SA was run with 9,999 permutations and bootstraps. Genetic distances were calculated from genotypes and converted to the autocorrelation coefficient $r$, which although not the same as, yields similar estimates to Queller & Goodnight's $R$ [45,46].

Spatial autocorrelation analyses are influenced by the distance class used [47]. We searched for biologically meaningful classes that would not be overly influenced by short-term movements of elephants during sampling sessions, or by distances between locations of recaptured individuals. We explored telemetry data used in Momont [20] from four adult female elephants in LNP. These elephants moved 2.8–4.4 km every 12 hours [20], but did not displace this distance. Therefore, we examined displacement distances of individuals by selecting one six-day period (the average number of days in week sampling sessions) from each month per individual, and recorded the longest distance between two location points. Average displacement was 4.8±2.3 km with a range of 2.8–7.2 km. Therefore, we chose distance classes of 5.0 km.

Significance in SA can be detected if (1) $r$ exceeds the upper and lower bounds of the 95% confidence interval generated from random permutations, or (2) the 95% error about $r$ generated from bootstrap tests does not intercept the x-axis at $r = 0$. The latter is more conservative and will favor the null hypothesis more frequently [47]. When positive significant genetic structure is found, $r$ will decrease as distance size classes increase. The first distance class where $r$ is not significant designates the spatial extent of genetic structure in the population [47]. This can depend on the size of the distance class, which is chosen by the user. To overcome this, we conducted analyses that plotted pairwise genetic distances against increasing inclusive distance classes (999 bootstraps) of 1.0 km intervals to determine the distance class at which $r$ was no longer significant [47].

No relatedness estimator outperforms others, and an estimator is data dependent [48]. We used COANCESTRY version 1.0 [49] to perform Monte Carlo simulations that calculated correlation coefficients between seven relatedness estimators and the values of known relatedness categories generated through simulations using observed allele frequencies and missing genotype rates. We simulated eight relationship categories of 100 dyads, with 100 reference individuals, and 1,000 bootstraps. We chose the Queller-Goodnight moment estimator [45] because it resulted in a strong correlation between true and estimated values ($r = 0.911$). We calculated pairwise relatedness ($R$) in RELATEDNESS version 5.0.8 [45] using the bias correction. To test the effectiveness of this estimator, we used eight known mother-calf pairs whose samples were collected during the observational study. Average pairwise relatedness of mother-calf pairs was 0.490±0.083, consistent with expectations.

To assess the consequences of using centroids for recaptured locations from the same individual, we conducted permutation tests between matrices of dyads that compared spatial proximity with relatedness, but allowed for nonindependent data points. We used Hemelrijk's Tau $K_r$ test [50] in MATRIXTESTER (www.rug.nl/fmns-research/beso/_people/hemelrijk) with 10,000 permutations. MATRIXTESTER allows for matrices containing dyads of ≤100 individuals, and therefore the year analyses for 2008 and 2010 were excluded because recaptures increased matrix size.

To understand how kin are positioned in space, we calculated average distances between related adult females in week and single day sessions. Dyads with a relatedness value of at least 0.2 had their Euclidean distances measured and averaged in ARCGIS 9.2. We chose 0.2 to ensure the following relationships would be captured; mother-daughter ($R = 0.5$), full siblings ($R = 0.5$), half siblings ($R = 0.25$), and grandmother-granddaughter ($R = 0.25$), as the mean expected relatedness value on average is 0.25, but can be lower or higher. Mitochondrial DNA haplotypes were mapped into ARCGIS 9.2 to examine the distribution of matrilines.

To investigate associations between individuals, we examined relatedness within and between groups of dung samples, and created network models from samples collected in groups. Samples were considered to be from the same group if they were collected on the same day, within 250 m of each other for each pair of samples, and were of the same freshness. For most samples, identifying groups was apparent, as it was rare to find numerous dung piles that were less than 24 hours old in the same area and of varying freshness. To be conservative, if there was any doubt that dung samples belonged to the same group, they were assigned to different groups.

We tested if individuals within a group were more related compared to individuals from other groups using permutation tests in PERM version 1.0 with 10,000 permutations and 10 iterations [51]. We tested adult females in groups, and all individuals found together. Additionally, we tested for differences in mean pairwise relatedness between groups of male and female dyads in COANCESTRY using 10,000 bootstraps.

We created network models from the genetic information from dung samples collected together, excluding samples found solitarily, and therefore not associated with any group. Networks were created in UCINET version 6.403 [52] and NETDRAW version 2.120 [53]. We reported the number of nodes (individuals), edges (ties between individuals if samples were detected in groups), and the number of components (nodes connected to each other and not connected to the rest of the network). Edges were weighted by relatedness and mitochondrial haplotypes were added as an attribute. We calculated average relatedness for each component using RELATEDNESS version 5.0.8.

## Results

We collected 501 dung samples and identified 89 unique adult females, 22 juvenile females, 16 females of unknown age, 18 adult males, 22 juvenile males, and 10 males of unknown age from genotypes. On average, individuals were recaptured 2.249 times with a range of 0–19 recaptures. Between 2008 and 2010, 23 adult females were re-sampled.

After applying a standard Bonferonni correction for multiple tests, all loci except LaT05 conformed to Hardy-Weinberg equilibrium expectations. This locus showed evidence of large allelic dropout and null alleles, and because it did not amplify consistently, removing it from analyses did not affect the number of genotyped individuals. The average number of alleles per locus was 12±3.9 and the mean observed heterozygosity was 0.820±0.081 (Table S1). The average expected overall reliability after replication was 0.995.

Mitochondrial haplotype diversity was 0.805±0.017 and nucleotide diversity averaged over all loci was 0.008±0.004. Ten haplotypes were identified, all of which differed from those previously reported of the same length (GenBank accession #KF938592-938601). One haplotype (Lope10) was unique to males.

Sessions 2, 4, the single day analysis, and both years revealed significant positive genetic structure at 0–5 km and negative structure at 5–10 km for adult females (Figures 2, 3). Spatial autocorrelation analyses with inclusive distance classes revealed significance up to 5.0 km distances in sessions 2, 4, the single day, 2008, and 2010 (Table 2). However, preceding distance classes were not necessarily significant. No autocorrelation coefficients were significant in sessions 1 and 3 (Figure 2). The Tau $K_r$ tests revealed slightly different results with the significance of non-independent data between spatial proximity and relatedness occurring in session 1 (Tau $K_r = -0.106$, $p = 0.003$), 4 (Tau $K_r = -0.101$, $p = 0.006$), the single day (Tau $K_r = -0.121$, $p = 0.003$), approaching significance in session 2 (Tau $K_r = -0.029$, $p = 0.055$), but none in session 3 (Tau $K_r = -0.049$, $p = 0.180$).

The mean distance between adult females with a relatedness value of ≥0.2 was 2.984±2.105 km (session 1), 2.208±1.709 km (session 2), 2.599±0.747 km (session 3), and 1.688±1.775 (session 4). The single day had a smaller mean distance of 0.823±0.548 km. The north-south sampling extent of the single

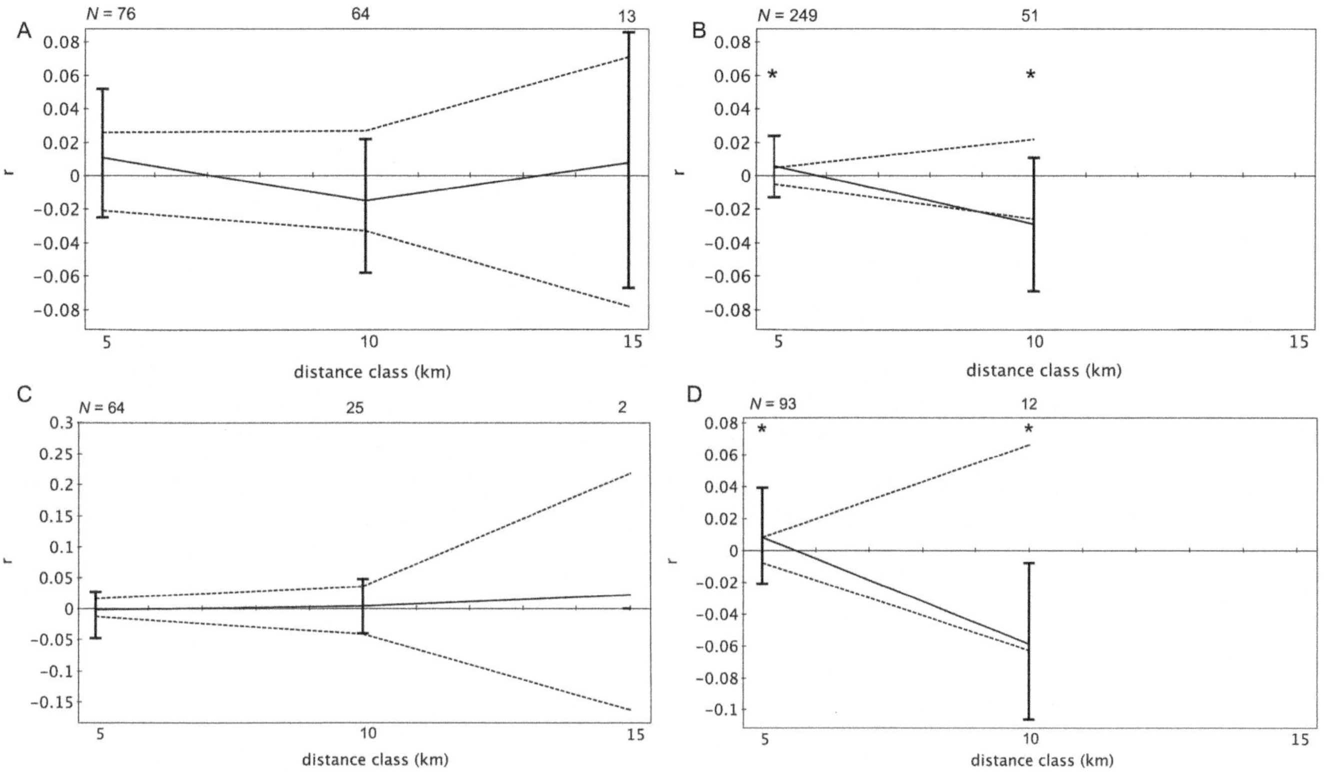


**Figure 2. Spatial autocorrelation correlograms for adult female pairs, by week.** (A) Represents session 1, (B) 2, (C) 3, and (D) 4. Significant distance classes are designated with an asterisk (*). Dotted lines represent the upper and lower bounds of the 95% confidence interval generated from random permutations, while bars represent 95% error generated from bootstrap tests.

day analysis was slightly smaller than the regular sampling sessions (9 km compared to a maximum of 13–14 km); however the east-west spatial extent was the same (7 km compared to 7–8 km). In all sessions, we found dyads that were closer to each other in distance, but less closely related.

We detected ten mtDNA haplotypes. Lope7 was the most common, but individuals with different haplotypes were found in close proximity (<1.0 km), and overall no spatial segregation was observed. The single day session yielded the same pattern, with samples collected <100 m from each other having different haplotypes.

We detected 70 groups and 26 included more than one adult female (Table S2). Individuals within groups were significantly more related to each other than to individuals from distant groups for adult females ($R = 0.236 \pm 0.187$, $p < 0.001$) and all individuals ($R = 0.255 \pm 0.194$, $p < 0.001$). In 84.6% of groups (22 of 26), adult females had the same mitochondrial haplotype, and in 82.9% of groups (58 of 70), all individuals shared the same mitochondrial haplotype (Table S2). Consistent with male sex-biased dispersal, females had a significantly higher mean pairwise relatedness than did males (males: $R = -0.045 \pm 0.146$; females: $R = -0.007 \pm 0.152$, $p < 0.05$).

The network included 106 individuals with 28 components and 83 edges (Figure 4). The largest component had 22 individuals. Eighteen of 131 dyads were detected two or more times together (13.7%), with six being the largest number of times two individuals were detected together. Average relatedness of network components was $0.155 \pm 0.197$ for all individuals, and $0.2058 \pm 0.167$ when including only adult females. Twenty-one components consisted of individuals that shared the same haplotype (75.0%).

## Discussion

This study revealed fine-scale genetic structure, albeit weak, and evidence of potentially complex, kin-based groups using a novel network model genetic approach derived from non-invasive dung sampling in African forest elephants. We tested the hypothesis that genetic similarity decreases with increasing geographic distance for two types of spatial genetic patterns; spatial autocorrelation (SA) analyses in which distance was separated into classes of 5 km, and the Tau $K_r$ test, which tests for the relationship between relatedness and distance, but accounts for non-independent repeat locations. Spatial autocorrelation analyses revealed that adult females are more closely related than expected by chance from 0–5 km in two of the four week sampling sessions, samples collected over four-months in 2008 and 2010, and in the single-day analysis. No genetic structure was found in session 1, but there was a significant correlation between distance and relatedness in the Tau $K_r$ test. Alternatively, significant genetic structure was detected in sampling session 2, but not in the Tau $K_r$ test, although it approached significance ($p = 0.055$). Conflicting results between the two methods may be due to this test's use of recaptures and not midpoints.

Significant SA coefficients were 0.014 or lower, and for inclusive distance classes, 0.083 or less within the first distance class. Although significance was detected, compared to other mammals (0.04, black rhinoceros *Diceros bicornis bicornis* [54]; 0.07, Eurasian badgers, *Meles meles* [55]; 0.08, tree-roosting bats [9]; 0.12 in Australian bushrats, *Rattus fuscipes* [47]), these values are low. The weaker genetic structure and inconsistent results between sampling sessions may be the result of using larger distance classes, which were necessary given the mobility of elephants and recaptures of

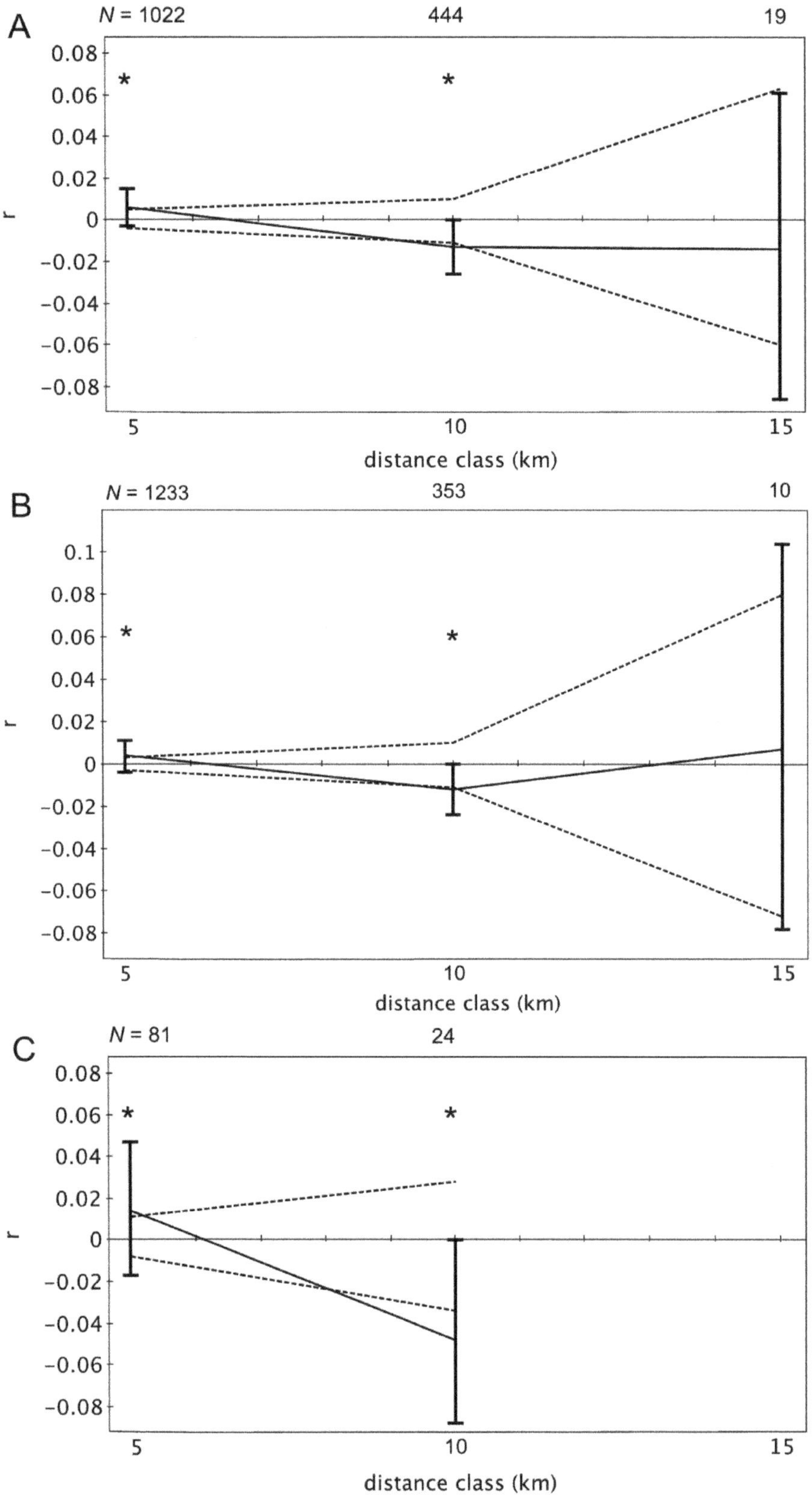


**Figure 3. Spatial autocorrelation correlograms for adult female pairs, by year and day.** (A) Represents correlogram from year 2008, (B) 2010, and (C) a single day. Significant distance classes are designated with an asterisk (*) and based on 95% confidence intervals from permutation analysis. Dotted lines represent the upper and lower bounds of the 95% confidence interval generated from random permutations, while bars represent 95% error generated from bootstrap tests.

**Table 2.** Summary of *r* coefficients from autocorrelation analyses with inclusive distance classes for adult female forest elephants in Lopé National Park, Gabon.

| | distance class (km) | | | | | | | | | | | | | |
|---|---|---|---|---|---|---|---|---|---|---|---|---|---|---|
| **sampling session** | **0 to 1** | **0 to 2** | **0 to 3** | **0 to 4** | **0 to 5** | **0 to 6** | **0 to 7** | **0 to 8** | **0 to 9** | **0 to 10** | **0 to 11** | **0 to 12** | **0 to 13** | **0 to 14** |
| session 1 AF | 0.083 | 0.044 | 0.013 | 0.022 | 0.011 | 0.015 | 0.006 | 0.002 | 0.001 | −0.001 | 0.000 | 0.000 | - | - |
| session 2 AF | **0.059** | **0.016** | 0.008 | 0.004 | **0.006** | −0.001 | 0.001 | 0.000 | - | - | - | - | - | - |
| session 3 AF | −0.011 | −0.004 | −0.016 | 0.000 | −0.002 | 0.002 | 0.003 | −0.002 | −0.001 | −0.001 | 0.000 | - | - | - |
| session 4 AF | 0.033 | 0.029 | 0.015 | 0.002 | **0.008** | 0.003 | 0.003 | 0.000 | - | - | - | - | - | - |
| single day | **0.075** | **0.025** | **0.013** | **0.014** | **0.014** | 0.007 | 0.002 | −0.001 | - | - | - | - | - | - |
| year 2008 AF | **0.026** | 0.005 | 0.003 | 0.004 | **0.006** | 0.002 | 0.000 | −0.001 | 0.000 | 0.000 | 0.000 | 0.000 | 0.000 | 0.000 |
| year 2010 AF | **0.029** | **0.012** | 0.004 | **0.006** | **0.004** | 0.001 | 0.000 | 0.000 | 0.000 | 0.000 | 0.000 | - | - | - |

Significant *r* coefficients from one-tailed test for positive autocorrelation P values are in bold type.

samples. This pooling of dyads into more inclusive distance classes may have biased SA coefficients downward. Results of the SA analyses with inclusive distance classes supports this (except session 3), as the autocorrelation coefficients in the smaller distance class of 0–1 km are higher than those in 0–5 km from regular SAs.

Interestingly, while SA analyses revealed significant positive genetic structure in the 0–5 km distance class for most sampling sessions, previous classes in the inclusive SA analyses were not necessarily significant. Related adult females were found at distances of less than one kilometer, and again at approximately five kilometers. As elephants have the ability to move over short time periods, the genetic structure may be diluted in some distance classes because the actual locations of individuals in relation to one another at the exact same time could not be captured. The dung samples are collected within a 24-hr window and therefore locations represent one position for the individual. Distance classes were pooled at larger intervals of five kilometers to account for this approximation. When distance classes are at a finer scale (inclusive SA analyses), individuals may not be placed in the distance class that more accurately represents their true positions relative to one another at that time. The only session to yield significant positive genetic structure with all analyses, and in all of the inclusive distance classes up to five kilometers, was the single day session. This suggests that the time involved in sampling is likely an important factor in capturing fine-scale spatial genetic patterns of forest elephants. The single day analysis includes recaptures, but likely reduced the effects of movements because of the shorter time scale. However, reducing sampling time also limits sample size. To examine these effects, we increased sample size by using data from dung samples collected for each year. Both 2008 and 2010 yielded

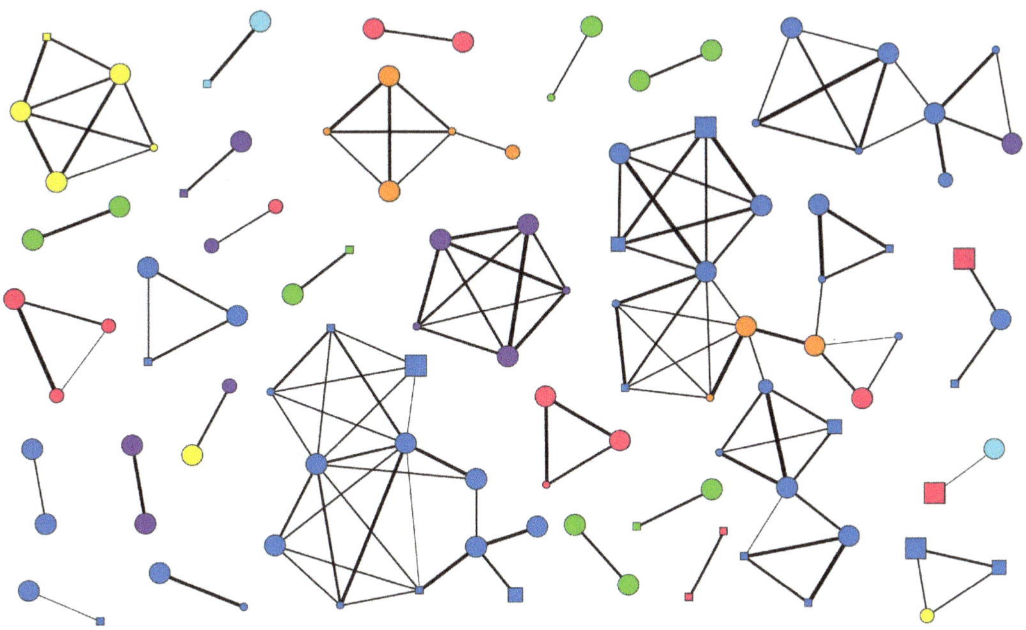

**Figure 4. Network constructed from dung sample group data using all individuals.** Dung samples that were collected outside of groups were not included. Nodes represent individuals and edges indicate individuals whose dung was collected as part of the group. Squares represent males, while circles represent females. The size of the node reflects the age category; adults are the largest, unknown ages are of medium size, and juveniles are the smallest. Colors represent mitochondrial DNA haplotype; pink, Lope1; orange, Lope3; yellow, Lope4; green, Lope5; aqua, Lope6; blue, Lope7; purple, Lope9. Edges are weighted according to relatedness; those with thicker lines representing more closely related dyads.

similar patterns with significant positive genetic structure in the 0–5 km distance class and negative in the 5–10 km class. Future studies should therefore concentrate on increasing sampling effort over shorter time periods.

When looking at the spatial patterns of related dyads (those with a relatedness value of ≥0.2), related adult females were found within several kilometers of each other. However, it was not uncommon to find more geographically distant dyads more closely related than individuals that were closer together, even for samples collected on the same day. There was no spatial structure associated with mitochondrial haplotypes, and based on dung locations and dates, individuals appeared to tolerate others with different haplotypes in close proximity. This may explain the weak genetic structure from the SA analyses and inconsistent results between sampling sessions. Although we detected a signal of positive genetic structure within five kilometers, it is likely diluted by the presence of unrelated individuals in the area. However, consistent with evidence of male-biased dispersal that typically occurs in mammals, females had a significantly higher mean pairwise relatedness than males in the population.

The SEGC study zone differs from most forest elephant habitat because it is a savanna-forest mosaic and also has slightly higher elephant densities [21]. Group sizes do not differ between forested and savanna habitats, and are comparable in size and composition to those found in other populations, with an adult female and her calves being the most common group type [16]. In Loango National Park, a protected area with similar habitat and elephant densities, mean relatedness in associated dung piles was not significantly higher than non-associated piles, which contradicts the pattern found in another population tested in that same study, and also results from this study [18]. This could be due to high elephant densities there, which could increase the chance of sampling different groups using the same area or seasonal migrations as unrelated groups would be in closer proximity when individuals migrate into the area. However, Schuttler et al. [56] found no evidence of seasonal migrations in the home ranges of forest elephants in Loango. White [57] demonstrated that elephant dung was seasonally correlated with the ripening of *Sacoglottis gabonensis*, with dung rates highest in September – November. Our study found genetic structure through SA in only one of the two of the sampling sessions that occurred in the months with both the most and fewest number of dung piles. Therefore, genetic structure can be detected, even when densities are high, however, higher densities also have the potential to dilute patterns, necessitating the importance of looking at individual association patterns.

We also did not find any relationship between rainfall and the results of spatial genetic analyses. Session 2 occurred during the month with the most rainfall, and we found significant SA, and an almost significant negative relationship between relatedness and spatial proximity in the Tau $K_r$ test. In contrast, session 3 had the second highest amount of rainfall, yet no spatial genetic structure was detected.

Other factors possibly contributing to inconsistent results may include social structure, poaching, and sample size. Evidence is mounting that forest elephants have fission-fusion sociality, where group sizes and composition change over time [15]. Sessions where genetic structure is detected could reflect periods when individuals group together. Poaching has had a severe impact on forest elephant populations [58] and may also influence social structure. Gobush et al. [29] found that non-kin grouped together in poached populations of savanna elephants. We found groups of associated dung piles to consist largely of individuals of the same matriline, therefore this is unlikely true for the LNP population. Another aspect is the small sample sizes in SA analyses. Although sampling sessions yielded adequate sample sizes of dung samples, removing juveniles, males, samples that would not reliably amplify using the PCR, and recaptures of individuals reduced sample sizes.

Despite weak genetic structure, we found relatedness in dung samples collected as groups to be consistent with family group expectations. Individuals within groups were significantly more related to each other than to individuals from other groups, and average pairwise relatedness between adult female elephants (0.236) was comparable to family groups in savanna elephants (0.150, 35,0.234, [59]). We linked samples over time using network models to investigate associations at the population level and found several components with more complex associations, despite behavioral studies revealing an adult female and dependent calves as the most common group type [16]. The largest component contained 22 individuals and six larger components (5–12 individuals) were found, with a mean of 3.786 for all components, and a median of two. Additionally, 75% of components consisted of individuals that shared a mtDNA haplotype and the average relatedness of individuals within components was 0.155, suggesting most associations are kin-based. However, 25% of associations were between different matrilines, which could represent group formation independent of kin [29], or cases where unrelated individuals used the same resources. Some individuals therefore had a larger number of associations than what is reflected from group sizes alone, and surprisingly most were based on the same matriline despite high haplotype diversity in the area and the close proximity between dung samples of different haplotypes. However, many components were small and consistent with group sizes from behavioral observations. Group sizes from dung collection may be underestimated as not all members may defecate, and not all samples collected were successfully genotyped. Therefore further research is still needed to address how common larger components are in forest elephant society.

Tracking associations from dung found together, combined with fine-scale spatial genetic sampling, allows for information about elephant sociality that cannot be gained from observations alone. Observational sampling is biased towards diurnal observations, as only groups visible or active during the day are observed. Using acoustic sampling, Wrege et al. [60] found that 79% of forest elephant visitations occurred at night. Although genetic methods have caveats including under-sampling groups, they can capture diurnal and nocturnal associations, as well as cryptic associations in forested habitat. Our network model created from the genetic information from non-invasive dung sampling revealed that individuals can associate with a larger number of individuals than what is reflected through group sizes, and that individuals are not always found with the same associates, which is consistent with the expectations of fission-fusion sociality. Although more information is needed to understand the full social repertoire of forest elephants, our results demonstrate that associations can be larger than what is observed in group sizes alone, and that forest elephants more often associate with individuals of the same matriline, even in genetically diverse populations. By combining results from observational studies to the association patterns detected with dung, we can get a clearer picture of forest elephant sociality.

## Supporting Information

**Table S1 Genetic diversity values for elephants at Lopé National Park, Gabon.** $N_a$ = allelic diversity, $H_e$ = expected and $H_o$ = observed heterozygosity. Multiplexes 1 and 4 had an annealing temperature of 60°C, while 2 and 3 were at 58°C.

**Table S2 Group composition, mitochondrial haplotypes, average pairwise relatedness (*R*) and 95% confidence intervals (CI) within groups.** A–adult, J–juvenile, U–unknown age category, F–female, and M–male.

## Acknowledgments

We thank the Gabonese government, CENAREST, and ANPN for permission to work in Lopé National Park, and SEGC and CIRMF for providing institutional and logistical support. J. Dibakou, J.T. Dikangadissi, E. Dimoto, and C. Nzotekoumie provided invaluable field support. We thank three anonymous reviewers for feedback in strengthening the manuscript and P. Smouse for advice on spatial autocorrelation analyses.

## Author Contributions

Conceived and designed the experiments: SS LE KJ. Performed the experiments: SS JP. Analyzed the data: SS JP. Contributed reagents/materials/analysis tools: LE SS. Wrote the paper: SS LS KJ JP.

## References


1. Johnstone RA, Cant MA (2008) Sex differences in dispersal and the evolution of helping and harming. The American Naturalist 172: 318–330.
2. Woxvold IA, Adcock GJ, Mulder RA (2006) Fine-scale genetic structure and dispersal in cooperatively breeding apostlebirds. Molecular Ecology 15: 3139–3146.
3. Hazlitt SL, Sigg DP, Eldridge MDB, Goldizen AW (2006) Restricted mating dispersal and strong breeding group structure in a mid-sized marsupial mammal (*Petrogale penicillata*). Molecular Ecology 15: 2997–3007.
4. Garant D, Kruuk LEB, Wilkin TA, McCleery RH, Sheldon BC (2005) Evolution driven by differential dispersal within a wild bird population. Nature 433: 60–65.
5. Postma E, van Noordwijk AJ (2005) Gene flow maintains a large genetic difference in clutch size at a small spatial scale. Nature 433: 65–68.
6. Greenwood PJ (1980) Mating systems, philopatry and dispersal in birds and mammals. Animal Behaviour 28: 1140–1162.
7. Storz JF (1999) Genetic consequences of mammalian social structure. Journal of Mammalogy 80: 553–569.
8. Hazlitt SL, Eldridge MDB, Goldizen AW (2004) Fine-scale spatial genetic correlation analyses reveal strong female philopatry within a brush-tailed rock-wallaby colony in southeast Queensland. Molecular Ecology 13: 3621–3632.
9. Rossiter SJ, Zubaid A, Mohd-Adnan A, Struebig MJ, Kunz TH, et al. (2012) Social organization and genetic structure: insights from codistributed bat populations. Molecular Ecology 21: 647–661.
10. Robinson SJ, Samuel MD, Lopez DL, Shelton P (2012) The walk is never random: subtle landscape effects shape gene flow in a continuous white-tailed deer population in the Midwestern United States. Molecular Ecology 21: 4190–4205.
11. Nussey DH, Coltman DW, Coulson T, Kruuk LEB, Donald A, et al. (2005) Rapidly declining fine-scale spatial genetic structure in female red deer. Molecular Ecology 14: 3395–3405.
12. Wittemyer G, Getz WM (2007) Hierarchical dominance structure and social organization in African elephants, *Loxodonta africana*. Animal Behaviour 73: 671–681.
13. Moss CJ (1988) Elephant memories. Chicago: University of Chicago Press.
14. Archie EA, Maldonado JE, Hollister-Smith JA, Poole JH, Moss CJ, et al. (2008) Fine-scale population genetic structure in a fission-fusion society. Molecular Ecology: 2666–2679.
15. Fishlock V, Lee PC (2012) Forest elephants: fission–fusion and social arenas. Animal Behaviour 85: 357–363.
16. White LJT, Tutin CG, Fernandez M (1993) Group composition and diet of forest elephants, *Loxodonta africana cyclotis* Matschie 1900, in the Lopé Reserve, Gabon. African Journal of Ecology 31: 181–199.
17. Turkalo A, Fay JM (1996) Studying forest elephants by direct observation: preliminary results from the Dzanga clearing, Central African Republic. Pachyderm 21: 45–54.
18. Munshi-South J (2011) Relatedness and demography of African forest elephants: inferences from noninvasive fecal DNA analyses. Journal of Heredity 102: 391–398.
19. Fishlock V, Lee PC, Breuer T (2008) Quantifying forest elephant social structure in Central African bai environment. Pachyderm 44: 17–26.
20. Momont L (2007) Sélection de l'habitat et organisation sociale de l'éléphant de forêt, *Loxodonta africana cyclotis* (Matschie 1900), au Gabon: Muséum National d'Histoire Naturelle.
21. White LJT (1994) Biomass of rain-forest mammals in the Lopé Reserve, Gabon. Journal of Animal Ecology 63: 499–512.
22. White LJT (1995) Factors affecting the duration of elephant dung piles in rain forest in the Lopé Reserve, Gabon. African Journal of Ecology 33: 142–150.
23. Eggert LS, Eggert JA, Woodruff DS (2003) Estimating population sizes for elusive animals: the forest elephants of Kakum National Park, Ghana. Molecular Ecology 12: 1389–1402.
24. Amos W, Whitehead H, Ferrari MJ, Glockner-Ferrari DA, Payne R, et al. (1992) Restrictable DNA from sloughed cetacean skin; its potential for use in population analysis. Marine Mammal Science 8: 275–283.
25. Archie EA, Moss CJ, Alberts SC (2003) Characterization of tetranucleotide microsatellite loci in the African savannah elephant (*Loxodonta africana africana*). Molecular Ecology Notes 3: 244–246.
26. Eggert LS, Maldonado JE, Fleischer RC (2005) Nucleic acid isolation from ecological samples: animal scat and other associated materials. Molecular Evolution: Producing the Biochemical Data Part B: 73–87.
27. Eggert LS, Ahlering M, Manka SG. Lessons from genetic censuses of forest elephants. In: Olson D, editor; 2008 November, 2007; International Elephant Foundation and Ringling Brothers Barnum and Bailey Center for Elephant Conservation. pp. 157–164.
28. Comstock KE, Wasser SK, Ostrander EA (2000) Polymorphic microsatellite DNA loci identified in the African elephant (*Loxodonta africana*). Molecular Ecology 9: 1004–1006.
29. Gobush K, Kerr BEN, Wasser S (2009) Genetic relatedness and disrupted social structure in a poached population of African elephants. Molecular Ecology 18: 722–734.
30. Ahlering MA, Maldonado JE, Fleischer RC, Western D, Eggert LS (2012) Fine-scale group structure and demography of African savanna elephants recolonizing lands outside protected areas. Diversity and Distributions 18: 952–961.
31. Frantz AC, Pope LC, Carpenter PJ, Roper TJ, Wilson GJ, et al. (2003) Reliable microsatellite genotyping of the Eurasian badger (*Meles meles*) using faecal DNA. Molecular Ecology 12: 1649–1661.
32. Hansen H, Ben-David M, McDonald DB (2008) Effects of genotyping protocols on success and errors in identifying individual river otters (*Lontra canadensis*) from their faeces. Molecular Ecology Resources 8: 282–289.
33. Waits LP, Luikart G, Taberlet P (2001) Estimating the probability of identity among genotypes in natural populations: cautions and guidelines. Molecular Ecology 10: 249–256.
34. Peakall ROD, Smouse PE (2006) GENALEX 6: genetic analysis in Excel. Population genetic software for teaching and research. Molecular Ecology Notes 6: 288–295.
35. Archie EA, Moss CJ, Alberts SC (2006) The ties that bind: genetic relatedness predicts the fission and fusion of social groups in wild African elephants. Proceedings of the Royal Society Biological Sciences Series B 273: 513–522.
36. Miller CR, Joyce P, Waits LP (2002) Assessing allelic dropout and genotype reliability using maximum likelihood. Genetics 160: 357–366.
37. McKelvey KS, Schwartz MK (2005) DROPOUT: a program to identify problem loci and samples for noninvasive genetic samples in a capture-mark-recapture framework. Molecular Ecology Notes 5: 716–718.
38. Munshi-South J, Tchignoumba L, Brown J, Abbondanza N, Maldonado JE, et al. (2008) Physiological indicators of stress in African forest elephants (*Loxodonta africana cyclotis*) in relation to petroleum operations in Gabon, Central Africa. Diversity and Distributions 14: 995–1003.
39. Ahlering MA, Hailer F, Roberts MT, Foley C (2011) A simple and accurate method to sex savannah, forest and asian elephants using noninvasive sampling techniques. Molecular Ecology Resources 11: 831–834.
40. Fernando P, Pfrender ME, Encalada SE, Lande R (2000) Mitochondrial DNA variation, phylogeography and population structure of the Asian elephant. Heredity 84: 362–372.
41. Eggert LS, Rasner CA, Woodruff DS (2002) The evolution and phylogeography of the African elephant inferred from mitochondrial DNA sequence and nuclear microsatellite markers. Proceedings of the Royal Society of London Series B 269: 1993–2006.
42. Sukumar R (2003) The Living Elephants: Evolutionary Ecology, Behavior, and Conservation. New York: Oxford University Press.
43. van Oosterhout C, Hutchinson WF, Wills DPM, Shipley P (2004) MICRO-CHECKER: software for identifying and correcting genotyping errors in microsatellite data. Molecular Ecology Notes 4: 535–538.
44. Raymond M, Rousset F (1995) GENEPOP (version 1.2): population genetics software for exact tests and ecumenicism. Journal of Heredity 86: 248–249.
45. Queller DC, Goodnight K (1989) Estimating relatedness using genetic markers. Evolution 43: 258–275.
46. Smouse PE, Peakall R (1999) Spatial autocorrelation analysis of individual multiallele and multilocus genetic structure. Heredity 82: 561–573.
47. Peakall R, Ruibal M, Lindenmayer DB (2003) Spatial autocorrelation analysis offers new insights into gene flow in the Australian bush rat, *Rattus fuscipes*. Evolution 57: 1182–1195.
48. Van De Casteele T, Galbusera P, Matthysen E (2001) A comparison of microsatellite-based pairwise relatedness estimators. Molecular Ecology 10: 1539–1549.

49. Wang J (2011) COANCESTRY: a program for simulating, estimating and analysing relatedness and inbreeding coefficients. Molecular Ecology Resources 11: 141–145.
50. Hemelrijk CK (1990) A matrix partial correlation test used in investigations of reciprocity and other social interaction patterns at group level. Journal of Theoretical Biology 143: 405–420.
51. Duchesne P, ÉTienne C, Bernatchez L (2006) PERM: a computer program to detect structuring factors in social units. Molecular Ecology Notes 6: 965–967.
52. Borgatti SP, Everett MG, Freeman LC (2002) Ucinet for Windows: software for social network analysis. Harvard, MA: Analytic Technologies.
53. Borgatti SP (2002) NetDraw software for network visualization. Lexington, KY: Analytic Technologies.
54. Van Coeverden de Groot P, Putnam A, Erb P, Scott C, Melnick D, et al. (2011) Conservation genetics of the black rhinoceros, *Diceros bicornis bicornis*, in Namibia. Conservation Genetics 12: 783–792.
55. Pope LC, Domingo-Roura X, Erven K, Burke T (2006) Isolation by distance and gene flow in the Eurasian badger (*Meles meles*) at both a local and broad scale. Molecular Ecology 15: 371–386.
56. Schuttler SG, Blake S, Eggert LS (2012) Movement patterns and spatial relationships among African forest elephants. Biotropica 44: 445–448.
57. White LJT (1994) *Sacoglottis gabonensis* fruiting and the seasonal movements of elephants in the Lopé Reserve, Gabon. Journal of Tropical Ecology 10: 121–125.
58. Maisels F, Strindberg S, Blake S, Wittemyer G, Hart J, et al. (2013) Devastating decline of forest elephants in Central Africa. PLoS ONE 8: e59469.
59. Wittemyer G, Okello JBA, Rasmussen H, Arctander P, Nyakaana S, et al. (2009) Where sociality and relatedness diverge: the genetic basis for hierarchical social organization in African elephants. Proceedings of the Royal Society B 276: 3513–3521.
60. Wrege PH, Rowland ED, Bout N, Doukaga M (2012) Opening a larger window onto forest elephant ecology. African Journal of Ecology 50: 176–183.

# Flying-Fox Species Density - A Spatial Risk Factor for Hendra Virus Infection in Horses in Eastern Australia

**Craig Smith[1]*, Chris Skelly[2,5], Nina Kung[1], Billie Roberts[3], Hume Field[1,4]**

**1** Queensland Centre for Emerging Infectious Diseases, Department of Agriculture, Fisheries and Forestry, Brisbane, Queensland, Australia, **2** Biosecurity Intelligence Unit, Department of Agriculture, Fisheries and Forestry, Brisbane, Queensland, Australia, **3** Griffith School of Environment, Griffith University, Brisbane, Queensland, Australia, **4** EcoHealth Alliance, New York, New York, United States of America, **5** GIS People Pty Ltd, Brisbane, Queensland, Australia

## Abstract

Hendra virus causes sporadic but typically fatal infection in horses and humans in eastern Australia. Fruit-bats of the genus *Pteropus* (commonly known as flying-foxes) are the natural host of the virus, and the putative source of infection in horses; infected horses are the source of human infection. Effective treatment is lacking in both horses and humans, and notwithstanding the recent availability of a vaccine for horses, exposure risk mitigation remains an important infection control strategy. This study sought to inform risk mitigation by identifying spatial and environmental risk factors for equine infection using multiple analytical approaches to investigate the relationship between plausible variables and reported Hendra virus infection in horses. Spatial autocorrelation (Global Moran's I) showed significant clustering of equine cases at a distance of 40 km, a distance consistent with the foraging 'footprint' of a flying-fox roost, suggesting the latter as a biologically plausible basis for the clustering. Getis-Ord Gi* analysis identified multiple equine infection hot spots along the eastern Australia coast from far north Queensland to central New South Wales, with the largest extending for nearly 300 km from southern Queensland to northern New South Wales. Geographically weighted regression (GWR) showed the density of *P. alecto* and *P. conspicillatus* to have the strongest positive correlation with equine case locations, suggesting these species are more likely a source of infection of Hendra virus for horses than *P. poliocephalus* or *P. scapulatus*. The density of horses, climate variables and vegetation variables were not found to be a significant risk factors, but the residuals from the GWR suggest that additional unidentified risk factors exist at the property level. Further investigations and comparisons between case and control properties are needed to identify these local risk factors.

**Editor:** Bradley S. Schneider, Metabiota, United States of America

**Funding:** The State of Queensland, through the Queensland Centre for Emerging Infectious Diseases, provided funding and logistical support for the project. Additional support was provided the by the State of Queensland, the State of New South Wales and the Commonwealth of Australia under the National Hendra Virus Research Program. The funders had no role in study design, data collection and analysis, decision to publish, or preparation of the manuscript.

**Competing Interests:** The employment of one of the authors (WCS) by a commercial company (GIS People P/L) does not affect the authors' Competing Interests.

* Email: craig.smith@daff.qld.gov.au

## Introduction

Hendra virus (genus *Henipavirus*, family *Paramyxoviridae*) was first described in September 1994 in Australia, when it caused an outbreak of typically fatal disease in horses and two close-contact humans [1–4]. Sporadic cases continue to occur in horses and humans, with some 80 confirmed or possible equine cases and seven confirmed human cases identified to 31 December, 2012 [5–9], and an increase in reported cases since 2006 [10]. Fruit-bats of the genus *Pteropus* (commonly known as flying-foxes) are the natural reservoir of the virus [11–16].

Effective treatment for Hendra virus (HeV) infection in horses and humans is lacking, and while a vaccine for horses has been recently released [17], minimising exposure risk remains a fundamental aspect of horse owner management strategies. It is well recognised that ecological processes influence the spatial distribution and patterns of disease risk and disease incidence [18]. Fundamental spatial risk factors comprise the geographic distribution of the pathogen, the natural host and potential spill-over hosts, overlaid by an environmental complexity of ecological and climatic variables that affect the behaviour of the above. The ability of climatic variables to influence host-pathogen interactions and spatial patterns of disease was highlighted in Australia recently with the identification of the role of relative humidity, maximum air temperature and wind speed in the spread of introduced equine influenza virus [19].

A limited number of studies have sought to elaborate the infection and transmission dynamics of Hendra virus in flying-foxes, and identify temporal and spatial risk factors for spill-over of infection from flying-foxes to horses. McFarlane et al (2011) identified a positive association with postcodes containing flying-fox roosts, and with geographic seasonal low rainfall, but found no evident association with horse density and vegetation primary productivity [20]. One of us (Smith, CS., unpublished data) previously found an association with radial proximity to flying-fox roosts, with a statistically significant increased risk of equine cases within 7 km of a known roost. Fogarty *et al* (2008) showed that Hendra virus survival in-vitro decreased with increasing temperature and desiccation [21], supporting the contention that climatic variables likely contribute to infection risk for horses. Correlates with the presence of anti-Hendra virus antibodies in flying-foxes are better understood, and include species, age, location, year, reproductive status (pregnancy or lactation) and season [6,15,22,23].

Reported equine Hendra virus cases to date have been restricted to the adjoining eastern Australia states of Queensland and New South Wales. The limited case numbers prior to 2011 has precluded robust spatial analysis to date, however the unprecedented cluster of 18 separate incidents involving 23 cases in 2011 somewhat relieves this constraint, and with our coincident improved knowledge of horse and flying-fox geographic distribution, invites further spatial analysis. In this paper, we model the spatial occurrence of reported Hendra virus infections in horses, and seek to identify key spatial and environmental risk factors.

## Methods

### Spatial Data

**Hendra virus equine cases.** There were about 80 recorded cases of Hendra infection in horses from 40 discrete case properties in Queensland (n = 31) and New South Wales (n = 9) between September 1994 and 30 December 2012. These cases were reported to and investigated by the Queensland Department of Agriculture, Fisheries and Forestry (DAFF) and the New South Wales Department of Primary Industries (DPI) respectively. The 40 epidemiologically unrelated properties, each with one or more confirmed cases [9,10] are used in this study to denote individual 'spill-over events' of infection from flying-foxes to horses. Case property spatial data in the form of property centroid (latitude and longitude) was sourced in de-identified format from DAFF and DPI.

**Equine control properties.** We randomly selected 1189 control properties from 118,900 registered horse properties in Queensland, New South Wales and Victoria that had not had a confirmed or suspect case of Hendra virus between September 1994 and 31 December 2012. This provided a case-control ratio of nearly 1:30, and ensured that case properties were matched regardless of variability in property characteristics. Control property spatial data in the form of property centroid (latitude and longitude) was sourced in de-identified format from DAFF, DPI, and the Victorian Department of Primary Industries.

**Flying-fox occurrence.** Flying-fox data are recorded observations of the geographic occurrence of flying-foxes drawn from two sources. Spatial data for the distribution of *P. alecto* and *P. poliocephalus* was derived from Roberts *et al* (2011) [24] and *P. conspicillatus* and *P. scapulatus* occurrence was sourced from *WildNet*, an electronic database managed by the Queensland Department of Environment and Heritage Protection [25]. The data collectively constitute observations from sources including the Australian Bird and Bat Banding Scheme, the Australian National Wildlife Collection, state museums, state wildlife atlases, and published and reputable unpublished research. Records span the period 1843–2007, with around 75% of the observations recorded since 1994, and cover all or part of the species mainland Australian distribution.

**Environmental data.** Spatial climate data (annual rainfall, annual minimum, maximum and mean temperature, relative humidity) for eastern Australia based on the most recent standard 30-year period 1961–1990 as defined by the World Meteorological Organisation [26], and normalised difference vegetation index (NDVI) for the recent peak HeV case years of 2011 and 2012 were sourced from the Australian Bureau of Meteorology [27]. Vegetation mapping (dominant species of the tallest stratum) was obtained from Geoscience Australia [28].

### Spatial Analysis

Analysis focused on eastern Australia, comprising the states of Queensland and New South Wales (where all reported equine Hendra virus cases have occurred) and the adjoining state of Victoria. Analysis on state-based boundaries (rather than biological boundaries) facilitated data collection. To facilitate computation, we used a 1% subset of the 118,900 potential control properties, as indicated above. We also merged the datasets of P. alecto and P. conspicillatus because of their demonstrated paraphyly [29] and evident species tropism of viral infections in bats [30–32]. All data were standardised by scaling from 0 and 1 to allow comparison of the coefficients [33].

Spatial autocorrelation (Global Moran's I, zone of indifference conceptualisation) was used to identify spatial clustering and distances at which spatial processing was most pronounced [34]. Hot spot analysis (Getis-Ord Gi*, zone of indifference conceptualisation) identified areas with statistically significant clustering of Hendra virus spillovers [35]. Geographically weighted regression (GWS) was used to investigate the spatial pattern of spillovers in relation to environmental data. The latter were identified using ordinary least squares (OLS). Kernel density and inverse distance weighted were employed to interpolate data. All variables were included in the initial OLS model. A variable inflation factor (VIF) >7.5 suggested redundancy amongst variables and they were removed from the model in succession. Heteroscedasticity, or the relationship between predicted values and changes in variable magnitudes, was measured using the Koenker (BP) Statistic. OLS models that display statistically significant heteroscedasiticity ($P< 0.05$) should use the reported Robust P, and are good candidates for GWR. Variables that did not have a statistically significant Robust P ($P<0.1$) were removed from the model in succession. Table 1 indicates the variables used in OLS and GWR. All spatial analysis was undertaken in Arcmap 10, Spatial Statistics Tools (ESRI, USA).

## Results

The 40 equine case properties reported up to December 2012 and the 1,189 randomly selected control horse properties are identified in Figure 1A.

The Global Moran's I test identified that spatial clustering of spill-over events was evident, and was statistically significant at a distance of 40 km ($P<0.001$) (Figure 2). This distance was then used as the 'distance band' in the Getis-Ord Gi* analysis (below) and as the search radius for interpolation.

Hot spot (Getis-Ord Gi*) analysis identified multiple areas of significant clustering along the east coast of Australia (Z Score > 1.96 SD) (Figure 1B). Also interesting is the suggestion of 'cold spots', clustering of control properties on the Southern Downs of south-east Queensland and along the central coast of NSW (Z Score <1.00 SD).

Of the 15 spatial and environmental variables included in the study (Table 1), six were identified as being statistically significant in the initial and final OLS model (Table 2). Of these six, three had an association in the GWR, with the strength and direction of the correlation varying (Table 2). The density of flying-foxes *P. alecto* and *P. conspicillatus* had the strongest positive correlation (> 0.5) with reported Hendra virus equine cases. The annual minimum temperature had a weak positive correlation (<0.2) and the density of *P. scapulatus* had a weak negative correlation (>−0.2). Negligible correlations (−0.1<0.1) were identified for *P. poliocephalus*, relative humidity (as measured at 0900) and horse population. The predicted (P) and residual (SD) values for a

**Table 1.** Description of spatial and environmental variables used in OLS and GWR.

| Variable | Description |
|---|---|
| P. alecto | Kernel density analysis of the recorded sightings of flying-foxes/$km^2$ |
| P. conspicillatus | " |
| P. poliocephalus | " |
| P. scapulatus | " |
| P. alecto/conspicillatus | Kernel density analysis of the merged recorded sightings of both *P. alecto* and *P. conspicillatus*/$km^2$ |
| Distance to roost | Distance to the nearest flying-fox roost |
| Horse population | Kernel density analysis of the population of horses/$km^2$ |
| Horse properties | Kernel density analysis of the number of horses properties/$km^2$ |
| Annual Rainfall | Average annual rainfall over the period 1961 to 1990 |
| Annual Mean Temp. | Average annual mean temperature over the period 1961 to 1990 |
| Annual Min. Temp. | Average annual minimum temperature over the period 1961 to 1990 |
| Annual Max. Temp. | Average annual maximum temperature over the period 1961 to 1990 |
| Relative Humidity 0900 | Average relative humidity at 0900 hours over the period 1961 to 1990 |
| Relative Humidity 1500 | Average relative humidity at 1500 hours over the period 1961 to 1990 |
| Vegetation | The dominant species of the tallest stratum |

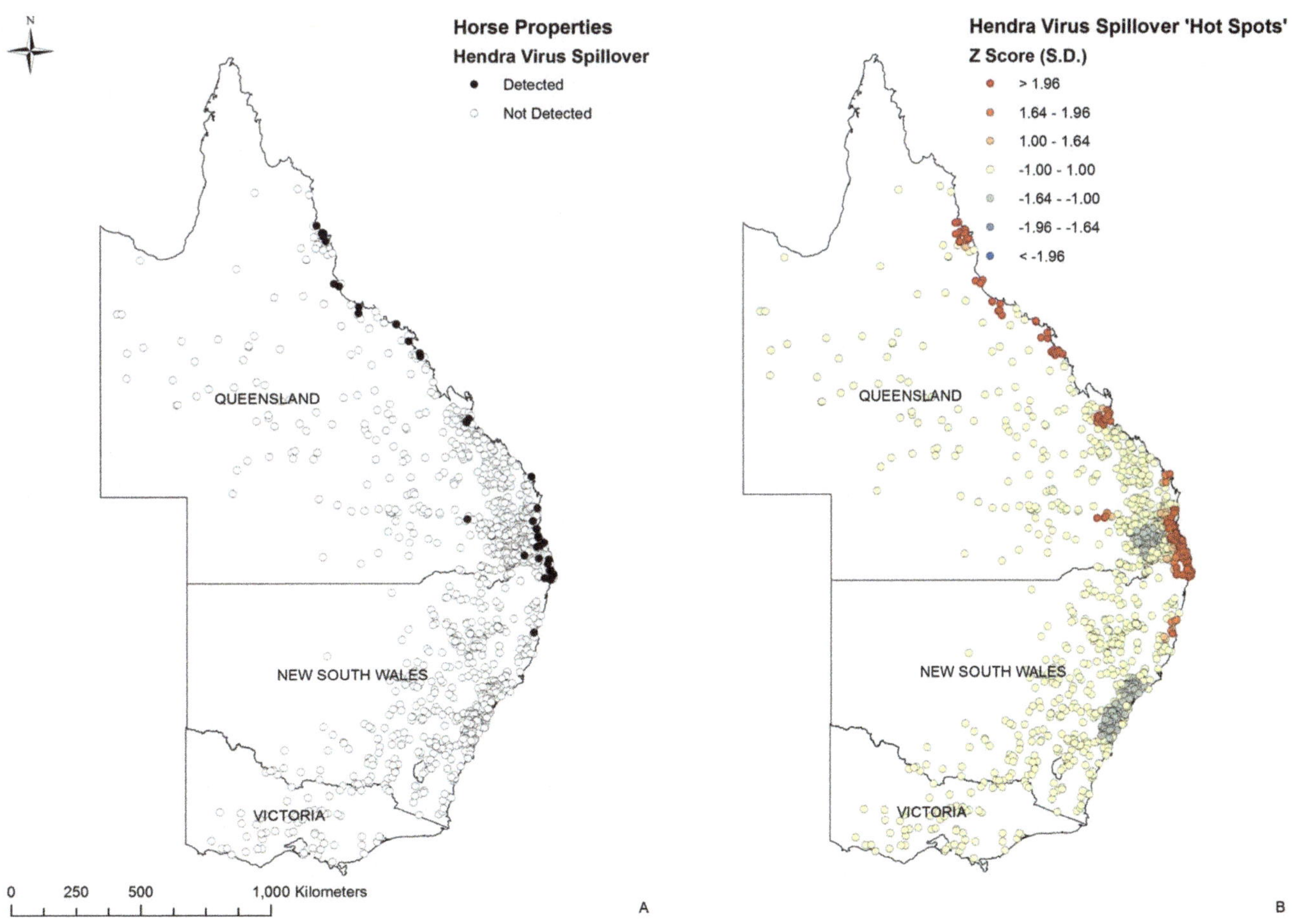


**Figure 1. Equine property locations and Hendra virus spillover hot spots.** (A) Forty reported Hendra virus equine cases September 1994 to December 2012 and 1,189 randomly selected control horse properties. (B) Hot spot analysis (Getis-Ord Gi*) identified areas of significant clustering of spill-overs (Z Score>1.96 SD) along the central and northern coasts of eastern Australia.

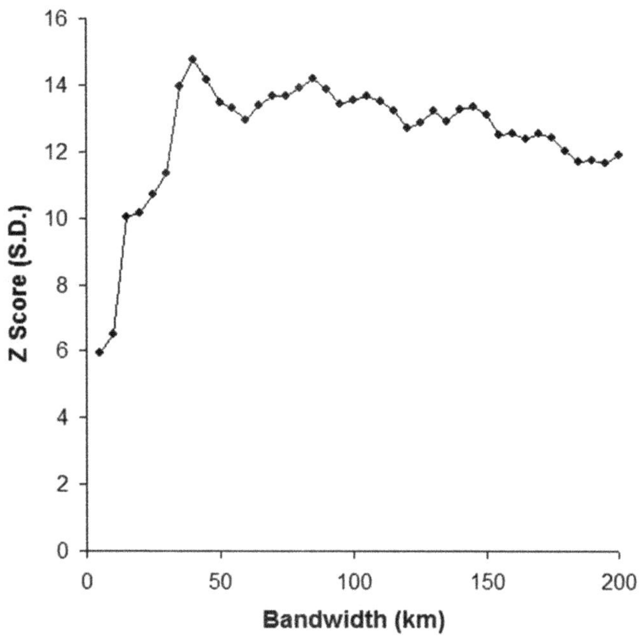


**Figure 2. Spatial autocorrelation (Global Moran's I) of Hendra virus spill-overs by distance.** A peak Z Score at 40 km suggests that spatial processes exist at this distance to produce pronounced spatial clustering.

Hendra virus spill-over on horse properties are mapped in Figures 3A and B.

## Discussion

Hendra virus was initially perceived as a Queensland problem, with the first six reported equine incidents (1994–2004) occurring in this state. That perception changed in 2006 when the first equine case was reported in the neighbouring state of New South Wales. While it is probable that under-reporting of cases occurred historically, it is less likely post-2008, with increased industry and public awareness following the fatal infection of a veterinarian. However, any reporting bias is likely to be constant geographically, and thus unlikely to significantly impact our analysis. A lack of serological surveillance data precludes inclusion of sub-clinical infections. To date, all reported cases have occurred in eastern Queensland and New South Wales, within the variably overlapping geographic ranges of all four mainland Pteropus species [10]. We analysed the spatial distribution of reported Hendra virus infections in horses and identified that spatial clustering existed. Spatial clustering in the landscape is evidence of underlying spatial processes, and the pronounced clustering at 40 km, a distance consistent with the nocturnal foraging range of flying-foxes [36,37] suggests the foraging 'footprint' of a roost as a biologically plausible underlying spatial dynamic.

Hot spot analysis (Getis-Ord GI*) is a useful tool to identify spatial clusters of both high and low values, and has previously been used to model disease occurrence [38–41]. Our analysis of Hendra virus spill-over events suggests multiple hot spots along the eastern Australia coast from far north Queensland to central New South Wales. The largest extends for nearly 300 km, and straddles the Queensland-New South Wales border (Fig. 1B). This hot spot includes almost half of all known events (19/40) to 31 December 2012. While it might be argued that the spatial significance is

**Table 2.** Identification and modelling of spatial and environmental variables for Hendra virus spill-overs using OLS[a] and GWR[b].

| | OLS Model (Initial) | OLS Model (Final) | GWR Model |
|---|---|---|---|
| P. alecto | 0.57 (0.06)*** | | |
| P. conspicillatus | 0.35 (0.07)* | | |
| P. poliocephalus | −0.07 (0.05)* | −0.09 (0.04)*** | −0.08 (0.08) |
| P. scapulatus | −0.21 (0.05)* | −0.23 (0.05)** | −0.20 (0.07) |
| P. alecto/conspicillatus | 0.48 (0.06)*** | 0.53 (0.05)*** | 0.55 (0.06) |
| Distance to roost | −0.01 (0.05) | | |
| Horse population | −0.09 (0.04)** | −0.07 (0.03)** | −0.05 (0.05) |
| Horse properties | 0.01 (0.04) | | |
| Annual Rainfall | 0.02 (0.08) | | |
| Annual Mean Temp. | 0.20 (0.22) | | |
| Annual Min. Temp. | −0.07 (0.16) | 0.21 (0.03)*** | 0.14 (0.06) |
| Annual Max. Temp. | 0.01 (0.16) | | |
| Relative Humidity 0900 | −0.06 (0.09) | 0.14 (0.03)*** | 0.06 (0.06) |
| Relative Humidity 1500 | 0.21 (0.11) | | |
| Vegetation | 0.01 (0.01) | | |
| Adjusted $R^2$ | 0.18 | 0.18 | 0.21 |
| AIC | −991.78 | −994.07 | −1037.00 |
| Global Moran's I (P) | 0.53 | 0.18 | 0.62 |

[a]For the OLS models, estimates correspond to the standardised coefficient and the standard error in parentheses.
[b]For the GWR model, estimates correspond to the standardised mean coefficient and the standard error in parentheses.
*P<0.10,
**P<0.05,
***P<0.01 are statistically significant levels of the OLS model.

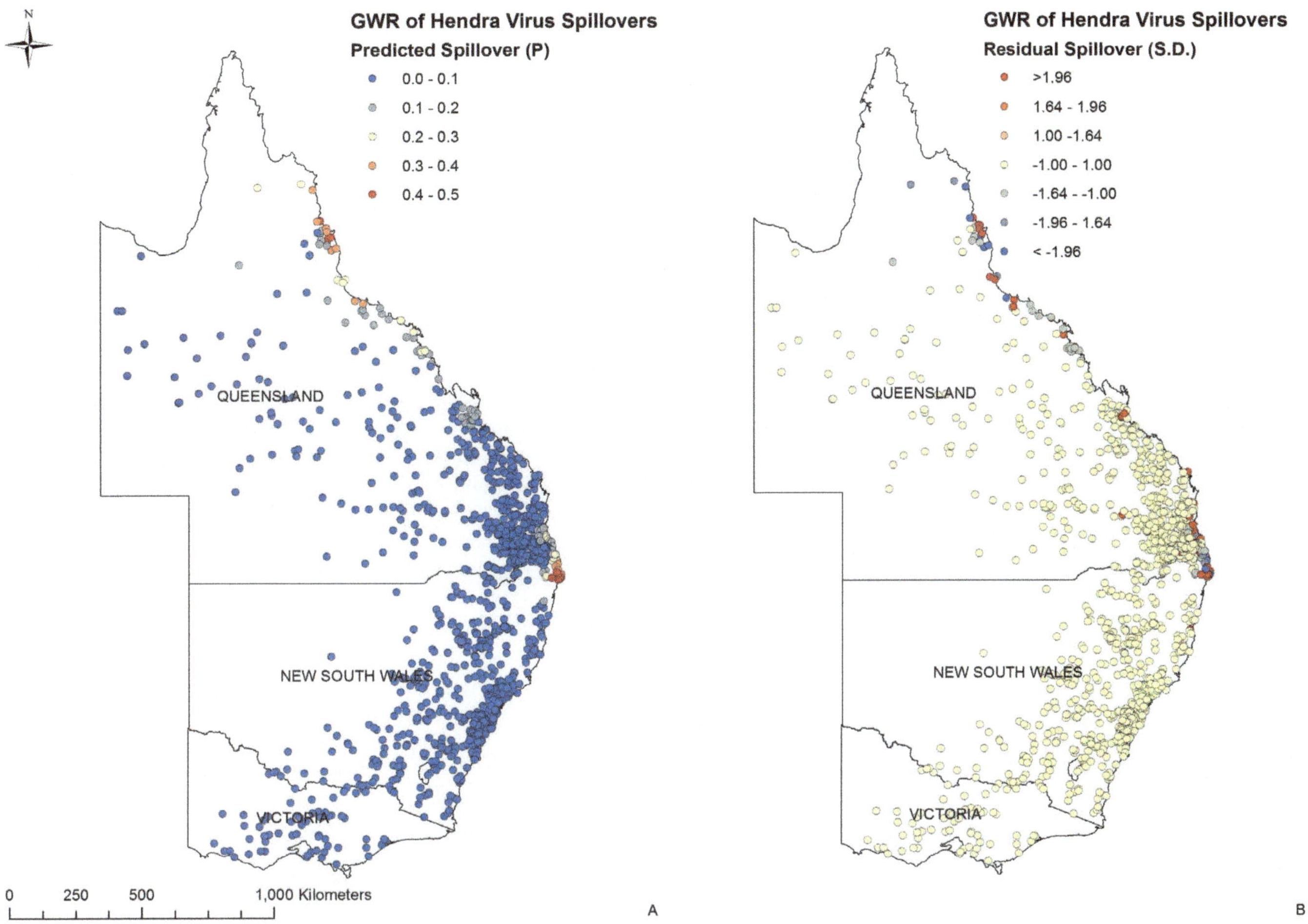


**Figure 3. The final GWR model for the spill-over of Hendra virus in eastern Australia, with predicted (A) and residual (B) values.** The density of flying-foxes *P. alecto* and *P. conspicillatus* had the strongest positive correlation with reported Hendra virus spill-overs (A). An absence of spatial autocorrelation of the residuals suggests additional (as yet unidentified) local risk factors play a role in Hendra virus spill-over from flying-foxes to horses (B).

confounded by the unprecedented number of cases that occurred in 2011, when this year is excluded, this hot spot still accounts for nearly 30% of the events, and any effect of the temporal cluster would not fundamentally impact our analyses. From an epidemiological perspective, the preponderance of spillover events in this region suggests either the existence of a predisposing causal factor or better detection in this region. Combined, the identified hot spots contain about 7% of equine control properties, translating to around 8,000 equine properties at putative elevated risk of Hendra virus infection in eastern Australia, and suggesting advantage in spatially targeted risk mitigation approach.

In addition to identifying areas of evident increased spillover risk, the hot spot analysis also suggested areas of decreased risk, or 'cold spots'. While the findings are not statistically significant, the putative 'cold spots' do correspond geographically to spatial clusters of equine control properties, and support the concept of a spatial risk gradient.

The geographically weighted regression analysis found the combined density of *P. alecto* and *P. conspicillatus* to be the strongest positive correlation with reported equine case locations in both the OLS and GWR models (Table 2). While the flying-fox dataset likely reflects observations of varying sampling intensity, the large number of observations and the extended timescale over which they were made supports its use to derive relative flying-fox species density. The spatial distribution of Hendra virus spill-over events and their association with flying-foxes has previously been reported [20,42]. However, Field and Kung (2011) argue that the relationship between horses and flying-foxes alone does not explain the observed spatial pattern [43]. Field *et al* (2011) suggest that some species of flying-foxes may play a greater role in the transmission of infection to horses [44], implying that the pattern of spill-overs may better reflect species preponderance. This prospect is supported by the occurrence of multiple species of flying-fox on mainland Australia, by their differing geographic ranges, by the evident species-specific tropism of some bat viruses, and by the spatial occurrence of Hendra virus spill-overs. Our positive correlation of combined *P. alecto* and *P. conspicillatus* density with equine case locations supports this thinking, and suggests that these species play a primary role in the infection of horses. The frequent detection of Hendra virus genome in the urine of these two species further supports this contention [44] (Edson, D. *et al*, unpublished).

Roberts *et al* (2011) [24] reported the southward expansion of the range of *P. alecto*. Their reported rate of southern range expansion (100 km per decade) plus our finding of the strong correlation between *P. alecto*/*P. conspicillatus* density and equine case location, suggests the likelihood of a future southward expansion of Hendra virus equine cases, unless currently unidentified moderating factors exist. The absence of reported equine cases in NSW in the 1990s, the single case in 2006, and the cluster of cases in 2011 is consistent with this scenario.

The densities of *P. scapulatus* and *P. poliocephalus* have weak negative or negligible correlations with equine cases, reflecting the absence of equine cases in that part of their range not overlapping *P. alecto* or *P. conspicillatus*. This lack of positive correlation suggests *P. scapulatus* and *P. poliocephalus* play a limited role in the infection of horses, and is consistent with the infrequent detection of Hendra virus genome in the urine of these species [44] (Edson, D. *et al*, unpublished), despite the presence of neutralising antibodies to Hendra virus. While it might be suggested that the spatial correlation between equine case locations and the combined density of *P. alecto* and *P. conspicillatus* reflects an underlying latitudinal trend rather than a biological correlation with these two species, two observations argue against this: firstly, the absence of a positive gradient between far north Queensland and southeast Queensland (indeed the reverse is true); and secondly, the evident negative correlation between Hendra virus excretion in *P. scapulatus* and *P. alecto* at corresponding latitudes (Field, H. *et al*, unpublished).

Thus, an increasingly plausible scenario is the occurrence of species-specific Hendra virus 'strains' in Australian flying-foxes with varying infectivity and/or pathogenicity to horses. The related Nipah virus provides a relevant precedent, with the 'Malaysian' strain found in *P. vampyrus*, the 'Bangladesh' strain found in *P. giganteus*, and both strains detected in the geographically intermediate *P. lylei* in Thailand [45]. This thinking prompts science-based risk mitigation, and poses an alternative to the current approach of assessing risk on the basis of flying-fox occurrence collectively.

While the presence of horses is obviously a necessary component of equine case causality, we found that the density of horses had negligible correlation with equine case locations, and is evidently not a significant risk factor. McFarlane *et al* (2011) found the same [20]. This scenario plausibly reflects the ubiquitous presence of horses throughout eastern Australia, albeit at varying densities, and thus the *a priori* potential for spill-over. But given the presence of horses, our findings show that the primary spatial risk factor for spill-over and consequent equine cases is the occurrence of *P. alecto* or *P. conspicillatus*. However, the lack of clustering of the residuals in both the OLS and GWR models (Global Moran's I, $P = 0.32$ and Figure 3B) suggests additional unidentified spatial risk factors likely play a role in Hendra virus spill-over. Visual inspection of the residuals indicates that case properties have a high positive residual ($SD > 1.96$), while neighbouring properties have a negative residual ($SD < -1.00$). These findings suggest that while the distribution of *P. alecto* and *P. conspicillatus* is a significant global risk factor for spill-over, additional local risk factors play a significant, but as yet undetermined role. These factors might include property attributes, and husbandry and management practices that reduce the likelihood of flying-fox-horse interaction, such as excluding horses from paddocks where flying-foxes are active, and covering horse food and water points were possible [9]. Further investigations and comparisons between case and control properties are needed to confirm these local risk factors.

Our analysis of climate data sought to identify correlations consistent with the experimental findings of Fogarty *et al* (2008) who reported that Hendra virus survivability decreases dramatically as temperature increases [21]. Similarly, Scanlan *et al* (2014) modelled the effect of temperature on Hendra virus survival in the environment, finding a good temporal fit with the cumulative annual clustering of spill-overs in the Australian winter [46]. Our finding of a weak positive correlation ($0.14 \pm 0.06$) between a higher annual minimum temperature and spill-over is incongruent, and is likely confounded by the northern geographic distribution (higher minimum temperatures) of *P. alecto* and *P. conspicillatus*. More broadly, the absence of any identifiable climatic risk factor does not preclude its existence. It might be argued that the period of the current standard 30-yr climate dataset constrained our ability to identify any emerging climatic factors, but more fundamentally there are inherent challenges in terms of which variables to include in an analysis, and on what spatial scale. With a nomadic natural host species able to move hundreds of kilometres in a week, it is entirely plausible that a climatic risk factor might manifest remote to a subsequent spill-over location. Additionally, a climatic risk factor that acts indirectly and/or after a time-lag (for example, annual rainfall on subsequent food resource) poses confronting analytical complexity.

## Conclusion

The threat of Hendra virus causes particular anxiety and concern for horse owners in eastern Australia because of the very high case fatality rates in infected horses and close contact humans. Despite the recent availability of a vaccine for horses, it remains important that horse owners continue to implement risk management strategies that minimise horse exposure to Hendra virus. Our findings validate anecdotal assessments of a spatial risk component to Hendra virus infection in horses. They indicate that equine incidents are indeed spatially clustered, indicating underlying spatially occurring risk factors, and that multiple equine infection hot spots exist along the eastern Australia coast. Furthermore, we have shown that the density of two related species of flying-fox, *P. alecto* and *P. conspicillatus*, has a strong positive correlation with equine case locations. This finding suggests that these species play a primary role in Hendra virus infection of horses. We continue to seek other unidentified risk-contributing variables and to refine our model accordingly.

## Acknowledgments

We acknowledge and thank the following people and institutions: Barbara Moloney (DPI NSW) and Roger Paskin (then DPI Victoria) for the respective equine property datasets, and Brendon Cowled (then DAFF Australia, now AusVet) and Simon Firestone (then The University of Sydney Faculty of Veterinary Science) for the 'cleaned' version of Queensland and New South Wales datasets; Carol de Jong and Amanda McLaughlin (Queensland Centre for Emerging Infectious Diseases) for advice and assistance with the manuscript.

## Author Contributions

Conceived and designed the experiments: C. Smith C. Skelly NK HF. Performed the experiments: C. Smith C. Skelly. Analyzed the data: C. Smith C. Skelly. Contributed reagents/materials/analysis tools: C. Smith C. Skelly NK BR. Wrote the paper: C. Smith HF C. Skelly NK BR.

## References

1. Murray K, Rogers R, Selvey L, Selleck P, Hyatt A, et al. (1995) A novel Morbillivirus pneumonia of horses and its transmission to humans. Emerging Infectious Diseases 1: 31–33.
2. Murray K, Selleck P, Hooper P, Hyatt A, Gould A, et al. (1995) A Morbillivirus that caused fatal disease in horses and humans. Science 268: 94–97.

3. Selvey LA, Wells RM, McCormack JG, Ansford AJ, Murray K, et al. (1995) Infection of Humans and Horses by a Newly Described Morbillivirus. Medical Journal of Australia 162: 642–645.
4. Baldock FC, Douglas IC, Halpin K, Field H, Young PL, et al. (1996) Epidemiological investigations into the 1994 Equine Morbillivirus outbreaks in Queensland, Australia. Singapore Veterinary Journal 20: 57–61.
5. Field H, Schaaf K, Kung N, Simon C, Waltisbuhl D, et al. (2010) Hendra Virus Outbreak with Novel Clinical Features, Australia. Emerging Infectious Diseases 16: 338–340.
6. Young PL, Halpin K, Selleck PW, Field H, Gravel JL, et al. (1996) Serologic evidence for the presence in Pteropus bats of a paramyxovirus related to equine morbillivirus. Emerging Infectious Diseases 2: 239–40.
7. Field HE, Breed AC, Shield J, Hedlefs RM, Pittard K, et al. (2007) Epidemiological perspectives on Hendra virus infection in horses and flying foxes. Australian Veterinary Journal 85: 268–270.
8. Playford EG, McCall B, Smith G, Slinko V, Allen G, et al. (2010) Human Hendra Virus Encephalitis Associated with Equine Outbreak, Australia, 2008. Emerging Infectious Diseases 16: 219–223.
9. Anon. What is Hendra virus? The Dept of Agriculture, Forestry & Fisheries. Queensland Government, Brisbane, Australia. Available: http://www.daff.qld.gov.au/animal-industries/animal-health-and-diseases/a-z-list/hendra-virus/general-information/what-is-hendra-virus. Accessed: 26 Feb 2014.
10. Field H, Crameri G, Kung N, Wang L (2012) Ecological aspects of Hendra virus. Curr Top Microbiol Immunol 359: 11–23.
11. Young PL, Halpin K, Selleck PW, Field H, Gravel JL, et al. (1996) Serologic evidence for the presence in pteropus bats of a paramyxovirus related to equine morbillivirus. Emerging Infectious Diseases 2: 239–240.
12. Halpin K, Young P, Field H (1996) Identification of likely natural hosts for equine morbillivirus. Communicable Diseases Intelligence 20: 476.
13. Halpin K, Young PL, Field HE, Mackenzie JS (2000) Isolation of Hendra virus from pteropid bats: a natural reservoir of Hendra virus. Journal of General Virology 81: 1927–1932.
14. Field H, Young P, Yob JM, Mills J, Hall L, et al. (2001) The natural history of Hendra and Nipah viruses. Microbes and Infection 3: 307–314.
15. Field H (2005) The ecology of Hendra virus and Australian bat lyssavirus. PhD thesis, The University of Queensland, Brisbane.
16. Halpin K, Hyatt AD, Fogarty R, Middleton D, Bingham J, et al. (2011) Pteropid Bats are Confirmed as the Reservoir Hosts of Henipaviruses: A Comprehensive Experimental Study of Virus Transmission. American Journal of Tropical Medicine and Hygiene 85: 946–951.
17. Middleton D, Pallister J, Klein R, Feng YR, Haining J, et al. (2014) Hendra virus vaccine, a one health approach to protecting horse, human, and environmental health. Emerging Infectious Disease. 20: 372–9. doi:10.3201/eid2003.131159.
18. Ostfeld R, Glass G, Keesing F (2005) Spatial epidemiology: an emerging (or re-emerging) discipline. Trends in Ecology & Evolution 20: 328–336.
19. Firestone S, Cogger N, Ward M, Toribio J-A, Moloney B, et al. (2012) The Influence of Meteorology on the Spread of Influenza: Survival Analysis of an Equine Influenza (A/H3N8) Outbreak. PLoS One 7.
20. McFarlane R, Becker N, Field H (2011) Investigation of the Climatic and Environmental Context of Hendra Virus Spillover Events 1994–2010. Plos One 6.
21. Fogarty R, Halpin K, Hyatt AD, Daszak P, Mungall BA (2008) Henipavirus susceptibility to environmental variables. Virus Research 132: 140–144.
22. Plowright RK, Field HE, Smith C, Divljan A, Palmer C, et al. (2008) Reproduction and nutritional stress are risk factors for Hendra virus infection in little red flying foxes (Pteropus scapulatus). Proceedings of the Royal Society B-Biological Sciences 275: 861–869.
23. Breed A, Breed M, Meers J, Field HE (2011) Evidence of Endemic Hendra Virus Infection in Flying-Foxes (Pteropus conspicillatus)-Implications for Disease Risk Management. PLoS One 6.
24. Roberts BJ, Catterall CP, Eby P, Kanowski J (2011) Latitudinal range shifts in Australian flying-foxes: A re-evaluation. Austral Ecology 37: 12–22.
25. Department of Environment & Heritage Protection. WildNet. Queensland Government, Brisbane, Australia. Available: http://www.ehp.qld.gov.au/wildlife/wildlife-online/index.html. Accessed: 26 Feb 2014.
26. Bureau of Meteorology. Periods used in climate maps and statistics. Australian Government, Canberra, Australia. Available: http://www.bom.gov.au/climate/how/newproducts/map-periods.shtml. Accessed 5 May 2014.
27. Bureau of Meteorology. Maps of average conditions. Australian Government, Canberra, Australia. Available: http://www.bom.gov.au/climate/averages/maps.shtml Accessed: 26 Feb 2014.
28. Geoscience Australia. Vegetation-post-European settlement (1988). Australian Government, Canberra, Australia. Available: http://www.ga.gov.au/meta/ANZCW0703005426.html. Accessed: 26 Feb 2014.
29. Almeida FC, Giannini NP, Simmons NB, Helgen KM (2014) Each flying fox on its own branch: A phylogenetic tree for *Pteropus* and related genera (Chiroptera: Pteropodidae). Mol Phylogenet Evol http://dx.doi.org/10.1016/j.ympev.2014.03.009.
30. Nadin-Davis SA, Huang W, Armstrong J, Casey GA, Bahloul C, et al. (2001) Antigenic and genetic divergence of rabies viruses from bat species indigenous to Canada. Virus Research 74: 139–156.
31. Chu DK, Poon LL, Chan KH, Chen H, Guan Y, et al. (2006) Coronaviruses in bent-winged bats (Miniopterus spp.). J Gen Virol 87: 2461–2466.
32. Gloza-Rausch F, Ipsen A, Seebens A, Gottsche M, Panning M, et al. (2008) Detection and Prevalence Patterns of Group I Coronaviruses in Bats, Northern Germany. Emerg Infect Dis 14: 626–631.
33. BioMedware (2012) Methods for data standardisation. Avialable: http://www.biomedware.com/files/documentation/boundaryseer/Preparing_data/Methods_for_data_standardization.htm. Accessed: 26 Feb 2014.
34. Li H, Calder C, Cressie N (2007) Beyond Moran's I: Testing for Spatial Dependence Based on the Spatial Autoregressive Model. Geographical Analysis 36: 357–375.
35. Getis A, Ord J (1992) The Analysis of Spatial Association by Use of Distance Statistics. Geographical Analysis 24.
36. Markus N, Hall L (2004) Foraging behaviour of the black flying-fox (*Pteropus alecto*) in the urban landscape of Brisbane, Queensland. Wildlife Research 31: 345–355.
37. Fox S (2011) The Spectacled Flying-fox-review of past and present knowledge. In: B. Law PE, D Lunney and L Lumsden, editor. The Biology and Conservation of Australasian Bats. Sydney, Australia.: Royal Zoological Society of NSW, Australia. 136–145.
38. Kitron U, Kazmierczak J (1997) Spatial analysis of the distribution of Lyme disease in Wisconsin. Am J Epidemiol 145: 558–566.
39. Brownstein J, Rosen H, Purdy D, Miller J, Merlino M, et al. (2002) Spatial analysis of West Nile virus: rapid risk assessment of an introduced vector-borne zoonosis. Vector Borne Zoonotic Dis 2: 157–164.
40. Martin V, Pfeiffer D, Zhou X, Xiao X, Prosser D, et al. (2011) Spatial Distribution and Risk Factors of Highly Pathogenic Avian Influenza (HPAI) H5N1 in China. PLoS Pathog 7: e1001308.
41. Firestone S, Christley R, Ward M, Dhand N (2012) Adding the spatial dimension to the social network analysis of an epidemic: investigation of the 2007 outbreak of equine influenza in Australia. Prev Vet Med 106: 123–135.
42. Plowright R, Foley P, Eby P, Dobson A, Field H, et al. (2011) Urban habituation, ecological connectivity and epidemic dampening: the emergence of Hendra virus from flying foxes (Pteropus spp.). Proceedings of The Royal Society B 278: 3703–3712.
43. Field H, Kung N (2011) Henipaviruses-unanswered questions of lethal zoonoses. Current Opinion in Virology 1: 658–661.
44. Field HE, De Jong C, Melville D, Smith C, Smith I, et al. (2011) Hendra virus infection dynamics in Australian fruit bats. Plos One 6: e28678.
45. Wacharapluesadee S, Boongird K, Wanghongsa S, Ratanasetyuth N, Supavonwong P, et al (2010) A longitudinal study of the prevalence of Nipah virus in Pteropus lylei bats in Thailand: evidence for seasonal preference in disease transmission. Vector Borne Zoonotic Dis 10: 183–90.
46. Scanlan J, Kung N, Selleck P, Field H (2014) Survival of Hendra virus in the environment-modelling the effect of temperature. EcoHealth DOI:10.1007/s10393-014-0920-4.

# 12

# Joint Effects of Habitat Heterogeneity and Species' Life-History Traits on Population Dynamics in Spatially Structured Landscapes

**Xinping Ye[1,2]*, Andrew K. Skidmore[1], Tiejun Wang[1]***

**1** Department of Natural Resources, Faculty of Geo-Information Science and Earth Observation (ITC), University of Twente, Enschede, The Netherlands, **2** Foping National Nature Reserve, Shaanxi, China

## Abstract

Both habitat heterogeneity and species' life-history traits play important roles in driving population dynamics, yet there is little scientific consensus around the combined effect of these two factors on populations in complex landscapes. Using a spatially explicit agent-based model, we explored how interactions between habitat spatial structure (defined here as the scale of spatial autocorrelation in habitat quality) and species life-history strategies (defined here by species environmental tolerance and movement capacity) affect population dynamics in spatially heterogeneous landscapes. We compared the responses of four hypothetical species with different life-history traits to four landscape scenarios differing in the scale of spatial autocorrelation in habitat quality. The results showed that the population size of all hypothetical species exhibited a substantial increase as the scale of spatial autocorrelation in habitat quality increased, yet the pattern of population increase was shaped by species' movement capacity. The increasing scale of spatial autocorrelation in habitat quality promoted the resource share of individuals, but had little effect on the mean mortality rate of individuals. Species' movement capacity also determined the proportion of individuals in high-quality cells as well as the proportion of individuals experiencing competition in response to increased spatial autocorrelation in habitat quality. Positive correlations between the resource share of individuals and the proportion of individuals experiencing competition indicate that large-scale spatial autocorrelation in habitat quality may mask the density-dependent effect on populations through increasing the resource share of individuals, especially for species with low mobility. These findings suggest that low-mobility species may be more sensitive to habitat spatial heterogeneity in spatially structured landscapes. In addition, localized movement in combination with spatial autocorrelation may increase the population size, despite increased density effects.

**Editor:** Lalit Kumar, University of New England, Australia

**Funding:** This work was supported by the following grants: EU Erasmus Mundus External Co-operation Window Programme (L14a0900029 XY) and ITC Research Fund (RC201003/PD XY). The funders had no role in study design, data collection and analysis, decision to publish, or preparation of the manuscript.

**Competing Interests:** The authors have declared that no competing interests exist.

* Email: ye17188@itc.nl (XY); t.wang@utwente.nl (TW)

## Introduction

Habitat heterogeneity is increasingly recognized as a significant factor affecting the dynamics and persistence of populations [1–5]. For mobile species inhabiting complex landscapes, spatial heterogeneity can influence their behaviour, thereby affecting survival and reproductive performance [6–8]. Most frequently, the effect of spatial heterogeneity has been studied either under a so-called binary 'habitat-matrix' framework [9–11] or under the assumption of stochastic variability wherein local variation is random in space [5,6]. However, defining a landscape as discrete habitat/matrix patches overlooks the spatial pattern of continuously varying habitat quality that might facilitate or constrain the performance of individuals and therefore population dynamics [12,13]. In addition, predictions of population dynamics are also influenced by how the spatial variation in habitat quality within a landscape is calculated [14]. In this sense, a continuous representation of habitat quality is expected to more accurately reflect how a species experiences a heterogeneous landscape [15].

One general property of landscape heterogeneity of particular importance for ecological dynamics is its spatial autocorrelation structure, which can be defined as the property of random variables (e.g., habitat quality) taking values over distance that are more (or less) similar than expected for randomly associated pairs of observations due to geographic proximity [16–18]. The spatial autocorrelation in habitat quality is known to affect species persistence, for instance, increased spatial autocorrelation could reduce the persistence of small populations [19–21]. Yet the effect of habitat spatial structure on population dynamics has rarely been tested for species in spatially structured landscapes. Bolker [4] incorporated spatially correlated heterogeneity into simulation models of sessile organisms and found that spatial autocorrelation generally improves population viability. For mobile species, however, the performance of a population may become different if spatial autocorrelation occurs within the range of dispersal [22,23]. From an ecological standpoint, the effects of habitat spatial structure on mobile species depend not only on the scale of spatial autocorrelation, but also on how this scale relates to the species movement capacity and habitat requirements [1,23]. If the scale of spatial autocorrelation is low, the likelihood of an individual encountering a different environment will increase quickly as it moves away from its present location [22]. If the movement range is large, the population becomes well-mixed and hence spatially unstructured [24,25]. Moreover, species differ in their environmental tolerance, thereby exhibiting differential responses to habitat conditions [26–28]. Therefore, we anticipate that habitat spatial structure will interact with species' environmental tolerance and movement capacity in determining species' demographic performance as well as population dynamics [29,30]. For instance, spatial autocorrelation in habitat quality may lead to spatial aggregation of individuals, thereby affecting local population density or competitive pressure on individuals [31–33]. Under such circumstances, individual-level variation can also provide underlying mechanisms of population regulation [34]. However, analytical models do not take into account the complexity of the multiple concurrent interactions (e.g., competition between individuals) which may influence population dynamics [35,36]. The key to improve understanding of the role of habitat spatial structure in driving the population dynamics of mobile species is therefore to link individual behaviours to population phenomena in a spatially explicit modelling framework [12,37,38].

In this paper, we explored how habitat spatial structure interact with species' environmental tolerance and movement capacity to affect the population dynamics of mobile species using a spatially explicit agent-based modelling approach. We compared the responses of four virtual species to four landscape scenarios differing in the scale of spatial autocorrelation in habitat quality. The ecological motivation for this simulation was to understand the combined impacts of these factors on mobile species that are dispersal limited and experience habitat heterogeneity (e.g., animals in mountainous forests). We hypothesized that for species in landscapes with small-scale spatial autocorrelation, the movement-distance does not have a strong effect on population dynamics, whereas for species in landscapes with large-scale spatial autocorrelation, short distance movement would increase the viability of the population.

## Methods

### The Agent-Based Model

The present spatially explicit, agent-based model simulates the population dynamics of a single virtual species in a spatially structured landscape. The description of the model follows the ODD protocol (overview, design concepts, and details) for describing individual- and agent-based models proposed by Grimm *et al.* [39,40]. The model was implemented in the R environment [41] with the package 'gstat' [42].

**Purpose.** The purpose of this model is (i) to understand how habitat spatial structure (depicted by the scale of spatial autocorrelation in habitat quality) influences the population dynamics of mobile species, and (ii) to explore how species' life-history traits shape the population responses in relation to habitat spatial structure. Since this is a purely theoretical study, the model simulates population performance of four hypothetical species that differ in aspects of life-history traits (low-tolerance vs. high-tolerance; low-mobility vs. high-mobility), rather than predict the population dynamics of any specific species. In other words, the model's formulation tends to increase generality in a trade-off for realism and precision.

**Entities, state variables, and scales.** The model contains two types of entities: individuals and landscape. The landscape is a grid of 30×30 habitat cells, in which each cell represents a discrete space of the environment, and is characterised by spatial coordinates and habitat quality ($Q$; range 0–1), where habitat quality simply reflects resource conditions essential for an individual's performance. Cell size is not spatially specified, but is assumed to fulfil the space requirement of an individual. The simulated landscape is, therefore, a continuous representation of habitat quality rather than a binary mosaic of habitat and matrix. The individuals are characterized by the following state variables: identity number, age, spatial location (cell coordinates), resource share, reproduction rate, mortality rate, and moving distance per time step. Only females are simulated in the model, a strategy common to many population models [11,12]. No age-structure is included in the model. Values of the demographic parameters (see Table 1) are selected in order to meet reasonable assumptions about the hypothetical species (i.e. long-lived animals which generally have low mortality rate and high probability of reproductive success), thereby ensuring this generic model to generate plausible projections. Time runs in discrete steps and the simulation lasts 100 time-steps. The length of a time step is not explicitly specified, but is assumed to be long enough for each individual to accomplish all demographic processes (i.e., movement, reproduction, and mortality).

**Process overview and scheduling.** Individuals may move, reproduce, and/or die at each time step. The order of individuals is randomized per time step, and the order of the three events (i.e. move, reproduce, and/or die) is randomized for each individual per time step. Probabilities of mortality, reproduction, and movement of an individual per time step depend on its resource share, which is updated before each event to take into account changes in the cell's density due to mortality and movement of other individuals within the time step. Only individuals with a resource share beyond 0.5 are capable of reproducing. The newly produced offspring join the population only at the end of the time step, and become adults upon commencement of the following time-step. The model is illustrated in Figures 1 and 2.

**Design concepts.** Population dynamics emerge from individual and species traits. Individuals' life cycles and movements are imposed by stochastic rules.

Individuals within the same cell affect each other through the "scramble competition" [43] that the resource share ($F$) per individual declines with an increasing number of individuals within a cell. Therefore, an individual's resource share is determined by habitat quality of the inhabited cell and crowding in that cell, i.e., $F = Q/n$, where $Q$ is the habitat quality of the cell, and $n$ is the number of individuals in the cell. Direct interaction between individuals of different cells is not considered in the model.

All modelled processes are implemented stochastically via a random order of individuals in the main routine, a random order of events, and moving in a random direction. Movement, reproduction, and mortality events are implemented based on probabilities.

Population dynamics are quantified by recording the following variables throughout the last 30 time steps of each run: 1) population size; 2) mean resource share of individuals; 3) mean mortality rate of individuals; 4) proportion of individuals in high-quality cells ($Q \geq 0.5$); and 5) proportion of individuals experiencing competition. By using the proportion of individuals under different conditions (e.g., residing in high-quality cells and/or experiencing competition), the absolute number of individuals are transformed to a range 0–1, allowing direct comparison between population dynamics of different species, and thereby correcting for differences in their population sizes. For model analysis, each response variable is averaged over the last 30 time steps of each run.

**Table 1.** Variables, parameters, and initial conditions used in the model.

| Parameter | Symbol | Values |
|---|---|---|
| Landscape size | - | 30×30 cells |
| Habitat quality of cells | $Q$ | 0–1 |
| Mean habitat quality of landscape | - | 0.5 |
| Scale of spatial autocorrelation in habitat quality | $S$ | 0.01, 2, 4, 8 |
| Individual's resource share | $F$ | 0–1 |
| Birth probability | $B$ | 0.7 |
| Basic mortality rate | $M_0$ | 0.15 |
| Level of environmental tolerance | $C_{envir}$ | 2, 3 |
| Mean moving distance | $D_{mean}$ | 1, 4 cells |
| Litter size | - | 1 |
| Max. lifespan | - | 10 time steps |
| Initial population size | - | 100 |
| Length of simulation period | - | 100 time steps |

There are no external environmental variables that drive the internal dynamics of the model.

**Initialisation.** At the beginning of each run, a two-dimensional landscape with 30×30 grid cells is generated using an unconditional Gaussian variogram model (*sill* = 0.025, *nugget* = 0, and *range* = $S$), where the *range* parameter defines the scale of spatial autocorrelation in habitat quality. The mean habitat quality of the landscape is fixed at 0.5. After generating a landscape, one hundred individuals are distributed at random through the landscape. The resource share of each individual is calculated based on the habitat quality and crowdedness of the occupied cell. Individuals' ages are independently drawn from the uniform distribution in the interval (0, lifespan), because preliminary tests based on different distributions (e.g., Poisson distribution) revealed that different initial age structure of individuals had no effect on population dynamics after the first 30 time-steps (see Text S1).

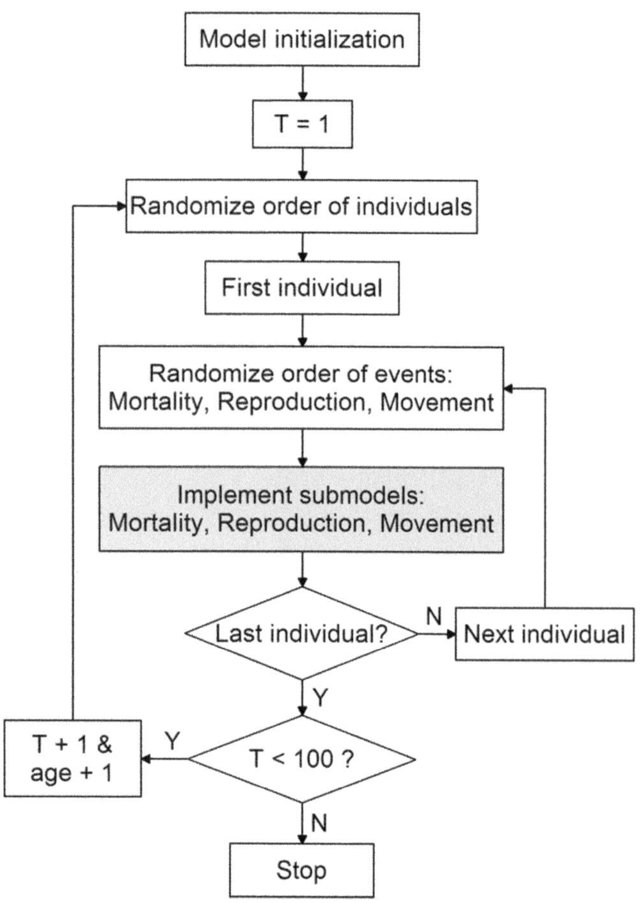


**Figure 1. Flow diagram of the main routine used in the simulation model.**

**Mortality submodel.** Probabilities of mortality are individual-specific and depend on their resource share per time step. An individual's mortality rate is calculated as: $M = M_0 \times \left(1 + \exp\left(-(F \times C_{envir})^2\right)\right)$, where $M_0$ is the basic mortality rate, $F$ is the resource share of the individual, and $C_{envir}$ is the level of environmental tolerance. In biological terms, $C_{envir}$ regulates the sensitivity of individual mortality rate to environmental variability (Figure 3a). Individuals that reach the maximal lifespan are "killed" from the population.

**Reproduction submodel.** Only adult individuals with a high resource share (i.e., $F \geq 0.5$) can reproduce, giving birth to one offspring with a constant probability of $B$. More complex relations, e.g., making reproduction probability and number of offspring function of habitat quality, would by virtue of Jensen's inequality prohibit comparison between landscapes [15,44]. The newly produced offspring are attached to their mothers (i.e., in the same cells), and join the population at the end of the current time step. They become adults able to move and reproduce at the following time step.

**Movement submodel.** Individual movement is a conditional response to local habitat quality and crowdedness, i.e., the condition-dependent dispersal [25,45], and the probability of movement per time step is calculated as: $P_{move} = 1 - Q/n$, where $Q$ is the habitat quality of the cell, and $n$ is the number of individuals in the cell. Ecologically, it means that individuals residing in cells of low habitat quality and/or high density tend to depart the cell, while those residing in cells of high quality and/or low density are more likely to stay in the current cell. The

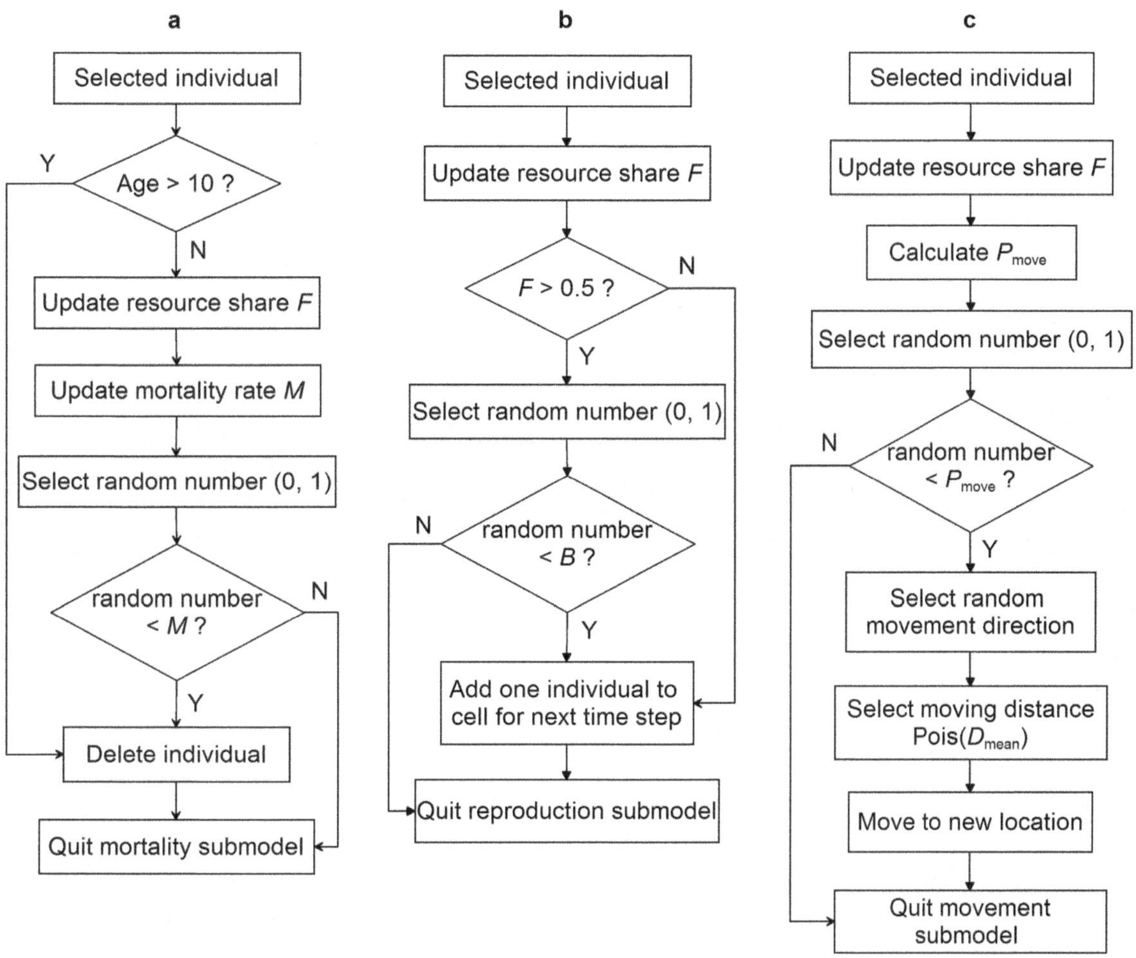


**Figure 2. Flow diagrams of the submodels in the simulation model that determine mortality (a), reproduction (b), and (c) movement for a single individual within a time step.**

movement direction per time step is randomly determined, and the movement distance per time step is drawn from a Poisson distribution with mean $\lambda = D_{mean}$ (Figure 3b). The assumption of random walk seems to be unrealistic for species that may actively search the landscape, but is conservative in the sense that effects of spatial heterogeneity are easy to detect [46]. From the moving individuals' perspective, the landscape is "wrapped", meaning that an individual that crosses the edge of the landscape continues in the same direction on the opposite edge [47].

## Simulation Experiments

To investigate how spatial autocorrelation of habitat quality interacts with species traits to affect population dynamics, we defined four landscape scenarios that reflect different levels of

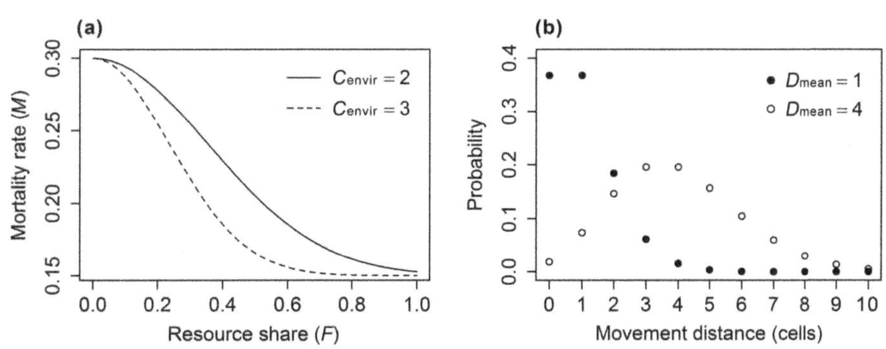


**Figure 3. Movement distance distributions (a) and mortality rate functions (b) used to define hypothetical species in the model.** $C_{envir} = 2$ and $C_{envir} = 3$ represent low-tolerant and high-tolerant species, respectively. $D_{mean} = 1$ and $D_{mean} = 4$ represent low-mobility and high-mobility species, respectively.

spatial autocorrelation in habitat quality: apparently uncorrelated ($S = 0.01$), small-scale ($S = 2$), intermediate-scale ($S = 4$), and large-scale ($S = 8$) spatial autocorrelation in habitat quality (Figure 4). For each landscape scenario, we compared the population dynamics of four virtual species differing in the level of environmental tolerance ($C_{envir} = 2$ or 3; Figure 3a) and mean movement distance ($D_{mean} = 1$ or 4 cells; Figure 3b). The combination of landscape scenarios and species classes resulted in a total of 16 treatments ($4 \times 4 = 16$) in the experiments. For each treatment, we conducted 50 replicates, each lasting for 100 time steps. This is based on preliminary runs using 200 time-steps, in which we found that any population reached a state of dynamic equilibrium within the first 50 time-steps.

The flexibility of agent-based models means that they typically have a large number of parameters. Ideally, parameter values would be estimated from empirical data. Unfortunately, such data are scarce, and the number of parameters is so large that it is not possible to investigate every combination of parameter values. Therefore, following a previous sensitivity analysis (data not shown), parameter values were chosen arbitrarily to yield an overall rate of population increase of ~1 over the first 50 time steps for the species with $C_{envir} = 2$ and $D_{mean} = 1$ in landscapes with apparently uncorrelated habitat quality (i.e., landscape scenario $S = 0.01$). Real-world species would experience life-history trade-offs in demographic and movement rates which influence population growth [48]. To permit a more direct comparison of the combined impacts of changing life-history traits and habitat spatial structure, we assumed that all virtual species had the same basic mortality rate ($M_0$) and birth probability ($B$). Since we were interested in combining landscape spatial autocorrelation with species traits, these assumptions were considered acceptable. The detailed parameter values are given in Table 1.

### Statistical Analysis

We first performed a Shapiro-Wilks normality test to examine data normality and found that data were normally distributed. To assess the main effects and interactions among the three experimental factors (i.e., scale of spatial autocorrelation in habitat quality $S$, species' environmental tolerance $C_{envir}$, and mean moving distance $D_{mean}$), we estimated the magnitude of effects of the variance sources using a three-way analysis of variance (ANOVA), as suggested by White et al. [49]. We performed three-way ANOVA for each of the five response variables (cf. *Observation* section). The three experimental factors were treated as categorical variables in accordance with the experimental design. The effect sizes of the factors and interactions were measured by percentage of variance explained [50] and generalized eta-squared [51,52] (see Text S2 for the calculation). For each treatment, we also investigated Spearman correlations of the mean resource share of individuals with the proportion of individuals in high-quality cells and with the proportion of individuals experiencing competition. By doing so, we aimed at increasing our understanding of the mechanisms behind observed population dynamics in different landscape scenarios. All data analyses were conducted in the R environment with the 'ez' package [53].

## Results

All hypothetical species exhibited a constant increase in population size as the scale of spatial autocorrelation in habitat quality increased (Figure 5a). However, the pattern of the increase in population size varied greatly between the species with different movement capacities. Low-mobility species ($D_{mean} = 1$) displayed a logarithmic-like increase in response to increased spatial autocorrelation, contrasting with the exponential-like increase of high-mobility species ($D_{mean} = 4i$). Under the same level of species' environmental tolerance, low-mobility species showed greater population size than high-mobility species in all landscape scenarios except for the scenario of $S = 0.01$ (Figure 5a).

The analysis of variance revealed that both species' environmental tolerance and the scale of spatial autocorrelation in habitat quality had strong effects on population size, accounting for ~47% and ~33%, respectively, of the variation in population size (Table 2, also see Table S1). Their interaction also had a significant effect ($S \times D_{mean}$ $F_{7,792} = 487.96$, $p<0.001$), but accounted only for ~1% of the variation in population size (Table 2). Species' movement capacity and its interaction with scale of spatial autocorrelation in habitat quality also had appreciable effects and explained ~10% of the variation in population size ($D_{mean}$ $F_{1,798} = 23.04$, $p<0.001$; $S \times D_{mean}$ $F_{7,792} = 85.29$, $p<0.001$; Table 2).

All virtual species also showed an exponential increase in mean resource share as the scale of spatial autocorrelation in habitat quality increased (Figure 5b), and low-tolerant species had appreciably higher mean resource shares than high-tolerant populations (Mann–Whitney test, $p<0.01$). Species' environmental tolerance had the greatest effect on individual resource share, explaining ~62% of the variation in mean resource share (Table 2, also see Table S1). Scale of spatial autocorrelation in habitat quality accounted for ~15% of the variation in mean resource share of individuals (Table 2). In addition, the effect of movement capacity on mean resource share was appreciably greater when habitat quality was spatially autocorrelated at a broad scale ($S \times D_{mean}$ $F_{7,792} = 21.18$, $p<0.001$).

For the mean mortality rate of individuals, species' environmental tolerance had a dominant effect and accounted for ~99% of the variation (Table 2, also see Table S1). In contrast, the scale of spatial autocorrelation in habitat quality and species' movement capacity had little effect on the mean mortality rate of individuals,

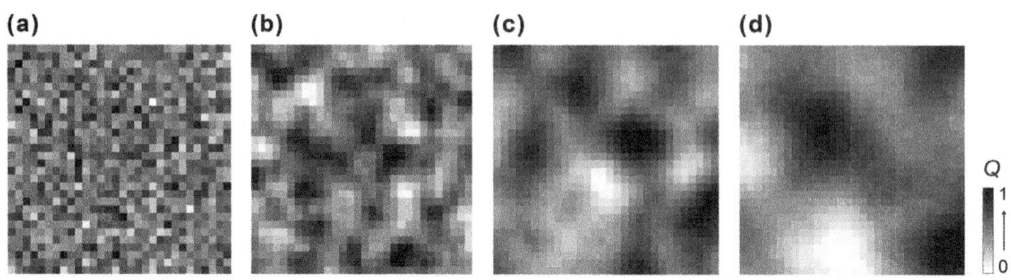


**Figure 4. Examples of artificial landscapes differing in the scale of spatial autocorrelation in habitat quality *S*.** (a) nearly random, $S = 0.01$; (b) small-scale, $S = 2$; (c) intermediate-scale, $S = 4$; and (d) large-scale, $S = 8$. The mean habitat quality of the landscapes is fixed at 0.5.

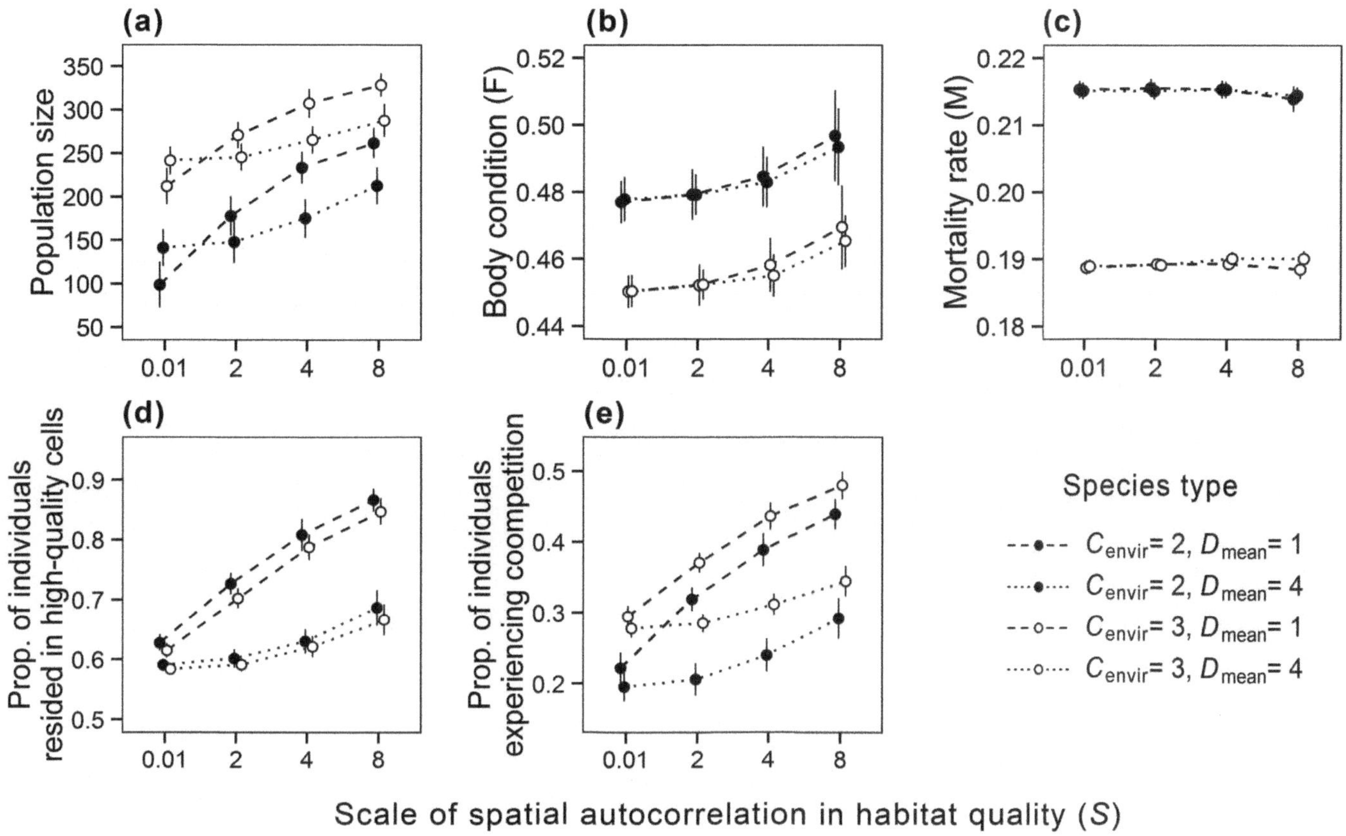


**Figure 5. Population responses of four hypothetical species to the scale of spatial autocorrelation in habitat quality *S*.** (a) population size; (b) mean resource share of individuals; (c) mean mortality rate of individuals; (d) proportion of individuals in high-quality cells ($Q \geq 0.5$); (e) proportion of individuals experiencing competition. The figure shows mean±1 SD for 50 replicates for each variable.

due to the dominant effect of species' environmental tolerance (Figure5c and Table 2).

The proportion of individuals residing in high-quality cells and the proportion of individuals experiencing competition reflect the spatial distribution of individuals in response to the spatial structure of habitat quality (Figure 6, also see Figures S1 and S2). Increasing the scale of spatial autocorrelation in habitat quality increased the proportion of individuals residing in high-quality cells, especially for low-mobility species (Figure 5d, also see Figure S1). Species' movement capacity and the scale of spatial autocorrelation in habitat quality both had strong effects, accounting for ~45 and 41%, respectively, of the variation in the proportion of individuals in high-quality cells (Table 2, also see Table S1). Their interaction also had a significant effect and

**Table 2.** Summary of the percentages of the variation in response variables explained by factors scale of spatial autocorrelation in habitat quality (*S*), species environmental tolerance ($C_{envir}$), and mean moving distance ($D_{mean}$). Detailed ANOVAs for each response variable are presented in Table S1.

| Source | Explained variation (%) in response variable | | | | |
|---|---|---|---|---|---|
| | Population size | Mean resource share of individuals | Mean mortality rate of individuals | Prop. of individuals in high-quality cells | Prop. of individuals experiencing competition |
| $S$ | 33.16 | 15.35 | 0.05 | 41.18 | 38.57 |
| $C_{envir}$ | 46.92 | 62.41 | 98.96 | 0.67 | 13.37 |
| $D_{mean}$ | 2.81 | 0.16 | 0.01 | 44.60 | 34.13 |
| $S \times C_{envir}$ | 1.09 | 0.01 | 0.05 | 0.04 | 0.44 |
| $S \times D_{mean}$ | 7.01 | 0.26 | 0.04 | 9.48 | 7.88 |
| $C_{envir} \times D_{mean}$ | 0.03 | 0.01 | 0.01 | 0.04 | 0.29 |
| $S \times C_{envir} \times D_{mean}$ | 0.02 | 0.01 | 0.00 | 0.02 | 0.05 |
| Total variation explained by the full model ($R^2$) | 0.91 | 0.78 | 0.99 | 0.96 | 0.95 |

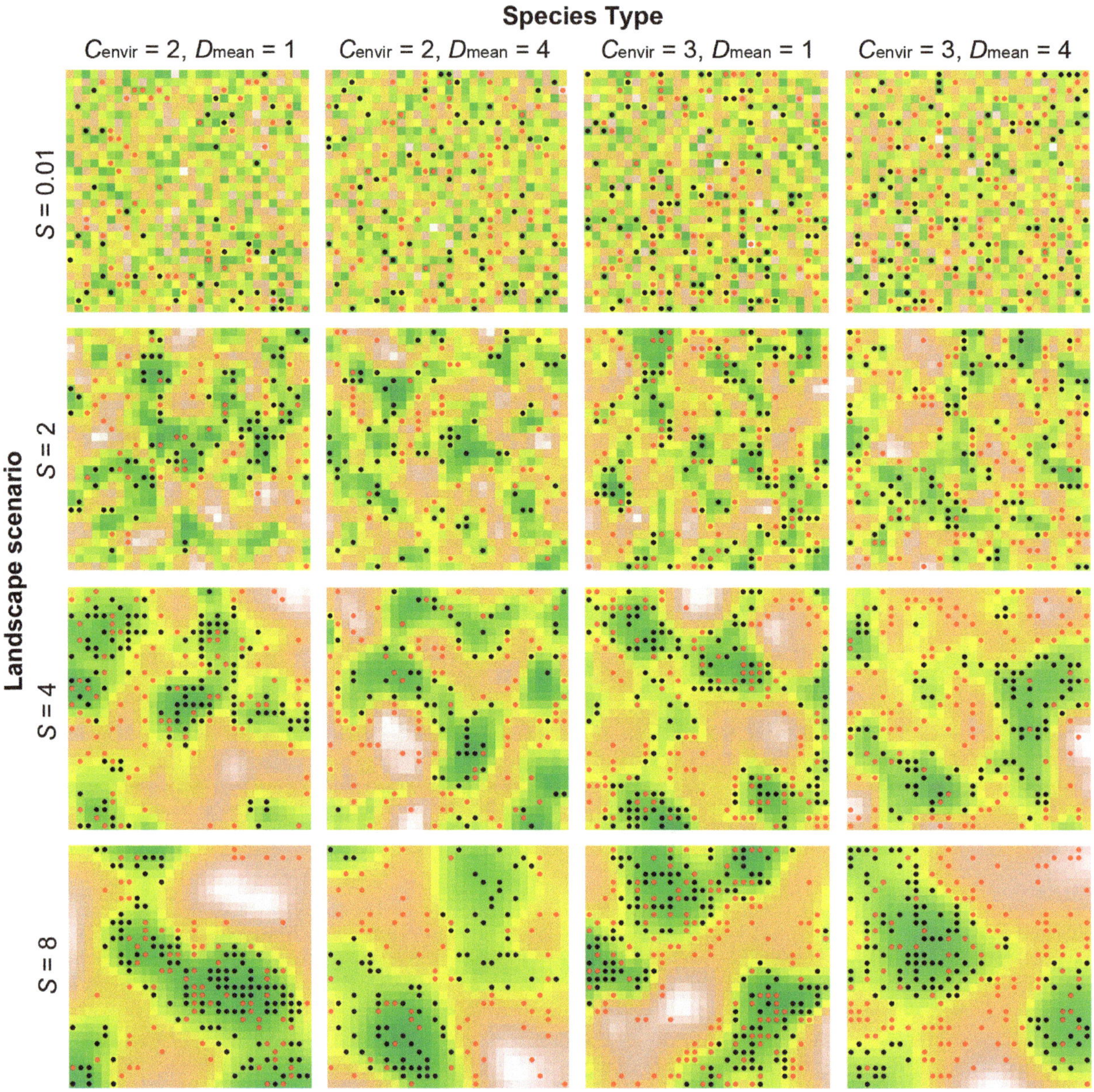


**Figure 6. Sample patterns of spatial distribution of individuals under different landscape scenarios.** Hypothetical species are parameterized by environmental tolerance $C_{envir}$ and mean moving distance $D_{mean}$. Black dots represent individuals residing in cells of $Q \geq 0.5$ and free of competition, while red dots are individuals residing in cells of $Q < 0.5$ or experiencing competition. Greener colour indicates higher habitat quality.

explained ~9% of the variation ($S \times D_{mean}$ $F_{7,792} = 2273.5$, $p < 0.001$).

Likewise, the proportion of individuals experiencing competition increased as the scale of spatial autocorrelation in habitat quality increased, and low-mobility species had higher proportions when habitat quality was spatially autocorrelated (Figure5e, also see Figure S2). The scale of spatial autocorrelation in habitat quality had the greatest effect on the proportion of individuals experiencing competition, followed by species' movement capacity and environmental tolerance (Table 2, also see Table S1). The interaction between the scale of spatial autocorrelation in habitat quality and species' movement capacity also had a significant effect, accounting for ~8% of the variation in the proportion of individuals experiencing competition ($S \times D_{mean}$ $F_{7,792} = 469.54$, $p < 0.001$).

In general, the mean resource share of individuals was positively associated with the proportion of individuals in high-quality cells when habitat quality was spatially autocorrelated at a broad scale (Figure 7a). In contrast, the associations between mean resource share of individuals and the proportion of individuals experiencing competition were differential, depending on the spatial structure of habitat quality. When the spatial structure of habitat quality shifted from spatially random to autocorrelated at a broad scale, the correlation changed from negative to positive (Figure 7b), indicating that local movement in combination with spatial autocorrelation can actually increase population size and resource share despite of increased density effects.

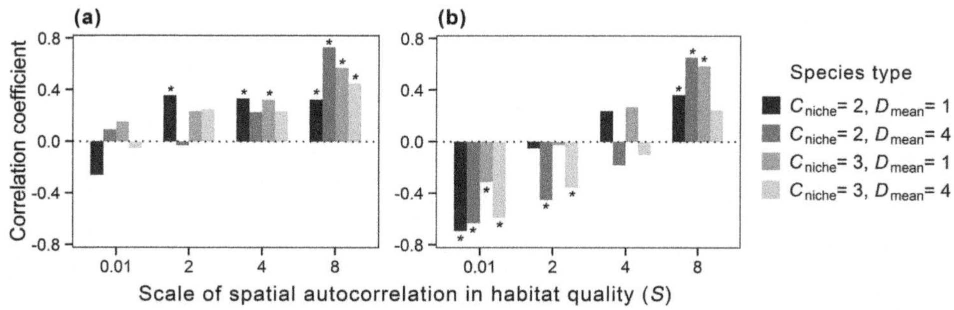


**Figure 7. Coefficients of Spearman correlations of the mean resource share of individuals with (a) the proportion of individuals in high-quality cells and (b) the proportion of individuals experiencing competition for four hypothetical species under different landscape scenarios.** Bars marked with an asterisk (*) indicate the coefficients are statistically significant at $P<0.05$.

## Discussion

In this study, we explored how habitat spatial autocorrelation interacts with species' life-history attributes to influence population dynamics within spatially structured landscapes. We found that increasing the scale of spatial autocorrelation in habitat quality can greatly increase the population size as well as mean resource share of virtual species and in line with common expectation, high-tolerant species had an appreciably higher population size than low-tolerant species. These results are generally consistent with previous findings that more specialized species are more affected by habitat spatial heterogeneity [4,8,14,54]. These observed patterns may be partly explained by edge effects, which is one of the factors influencing population dynamics [55]. As the scale of spatial autocorrelation increases, habitat quality becomes less rugged across space, forming fewer but larger suitable areas in the landscape, and consequently reduces changes in abundance or fitness of low-tolerant species with respect to low quality of "edge" areas. In contrast, the high-tolerant species, due to higher tolerances to variation in habitat quality, may be insensitive to such edge effects [48]. Moreover, our results are comparable to the population dynamics in the "source-sink" context. Spatial autocorrelation in habitat quality can result in neighbourhoods of high-quality and low-quality areas, thereby creating source-sink dynamics in relation to habitat heterogeneity [31,32]. If a landscape is spatially autocorrelated (particularly at large extents), individuals dispersing away from "source" habitats are more likely to encounter higher quality habitats rather than disperse into a demographic sink.

Species' movement capacity can interact with habitat heterogeneity to influence population dynamics and distribution of individuals [56,57]. Our simulation demonstrated that low-mobility species had greater population sizes than high-mobility species in the presence of spatial autocorrelation in habitat quality. The findings also suggest that species' movement capacity plays an important role in shaping the increase in population size in response to an increased spatial autocorrelation in habitat quality, where low-mobility species had a logarithmic-like increase in response to increased scale of spatial autocorrelation, contrasting with the exponential-like increase for high-mobility species. This may be because, under the assumption of random walk, individuals with short-distance movement are likely to stay in suitable habitats, while long-distance movement will increase the risk of landing in an unsuitable habitat when habitat quality is autocorrelated at fine scales [4,58,59]. Such contrasting patterns of population increase in relation to movement capacity indicate that the effect of distance-based movement capacity may function in highly scale-dependent ways, and its relative strength is determined by interactions among movement distance, the scale of spatial autocorrelation, and the effects of species competition [56,57,60]. For instance, the logarithmic-like increase in population size of low-mobility species in response to increased spatial autocorrelation indicates that the benefit from short-distance movement was buffered by increasing effects of population density, whereas the exponential-like increase in population size for high-mobility species may suggest that long-distance movement will reduce competitive interactions between individuals, thereby increasing the net population. Skelsey and Garrett [61] predicted that the magnitude of dispersal (number of individuals) should be maximized at intermediate scales of heterogeneity, i.e., when the scale of spatial heterogeneity is neither too fine nor too coarse relative to the movement capacity of a species. This potentially provides an additional explanation for the observed patterns in population dynamics in response to spatial autocorrelation in habitat quality.

The population dynamics of species inhabiting complex landscapes generally involve two major components: the dispersal of individuals and habitat-specific mortality rates [30]. The interplay between spatial autocorrelation in habitat quality and species life-history traits can lead to spatial aggregation of individuals, creating variation in demographic performance among individuals [31,32]. This is evident by our model, where the proportions of individuals in high-quality cells and the proportion of individuals experiencing competition increased differentially in response to increased spatial autocorrelation between species with different movement capacity (see Figure S1 and Figure S2). These results indicate that localized movement in combination with habitat spatial structure may increase the population size despite of increased density effects in the presence of spatial autocorrelation in habitat quality. Furthermore, we found that the mean resource share was positively correlated with the proportion of individuals experiencing competition when habitat quality was spatially autocorrelated. This may sound contradictory, because an individual's resource share would be, by definition, reduced by increased local crowding. However, increasing spatial autocorrelation in habitat quality could also improve the quality of individuals by increasing the probability of finding high-quality habitat within the range of movement, which may buffer the density-dependent effects on the population [21]. Such complex relationships indicate how the interplay of movement capacity and environmental autocorrelation can jointly influence the outcome of competitive interactions [62].

The results of this simulation are contingent upon numerous simplifying model assumptions, and it remains to be seen whether the same or similar effects also occur for species living in real landscapes. For example, the model used a simplified landscape with the edge being "wrapped" so that an individual crossing the edge of the landscape continues in the same direction on the opposite edge [47]. However, moving individuals would experience sharp discontinuities as they cross the edges if the simulated landscape boundaries are not periodic. To avoid this, one can generate landscapes with either periodic boundary conditions or with a larger spatial extent, depending on the study of interest [63]. Furthermore, we assumed individuals move in a random manner. The random movement assumption seems to be unrealistic for species that may actively search the landscape within their home range (e.g. birds and mammals). However, if individuals can perceive and orient new habitat location from some distance, it may result in a reduced effect of habitat spatial structure on population dynamics. Therefore, the random movement assumption is conservative in the sense that effects of landscape heterogeneity are easy-to-detect [46]. Moreover, male individuals were not modelled based on the perspective that females form the reproductive unit and thereby determine the population dynamics [11,12]. We also excluded life-history trade-offs in demographic and movement rates by assuming that only mortality rate is associated with local habitat quality and density. However, more complex relations, e.g., making both mortality and reproduction functions of habitat quality, would increase the unpredictability in model outputs and prohibit comparison between landscapes due to the Jensen's inequality [15,44].

Although the model incorporates numerous simplifying assumptions, the general prediction that low-mobility species will have greater sensitivity to changing scale of spatial autocorrelation in habitat quality strongly suggests that mechanisms shaping population dynamics are complex and depend on both habitat spatial heterogeneity and species' life history traits considered. We have yet to confirm them in the field, but if the spatial autocorrelation favouring species exists, we may expect a nonlinear relationship between local population dynamics and habitat heterogeneity in the sense that species movement limitation represents a threshold [22]. Nevertheless, understanding which mechanisms in spatially structured landscapes regulate a population is important for explaining species abundances and designing management plans for species conservation.

## Supporting Information

**Figure S1 Sample patterns of spatial distribution of individuals residing in high-quality cells under different landscape scenarios.**

**Figure S2 Sample patterns of spatial distribution of individuals experiencing competition under different landscape scenarios.**

**Table S1 Factorial ANOVAs on each of the five response variables under different scale of spatial autocorrelation in habitat quality and species' environmental tolerance and movement capacity.**

**Text S1 Testing for the effect of different initial age-structures on model output.**

**Text S2 Calculation of the generalized eta-squared in factorial ANOVA.**

## Acknowledgments

The authors would like to thank Lalit Kumar and two anonymous reviewers for suggestions for improvements on this article.

## Author Contributions

Conceived and designed the experiments: XY AKS TW. Performed the experiments: XY. Analyzed the data: XY. Contributed reagents/materials/analysis tools: XY TW. Contributed to the writing of the manuscript: XY.

## References

1. Lovett GM, Jones CG, Turner MG, Weathers KC (2005) Ecosystem function in heterogeneous landscapes. In: Lovett GM, Turner MG, Jones CG, Weathers KC, editors. Ecosystem function in heterogeneous landscapes: Springer New York. 1–4.
2. Wu JG, Hobbs RJ (2007) Key topics in landscape ecology. Cambridge: Cambridge University Press. 297 p.
3. Pike N, Tully T, Haccou P, Ferriere R (2004) The effect of autocorrelation in environmental variability on the persistence of populations: an experimental test. Proceedings of the Royal Society of London Series B: Biological Sciences 271: 2143–2148.
4. Bolker BM (2003) Combining endogenous and exogenous spatial variability in analytical population models. Theoretical Population Biology 64: 255–270.
5. Lou Y (2006) On the effects of migration and spatial heterogeneity on single and multiple species. Journal of Differential Equations 223: 400–426.
6. Kendall BE, Fox GA (1998) Spatial structure, environmental heterogeneity, and population dynamics: analysis of the coupled logistic map. Theoretical Population Biology 54: 11–37.
7. Burgess MD, Nicoll MA, Jones CG, Norris K (2011) Multiple environmental gradients affect spatial variation in the productivity of a tropical bird population. Journal of Animal Ecology 80: 688–695.
8. Zając T, Solarz W, Bielański W (2008) Site-dependent population dynamics: the influence of spatial habitat heterogeneity on individual fitness in the sedge warbler Acrocephalus schoenobaenus. Journal of Avian Biology 39: 206–214.
9. Fleishman E, Ray C, Sjögren-Gulve P, Boggs CL, Murphy DD (2002) Assessing the roles of patch quality, area, and isolation in predicting metapopulation dynamics. Conservation Biology 16: 706–716.
10. Pulliam HR (1988) Sources, sinks, and population regulation. American Naturalist 132: 652–661.
11. Pulliam HR, Dunning JB, Liu J (1992) Population dynamics in complex landscapes: a case study. Ecological Applications 2: 165–177.
12. Wiegand T, Moloney K, Naves J, Knauer F (1999) Finding the missing link between landscape structure and population dynamics: a spatially explicit perspective. The American Naturalist 154: 605–627.
13. Humphries NE, Weimerskirch H, Sims DW (2013) A new approach for objective identification of turns and steps in organism movement data relevant to random walk modelling. Methods in Ecology and Evolution 4: 930–938.
14. Ye X, Wang T, Skidmore A (2013) Spatial pattern of habitat quality modulates population persistence in fragmented landscapes. Ecological Research 28: 949–958.
15. Stoddard ST (2010) Continuous versus binary representations of landscape heterogeneity in spatially-explicit models of mobile populations. Ecological Modelling 221: 2409–2414.
16. Legendre P (1993) Spatial autocorrelation: trouble or new paradigm? Ecology 74: 1659–1673.
17. Dutilleul P, Legendre P (1993) Spatial heterogeneity against heteroscedasticity: an ecological paradigm versus a statistical concept. Oikos 66: 152–171.
18. Harrison S, Bruna E (1999) Habitat fragmentation and large-scale conservation: what do we know for sure? Ecography 22: 225–232.
19. Clarke RT, Thomas JA, Elmes GW, Hochberg ME (1997) The effects of spatial patterns in habitat quality on community dynamics within a site. Proceedings of the Royal Society of London B: Biological Sciences 264: 347–354.
20. Petchey OL, Gonzalez A, Wilson HB (1997) Effects on population persistence: the interaction between environmental noise colour, intraspecific competition and space. Proceedings of the Royal Society of London B: Biological Sciences 264: 1841–1847.
21. Schiegg K (2003) Environmental autocorrelation: curse or blessing? Trends in Ecology & Evolution 18: 212–214.
22. Schooley RL, Branch LC (2007) Spatial heterogeneity in habitat quality and cross-scale interactions in metapopulations. Ecosystems 10: 846–853.

23. Jacobson B, Peres-Neto P (2010) Quantifying and disentangling dispersal in metacommunities: how close have we come? How far is there to go? Landscape Ecology 25: 495–507.
24. Brachet S, Olivieri I, Godelle B, Klein E, Frascaria-Lacoste N, et al. (1999) Dispersal and metapopulation viability in a heterogeneous landscape. Journal of Theoretical Biology 198: 479–495.
25. Bowler DE, Benton TG (2005) Causes and consequences of animal dispersal strategies: relating individual behaviour to spatial dynamics. Biological Reviews 80: 205–225.
26. Kolasa J, Li B-L (2003) Removing the confounding effect of habitat specialization reveals the stabilizing contribution of diversity to species variability. Proceedings of the Royal Society of London B: Biological Sciences 270: S198–S201.
27. Pigliucci M (2001) Environmental heterogeneity: temporal and spatial. eLS: John Wiley & Sons, Ltd.
28. Jonsen ID, Fahrig L (1997) Response of generalist and specialist insect herbivores to landscape spatial structure. Landscape Ecology 12: 185–197.
29. Mortelliti A, Amori G, Boitani L (2010) The role of habitat quality in fragmented landscapes: a conceptual overview and prospectus for future research. Oecologia 163: 535–547.
30. Pulliam HR, Danielson BJ (1991) Sources, sinks, and habitat selection: a landscape perspective on population dynamics. American Naturalist: S50–S66.
31. Rodenhouse NL, Sherry TW, Holmes RT (1997) Site-dependent regulation of population size: a new synthesis. Ecology 78: 2025–2042.
32. Dhondt AA, Kempenaers B, Adriaensen F (1992) Density-dependent clutch size caused by habitat heterogeneity. Journal of Animal Ecology 61: 643–648.
33. Boeye J, Kubisch A, Bonte D (2014) Habitat structure mediates spatial segregation and therefore coexistence. Landscape Ecology 29: 593–604.
34. Jager HI (2001) Individual variation in life history characteristics can influence extinction risk. Ecological Modelling 144: 61–76.
35. Getz WM, Saltz D (2008) A framework for generating and analyzing movement paths on ecological landscapes. Proceedings of the National Academy of Sciences 105: 19066–19071.
36. Parrott L, Kok R (2000) Incorporating complexity in ecosystem modelling. Complexity International 7: Online.
37. Wiens JA, Stenseth NC, Vanhorne B, Ims RA (1993) Ecological Mechanisms and Landscape Ecology. Oikos 66: 369–380.
38. Wu JG, Hobbs R (2002) Key issues and research priorities in landscape ecology: An idiosyncratic synthesis. Landscape Ecology 17: 355–365.
39. Grimm V, Berger U, Bastiansen F, Eliassen S, Ginot V, et al. (2006) A standard protocol for describing individual-based and agent-based models. Ecological Modelling 198: 115–126.
40. Grimm V, Berger U, DeAngelis DL, Polhill JG, Giske J, et al. (2010) The ODD protocol: A review and first update. Ecological Modelling 221: 2760–2768.
41. R Development Core Team (2011) R: A language and environment for statistical computing, version 2.13.0. R Foundation for Statistical Computing, Vienna, Austria.
42. Itakura J, Kurosaki M, Itakura Y, Maekawa S, Asahina Y, et al. (2010) Reproducibility and usability of chronic virus infection model using agent-based simulation; comparing with a mathematical model. Biosystems 99: 70–78.
43. Hassell MP (1975) Density Dependence in Single Species Populations. Journal of Animal Ecology 44: 283–295.
44. Ruel JJ, Ayres MP (1999) Jensen's inequality predicts effects of environmental variation. Trends in Ecology and Evolution 14: 361–366.
45. Armsworth P (2009) Conditional dispersal, clines, and the evolution of dispersiveness. Theoretical Ecology 2: 105–117.
46. Fahrig L (1998) When does fragmentation of breeding habitat affect population survival? Ecological Modelling 105: 273–292.
47. Haefner JW, Poole GC, Dunn PV, Decker RT (1991) Edge effects in computer models of spatial contagion. Ecological Modelling 56: 221–244.
48. Pearson SM, Fraterrigo JM (2011) Habitat quality, niche breadth, temporal stochasticity, and the persistence of populations in heterogeneous landscapes. In: Liu J, Hull V, Morzillo AT, Wiens JA, editors. Sources, sinks and sustainability: Cambridge University Press.
49. White JW, Rassweiler A, Samhouri JF, Stier AC, White C (2014) Ecologists should not use statistical significance tests to interpret simulation model results. Oikos 123: 385–388.
50. Cohen J (1988) Statistical power analysis for the behavioral sciences. Hillsdale, NJ: Lawrence Erlbaum Associates.
51. Olejnik S, Algina J (2003) Generalized eta and omega squared statistics: measures of effect size for some common research designs. Psychological Methods 8: 434–447.
52. Bakeman R (2005) Recommended effect size statistics for repeated measures designs. Behavior Research Methods and Instrumentation 37: 379–384.
53. Lawrence M (2011) ez: Easy analysis and visualization of factorial experiments. R package version 3.0-0. Available: http://CRAN.R-project.org/package=ez. Accessed March 2013.
54. Gonzalez A, Holt RD (2002) The inflationary effects of environmental fluctuations in source–sink systems. Proceedings of the National Academy of Sciences 99: 14872–14877.
55. Ries L, Fletcher RJ Jr, Battin J, Sisk TD (2004) Ecological responses to habitat edges: mechanisms, models, and variability explained. Annual Review of Ecology, Evolution, and Systematics 35: 491–522.
56. Snyder RE, Chesson P (2003) Local dispersal can facilitate coexistence in the presence of permanent spatial heterogeneity. Ecology Letters 6: 301–309.
57. Pinto SM, MacDougall AS (2010) Dispersal limitation and environmental structure interact to restrict the occupation of optimal habitat. The American Naturalist 175: 675–686.
58. With KA, King AW (1999) Extinction thresholds for species in fractal landscapes. Conservation Biology 13: 314–326.
59. Hiebeler D (2004) Competition between near and far dispersers in spatially structured habitats. Theoretical Population Biology 66: 205–218.
60. Gilbert B, Lechowicz MJ (2004) Neutrality, niches, and dispersal in a temperate forest understory. Proceedings of the National Academy of Sciences of the United States of America 101: 7651–7656.
61. Skelsey P, Garrett KA (2013) Why dispersal should be maximized at intermediate scales of heterogeneity. Theoretical Ecology 6: 203–211.
62. Nevoux M, Gimenez O, Arlt D, Nicoll M, Jones C, et al. (2010) Population regulation of territorial species: both site dependence and interference mechanisms matter. Proceedings of the Royal Society B: Biological Sciences 278: 2173–2181.
63. Chipperfield JD, Dytham C, Hovestadt T (2011) An Updated Algorithm for the Generation of Neutral Landscapes by Spectral Synthesis. PLoS ONE 6: e17040.

# 13

# Application of Large-Scale Parentage Analysis for Investigating Natal Dispersal in Highly Vagile Vertebrates: A Case Study of American Black Bears (*Ursus americanus*)

**Jennifer A. Moore[1,4]*, Hope M. Draheim[2], Dwayne Etter[3], Scott Winterstein[4], Kim T. Scribner[4]**

1 Biology Department, Grand Valley State University, Allendale, Michigan, United States of America, 2 Department of Zoology, Michigan State University, East Lansing, Michigan, United States of America, 3 Wildlife Division, Michigan Department of Natural Resources, East Lansing, Michigan, United States of America, 4 Department of Fisheries and Wildlife, Michigan State University, East Lansing, Michigan, United States of America

## Abstract

Understanding the factors that affect dispersal is a fundamental question in ecology and conservation biology, particularly as populations are faced with increasing anthropogenic impacts. Here we collected georeferenced genetic samples (n = 2,540) from three generations of black bears (*Ursus americanus*) harvested in a large (47,739 km$^2$), geographically isolated population and used parentage analysis to identify mother-offspring dyads (n = 337). We quantified the effects of sex, age, habitat type and suitability, and local harvest density at the natal and settlement sites on the probability of natal dispersal, and on dispersal distances. Dispersal was male-biased (76% of males dispersed) but a small proportion (21%) of females also dispersed, and female dispersal distances (mean $\pm$ SE = 48.9$\pm$7.7 km) were comparable to male dispersal distances (59.0$\pm$3.2 km). Dispersal probabilities and dispersal distances were greatest for bears in areas with high habitat suitability and low harvest density. The inverse relationship between dispersal and harvest density in black bears suggests that 1) intensive harvest promotes restricted dispersal, or 2) high black bear population density decreases the propensity to disperse. Multigenerational genetic data collected over large landscape scales can be a powerful means of characterizing dispersal patterns and causal associations with demographic and landscape features in wild populations of elusive and wide-ranging species.

**Editor:** João Pinto, Instituto de Higiene e Medicina Tropical, Portugal

**Funding:** Michigan Department of Natural Resources through the Wildlife and Sportfish Restoration Program F11AF00640, Michigan State University Department of Fisheries and Wildlife. The funders had no role in study design, data collection and analysis, decision to publish, or preparation of the manuscript.

**Competing Interests:** The authors have declared that no competing interests exist.

* E-mail: moorejen@gvsu.edu

## Introduction

Dispersal is an important ecological process that allows individuals to exploit temporally and spatially variable resources, and has implications for population dynamics and population viability through the spatial redistribution of individuals [1]. Understanding the mechanisms that motivate individuals to disperse from, and to settle in specific locales is a fundamental question in ecology and conservation biology [1] particularly as populations are faced with increasing anthropogenic impacts and rapidly changing and fragmented environments. Animal dispersal is typically non-random, as informed dispersal decisions (i.e., those that are based on ecological or social cues) should confer an evolutionary advantage over uninformed decisions [2]. Dispersal behavior can thus be based on factors intrinsic to the individual (i.e., phenotype-dependent) and/or extrinsic social or ecological factors (i.e., condition-dependent) [3].

Numerous hypotheses have been proposed to explain the evolution of, and variation in dispersal behavior, yet most explanations involve competition, inbreeding avoidance, and/or environmental stochasticity [3,4]. Individuals are theoretically expected to disperse from areas of high to low population density, to exploit areas with comparatively more abundant resources (e.g., space, food, mates) and areas with fewer competitors [3]. Results from the few empirical studies that have investigated density-dependent dispersal directly have found equivocal support for theoretical expectations, which may reflect context-specific probabilities of immigration and emigration in response to density [4–6]. Alternatively, when the relationship between population density and resource availability is not closely coupled, individuals may be better able to survive and reproduce by dispersing into locales characterized by habitat that is similar to habitat in their natal areas (i.e., natal habitat-biased dispersal [7,8] or natal habitat imprinting [9,10]).

Dispersal probability may also be inversely related to maternal age. Whether or not a female offspring disperses or exhibits philopatry is dependent upon the costs of dispersal as well as the costs of kin competition. Theory predicts that as a mother ages, her investment in future reproductive events (i.e., her residual reproductive value) declines [11–13]. Therefore, older mothers should invest more in the growth and survival of each current offspring (i.e., by promoting offspring philopatry) than younger mothers, who should invest more in their own growth and survival and thus future reproduction (i.e., by promoting offspring dispersal) [11,13]. Dispersal is also commonly biased toward one sex, which reduces the risk of mating with close relatives, regardless of whether or not individuals are capable of recognizing and deliberately avoiding mating with kin [14]. In mammals, males are typically the dispersing sex while females remain philopatric [15,16], yet there is variation among individuals of the same sex associated with phenotype and/or condition [17].

The above hypotheses are not mutually exclusive, and thus patterns in data, particularly when based on small sample sizes, may not characterize underlying processes unambiguously. Traditional studies of dispersal require marking and tracking a large number of individuals over time, and complications can arise due to difficulty obtaining adequate sample sizes and tracking long distance dispersers. Further, the limited duration over which observations are collected, particularly for long-lived, iteroparous species, can affect the accuracy and power of inferences [18]. For highly mobile vertebrates, long distance dispersers are difficult to track or recapture and differences in detectability between dispersers and non-dispersers may be large and could bias results [19]. Parentage analysis provides an alternative to traditional field-based methods (e.g., mark-recapture, radio telemetry), and has great potential to inform our understanding of dispersal [20,21]. However, the logistical challenges associated with genetic sampling of a sufficient number of wide-ranging vertebrates have recently lead some researchers to question the feasibility of applying parentage analysis to investigate dispersal in these species [18].

In this study, we use a large number of harvest samples collected from an isolated, closed population of American black bears (*Ursus americanus*) inhabiting a large geographical area in the Northern Lower Peninsula (NLP) of Michigan, USA (Figure 1, inset). Black bears are solitary omnivores with a promiscuous mating system, male-biased dispersal [19,21], and home ranges that can overlap considerably depending on resource availability [19,22–24]. Annual population size estimates of NLP black bears over the last decade are stable and have ranged from 1,500–1,900 individuals, and local density varies. Between 13 and 29% of the population is harvested annually ([25]; unpublished data, Michigan Department of Natural Resources). Harvest quotas fluctuate annually, and are largely based upon the population estimates.

Parentage analysis using harvest samples has provided us with a unique opportunity to conduct a large-scale investigation of dispersal in a highly vagile species by quantifying sex-specific dispersal probabilities and distances and factors influencing these behaviors, like habitat quality and harvest density. We tested the following hypotheses:

(1) Dispersal depends on sex and age of the offspring. Radio telemetry data indicate that dispersal in male NLP black bears usually has occurred by the time a bear reaches two years of age [26]. Thus, we predict that NLP black bears will follow the typical mammalian pattern of male-biased dispersal, and that the probability that a male has dispersed will increase after two years of age.

(2) Female dispersal is inversely related to the age of the mother. We predict that females will be largely philopatric [16], and the mother will be increasingly more tolerant of offspring philopatry as she ages and her residual reproductive value declines.

(3) Dispersal probability and dispersal distance are dependent upon harvest density. We predict that dispersal probability will be greatest in areas of high harvest density, where bears are more often displaced by hunting practices. Further, if harvest density is correlated with bear population density (e.g., for red grouse, [27]), we expect conspecific competition to be greatest at high density, thus increasing dispersal and causing dispersers to seek out areas with lower harvest density and lower bear density [28].

(4) Black bears exhibit natal habitat-biased dispersal. Familiarity with a particular habitat type during formative early ontogenetic periods should increase an individual's ability to acquire resources in that habitat type. We predict that there will be a positive relationship between the habitat type at an individual's natal site and the habitat type into which that individual settles.

## Methods

### Study Area

Our study area covered the northern two-thirds of the lower peninsula of Michigan (~47,739 km$^2$). This area is considered an insular population, being bounded on three sides by the Great Lakes, and to the south by a landscape composed of intensive agricultural and expansive urban areas. The NLP landscape is a largely forested mix of northern and mixed hardwoods, pines, and forested and non-forested wetlands. The NLP is also highly fragmented by roads, forestry activities, and agriculture [29].

### Field Sampling

During the annual NLP bear hunting season (September and October), hunters must register harvested bears at registration stations facilitated by the Michigan Department of Natural Resources (MDNR). At the check stations, hunters report the bear's sex and harvest location (to a township, range, and section; 2.6 km$^2$). A pre-molar tooth is extracted for aging and DNA extraction. Bears are aged by the MDNR using the cementum annuli method [30], which has been estimated to be 96% accurate (unpublished data, D. Etter). In seven years (2002 and 2003, and 2006–2010), MDNR personnel collected samples from 2,580 harvested bears. All samples were collected from bears that were legally harvested under bear hunting licenses issued by the MDNR to individual hunters. All harvest took place in one of three bear management units, and black bears are not threatened or endangered in the state of Michigan. No vertebrate Institutional Animal Care and Use Committee approval was sought because the harvest season takes place annually, it is strictly regulated by the MDNR, and no bear harvest was conducted specifically for the purpose of this study. Rather, harvest samples were provided by MDNR cooperators.

### Laboratory Analysis

We extracted DNA from bear teeth using Qiagen DNEasy Tissue Kits following manufacturer protocols (Qiagen Inc., Valencia, CA). DNA was quantified using a Nanodrop spectrophotometer (Thermo Scientific, Waltham, MA) and diluted to a 20 ng/µl working concentration. Using polymerase chain reaction (PCR), we amplified 12 microsatellite loci including *G10X*, *G10L*, *G10D*, *G10M*, *G10B* [31], *UarMU59*, *UarMU50* [32], *UT29*, *UT35*, *UT38* [33], *ABB1*, and *ABB4* [34]. 10 µl PCR reactions included 40 ng of DNA in 10 mM Tris-HCl, 50 mM KCl, 2.0 mM $MgCl_2$, 0.2 mM of each dNTP, 1 pmol of each primer, and 0.5 units of *Taq* polymerase. Thermocycler conditions included a 4 min denaturation at 94°C, followed by 25–42 cycles of 30s at 94°C (1 min for *G10L*), 30s at the locus-specific annealing temperature (58°C for *G10X, G10L, G10D, G10B, UarMU59, UarMU59;* 54°C for *UT29, UT35, UT38, ABB1, ABB4*), 1 min at 72°C, and a final 10 min extension step at 72°C. We used 6.5%

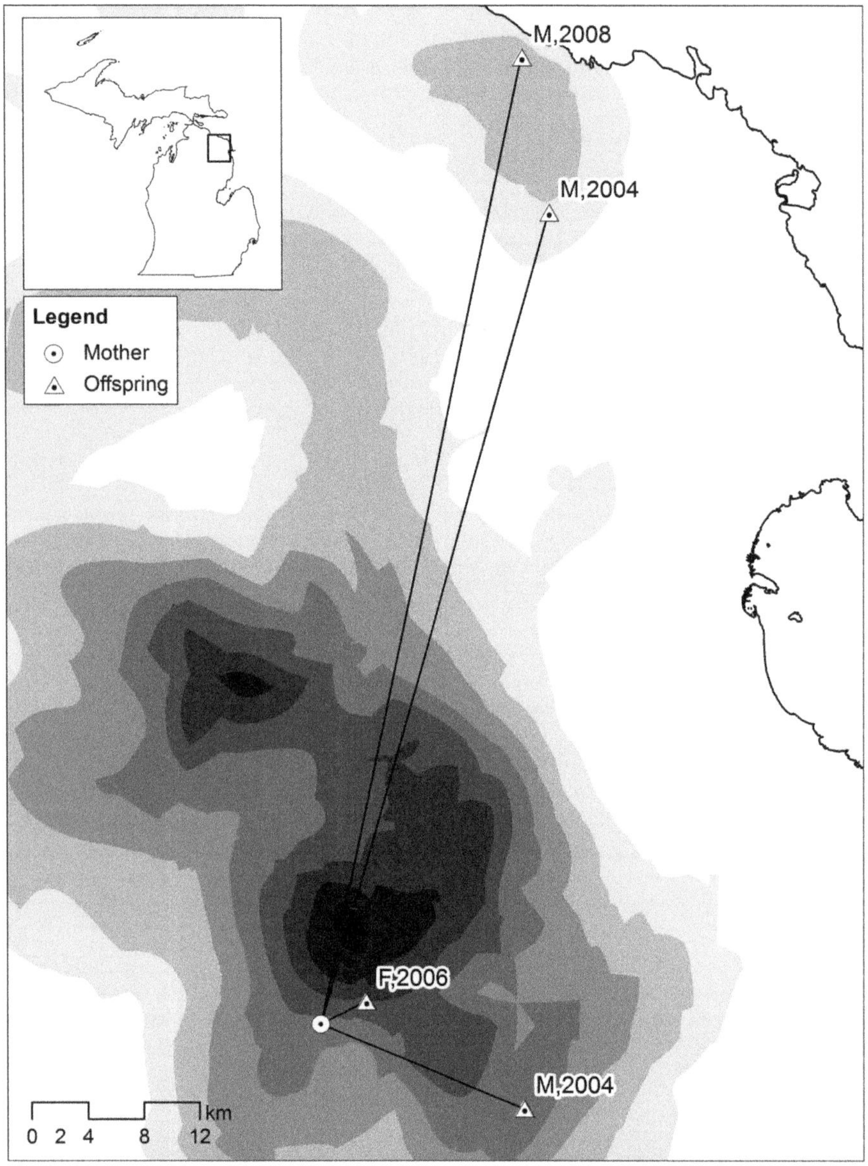


**Figure 1. Example of black bear dispersal patterns.** Example illustrating three cohorts of offspring from one black bear mother, showing patterns of sex-specific distances, and dispersal patterns in relation to harvest density (low = white, high = black). Offspring are labeled according to sex and year of birth. Square on inset indicates approximate area in Michigan.

denaturing acrylamide gels for electrophoresis visualized on a LI-COR 4200 Global IR2 System (LI-COR Inc., Lincoln, NE). For each gel, we included molecular weight standards and individual bears with known genotypes. All alleles were scored independently by two experienced lab personnel using Saga genotyping software (LI-COR Inc., Lincoln, NE). 10% of samples were randomly selected and genotyped twice to provide a genotyping error rate of <2%.

Using MICRO-CHECKER [35], no loci were found to deviate significantly from Hardy-Weinberg and linkage equilibrium, so all 12 were retained for further analyses.

To minimize potential errors in field sexing, we genetically sexed 1) all females that were identified as having had cubs in consecutive years (female black bears typically produce young every other year [26]), 2) all female offspring that were identified as dispersers, and their mothers, and 3) all males that were identified as non-dispersers, and their mothers. Genetic sexing followed the protocol outlined in [36] using the *SE47* and *SE48* amelogenin primers [37].

## Parentage Analysis

Parentage analysis was conducted using the program FRANz [38] to identify mother-offspring (MO) dyads. FRANz is a Bayesian pedigree reconstruction program that allows for the incorporation of prior information about sex, birth year, and death year. Incorporating age priors of putative parents and offspring enables the program to handle multi-generational data, without *a priori* separation of individuals into cohorts. For each parent-offspring relationship identified, FRANz estimates a posterior probability of the identified parent being the true parent and a parent-pair log odds ratio (LOD) score [38,39]. To assess the accuracy of parentage assignment and to set the threshold

**Table 1.** Summary of pairwise Euclidean distances for black bear mother-offspring dyads, by offspring sex, including samples sizes (N), and ages of offspring (at time of death), and mothers (at time of offspring birth).

| | | Euclidean distance (km) | | Offspring age (years) | | Maternal age (years) | |
|---|---|---|---|---|---|---|---|
| | N | Mean (± SE) | Range | Mean (± SE) | Range | Mean (± SE) | Range |
| Males | 178 | 41.1±2.5 | 0–251.2 | 2.5±0.2 | 1–14 | 5.8±0.3 | 2–19 |
| Dispersers | 106 | 59.0±3.2 | 30–251.2 | 2.8±0.2 | 1–14 | 5.6±0.3 | 2–17 |
| Residents | 72 | 14.9±0.9 | 0–28.8 | 2.1±0.2 | 1–11 | 6.0±0.5 | 2–19 |
| Females | 159 | 14.2±1.8 | 0–187.2 | 2.9±0.2 | 1–12 | 5.9±0.3 | 2–22 |
| Dispersers | 26 | 48.9±7.7 | 20–187.2 | 3.5±0.5 | 1–9 | 4.9±0.7 | 2–13 |
| Residents | 133 | 7.1±0.4 | 0–19.4 | 2.9±0.2 | 1–12 | 6.1±0.4 | 2–22 |

(Dispersers are defined by distances between natal and harvest locations (>30 km for males, and >20 km for females) based on results of an independent genetic spatial autocorrelation analysis)

posterior probability for true parentage assignments, we first performed a simulation by assigning parentages to known (simulated) offspring (see File S1 for detailed methodology). In the simulation, FRANz identified the correct parents for 98.5% of the simulated offspring with posterior probabilities ranging from 0.41–1 (mean = 0.96). We then performed a parentage analysis in FRANz using the real bear genotypes, and their sexes and birth and death years as priors, with the default parameter settings, with the exception of: maximum number of candidate fathers (Nmax) = 800, our empirical estimate of genotyping error = 0.02, the increment in steady state distribution variational distance ($\delta$) = 0.01, and the convergence tolerance ($\varepsilon$) = 0.1. For further analyses, we retained only those MO dyads with posterior probabilities ≥0.9 (based on simulation results, see File S1).

## Spatial Analyses

Locations of mothers and offspring were georeferenced to the centroid of their reported harvest sections (a 2.6 $km^2$ area). Euclidean distances between MO dyads were determined using ETGeowizards and Hawth's Tools [40] in ArcGIS 9.3 (ESRI, Redlands, CA). We classified offspring as 'dispersers' or 'non-dispersers' (hereafter called 'residents') based on threshold Euclidean distances between mother and offspring locations. Male dispersers were ≥30 km from the mother's location, and female dispersers were ≥20 km from the mother's location. Distance thresholds were set based on genetic spatial autocorrelation analyses that were performed separately for each sex (H. Draheim, unpublished data). Genetic spatial autocorrelation analyses examine the relatedness of pairs of bears at different distance classes, and can provide a good indicator of the extent of effective gene flow (i.e., dispersal) across the study area [41]. Results of the genetic spatial autocorrelation analyses showed that beyond the 20 and 30 km distance classes, female-female and male-male pairs, respectively, were no more related than expected by chance. We also assumed that beyond these distances, bears were outside of the core home ranges of assigned mothers [19,22,26,29,42].

To extract habitat information for each mother and offspring location, we reclassified the 2006 NOAA Coastal Change Analysis Program Land Cover dataset (resolution = 30 m) into seven land cover classifications and ranked them according to bear habitat suitability (most (1) to least (5) suitable) based on an ecological model of bear habitat selection in Michigan (developed using radio telemetry and discrete-choice modeling [29]), as follows: mixed deciduous forest (MF, 1), forested wetland (FW, 1), evergreen forest (EF, 2), non-forested upland (NFU, 3), agriculture (AG, 4), non-forested wetland (NFW, 5), and developed (DEV, 5).

We determined localized harvest density using bear harvest locations. Although we do not have direct estimates of local population density, or harvest effort, we expect some degree of correlation between harvest density and bear density. One qualitative factor contributing to this supposition is that, in annual harvest surveys, hunters consistently cite bear density as the primary reason for selecting a hunting site [43]. Likewise, a previous study of NLP black bears found that the geographic distribution of harvest locations [44] (i.e., regions of relatively high and low harvest) does not show significant annual variation. As such, we used the harvest locations for each year from 2002–2010 to create annual kernel density function [45] grids and reclassified grids into categories ranging from 1–10 (low to high harvest density). We then created a median harvest density grid by calculating the median values over the nine annual harvest density grids. A 1.61 km diameter circular buffer (representing the approximate area of a section) was created around each harvest location, and we extracted harvest density and land cover grid cell values falling within each circular buffer. The value that constituted the majority of grid cells within a buffer was assigned for each individual's location.

## Statistical Analyses

We used mixed models (*Lme4* package in R [46]), to model the factors that were associated with dispersal probability (0, 1) and dispersal distance while accounting for repeated measures of mothers (random effect) with multiple offspring. We assumed that dispersal patterns did not change considerably over time, and thus pooled all data for analyses. To quantify associations between intrinsic and extrinsic factors affecting whether or not individuals disperse, we first modeled dispersal probability (with each offspring classified as a disperser (1) or resident (0), based on the distance thresholds presented above as a function of offspring sex (*offsex*), offspring age (*offage*), characteristics of the natal (mother's) site including habitat type (*momhab*), habitat suitability (*momhs*), and harvest density (*momharv*) and characteristics of the settled (offspring harvest) site including habitat type (*offhab*), habitat suitability (*offhs*), harvest density (*offharv*), and mother's age (at the time of offspring birth, *momage_corr*). Subsequently, using only those individuals classified as dispersers, we modeled dispersal distance as a function of offspring sex and age and characteristics of the settlement site (habitat type, habitat suitability, and harvest density) and the natal site (habitat type, habitat suitability, and

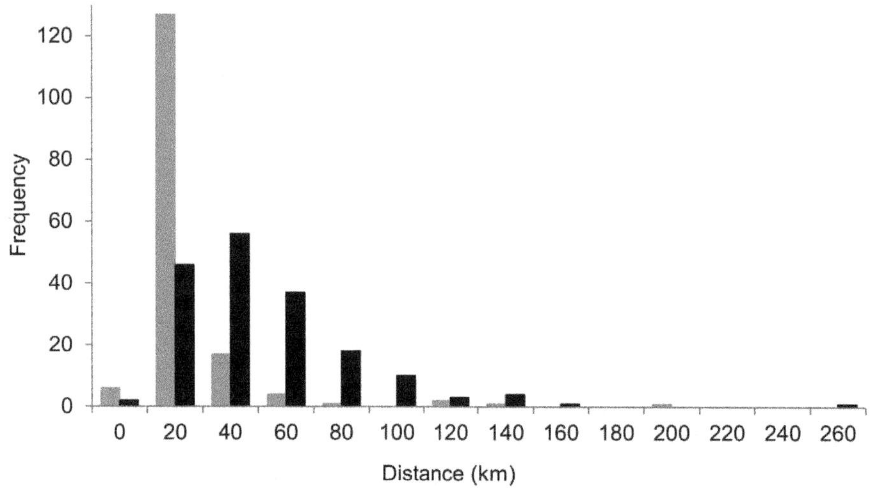


**Figure 2. Distribution of black bear dispersal distances.** Frequency histogram of pairwise distances between black bear mother and offspring dyads, for all individual males (black bars, n = 178) and females (grey bars, n = 159). Distances along the x-axis represent the upper bounds for each bin.

harvest density). Lastly, again using only those individuals classified as dispersers, we investigated whether natal habitat-biased dispersal occurred. We modeled the habitat type at the settlement site as a function of offspring sex, offspring age, and the habitat type at the natal site. We tested 64 *a priori* hypothetical models of dispersal probability, 53 models of dispersal distance, and 7 models of habitat type (see File S2 for a complete list of models and results). We used Akaike's Information Criterion (AIC) for model selection, with the lowest values representing the best supported model(s). Once the top models were selected, a sex*age interaction was added to test for improved model fit. The sex*age interaction was aimed at capturing age-specific dispersal differences that may have only been present in one sex. Models were rescaled relative to the model with the lowest AIC value in the set (ΔAIC), and Akaike weights ($w_i$) were calculated relative to the other models in each set. Models with ΔAIC values of ≤2 were considered to be supported.

## Results

We genotyped 2,540 black bears (1,415 males, 1,125 females) for inclusion in the parentage analysis to identify MO dyads. Parentage analysis identified 337 MO dyads with posterior probabilities ≥0.9. Dyads were comprised of 159 female and 178 male offspring, from 214 mothers. The average age at time of death for offspring, and at the time of offspring birth for mothers was 3.0 years (range = 1–12), 2.5 years (range = 1–14), and 5.8 years (range = 2–22) for female offspring, male offspring, and mothers, respectively.

Euclidean distances averaged 14.2 km (median = 7.7 km) between mother-female offspring dyads, and 41.1 km (median = 34.9 km) for mother-male offspring dyads (Table 1, Figure 2). The maximum dispersal distance was 251.2 km for males and 187.2 km for females (Figure 2). 60.0% of male offspring and 17.0% of female offspring dispersed (Figure 3). For bears that were ≥3 years of age at the time of harvest (i.e., those that were most likely to have dispersed), the frequency of dispersal was 21% for

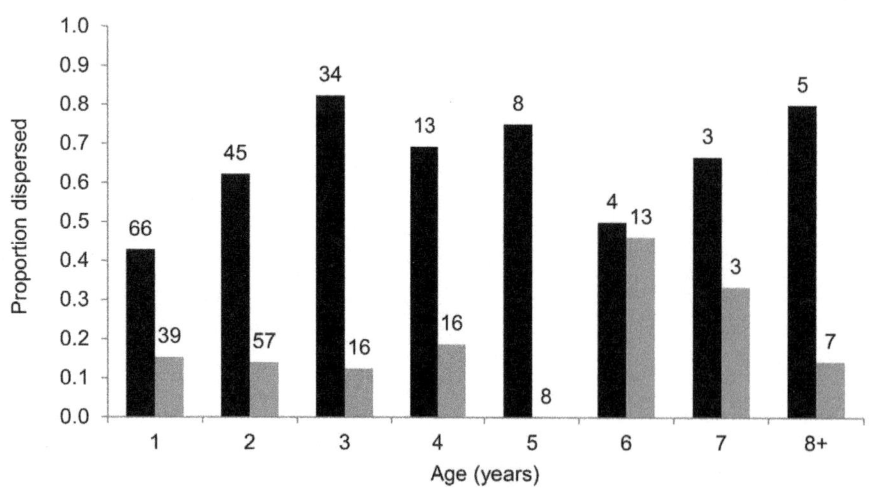


**Figure 3. Dispersal probabilities, by age.** Proportion of male (black bars) and female (grey bars) black bears that dispersed, by age. Sample sizes are presented above bars.

**Table 2.** Top ten linear mixed effects models of black bear dispersal probability and dispersal distance including Akaike's Information Criteria (AIC) rescaled to the lowest value (ΔAIC), and Akaike weights ($w_i$).

| Model parameters | | | |
|---|---|---|---|
| **Response variable = dispersal probability** | **AIC** | **ΔAIC** | $w_i$ |
| offsex + offage + offharv | 361.6 | 0.0 | 0.22 |
| offsex + offage + offharv + offhs | 361.9 | 0.3 | 0.19 |
| offsex + offage + offharv + momhs | 362.8 | 1.2 | 0.12 |
| offsex + offage + offharv + momhs + offhs | 363.3 | 1.7 | 0.10 |
| offsex + offage + momharv + offharv | 363.5 | 2.0 | 0.08 |
| offsex + offage + momharv + offharv + momage_corr | 363.8 | 2.3 | 0.07 |
| offsex + offage + momharv + offharv + offhs | 363.9 | 2.3 | 0.07 |
| offsex + offage + momharv + offharv + momhs | 364.8 | 3.2 | 0.04 |
| offsex + offage + momharv + offharv + momhs + offhs | 365.2 | 3.7 | 0.04 |
| offsex + offharv + offhs | 367.9 | 6.3 | 0.01 |
| **Response variable = dispersal distance** | | | |
| offsex * offage + offhab + momhab + offharv + momharv | 1251.71 | 0.00 | 0.46 |
| offsex * offage + offhab + momhab + offharv | 1253.35 | 1.64 | 0.20 |
| offsex + offage + offhab + momhab + offharv + momharv | 1254.03 | 2.32 | 0.15 |
| offsex * offage + offhab + momhab + momharv | 1255.65 | 3.94 | 0.06 |
| offsex + offage + offhab + momhab + offharv | 1255.67 | 3.96 | 0.06 |
| offsex * offage + offhab + momhab | 1257.43 | 5.71 | 0.03 |
| offsex + offage + offhab + momhab + momharv | 1258.00 | 6.29 | 0.02 |
| offsex + offage + offhab + momhab | 1259.78 | 8.07 | 0.01 |
| offsex + offhab + momhab | 1261.64 | 9.93 | 0.00 |
| offage + offhab + momhab | 1286.34 | 34.63 | 0.00 |

Mother ID was included as a random effect in all models. Models with ΔAIC ≤ 2 are supported. See File S2 for a complete list of model results. (Model parameters are sex of offspring (*offsex*), age of offspring (*offage*), maternal age at the time of offspring birth (*momage_corr*), habitat type at the natal site (*momhab*), habitat type at the settlement site (*offhab*), harvest density at the natal site (*momharv*), harvest density at the settlement site (*offharv*), habitat suitability rank at the settlement site (*offhs*), and habitat suitability rank at the natal site (*momhs*), * indicates an interaction).

females (n = 63) and 76% for males (n = 68). Dispersal distances for female dispersers averaged 48.9±7.7 km, while male dispersers averaged 59.0±3.2 km. Distances between mothers and female resident offspring averaged 7.1±0.4 km, while male resident offspring averaged 14.9±0.9 km (Table 1).

The best model characterizing dispersal probability included offspring sex ($\beta_{offsex}$ = 2.24), offspring age ($\beta_{offage}$ = 0.18), and harvest density at the offspring settlement site ($\beta_{offharv}$ = −0.39) (AIC = 361.56; $w_i$ = 0.22; Table 2, File S2). Competitive (ΔAIC<2) models of dispersal probability included harvest density at the natal site, as well as habitat suitability at both the natal and settlement sites (Table 2). Based on these models, males were more likely to disperse than females, and male dispersal probability increased considerably after 1–2 years of age (Figure 3). Dispersal probability was greatest for individuals dispersing from areas of high quality habitat and low harvest density and settling in areas of high quality habitat with low harvest density.

The best model of dispersal distance included an interaction between offspring age and sex ($\beta_{offage*offsex}$ = −0.7), harvest density ($\beta_{momharv}$ = −1.5, $\beta_{offharv}$ = −3.5) and habitat type at both the natal and the settlement sites (AIC = 1251.7; $w_i$ = 0.46, Table 2, Figure 4). A competitive model (ΔAIC<2) included all of the factors in the best model with the exception of natal site harvest density. Dispersal distance was greatest for young males dispersing from areas with low harvest density and settling in areas with low harvest density (Figure 4). Generally, dispersal distances tended to be greater for bears dispersing from areas of lower quality habitat like non-forested upland ($\beta_{NFU}$ = 7.52, mean distance = 56.1 km) and non-forested wetland ($\beta_{NFW}$ = 13.8, mean distance = 62.1 km) and for individuals settling in more suitable habitat types like mixed forest ($\beta_{MF}$ = −2.5, mean distance = 53.9 km), forested wetland ($\beta_{FW}$ = 2.1, mean distance = 58.61 km), and evergreen forest ($\beta_{EF}$ = −0.5, mean distance = 63.35 km).

The best model of settled habitat type included offspring sex ($\beta_{offsexM}$ = −0.2; AIC = 357.3; $w_i$ = 0.66, File S2), and the next best model was a null model that included only the random effect of maternal identity. Thus, we found no evidence of natal habitat imprinting, but this could have been due more to the fact that most bears were found in one of two habitat types (mixed deciduous forest, 44%, and forested wetland, 38%), which may have limited the power to detect a biologically meaningful relationship. A higher proportion of female than male dispersers was harvested in mixed deciduous forest (59% of females vs. 39% of males), while a higher proportion of male than female dispersers was harvested in forested wetlands (26% of females vs. 41% of males).

The majority of dispersers either showed no change in harvest density between the natal versus the settled site (24%) or settled in a site with lower harvest density than in their natal site (44%) (see Figure 1 for an example). Only 33% of dispersers settled in an area with a higher harvest density than was observed in their natal site. Likewise, the majority of dispersers (70%, n = 95) showed no change in habitat suitability from the natal to settled site. Similar proportions of dispersers settled in sites with higher habitat suitability (13%) and lower habitat suitability (17%) compared to their natal site.

## Discussion

We have shown that parentage analysis is a powerful means of characterizing condition-dependent dispersal patterns in elusive species, particularly when data are collected over large temporal and spatial scales. We found that black bear dispersal in Michigan's NLP is strongly male biased with 76% of males, but only 21% of females dispersing. Furthermore, harvest density and habitat suitability had strong effects on dispersal probability and dispersal distances, and black bears did not appear to exhibit natal habitat-biased dispersal. The probability of dispersal increased with decreasing harvest density and increasing habitat quality at both the natal and settlement sites. Likewise, black bears in areas with low harvest density tended to travel greater distances, and most commonly settled in areas of high quality habitat.

Harvest pressure could be driving our observed pattern of higher dispersal probability and distance in areas with lower densities of bear harvest locations. Bears are primarily hunted in the NLP using bait and dogs [41]. If harvest density was driving our observed patterns of dispersal, we might expect to see higher dispersal in areas of high harvest density, due to competing bears being lured by bait stations, or regularly harassed by dogs. On the other hand, in areas of high harvest density, individuals who disperse greater distances may be more susceptible to harvest. Intensive harvest can result in reduced abundance [47], changes in offspring survival [48], skewed sex ratios [49], age distributions

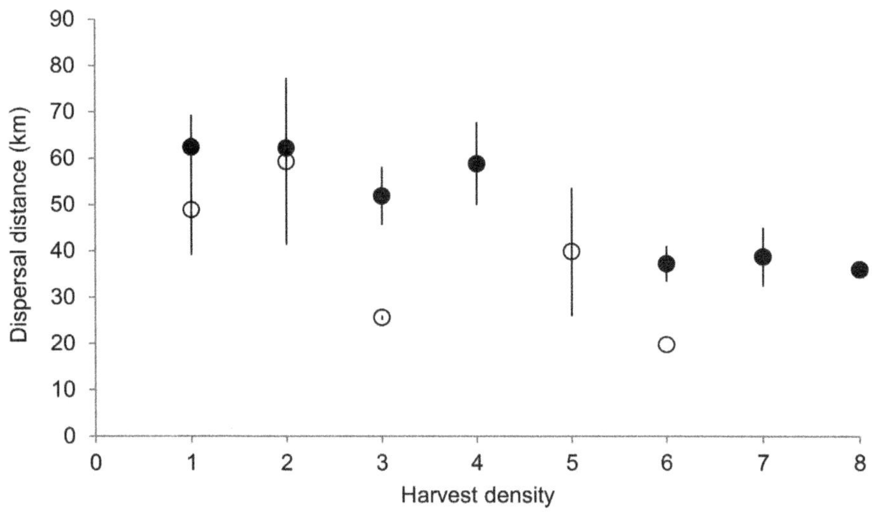


**Figure 4. Dispersal distances and harvest densities.** Average dispersal distances (±SE), based on harvest density at the settlement site (1–8, low to high, from rescaled kernel density function grid estimates of black bear harvest locations), for male (filled symbols) and female (open symbols) black bears.

[47,49], and distributions of phenotypic variation [50]. Anthropogenic selective pressures are imposed upon harvested populations that non-harvested populations are not subjected to [50,51]. We speculate that high harvest intensity may either select for reduced dispersal or cause individuals to behaviorally modify their movement patterns (but see [52]).

Alternatively, black bear population density may be a strong regulator of dispersal in high density areas, via conspecific competition. Most dispersal hypotheses predict positive density-dependent dispersal resulting from intraspecific competition, which is strongest in territorial species and species that incur high dispersal costs [28,53]. However, empirical studies, mostly based on small mammals [53–55], have shown that rates of immigration and emigration can also decrease at high density, due to saturation of aggressive and territorial individuals at high densities that restrict movements [5] of individuals into or out of these areas (i.e., the social fence hypothesis [56]). Black bears in areas of low density may not be limited by aggressive conspecifics, making them more able to move freely at lower cost [57,58]. Using population-level spatial autocorrelation analysis, Roy et al. [58] showed that genetic spatial autocorrelation was strongest in low density black bear populations, which these authors attributed to restricted dispersal in high density populations [58]. Habitat quality was not found to be the strongest predictor of dispersal patterns for NLP bears. However, this may be due to the fact that much of the bear habitat in the NLP is high quality, so resources are not the strongest limiting factor.

Dispersal distances of NLP black bears are comparable to other studies of black and grizzly bears, where average female distances range from 2.9 to 14.3 km, and average male distances range from 34.0 to 61 km [21,59,60]. Long-distance dispersers (e.g., those individuals that make up the right-hand tails of the distributions in Figure 2) are rare in our study population. However, long-distance dispersers can be extremely important for population dynamics and gene flow, and the reasons for long- and short-distance dispersal may differ [17,61]. Short-distance dispersal may be sufficient to avoid mating with kin, whereas long-distance dispersal may enable colonization of new areas, or escape from high density areas [17,28]. Based on the variables measured, we were unable to identify any intrinsic characteristics (e.g., based on age, family, maternal age, birth year) that would differentiate long-distance dispersers from short-distance dispersers. Because of the importance of long-distance dispersal, further investigation into the factors that drive certain individuals to exhibit this behavior is warranted.

Approximately 24% of adult male black bears (≥3 years of age) remained within 30 km of their natal ranges. The relatively high percentage of non-dispersing male bears is surprising considering black bears are thought to unequivocally exhibit strong male-biased dispersal, with 100% of males dispersing in some study systems [21,42,59]. The NLP black bears experience high turnover of individuals that are removed by harvest (approximately 25% of the population is harvested annually [26]). In areas of high harvest density, juvenile males may be able to utilize habitats adjacent to their natal ranges that are left vacant following harvest of neighboring males. High male turnover would also contribute to our observed lower probability of dispersal in high harvest areas. Further investigation is needed to determine whether remaining near the natal range increases the likelihood of a male mating with kin [21,62].

Comparisons of studies using ecological and genetic methods often reveal disparities in their results, even when conducted in the same study area. In a long-term radio telemetry study of NLP bears, Etter et al. (2002) found that 30 of 31 male bears (95%) dispersed, compared to the 52 of 68 male bears (76%) ≥3 years old that dispersed in our study (Fig. 3). The disparity in results may be attributed to differing methodology, data or sample sizes (i.e., differing thresholds defining dispersers vs. non-dispersers, different temporal span of the studies, order of magnitude difference in sample size), or the fact that the data from these two studies were collected a decade apart. Nevertheless, if differing methodology were the cause of the discrepancy, we would expect the telemetry data to provide an estimate of dispersal frequency that is biased low, due to the potential for dispersers to be lost from contact or incur higher mortality [18,19]. Alternatively, harvest regimes and population density may have changed enough in a decade to impact dispersal frequency. In a study of black bears in New Mexico, Costello et al. [19,21] found a similar pattern to our system, where almost all males tracked using radio telemetry dispersed [19], but a spatial analysis of genetic relatedness during the same time period was indicative of a much lower male dispersal rate (~75%) [21].

Female philopatry, like male dispersal, is not absolute. Approximately 21% of females ≥3 years old emigrated, often over large distances (Figure 2). A similar result was found for NLP females using radio telemetry (9 of 28, or 32% of females dispersed [26]). The frequency of dispersal in female black bears in the NLP is, on average, 5–15% higher than estimates from previous studies of black bear populations [19,23,42,63]. Relatively high female dispersal rates in our population may be indicative of overall low population density relative to other black bear populations, a low cost of dispersal for females, and/or abundant resources. Female black bears in the NLP are generally in good condition, which is reflected in a young age of first reproduction and high reproductive rates relative to other populations [26,42]. Dispersal probability increased with the quality of the habitat, presumably because bears in better condition are less likely to incur major dispersal costs related to resource acquisition.

Although maternal age was not a strong predictor of offspring dispersal probability in our models, this might be due more to the small proportion of females that did disperse (i.e., limiting our statistical power) or to the intensive harvest that reduces the median age of mothers in the population. In Scandinavian brown bears, 41% of females disperse, and female dispersal probability is strongly negatively correlated with maternal age [62]. Brown bears form matrilineal assemblages, and female kin compete to remain in or near the home ranges of their mothers, which means that smaller female offspring are more likely to disperse [62]. For black bears in the NLP, maternal age was lower (by an average of one year) for female dispersers than for female residents (Table 1). This corresponds to our prediction that younger mothers should promote offspring dispersal in favor of investing more in their own growth and survival, which would enhance future reproductive efforts. We identified seven mothers who produced female offspring in multiple years and where one offspring dispersed and the others were philopatric. For all but one of these mothers, the disperser was the first born among her offspring. Harvesting mothers before they are able to reach an advanced age may confound what might otherwise be a strong tendency toward the formation of multi-generational matrilineal assemblages in black bears.

The putative alteration of natural dispersal patterns by intensive harvest is a situation deserving of management attention. Several studies comparing harvested and non-harvested bear populations have shown that vital rates, like reproductive success and survival, can be density-dependent and can be altered by intensive harvest [48,64,65]. Intensive harvest that reduces the probability of dispersal, regardless of the mechanism (behavioral plasticity or selection) could reduce population connectivity even if abundances are not declining. Our study provides evidence for a link between harvest density, population density, and dispersal probability that warrants further understanding for the management of harvested species.

## Supporting Information

**File S1 Methods and results of parentage analysis based on simulated black bear offspring.**

**File S2 Full list of linear mixed effects models of black bear dispersal probability and dispersal distance.**

## Acknowledgments

We thank Michigan's bear hunters for providing genetic samples from harvested bears and the many Michigan Department of Natural Resources staff for collecting samples. We thank Jean Fierke, Jeannette Kanefsky, Rebecca Lust, Kristi Filcek, and Kristi Brown for assistance in the laboratory.

## Author Contributions

Conceived and designed the experiments: JAM HMD KTS. Analyzed the data: JAM HMD. Contributed reagents/materials/analysis tools: KTS DE SW. Wrote the paper: JAM.

## References


1. Clobert J, Le Galliard JF, Cote J, Meylan S, Massot M (2009) Informed dispersal, heterogeneity in animal dispersal syndromes and the dynamics of spatially structured populations. Ecology Letters 12: 197–209.
2. Ronce O, Olivieri I, Clobert J, Danchin E (2001) Perspectives on the study of dispersal evolution. In: Clobert J, Danchin E, Dhondt AA, Nichols JD, editors. Dispersal. New York: Oxford University Press. pp. 341–357.
3. Ronce O (2007) How does it feel to be like a rolling stone? Ten questions about dispersal evolution. Annual Review of Ecology Evolution and Systematics 38: 231–253.
4. Lambin X, Aars J, Piertney SB (2001) Interspecific competition, kin competition and kin facilitation: a review of empirical evidence. In: Clobert J, Danchin E, Dhondt AA, Nichols JD, editors. Dispersal. New York: Oxford University Press pp. 110–122.
5. Matthysen E (2005) Density dependent dispersal in birds and mammals. Ecography 28: 403–416.
6. Bitume EV, Bonte D, Magalhães S, San Martin G, Van Dongen S, et al. (2011) Heritability and artificial selection on ambulatory dispersal distance in *Tetranychus urticae*: effects of density and maternal effects PLoS One 6: e26927.
7. Stamps JA (1987) The effect of familiarity with a neighborhood on territory acquisition. Behavioral Ecology and Sociobiology 21: 273–277.
8. Davis JM, Stamps JA (2004) The effect of natal experience on habitat preferences. Trends in Ecology & Evolution 19: 411–416.
9. Stamps JA, Krishnan VV, Willits NH (2009) How different types of natal experience affect habitat preference. American Naturalist 174: 623–630.
10. Stamps JA, Davis JM (2006) Adaptive effects of natal experience on habitat selection by dispersers. Animal Behaviour 72: 1279–1289.
11. Pianka ER, Parker WS (1975) Age-specific reproductive tactics. American Naturalist 109: 453–464.
12. Clutton-Brock TH (1984) Reproductive effort and terminal investment in iteroparous animals. American Naturalist 123: 212–229.
13. Clutton-Brock TH (1991) The Evolution of Parental Care. Princeton, New Jersey: Princeton University Press.
14. Lehmann L, Perrin N (2003) Inbreeding avoidance through kin recognition: choosy females boost male dispersal. American Naturalist 162: 638–652.
15. Pusey AE (1987) Sex-biased dispersal and inbreeding avoidance in birds and mammals. Trends in Ecology & Evolution 2: 295–299.
16. Greenwood PJ (1980) Mating systems, philopatry and dispersal in birds and mammals. Animal Behaviour 28: 1140–1162.
17. Handley LJL, Perrin N (2007) Advances in our understanding of mammalian sex-biased dispersal. Molecular Ecology 16: 1559–1578.
18. Clutton-Brock TH, Lukas D (2012) The evolution of social philopatry and dispersal in female mammals. Molecular Ecology 21: 472–492.
19. Costello CM (2010) Estimates of dispersal and home-range fidelity in American black bears. Journal of Mammalogy 91: 116–121.
20. Jones AG, Small CM, Paczolt KA, Ratterman NL (2010) A practical guide to methods of parentage analysis. Molecular Ecology 10: 6–30.
21. Costello CM, Creel SR, Kalinowski ST, Vu NV, Quigley HB (2008) Sex-biased natal dispersal and inbreeding avoidance in American black bears as revealed by spatial genetic analyses. Molecular Ecology 17: 4713–4723.
22. Horner MA, Powell RA (1990) Internal structure of home range of black bear and analyses of home range overlap. Journal of Mammalogy 71: 402–410.
23. Schwartz CC, Franzmann AW (1992) Dispersal and survival of subadult black bears from the Kenai Penninsula, Alaska. Journal of Wildlife Management 56: 426–431.
24. Garshelis DL, Pelton MR (1981) Movements of black bears in the Great Smoky Mountain National Park. Journal of Wildlife Management 45: 912–925.
25. Dreher BP, Winterstein SR, Scribner KT, Lukacs PM, Etter DR, et al. (2007) Noninvasive estimation of black bear abundance incorporating genotyping errors and harvested bear. Journal of Wildlife Management 71: 2684–2693.
26. Etter DR, Visser LG, Schumacher CM, Carlson E, Reis T, et al. (2002) Black bear population management techniques. Federal Aid in Wildlife Restoration

Project W-127-R-20. Lansing, MI, USA: Michigan Department of Natural Resources.
27. Cattadori IM, Haydon DT, Thirgood SJ, Hudson PJ (2003) Are indirect measures of abundance a useful index of population density? The case of red grouse harvesting. Oikos 100: 439–446.
28. Travis JMJ, Murrell DJ, Dytham C (1999) The evolution of density-dependent dispersal. Proceedings of the Royal Society of London Series B-Biological Sciences 266: 1837–1842.
29. Carter NH, Brown DG, Etter DR, Visser LG (2010) American black bear habitat selection in northern Lower Peninsula, Michigan, USA, using discrete-choice modeling. Ursus 21: 57–71.
30. Willey CH (1974) Aging black bears from first premolar tooth sections. Journal of Wildlife Management 38: 97–100.
31. Paetkau D, Calvert W, Stirling I, Strobeck C (1995) Microsatellite analysis of population structure in Canadian polar bears. Molecular Ecology 4: 347–354.
32. Taberlet P, Camarra JJ, Griffin S, Uhres E, Hanotte O, et al. (1997) Noninvasive genetic tracking of the endangered Pyrenean brown bear population. Molecular Ecology 6: 869–876.
33. Shih CC, Huang CC, Li SH, Hwang MH, Lee LL (2009) Ten novel tetranucleotide microsatellite DNA markers from Asiatic black bear, *Ursus thibetanus*. Conservation Genetics 10: 1845–1847.
34. Wu H, Zhang SN, Wei FW (2010) Twelve novel polymorphic microsatellite loci developed from the Asiatic black bear (*Ursus thibetanus*). Conservation Genetics 11: 1215–1217.
35. Van Oosterhout C, Hutchinson WF, Wills DPM, Shipley P (2004) MICRO-CHECKER: software for identifying and correcting genotyping errors in microsatellite data. Molecular Ecology Notes 4: 535–538.
36. Peacock E (2004) Estimation of black bear population size on Kuiu Island, Alaska using tetracycline biomarking supplemented with genetic methods. Juneau, Alaska: Alaska Department of Fish & Game.
37. Ennis S, Gallagher TF (1994) PCR-based sex determination assay in cattle based on the bovine Amelogenin locus. Animal Genetics 25: 425–427.
38. Riester M, Stadler PF, Klemm K (2009) FRANz: reconstruction of wild multi-generation pedigrees. Bioinformatics 25: 2134–2139.
39. Meagher TR, Thompson E (1986) The relationship between single parent and parent pair genetic likelihoods in genealogy reconstruction. Theoretical Population Biology 29: 87–106.
40. Beyer HL (2004) Hawth's Analysis Tools for ArcGIS. http://www.spatialecology.com/htools.
41. Epperson BK (2005) Estimating dispersal from short distance spatial autocorrelation. Heredity 95: 7–15.
42. Rogers LL (1987) Effects of food supply and kinship on social behavior, movements, and population growth of black bears in northeastern minnesota. Wildlife Monographs 97: 1–72.
43. Frawley BJ (2009) 2008 Michigan Black Bear hunter survey. Lansing, USA: Michigan Department of Natural Resources.
44. Draheim H, Scribner KT, Winterstein S, Lopez V, Etter D, (In Review) Effects of spatial and temporal sampling scale on spatial genetic structure of a black bear population in Michigan. Wildlife Biology.
45. Silverman BW (1986) Density Estimation for Statistics and Data Analysis. New York: Chapman and Hall.
46. RDevelopmentCoreTeam (2006) R: A Language and Environment for Statistical Computing. Vienna, Austria: R Foundation for Statistical Computing.
47. Novak JM, Scribner KT, Dupont WD, Smith MH (1991) Catch-effort estimation of white-tailed deer population size. Journal of Wildlife Management 55: 31–38.
48. Czetwertynski SM, Boyce MS, Schmiegelow FK (2007) Effects of hunting on demographic parameters of American black bears. Ursus 18: 1–18.
49. Scribner KT, Wooten MC, Smith MH, Johns PE (1985) Demographic and genetic characteristics of white-tailed deer populations subjected to still or dog hunting. In: Beason SL, Robinsin SF, editors. Game Harvest Management. Kingsville, Texas, USA: Ceasar Kleberg Wildlife Research Institute. pp. 197–212.
50. Coltman DW, O'Donoghue P, Jorgenson JT, Hogg JT, Strobeck C, et al. (2003) Undesirable evolutionary consequences of trophy hunting. Nature 426: 655–658.
51. Allendorf FW, England PR, Luikary G, Ritchie PA, Ryman N (2008) Genetic effects of harvest on wild animal populations. Trends in Ecology & Evolution 23: 327–337.
52. Broseth H, Tufto J, Pedersen HC, Steen H, Kastdalen L (2005) Dispersal patterns in harvested willow ptarmigan. Journal of Applied Ecology 42: 453–459.
53. Wolff JO (1997) Population regulation in mammals: An evolutionary perspective. Journal of Animal Ecology 66: 1–13.
54. Lambin X (1994) Natal philopatry, competition for resources, and inbreeding avoidance in Townsend's voles (*Microtus towndsendii*). Ecology 75: 224–235.
55. Wolff JO (1992) Parents suppress reproduction and stimulation dispersal in opposite-sex juvenile white-footed mice. Nature 359: 409–410.
56. Hestbeck JB (1982) Population regulation of cyclic mammals - the social fence hypothesis. Oikos 39: 157–163.
57. Stoen OG, Zedrosser A, Saebo S, Swenson JE (2006) Inversely density-dependent natal dispersal in brown bears *Ursus arctos*. Oecologia 148: 356–364.
58. Roy J, Yannic G, Cote SD, Bernatchez L (2012) Negative density-dependent dispersal in the American black bear (*Ursus americanus*) revealed by noninvasive sampling and genotyping. Ecology and Evolution 2: 525–537.
59. Rogers LL (1987) Factors influencing dispersal in the black bear. In: Chepko-Sade BD, Halpin ZT, editors. Mammalian Dispersal Patterns. Chicago, IL: University of Chicago Press.
60. Proctor MF, McLellan BN, Strobeck C, Barclay RMR (2004) Gender-specific dispersal distances of grizzly bears estimated by genetic analysis. Canadian Journal of Zoology-Revue Canadienne De Zoologie 82: 1108–1118.
61. Murrell DJ, Travis JMJ, Dytham C (2002) The evolution of dispersal distance in spatially-structured populations. Oikos 97: 229–236.
62. Zedrosser A, Stoen O-G, Saebo S, Swenson JE (2007) Should I stay or should I go? Natal dispersal in the brown bear. Animal Behaviour 74: 369–376.
63. Elowe KD, Dodge WE (1989) Factors affecting black bear reproductive success and cub survival. Journal of Wildlife Management 53: 962–968.
64. Miller MA, Sellers RA, Keay JA (2003) Effects of hunting on brown bear cub survival and litter size in Alaska. Ursus 14: 130–152.
65. Obbard ME, Howe EJ (2008) Demography of black bears in hunted and unhunted areas of the boreal forest in Ontario. Journal of Wildlife Management 72: 869–880.

# 14

# The Spatial Analysis on Hemorrhagic Fever with Renal Syndrome in Jiangsu Province, China Based on Geographic Information System

**Changjun Bao[2☯], Wanwan Liu[1☯], Yefei Zhu[2], Wendong Liu[2], Jianli Hu[2], Qi Liang[2], Yuejia Cheng[1], Ying Wu[2], Rongbin Yu[1], Minghao Zhou[2], Hongbing Shen[1], Feng Chen[1], Fenyang Tang[2*], Zhihang Peng[1*]**

**1** Department of Epidemiology & Biostatistics, School of Public Health, Nanjing Medical University, Nanjing, Jiangsu, China, **2** Jiangsu Province Center for Disease Control and Prevention, Nanjing, Jiangsu, China

## Abstract

***Background:*** Hemorrhagic fever with renal syndrome (HFRS) is endemic in mainland China, accounting for 90% of total reported cases worldwide, and Jiangsu is one of the most severely affected provinces. In this study, the authors conducted GIS-based spatial analyses in order to determine the spatial distribution of the HFRS cases, identify key areas and explore risk factors for public health planning and resource allocation.

***Methods:*** Interpolation maps by inverse distance weighting were produced to detect the spatial distribution of HFRS cases in Jiangsu from 2001 to 2011. Spatio-temporal clustering was applied to identify clusters at the county level. Spatial correlation analysis was conducted to detect influencing factors of HFRS in Jiangsu.

***Results:*** HFRS cases in Jiangsu from 2001 to 2011 were mapped and the results suggested that cases in Jiangsu were not distributed randomly. Cases were mainly distributed in northeastern and southwestern Jiangsu, especially in Dafeng and Sihong counties. It was notable that prior to this study, Sihong county had rarely been reported as a high-risk area of HFRS. With the maximum spatial size of 50% of the total population and the maximum temporal size of 50% of the total population, spatio-temporal clustering showed that there was one most likely cluster (LLR = 624.52, P<0.0001, RR = 8.19) and one second-most likely cluster (LLR = 553.97, P<0.0001, RR = 8.25), and both of these clusters appeared from 2001 to 2004. Spatial correlation analysis showed that the incidence of HFRS in Jiangsu was influenced by distances to highways, railways, rivers and lakes.

***Conclusion:*** The application of GIS together with spatial interpolation, spatio-temporal clustering and spatial correlation analysis can effectively identify high-risk areas and factors influencing HFRS incidence to lay a foundation for researching its pathogenesis.

**Editor:** Xue-jie Yu, University of Texas Medical Branch, United States of America

**Funding:** This work was supported in part by Natural Science Foundation Key Project of Jiangsu Province, China (BK2010079), 'National S&T Major Project Foundation of China' (2011ZX10004-902). Jiangsu Province Health Development Project with Science and Education (NO. ZX201109), and the Priority Academic Program Development of Jiangsu Higher Education Institutions (PAPD). The funders had no role in study design, data collection and analysis, decision to publish, or preparation of the manuscript.

**Competing Interests:** The authors have declared that no competing interests exist.

* Email: zhihangpeng@njmu.edu.cn (ZHP); tfyepi@163.com (FYT)

☯ These authors contributed equally to this work.

## Introduction

Hemorrhagic fever with renal syndrome (HFRS) is caused by hantaviruses comprising 23 identified species, while more than 30 as-yet-undetermined species are rodent-borne pathogens with a global distribution [1–2]. Human beings are believed to be infected by hantaviruses through inhalation of contaminated aerosols shed in the excreta, saliva, and urine of infected rodents [3]. Old-world rodents carry viruses that cause HFRS, while new-world rodents carry viruses that cause hantavirus pulmonary syndrome (HPS) [4]. HFRS is highly epidemic in China, accounting for 90% of total reported cases worldwide [5]. Although integrated control measures such as rodent control, environmental management, and vaccination have currently been implemented, HFRS remains a severe public health problem in mainland China; more than 10,000 human cases are diagnosed annually [6]. The incidence of HFRS shows high variability at both rural and urban levels [7]. Jiangsu, a highly developed coastal province, is one of the most severely affected provinces in China [8]. HFRS has spread to almost every county in Jiangsu since the 1980s, when the first HFRS case in Jiangsu was found [9]. Previous epidemiological surveys revealed the distribution of HFRS cases differed considerably from place to place [9].

In recent years, experts at home and abroad have focused on the aetiology, epidemiology and pathogenic mechanisms of HFRS, and have made substantial progress. Clement Jan et al. found that

**Table 1**. The results of the analysis on HFRS cases by general spatial autocorrelation.

| Year | Z-score | P-value | Moran's Index | Cluster |
|---|---|---|---|---|
| 2001 | 4.065 | 0 | 0.391 | Highly clustered |
| 2002 | 4.069 | 0 | 0.426 | Highly clustered |
| 2003 | 4.063 | 0 | 0.396 | Highly clustered |
| 2004 | 3.829 | 0 | 0.389 | Highly clustered |
| 2005 | 7.679 | 0 | 0.806 | Highly clustered |
| 2006 | 4.099 | 0 | 0.419 | Highly clustered |
| 2007 | 3.749 | 0 | 0.342 | Highly clustered |
| 2008 | 3.299 | 0 | 0.318 | Highly clustered |
| 2009 | 3.762 | 0 | 0.361 | Highly clustered |
| 2010 | 3.142 | 0.002 | 0.302 | Highly clustered |
| 2011 | 3.568 | 0 | 0.363 | Highly clustered |

the infection rate of hantaviruses was closely associated with global warming [10]. A recent research report, which was the first to document an association between HFRS and $PM_{10}$ levels, indicated that $PM_{10}$ might be an important factor in HFRS infection [11]. The current study conducted an ArcGIS10.0-based spatial analysis involving the inverse distance weighting interpolation (IDW), spatio-temporal clustering and spatial correlation analysis to better comprehend space-time distribution patterns of HFRS cases in Jiangsu from 2001 to 2011, to find influencing factors of HFRS for public health interventions and to lay the groundwork for future research on these influencing factors.

## Materials and Methods

### Study area and data collection

The study area consisted of all 106 counties in Jiangsu province. Jiangsu is located at 116.60°~121.67° east longitude and 31.01°~34.89° north latitude on the central coast of China and has an area of 102.6 thousand square kilometers. Study samples were composed of all HFRS cases from 106 counties in Jiangsu from 2001 to 2011. Data that was not accessible to the general public were obtained from the Jiangsu provincial Center for Disease Control and Prevention (http://www.jshealth.com/). The work was approved by the ethical committee of Nanjing Medical University ("F", "CH", "Nanjing Med U", "FWA00001501", "NANJING", 11/21/2004), and IRB (Institutional Review Board) approval was obtained prior to initiating the study. The authors have read and abided by the statement of ethical standards for manuscripts submitted to PLoS One.

### Spatial autocorrelation

Spatial autocorrelation was applied to describe the similarity of geographically proximate units. In this study, general spatial autocorrelation was used to explore the characteristics of the spatial distribution of HFRS cases, the results of which determined the presence of spatial clustering independent of location.

### Creation of interpolation maps

The IDW was applied to estimate the value of every point according to the principle that proximate sample points have similar values. In other words, the authors used measured values of the sample points to predict estimated values of every point in the study, then evaluated the error and variability of those predictions [12]. The researchers obtained the spatial distribution of HFRS cases in the overall region of Jiangsu from the limited spatial sample points by IDW. In this study, IDW was used to make interpolated maps using the data of HFRS from 2001 to 2011 in Jiangsu.

Our aim was to detect whether or not the epidemic was aggregated at the provincial level. General Moran's $I$ was most often used, with the coefficient calculated as follows:

$$\mathrm{I} = \frac{n}{\sum_{i=1}^{n}\sum_{j=1}^{n} w_{ij}} \times \frac{\sum_{i=1}^{n}\sum_{j=1}^{n} w_{ij}(x_i-\bar{x})(x_j-\bar{x})}{\sum_{i=1}^{n}(x_i-\bar{x})^2}$$

where $n$ is the numbers of counties, $x_i$ and $x_j$ the observations from unit $i$ to unit $j$ with regard to the phenomenon $x$ (the indicator of autocorrelations). $w_{ij}$(d) represents the adjacent weight matrix from the distance d. If the unit $i$ for regional data is adjacent to unit $j$, then $w_{ij}$(d) is 1, otherwise it is 0. The $I$ index was compared with the critical value of $Z$. If $p<0.05$, then Moran's $I \neq 0$, which means that the area indeed had an aggregation of HFRS infections.

### Spatio-temporal clustering

A spatio-temporal clustering analysis is defined by a cylindrical window with a circular (or elliptical) geographic base and a height corresponding to time [13]. To formulate the dynamic size and position of the cylindrical window, the center of the window moves according to the geographic union center and the radius changes continuously from zero up to the maximum radius, never including more than 50% of the total population. Every moment the radius changes, the log likelihood ratio (LLR) is calculated according to the difference between HFRS cases inside and outside the window. The formula for calculating the LLR is as follows: $LLR = \log(c/n)^c[(C-c)/(C-n)]^{C-c}$: where C is the total number of HFRS cases, c is the observed number of HFRS cases inside the window and n is the expected number of HFRS cases inside the window.

The window with maximum LLR had the highest intensity of anomalies and was determined to be most likely to be a spatio-temporal cluster area. That is to say, the cluster was least likely to

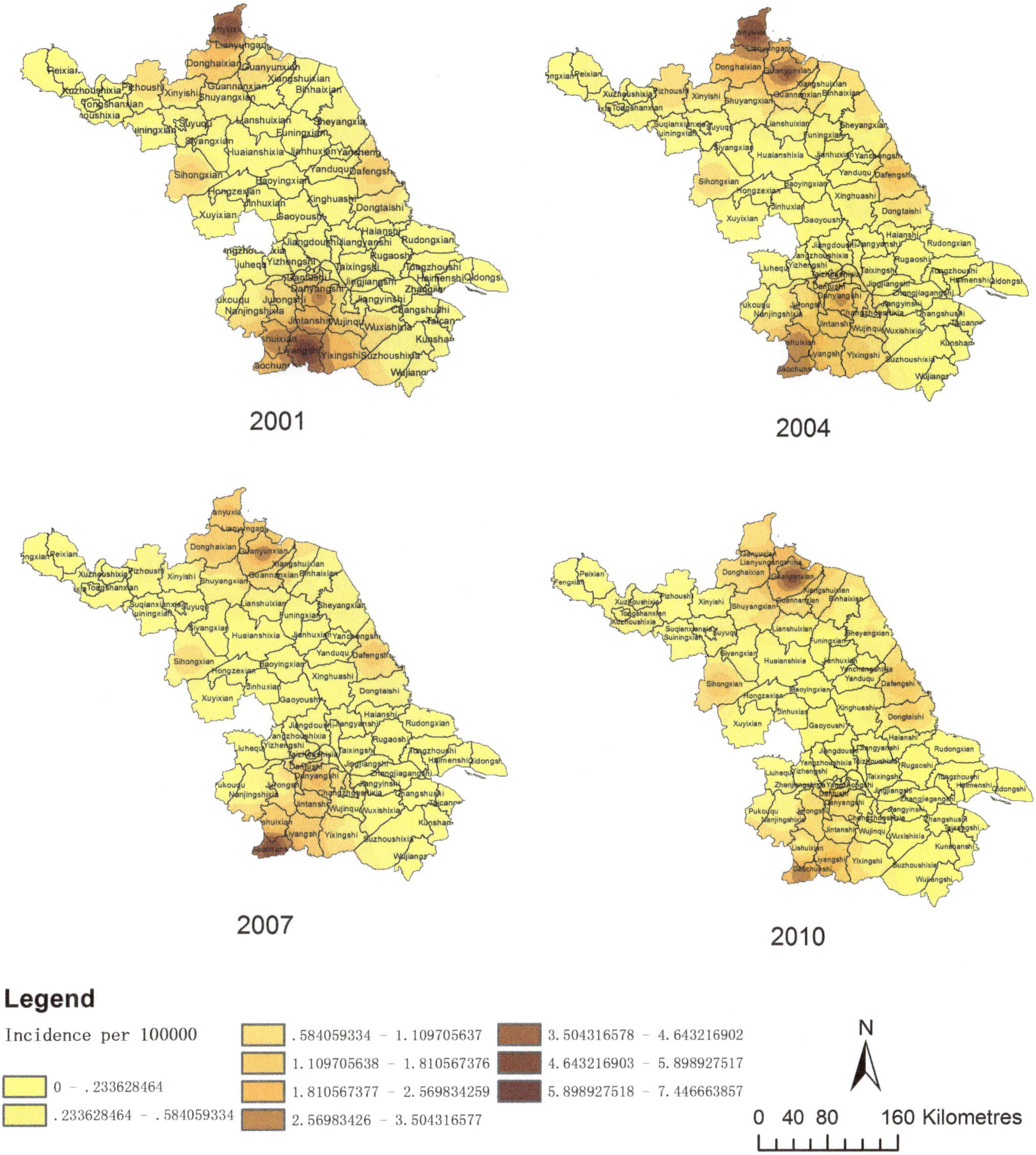


**Figure 1. Interpolated maps of HFRS by IDW in Jiangsu in 2001, 2004, 2007 and 2010.** The incidence of HFRS per 100,000 residents is shown in the map. The incidence of HFRS has a positive relationship with color depth.

be clustered due to chance. For each window, a Monte Carlo simulation was used to calculate p to test the null hypothesis that the relative risk (RR) of HFRS was the same between any counties or any group of counties and remaining counties [14]. If the p-value of LLR was less than 0.05, the differences in relative risk (RR) between inside and outside the scan window and the abnormality degree of the incidence were considered statistically significant. In this study, a retrospective spatial cluster analysis for higher incidence was used. The maximum window radius was set to be smaller than 50% of the total population to find possible clusters.

**Table 2.** Spatio-temporal clustering on HFRS in Jiangsu from 2001 to 2011.

| Spatio-temporal clusters | The most likely cluster | The second-most likely cluster |
|---|---|---|
| Center | Gaochun | Ganyu |
| Geographic coordinates of the center of clusters | (31.36N, 118.99E) | (34.90N, 119.01E) |
| Radius(Km) | 86.38 | 65.76 |
| Time | 2001~2004 | 2001~2004 |
| Number of counties included | 7 | 4 |
| LLR[a] value | 624.52 | 553.97 |
| P-value | <0.0001 | <0.0001 |
| RR[b] Value | 8.19 | 8.25 |

a. LLR: Log likelihood ratio. b. RR: Relative risk.

### Spatial correlation analysis

The spatial correlation analyses in this study included univariate analysis and a spatial regression model that combined all underlying factors. Spatial regression analysis was used to detect potential relationships between the spatial distribution of a disease and independent variables (environmental factors such as water, air, soil, socioeconomic factors, etc.) determined by geography [15]. In order to avoid including influencing factors with no statistical significance, potential influencing factors were analyzed first by univariate analysis. Then, to avoid collinearity among the influencing factors with statistical significance, a multi-collinearity test was conducted on them. There have been a few reports suggesting that the incidence of HFRS is likely to be influenced by a complex combination of factors, including environment factors and climate factors, rather than by a single foci pathogenic factor [16–18]. Eleven risk factors for the spatial correlation analysis were selected from the previous reports including the annual mean temperature; the amount of water vapor; elevation; the normalized difference vegetation indexes (NDVI) for spring, summer, autumn and winter; and the distribution of lakes, rivers, railways and highways.

## Results

### General spatial autocorrelation

The research group found that the general Moran indexes of HFRS cases in all counties in Jiangsu were all greater than zero (Table 1). In the normal distribution hypothesis, the result of the Moran's I test showed spatial autocorrelation was highly significant, which indicated that HFRS cases in Jiangsu from 2001 to 2011 were significantly spatially autocorrelated. This revealed that the similarity of the observations in any two counties showed a negative correlation with the distance between the corresponding counties. This corresponds to the precondition of IDW.

### The inverse distance weighted interpolation

Interpolated maps of HFRS incidence in Jiangsu from 2001~2011 were produced by IDW. Darker colors were associated with higher incidence. The maps revealed that high-risk areas were distributed in the following counties: Ganyu, Lianyungang, Ganyu of northeastern Jiangsu, Gaochun and Liyang of southwestern Jiangsu, Dafeng of central-eastern Jiangsu and Sihong of central-western Jiangsu (Figure 1).

**Table 3.** The results of single factor analysis between the incidence of HFRS and potential influencing factors in Jiangsu from 2001 to 2011.

| | Curve fitting equation | $R^2$ | F | P |
|---|---|---|---|---|
| Temperature | $y=-0.165x^2+6.338x-60.299$ | 0.036 | 1.219 | 0.302 |
| The amount of water vapor | $y=0.633x^3-13.661x+24.588$ | 0.126 | 4.623 | 0.013 |
| Elevation | $y=-4.43E\text{-}005x^3+0.002x^2-0.003x+0.309$ | 0.039 | 0.852 | 0.471 |
| Distance to railway | $y=-4.43E\text{-}005x^3+0.002x^2-0.003x+0.309$ | 0.998 | 2524.638 | 0.000 |
| Distance to highway | $y=9.98E\text{-}005x^2-0.006x+0.539$ | 1.000 | 9304.654 | 0.000 |
| Distance to river | $y=9.98E\text{-}005x^2-0.006x+0.539$ | 0.999 | 1766.491 | 0.000 |
| Distance to lake | $y=1.16E\text{-}005x^3-0.001x^2+0.022x+0.198$ | 0.999 | 3651.710 | 0.000 |
| NDVI[a] for spring | $y=-12.551x^3+10.481x^2\text{-}0.810x\text{-}0.044$ | 0.030 | 0.659 | 0.580 |
| NDVI for summer | $y=0.003x^3\text{-}0.023x^2+0.048x+0.451$ | 0.016 | 0.334 | 0.794 |
| NDVI for autumn | $y=-30.646x^3+20.580x^2\text{-}1.519x\text{-}.0136$ | 0.021 | 0.455 | 0.715 |
| NDVI for winter | $y=52.931x^3\text{-}53.296x^2+15.125x\text{-}0.732$ | 0.032 | 0.689 | 0.562 |

a. NDVI: normalized difference vegetation index.

**Table 4.** Collinearity diagnostics.

| Model Dimension | Eigenvalue | Condition Index | Variance Proportions | | | | |
|---|---|---|---|---|---|---|---|
| | | | (Constant) | Railway | Highway | River | Lake |
| 1 | 2.603 | 1.000 | 0.05 | 0.04 | 0.05 | 0.03 | 0.05 |
| 2 | 1.037 | 1.584 | 0.01 | 0.14 | 0.02 | 0.40 | 0.09 |
| 3 | 0.644 | 2.010 | 0.05 | 0.01 | 0.20 | 0.29 | 0.45 |
| 4 | 0.372 | 2.645 | 0.02 | 0.48 | 0.72 | 0.28 | 0.08 |
| 5 | 0.344 | 2.752 | 0.87 | 0.32 | 0.01 | 0.00 | 0.33 |

At the same time it was discovered that in addition to southwestern, northeastern and central-eastern Jiangsu, HFRS had also spread to central-western Jiangsu. However, the area of central-western Jiangsu where HFRS infections occurred was relatively small compared with the areas in southwestern, northeastern and central-eastern Jiangsu. This showed that the range of the region where HFRS cases were distributed had expanded, but in general southwestern and northeastern Jiangsu were the main areas affected and the general incidence of HFRS had dropped dramatically.

## Spatio-temporal clustering

Spatio-temporal clustering of HFRS cases in 2001~2011 in Jiangsu showed that HFRS was not distributed randomly in space-time. One most likely cluster and one second-most likely cluster were identified with a maximum spatial size of 50% of the total population and a maximum temporal size of 50% of the total population. The most likely cluster (LLR = 624.52, P<0.0001, RR = 8.19) was in an 86.38 km radius around Gaochun located at latitude 31.36 north, longitude 118.99 east from 2001 to 2004. It encompassed seven counties (Gaochun, Lishui, Liyang, Jintan, Jurong, Yixing, Danyang). A second-most likely cluster (LLR = 553.97, P<0.0001, RR = 8.25) was in a 65.76 km radius around Ganyu located at latitude 34.90 north, longitude 119.006311 east from 2001 to 2004. This cluster encompassed four counties: Ganyu, Lianyungang, Donghai and Ganyun (Table 2).

The most likely cluster and the second-most likely cluster were distributed in northeastern and southwestern Jiangsu from 2001 to 2004, which showed that HFRS cases were mainly distributed in northeastern and southwestern Jiangsu and also that the prevalence of HFRS had dropped since 2004. This is in line with the results of IDW. It seems practical, therefore, to apply IDW and spatio-temporal clustering to an analysis of spatio-temporal dynamic trends of HFRS in Jiangsu.

## Spatial correlation analysis

The results of the univariate analysis on the potential relationships between the incidence of HFRS and the annual mean temperature; average amount of water vapor; average elevation; distances to railways, highways, rivers and lakes; and NDVI values are shown in Table 3. Results indicated that the incidence of HFRS had a weak correlation with annual mean temperature, average amount of water vapor, elevation and seasonal NDVI; and a strong correlation with distances from railways, highways, rivers and lakes (Table 3). The results of multicollinearity tests of the four factors which were in strong correlation with the incidence of HFRS showed that the four factors had no linear relation (Table 4).

Assuming the incidence of HFRS as a dependent variable and the distance from railways, highways, rivers and lakes as independent variables, the regression equation could be established. The regression equation was as follows:

$$y = 0.4641x_1^2 + 1.21756 \times 10^{-5}x_2^2 - 7.32 \times 10^{-4}x_2 - 4.8734 \times 10^{-7}x_3^3 + 3.1388 \times 10^{-6}x_3 - 4 \times 10^{-6}x_4^2 + 8.8 \times 10^{-5}x_4 + 0.117181$$

$x_1$, $x_2$, $x_3$ and $x_4$ indicate the distances from railways, highways, rivers and lakes, respectively. The significance test and the goodness-of-fit test were applied to the regression equation. The results were $R^2 = 0.036, sig = 0.679$ and $F = 0.579$, indicating that

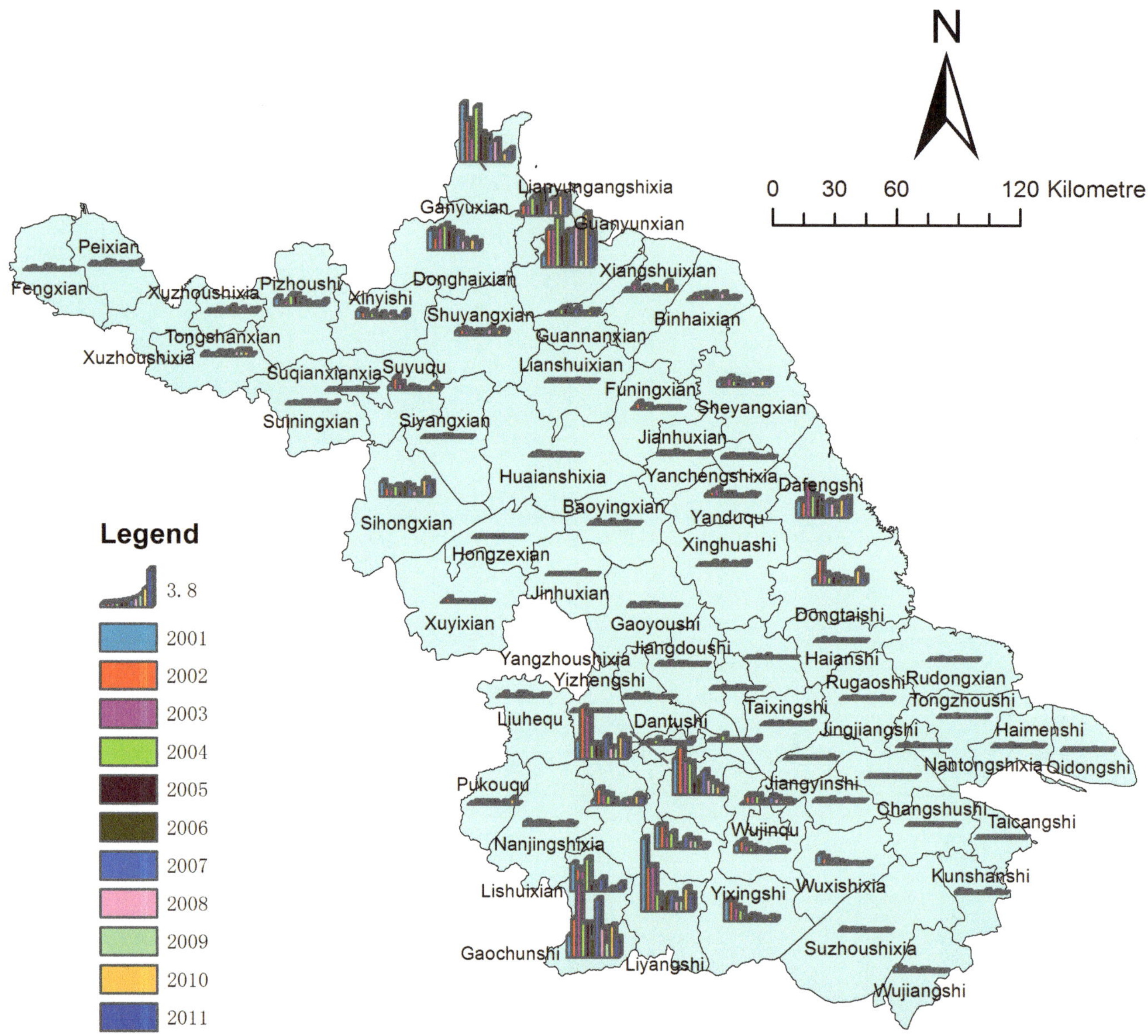


**Figure 2. The density of HFRS cases based on counties in Jiangsu from 2001 to 2011.** The mean annual incidence per 100,000 residents in each county of Jiangsu is shown. Different colors represent different years.

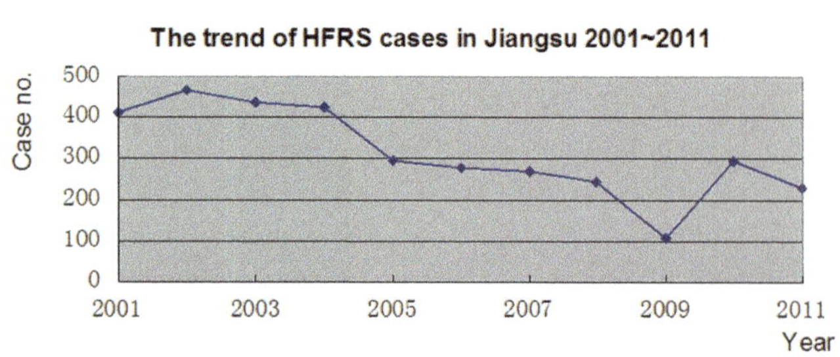


**Figure 3. The trend of the number of HFRS cases in Jiangsu from 2001 to 2011.** Each point represents the number of HFRS cases in a specific year. All of the points are lined to indicate the trend of the HFRS cases in Jiangsu.

the equation had a high level of statistical significance and a low level of goodness of fit.

## Discussion

Geographic information systems (GIS) are widely used in epidemiology research for their advantages in accurately showing the spatial distribution of diseases, revealing spatial clusters of diseases and detecting potential risk factors contributing to disease transmission [19–20]. However, the application of IDW must meet the condition that the sample sites are efficiently distributed densely enough to cover all areas for which predictions are sought [21]. In this study, the sample of 106 sites appears to be large enough to cover all areas, and the sites were distributed evenly. Therefore, it was suitable to apply IDW. In figure 1, IDW interpolation maps show that HFRS cases are mainly distributed in Ganyu, Lianyungang, and Guanyun counties of northeastern

Jiangsu, Gaochun and Liyang counties of southwestern Jiangsu, and Dafeng in central-eastern Jiangsu and Sihong in central-western Jiangsu. This finding is almost entirely in line with the results of Rongqiang Zu's study [9], with the exception of Sihong and Dafeng. Among the high-risk areas, Sihong was the only one that had rarely been reported in previous studies. Over the past several years, more land in Sihong has become farmland [22–23], which provides an abundance of food and suitable habitats for rodents and has resulted in an increasing number of rodents. The growth in the rodent population increases the chances of human contact with rodents and their excreta. Therefore, the increased area of farmland may be an important factor that contributed to the prevalence of HFRS in Sihong.

The application of spatio-temporal clustering to computing the difference between incidence inside and outside the circular window lowers the possibility of pre-selection bias, which occurs when researchers determine the range of the study area and the center of the circle [24]. In this study, spatio-temporal clustering was used to determine the distribution pattern of HFRS cases and to identify key areas for future planning and resource allocation. Jiangsu can be divided into four parts according to its terrain: the western hilly region, the northwestern low hilly region, the southwestern low hilly region and the plains region. The spatio-temporal clustering results showed two cluster regions distributed in the northwestern and southwestern low hilly regions, respectively. The findings of HFRS clustering are consistent with an earlier study by Hualiang Lin, et al. [25]. The low hilly regions have more forest than the plains regions and more crops than the hilly regions. There is more forest in the northwestern and southwestern hilly regions, which provides a habitat for rodents, and the abundant crops provide them with sufficient food. Therefore, the chances of human contact with rodents and their excreta increase. At the same time, it was found that both cluster regions occurred in 2001~2004 rather than in different periods of time. There were two clusters from 2001 to 2004 whereas there were no clusters from 2005 to 2011, which showed that the prevalence of HFRS decreased from 2005 to 2011 compared to the period from 2001 to 2004. This corresponds to the trend that the number of people infected with HFRS and patient density based on counties in Jiangsu declined from 2001 to 2011 (Figure 2 and Figure 3).

The spatial correlation analysis found that the incidence of HFRS was highly correlated with distances to highways, railways, rivers and lakes. Research has shown that the distance to a railway may cause a change in rodent density to a certain extent [26]. It is possible that railway transportation provides a suitable living environment and ample food for rodents.

In spite of the insights gained, the limitations of the study should also be acknowledged. Although spatio-temporal scan statistics can avoid a pre-selection bias successfully, a population shift bias may occur due to the lack of consideration of population growth. Therefore, both the number of HFRS cases and the unevenness of the population distribution among counties at different times can contribute to the occurrence of clusters [27]. If the population of one region increases faster than that of other regions, population shift bias may occur. The time span of this study is 11 years, during which the rate of population growth may have varied in different counties, so it is likely that population shift bias has occurred. In addition, this study only analyzed the causes of clusters of HFRS cases qualitatively instead of studying the quantitative relationship between possible influencing factors and the incidence of HFRS, due to a shortage of sufficient data about the environment, economy and geography. This may have reduced the study's rigor.

## Author Contributions

Conceived and designed the experiments: ZHP FYT CJB HBS JLH MHZ. Performed the experiments: WDL YW RBY FC. Analyzed the data: WWL YJC QL ZHP. Contributed reagents/materials/analysis tools: WWL YFZ WDL ZHP. Wrote the paper: WWL FYT ZHP.

## References


1. Zuo SQ, Fang LQ, Zhan L, Zhang PH, Jiang JF, et al. (2008) Geo-spatial hotspots of hemorrhagic fever with renal syndrome and genetic characterization of Seoul variants in Beijing, China. PLoS Negl Trop Dis 5: e945. doi: 10.1371/journal. pntd. 0000945
2. Lin JL, Ling JX, Yang ZQ (2010) The research progress in HFRS, HPS, and HV. Chin J Epidemiol 31: 1434–1435.
3. Lähdevirta J (1971) Nephropathia epidemica in Finland. A clinical histological and epidemiological study. Ann Clin Res 3: 1–154.
4. Schmaljohn CS (2012) Vaccines for hantaviruses: progress and issues. Expert Rev.Vaccines 11: 511–513.
5. Wei L, Qian Q, Wang Z-Q, Gregory E.G, Song SX, et al. (2011) Using geographic information system-based ecologic niche models to forecast the risk of hantavirus infection in Shandong province, China. Trop Med Hyg 84: 497–503.
6. Zhang YZ, Zou Y, Fu ZF, Plyusnin A, et al. (2011) Hantavirus infection in human and animals, China. Emerg Infect Dis 16: 1195–1203.
7. Fang LQ, Yan L, Liang S, Sake J de Vlas, Feng D, et al. (2006) Spatial analysis of hemorrhagic fever with syndrome in China. BMC Infect Dis 6: 77. Available: http://www.biomedcentral.com/1271-2334/6/77. Accessed 2012 July 8.
8. Fang LQ, Cao WC, Chen HX, Wang BG, Wu XM, et al. (2003) Study on the application of geographic information system in spatial distribution of hemorrhagic fever with renal syndrome in China. Chin J Epidemiol 24: 265–268.
9. Zu RQ, Wu YS, Zhu FC, Liu GZ (2003) Surveillance on hemorrhagic fever with renal syndrome in Jiangsu province from 1996 to 2000. Chin J Epidemiol 24: 109–112.
10. Clement J, Vercauteren J, Verstraeten WW, Ducoffre G, Barrios JM, et al. (2009) Relating increasing hantavirus incidences to the changing climate: the mast connection. Int J Health Gooqr 8: 1. Available: http://www.ij-healthgeographics.com/content/8/1/1. Accesssed 2012 July 7.
11. Han SS, Kim S, Choi Y, Kim S, Kim YS (2013) Air pollution and hemorrhagic fever with renal syndrome in South Korea: an ecological correlation study. BMC public health 13: 347.
12. Fang LQ, Li CY, Yang H, Wu XM, Yang H, et al. (2004) Using geographic information system to study on the association between epidemic areas and animal hosts of hemorrhagic fever with renal syndrome in China. Chin J Epidemiol 25: 929–933.
13. Wu W, Guo JQ, Guan P, Sun YW, Zhou BS (2011) Cluster of spatial, temporal, and space-time distribution of hemorrhagic fever with renal syndrome in Liaoning Province, Northeastern China. BMC Infect Dis 11: 229. Available: http://www.biomedcentral.com/1471-2334/11/229. Accessed 2012 July 7.
14. Qian HK, Yang P, Zhang Y, Wang XL, Duan W, et al. (2011) Spatial-temporal scan statistic on scarlet fever cases in Beijing, 2005–2010. Disease Surveillance 26: 435–438.
15. Cheng YJ, Ding KQ, Bao CJ, Liang Q, Hu JL, et al. (2011) GIS-based spatial distribution of typhoid and paratyphoid fever in Jiangsu province. Chin J of Health Statistics 28: 632–637.
16. Chen HX, Qiu FX (1994) Study on the environment structure of natural nidi and epidemic areas of hemorrhagic fever with renal syndrome in China. Chin Med J 107: 107–112.
17. Zhang WY, Guo WD, Fang LQ, Li CP, Bi P, et al. (2010) Climate variability and hemorrhagic fever with renal syndrome transmission in northeastern China. Environment Health Perspectives 118: 915–920.
18. Liu XD, Jiang BF, Gu WD, Liu Q (2011) Temporal trend and climate factors of hemorrhagic fever with renal syndrome epidemic in Shenyang City, China. BMC Infect Dis 11: 331. Available: http://www.biomedcentral/com/1471-2334/11/331. Accessed 2012 July 8.
19. Ding KQ, Bao CJ, Hu JL, Liang Q, Lei N, et al. (2011) Spatial analysis of tuberculosis epidemic in Jiangsu province based on Geographic Information System. Chin J Public Health 27: 1097–1099.
20. Cheng YJ, Norris J, Bao CJ, Liang Q, Hu JL, et al. (2012) Geographical information system-based spatial analysis and implications for syphilis interventions in Jiangsu province, People's Republic of China. Geospatial Health 7: 63–72.
21. Wang Y, Chen XD, Wang CS. (2009) Using inverse distance weighting in studying the distribution of endemic fluorosis in Jiangsu Province. Chin J Endemiol 28: 97–100.
22. Admi (2007) Report on land use change in Sihong in 2007. [online] Available: http://www.shmlr.gov.cn/Article/jsfc/niandtjzl/200711/677.html. Accessed 2012 July.

23. Dijigu (2008) Report on land use change in Sihong in 2008. [online] Available: http://www.shmlr.gov.cn/Article/jsfc/niandtjzl/201010/3028.html. Accessed 2012 July.
24. Chaput EK, Meek JI, Heimer R (2002) Spatial analysis of human granulocytic ehrilichiosis near Lyme, Connecticut. Emerg Infect Dis 8: 943–948.
25. Lin HL, Liu QY, Guo JQ, Zhang JB, JF, et al. (2007) Analysis of the geographic distribution of HFRS in Liaoning Province between 2000 and 2005. BMC Public Health 7: 207. Available: http://www.biomedcentral.com/1471-2458/7/207. Accessed 2012 July 9.
26. Yang SM, Huai HY, Zhang YL, Yin BF, Zhou L, et al. (2006) Effects of railway traffic on the community structure of rodents in warm steppe along the Qinghai-Tibet Railway. Acta Theriologica Sinica 26: 267–273.
27. Tang XY, Gan WY, Xu B, Chen CY, Chu Min, et al. (2011) Study on the spatiotemporal trend of Japanese encephalitis in Guangxi,based on geographic information system and space-time permutation scan statistic. Chin J of Epidemiology 32: 274–278.

15

# Virio- and Bacterioplankton Microscale Distributions at the Sediment-Water Interface

**Lisa M. Dann[1]*, James G. Mitchell[1], Peter G. Speck[1], Kelly Newton[1], Thomas Jeffries[2], James Paterson[1]**

**1** School of Biological Sciences, Flinders University, Adelaide, South Australia, Australia, **2** School of the Environment, University of Technology Sydney, Sydney, New South Wales, Australia

## Abstract

The marine sediment-water interface is an important location for microbially controlled nutrient and gas exchange processes. While microbial distributions on the sediment side of the interface are well established in many locations, the distributions of microbes on the water side of the interface are less well known. Here, we measured that distribution for marine virio- and bacterioplankton with a new two-dimensional technique. Our results revealed higher heterogeneity in sediment-water interface biomass distributions than previously reported with a greater than 45– and 2500-fold change $cm^{-1}$ found within bacterial and viral subpopulations compared to previous maxima of 1.5- and 1.4-fold $cm^{-1}$ in bacteria and viruses in the same environments. The 45-fold and 2500-fold changes were due to patches of elevated and patches of reduced viral and bacterial abundance. The bacterial and viral hotspots were found over single and multiple sample points and the two groups often coincided whilst the coldspots only occurred over single sample points and the bacterial and viral abundances showed no correlation. The total mean abundances of viruses strongly correlated with bacteria ($r = 0.90$, $p < 0.0001$, $n = 12$) for all three microplates ($n = 1350$). Spatial autocorrelation analysis via Moran's I and Geary's C revealed non-random distributions in bacterial subpopulations and random distributions in viral subpopulations. The variable distributions of viral and bacterial abundance over centimetre-scale distances suggest that competition and the likelihood of viral infection are higher in the small volumes important for individual cell encounters than bulk measurements indicate. We conclude that large scale measurements are not an accurate measurement of the conditions under which microbial dynamics exist. The high variability we report indicates that few microbes experience the 'average' concentrations that are frequently measured.

**Editor:** A. M. Ibekwe, U. S. Salinity Lab, United States of America

**Funding:** Funding provided to JGM by the Australian Research Council. The funders had no role in study design, data collection and analysis, decision to publish, or preparation of the manuscript.

**Competing Interests:** The authors have declared that no competing interests exist.

* Email: lisa.dann@flinders.edu.au

## Introduction

Marine ecosystems are dependent on the microbial loop to carry out critical marine biogeochemical processes such as carbon and nitrogen fixation [1–8]. These marine biogeochemical processes often occur at or near cell surfaces with interactions occurring over micrometres to centimetres [9–17]. However, marine microbe distributions are commonly measured in large volumes and then that value is extrapolated over larger areas or volumes in what is sometimes termed a 'mean field' approach [12,18–24].

There is growing awareness of deficiencies in this mean field approach for productivity, biomass and process estimates [12,25–28]. Studies on patch dynamics in terrestrial and aquatic ecosystems show most organisms have heterogeneous spatial distributions. This patchiness would be missed if the sampling resolution used was not appropriate for the size of the organism studied [29–32]. For instance, Brentnall *et al.* [33] found that the mean field production of planktonic communities greatly underestimated actual production due to small scale patchiness. The little work on picoplankton distributions indicates that there is microscale spatial heterogeneity of bacteria and viruses, and that 'hotspots' and 'coldspots' are common, with bacterial and viral abundance changing 5-fold and 1.4-fold $cm^{-1}$ [15,34].

Hotspots are areas of high microbial abundance which are believed to result from interactions between microbes and organic matter [9,12,13,15,35]. These viral and bacterial hotspots were often found to be due to bacterial response to nutrient patches, as marine bacteria have been found to accumulate around areas of high nutrient concentration via chemotaxis [9,31,36–40], This aggregation of bacteria then results in bacterial hotspots that are often followed by increased viral production via lysis due to close host proximity [15]. In contrast, coldspots are areas of low microbial abundance that are believed to be a result of lysis or grazing events in the case of lowered bacterial abundance and grazing or attachment in the case of lowered viral abundance [17].

The presence of these hotspots and coldspots are the units of microscale patchiness. The lower size limit of patches is controlled by the Batchelor scale, which is the smallest scale that nutrient gradients can occur before being dispersed by diffusion [41,42]. Chemotactic bacteria can exploit ephemeral nutrient gradients above the Batchelor scale. This exploitation leads to hotspots in bacterial abundance [37,38]. Patchiness in bacterial and viral distributions are also dependent on the season and time of sampling [13,39]. For instance, previous studies have shown bacterial abundance and activity were higher in the late afternoon [13]. This study ensured sampling was taken at the same time of day and within the same season to avoid confounding from these low frequency signals.

Although hotspots are a ubiquitous feature of microbial distributions they are often vaguely defined in studies, primarily being classified as patches of high microbial abundance that 'exceed' or are 'elevated above' background variation across one or two sampling points [12,13,43]. The methods for determining background are not given and the amount of elevation is rarely stated and differs considerably between studies making comparative analysis difficult. For instance, in the case of Seymour *et al.* [12] the data shown for hotspots suggests that elevation above background is less than one order of magnitude whereas in Seymour *et al.* [43] a conservative value of ≥4-fold higher is considered as a hotspot. In addition to this, coldspots have also lacked a definitive characterisation despite being observed in the microscale distributions of microbial communities previously [12,15,17].

As well as hotspots and coldspots, surface gradients are a common pattern seen in microbial distributions, and are seen when bacterial and viral abundance is highest at an interface and then dissipates into the water column as the distance from the interface increases. These surface gradients arise from the sinking of organic matter and its incorporation into the benthos followed by degradation and transformation by microbes leading to high nutrient concentrations and hence high bacterial and viral concentrations directly above the sediment-water interface [35]. The microscale distributions of microbial communities on the sediment side of the interface are well established, whilst the distributions of microbes on the water side of the interface are less well known and where this has been investigated the focus has been largely on bacteria [12,15,34]. However, of the studies analysing the water side, surface gradients have been seen in the distributions of bacterial and viral subpopulations above coral colonies [14], in the water column above an anchialine sinkhole [44] and in coastal [12] and estuary environments [15,35]. Viral studies show the greatest level of heterogeneity is found 1–2 cm above the sediment-water interface with abundance at St Kilda mangroves in South Australia varying by more than 2-fold, over a 15 cm sampling distance, about 1.13-fold $cm^{-1}$, with the highest concentrations found 1.5 cm from the sediment-water interface [15,35]. Similar patterns were seen in the vertical profiles in Seymour *et al.* [35] which showed high viral and bacterial abundance closest to the sediment-water interface with approximate increases of 1.4- and 1.5-fold $cm^{-1}$ in total virus and total bacteria. However, due to one-dimensional sampling and lack of replication, determining whether these changes in abundance were single or multiple point maxima or surface gradients was unknown.

The null hypothesis tested in this study was that patchiness has no effect on mean biomass estimates and therefore the mean field approach is an accurate estimate of biomass. To test this hypothesis, two-dimensional sampling with extensive replication was used which builds upon previous work by Seuront *et al.* [45], Seymour *et al.* [13–15,35,43] and Waters *et al.* [17]. Here, a new sampler with 9 mm resolution provided replication via the collection of 8 vertical profiles per sampler or 24 profiles per environment, which is much larger than the previous maxima of 5 single vertical profiles per environment [35]. This replication allowed a detailed picture of microbial distributions immediately above the sediment water interface. The value of this design is that it shows how much the 'mean field' approach underestimates absolute concentration and misses the resources, competition and viral exposure gradients that microbes experience.

## Materials and Methods

### Study site

Samples were collected from Saint Kilda mangroves and Port Noarlunga in South Australia. These environments were chosen to build on previous work using these sites [12,13,15,35]. The site at Port Noarlunga is located near the mouth of the Onkaparinga River and is bordered by sand dunes covered in remnant vegetation. Previous studies have described the water at Port Noarlunga as oligotrophic [12]. Samples were collected from Port Noarlunga (35°16′S, 138°47′E) on the $12^{th}$ of July 2011 at 11:20 am; the temperature of the water was 12.3°C with a pH of 9.4, a dissolved $O_2$ level of 14.3 ppm and a conductivity value of 40.7 ppt.

The St Kilda site was characterised by hypersaline lagoons less than 1 m deep in dense mangrove forests. Previous studies have shown St Kilda mangroves host a microbial community that is characterised by highly productive microbial mats and sulphur-oxidising bacteria [12,13,35,46]. Samples were collected from St Kilda (34°74′S, 138°54′E) on the $6^{th}$ of July 2011 at 10:00 am; the temperature of the water was 11.5°C with a pH of 8.5, a dissolved $O_2$ level of 14.3 ppm and a conductivity value of 41.5 ppt.

Environmental conditions data was collected using a Hydrolab Datasonde 4a sensor. Permission to access the sampling site from the St Kilda mangrove trail site was provided by the National Parks and Wildlife Rangers of the City of Salisbury Council. Specific permission to access the Noarlunga sampling site was not required. The field studies did not involve endangered or protected species.

### Sampling device (and supplementary methods)

Microplate triplicates were used to measure the microscale spatial distribution of marine viral and bacterial populations. Corning Costar TC-treated cell culture cluster microplates were used. They were 8 cm×12.3 cm, nonpyrogenic, polystyrene and sterile, consisting of 96 flat bottom 8 mm×8 mm wells which held 360 µl per well. The centre to centre distance between wells was 9 mm and the well perimeters were 7.2 cm at the ends and 10.8 cm on the sides resulting in a collection area of about 77.8 $cm^2$. This allowed the collection of 8 vertical profiles per microplate, which each consisted of 12 sampling points (Figure S1). Microplate triplicates gave 24 vertical profiles. The sampler was placed vertically against the respective surface and removed in a vertical motion with a 16 cm×16 cm glass plate used as a cover to minimise mixing and turbulence when collecting samples.

### Sample collection and preservation

All samples were transferred into 2 ml cryovials containing 4 µl of glutaraldehyde (0.5% final concentration) using a pipette and stored at 4°C in the dark for 15 minutes. Samples were then snap frozen in liquid nitrogen and stored in a −80°C laboratory freezer [47]. Flow cytometric analysis was performed within three weeks to avoid deterioration [47].

### FCM sample preparation and analysis

Many previous studies have used flow cytometry for the enumeration of bacteria and viruses due to its rapidity and accuracy [48,49]. Here we use the same flow cytometry technique as previous studies [12–15,34,35,43] and built on these studies by adding extensive replication. Briefly, samples were defrosted and prepared for flow cytometric viral and bacterial enumeration by diluting each sample 1:10 with Tris-EDTA buffer (pH 8.0, 0.2 µm filtered, 10 mM Tris, 1 mM EDTA) and staining with a nucleic acid-specific dye, SYBR Green I (1:500 dilution commercial stock; Molecular Probes). Samples were then incubated at 80°C in the dark for 10 minutes to optimise viral counts [35,47–51]. Reference beads (1 µm diameter, Molecular Probes) were added as an internal concentration and size standard with a final concentration of approximately $10^5$ beads $ml^{-1}$ in each sample. Measured flow cytometry parameters were normalised to the fluorescence and concentration of these beads [4,47–51].

Flow cytometry was conducted on a FACSCanto II flow cytometer (BD) equipped with a blue (488 nm, 20 mW, air-cooled), red (633 nm, 17 mW) and violet (405 nm, 30 mW) laser. Forward-angle light scatter (FSC), right-angle light scatter (SSC) and green fluorescence (SYBR I) were acquired for each sample and a phosphate-buffered saline (PBS) solution was used as a sheath fluid. The flow cytometer settings were normalised to fluorescence and bead concentration [35,50]. Each sample was run for two minutes at a medium flow rate setting for Noarlunga samples and a low flow rate setting for St Kilda samples due to more suspended particulate matter being present in St Kilda samples. In each flow cytometry session, triplicate blank control samples were analysed which consisted of 500 µl of 0.2 µl filtered Tris-EDTA buffer stained with 12.5 µl of SYBR Green I to eliminate any background noise that may have been created during the sample preparation or from flow cytometer artifacts.

## Sample processing

The resulting cytograms, density plots and histograms acquired from flow cytometry were exported as listmode files and analysed using Win Midi 2.9 (Joseph Trotter) to enumerate the bacterial and viral populations present [35]. Viral and bacterial populations were discriminated via side-scatter (SSC), which indicates cell size; and SYBR Green fluorescence, which is indicative of nucleic acid content [47,48,51].

## Data analysis and representation

For this study, rank abundance graphs were used to discriminate between hotspots, coldspots and background values. The background was determined according to Weibe [52] where the median value of the dataset is used as the background due to inclusion or exclusion of hotspots or coldspots not overly affecting the values obtained.

Surfer 10 (Golden Software, Inc.) was used to create two dimensional contour plots of the spatial distribution of viral and bacterial populations. A minimum contour interval value of at least 1000 was chosen when constructing plots due to this being larger than the maximum FACSCanto II flow cytometer error observed. This error refers to the background noise seen within triplicate blank control samples during flow cytometric analysis. This value is conservative as the maximum machine error was less than 36 events $\mu l^{-1}$.

Correlations were determined via Pearson's coefficient with the $\alpha$ of 0.05 being reduced by sequential Bonferroni [53]. Spatial autocorrelation statistics were used to determine whether correlations were present two-dimensionally and multi-directionally between sample points that were proximate and had similar values. From this, values were derived that illustrated the spatial autocorrelation present in the dataset. The two most common geospatial statistical tests are Moran's I and Geary's C statistics and both of these were employed as they allow statistical testing on smaller datasets [17].

**Moran's I Statistic.** Moran's *I* spatial autocorrelation statistics test within CrimeStat 3.3 (courtesy of Ned, Levine software) was used to determine the level of randomness within each population. The equation is as follows:

$$I = \frac{N \sum i \sum j W_{ij}(X_i - \overline{X})(X_j - \overline{X})}{(\sum i \sum j W_{ij}) \sum i (X_i - \overline{X})^2}$$

Where N is the sample number, $X_i$ and $X_j$ represent the variable values at a specific locations, i and j (where $i \neq j$), $\overline{X}$ is the mean of the variable and $W_{ij}$ is the weight applied to the i and j comparison. A weighted Moran's I test was used which gave a weight value of 1 to sample points adjacent and a weight value of 0 to sample points not adjacent.

The Moran's I statistical test [54] is a global statistical test used to determine spatial autocorrelation within a set of values. It is multi-directional, being able to use vertical, horizontal and diagonal directional analysis to consider all position correlations. The Moran's I value ranges from +1 to −1, with +1 indicating perfect clustering where high values are proximate and low values are proximate, −1 indicating perfect dispersion where high values are found far apart and low values are found far apart and zero being indicative of a random distribution. Critical cut-off values for most statistical tests are dependent on the scale and values used and are often well established through previous literature using the same or similar types of studies. However, in the instance of Moran's I values for small-scale spatial studies of bacteria and viruses there is only 1 study to compare which I values are indicative of significance. Waters et al. [17] looked at phytoplankton distributions on a 2 cm and 4 cm scale. From looking at the I values obtained in the Waters et al. [17] study, it appears I values between 0.08 and 0.18 indicate a considerable level of clustering, as opposed to a random distribution, which is far less than the general clustering I value of +1.

**Moran Correlograms.** Moran correlograms were constructed using the Moran's I statistic, which is applied to pairs of sample values separated by a lag distance, which in this case was the distance between each sampling well (0.9 cm). Geostatistical analysis requires each lag distance to have ≥30 pairs of sample points to enable statistical reliability; therefore sample intervals with less than this were not included in the correlograms [17,55]. The standard error obtained via Moran's I analysis was used to determine if correlograms and lag distance values were significant. Output values were used to determine whether spatial autocorrelation or independence was present at each distance interval from the sediment-water interface.

**Geary's C Statistic.** Geary's C statistics [56] test within CrimeStat 3.3 (courtesy of the Ned, Levine software) is a global statistic that can be used in conjunction with Moran's I as it is more sensitive to local clustering and is able to identify deterministic patterning of extreme values and non-spatially related chance phenomena within the dataset, which cannot be identified using Moran's I alone. Geary's C is similar to Moran's I, however it calculates the spatial autocorrelation of a dataset by the deviation in intensity of each sample value's location compared to one another, whereas in Moran's I spatial autocorrelation is calculated by the cross-product of the deviations from the mean within the dataset. The equation is as follows:

$$C = \frac{(N-1)\left[\sum i \sum j W_{ij}(X_i - X_j)^2\right]}{2(\sum i \sum j W_{ij}) \sum i (X_i - \overline{X})^2}$$

All terms are the same as Moran's I. For Geary's C, the values range from 0 to approximately 2, however there is no definitive upper limit [57]. A value of 1 indicates spatial independence, whilst values $<1$ indicate positive spatial autocorrelation and values $>1$ indicate negative spatial autocorrelation. Therefore this statistical test is inversely related to the Moran's I test [58]. The Moran's I and Geary's C can be used in conjunction to identify deterministic patterning of extreme values [57].

## Results

### FCM analysis

Flow cytometric analysis revealed three distinct bacterial subpopulations and two distinct viral subpopulations at both sites which were comparable to those identified previously in marine systems [12,15,48]. Subpopulation discrimination was based on the presence of aggregated regions on biparametric cytograms of side scatter and SYBR green fluorescence and discrete peaks in monoparametric histograms of SYBR green fluorescence (Figure. 1A–D). Bacterial subpopulations were differentiated and separated into high-DNA (HDNA) and low-DNA (LDNA) groups (Figure. 1A–D). The HDNA 1 bacterial population is the intermediate bacterial population that contains bacteria with a size and nucleic acid content that is intermediate to the LDNA and HDNA 2 populations. The two viral subpopulations were identified and separated into two virus-like particle subpopulations (VLP 1 and VLP 2).

At Noarlunga, the total mean abundance of the VLP 1 subpopulation was $3.7\times10^6$ cells $ml^{-1}$ (95%CI $=0.3\times10^6$ cells $ml^{-1}$, $n=269$), whereas it was $1.3\times10^6$ cells $ml^{-1}$ for the VLP 2 subpopulation (95%CI $=0.1\times10^6$ cells $ml^{-1}$, $n=269$) (Figure. 1A and 1B). Hence, the total mean abundance of VLP 2 was 2.8-fold less than the total mean abundance of VLP 1. At St Kilda, the total mean abundance of VLP 2 was also less than VLP 1, but to a greater extent as the total mean abundance of VLP 1 was $75\times10^5$ cells $ml^{-1}$ (95%CI $=5.1\times10^5$ cells $ml^{-1}$, $n=209$) compared to $6.1\times10^5$ cells $ml^{-1}$ (95%CI $=0.45\times10^5$ cells $ml^{-1}$, $n=209$) for VLP 2 (Figure. 1C and 1D). Hence, the total mean abundance of VLP 2 was 12.3-fold less than the total mean abundance of VLP 1.

For the bacteria, the LDNA subpopulation at Noarlunga was less abundant than the HDNA 1 and the HDNA 2 subpopulations. The total mean abundance for LDNA was $0.4\times10^5$ cells $ml^{-1}$ (95%CI $=0.03\times10^5$ cells $ml^{-1}$, $n=269$) compared to total mean abundances of $3.1\times10^5$ cells $ml^{-1}$ (95%CI $=0.2\times10^5$ cells $ml^{-1}$, $n=269$) for HDNA 1 and $2.7\times10^5$ cells $ml^{-1}$ (95%CI $=0.2\times10^5$ cells $ml^{-1}$, $n=269$) for HDNA 2 (Figure. 1A and 1B). Hence, the total mean abundance of LDNA was up to 7-fold less than the total mean abundance of HDNA 1 and HDNA 2. Whereas at St Kilda, HDNA 1 was the least abundant bacterial subpopulation (Figure. 1C and 1D) with total mean abundances of $3.0\times10^5$ cells $ml^{-1}$ (95%CI $=0.32\times10^5$ cells $ml^{-1}$, $n=209$) for LDNA, $2.1\times10^5$ cells $ml^{-1}$ (95%CI $=0.47\times10^5$ cells $ml^{-1}$, $n=209$) for HDNA 1 and $5.9\times10^5$ cells $ml^{-1}$ (95%CI $=1.7\times10^5$ cells $ml^{-1}$, $n=209$) for HDNA 2. Hence, the total mean abundance of HDNA 1 was 2-fold less than the total mean abundances of LDNA and HDNA 2 (Table S1 and S2).

### Vertical profiles

At Noarlunga, the mean of 24 vertical profiles had an r value of 0.9 ($p<0.0001$, $n=12$) and showed almost identical patterns between the total mean bacterial and total mean viral population (Figure. 2A), as well as the total mean LDNA, HDNA 1 and HDNA 2 subpopulations (Figure. S2A), the total mean VLP 1 and VLP 2 subpopulations (Figure. S2B) and the mean vertical profile of VLP 1 and LDNA from microplate 1, which all had an r value of 0.99 ($p<0.0001$, $n=12$) (Figure. S3). In addition, 145 of the 240 possible bacterial and viral single vertical profile subpopulation correlations were significantly correlated with a p-value generally less than 0.0001, such as the single vertical profile of VLP 1 and LDNA from microplate 1 which had an r value of 0.98 ($p<0.0001$, $n=12$) (Figure. 3B). However, 95 of the single vertical profiles were not significantly correlated, such as the single vertical profile of VLP 1 and LDNA from microplate 3 which had an r value of $-0.33$ ($p=0.12$, $n=12$) (Figure. 3A). At St Kilda, the mean of 24 vertical profiles was not significantly correlated (Figure. 2B). However, the total mean LDNA, HDNA 1 and HDNA 2 populations were correlated ($r=0.98$, $p<0.0001$, $n=12$), as well as the total VLP 1 and total VLP 2 populations ($r=0.64$, $p=0.02$, $n=12$). Of the 240 possible bacterial and viral single vertical profile subpopulation correlations, only 64 were significantly correlated with a p-value generally less than 0.0001, whilst 176 were not correlated.

### Determination of hotspots and coldspots

To determine the presence of hotspots and coldspots in the viral and bacterial distributions, rank abundance graphs were created (Figure. 4A–D). In some instances only one linear trend was seen within the background values and in others there were breaks in the fits that indicated two or three separate linear trends. St Kilda showed primarily one linear trend, for instance in the VLP 1 and HDNA 1 subpopulations (Figure. 4A–B) whilst Noarlunga showed primarily two linear trends within their background values, for instance in the VLP 1 and HDNA 2 subpopulations (Figure. 4C–D). The hotspots were identified as the sample points that exceeded this linear fit (Figure. 4A–D). Exponent values for these hotspots ranged from $-0.20$ to $-1.1$ for the bacteria and $-0.14$ to $-0.63$ for the viruses at Noarlunga, whilst exponent values at St Kilda ranged from $-0.27$ to $-1.4$ for the bacteria and $-0.23$ to $-0.66$ for the viruses (Figure. 4A–D). Lastly, the coldspot sample values were identified as sample points that fit a linear trend but exhibited steeper slopes than what were seen in the background values due to large differences between adjacent sample points (Figure. 4A–D).

### Two dimensional distributions

Two-dimensional contour plots of both sampling sites showed heterogeneous distributions with patches or hotspots of high abundance that occurred over single or multiple sample points as well as gaps or coldspots of low abundance that occurred only over single sample points (Figure. 5A–D).

At Noarlunga, high levels of heterogeneity were seen, with the VLP 2 subpopulation showing the highest level of heterogeneity of the two viral subpopulations. It was due to the presence of coldspots in the VLP 2 subpopulation that resulted in a 2585-fold change in abundance over 0.9 cm, with a coldspot of $6.64\times10^2$ particles $ml^{-1}$ below a background of $1.07\times10^6$ particles $ml^{-1}$ (Figure. 5A). For the bacterial subpopulations, the highest level of heterogeneity was seen in the LDNA subpopulation with a maximum hotspot of $2.59\times10^5$ cells $ml^{-1}$ above a background of $3.47\times10^4$ cells $ml^{-1}$ resulting in a 12.9-fold change in abundance over 0.9 cm (Figure. 5B).

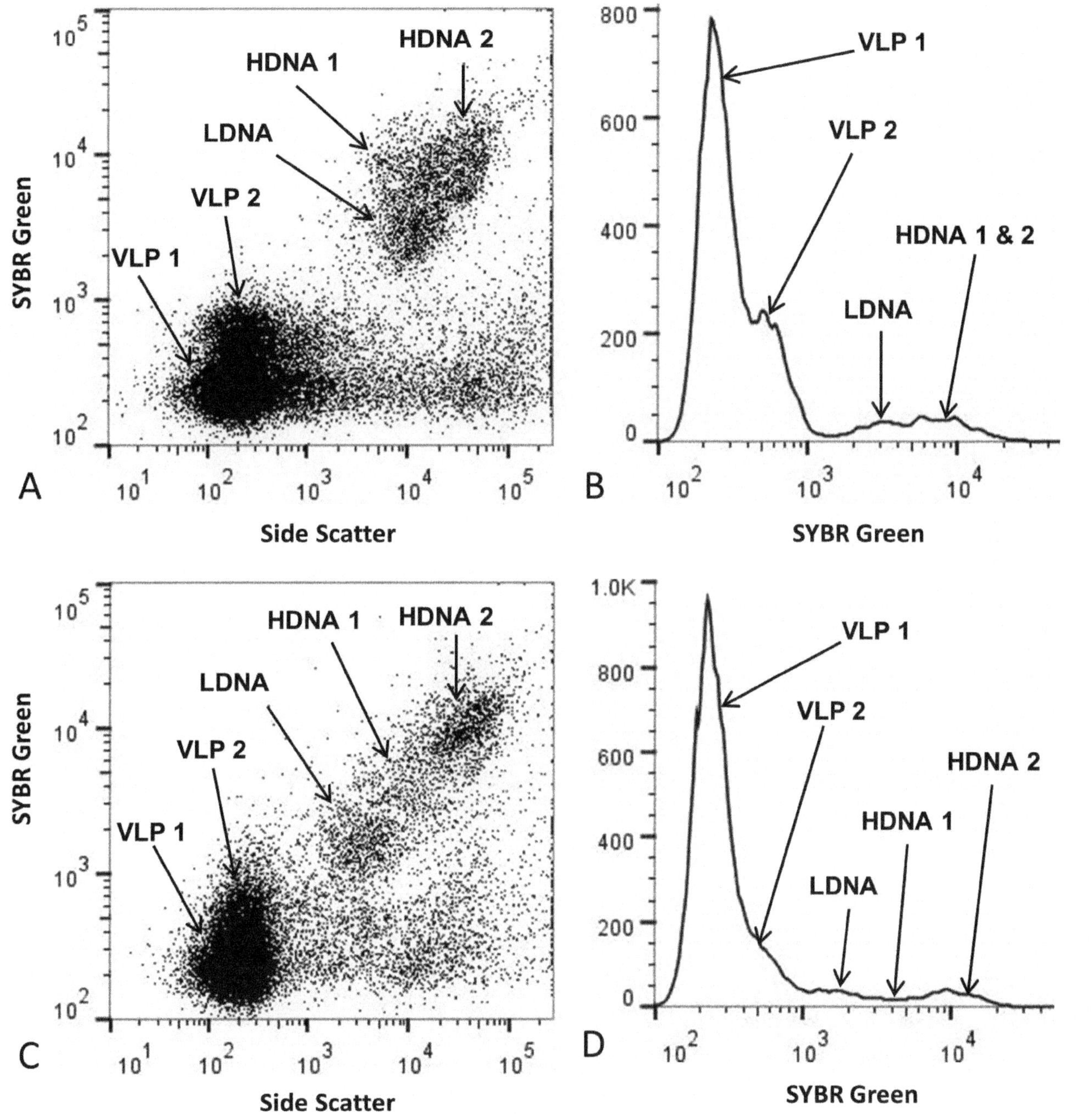


**Figure 1. Identification of bacterial and viral subpopulations via flow cytometry.** Flow cytometric cytograms of side-scatter light versus green fluorescence (SYBR Green) and histograms of green fluorescence (SYBR Green). **Noarlunga: A** cytogram **B** histogram; **St Kilda: C** cytogram **D** histogram showing two distinct viral populations (VLP 1 and VLP 2) and three distinct bacterial populations (LDNA, HDNA 1 and HDNA 2).

At St Kilda, the highest viral heterogeneity was present in the VLP 1 subpopulation which had a maximum hotspot of $1.62\times10^7$ particles $ml^{-1}$ above a background of $6.12\times10^6$ cells $ml^{-1}$ which resulted in a 10.52-fold change in heterogeneity seen over 0.9 cm (Figure. 5C). For the bacterial subpopulations at St Kilda a 45.2-fold change in heterogeneity was seen in the HDNA 2 subpopulation due a maximum hotspot of $1.45\times10^7$ cells $ml^{-1}$ above a background of $3.7\times10^5$ cells $ml^{-1}$ (Figure. 5D). There were a range of heterogeneities over 0.9 cm (Figure. S4) indicating a variety of intensities for hotspots and coldspots. Representative adjacent orthogonal gradient distributions as rank abundances show the distribution of abundance values for selected bacterial and viral populations at St Kilda and Noarlunga (Figure. S4).

## Spatial autocorrelation

**Moran's I.** At Noarlunga, the Moran's I values for all of the bacterial subpopulations were non-significant. Whereas at St Kilda, significant Moran's I values were found in 33% of the bacterial subpopulations; the LDNA subpopulation of microplate 2 and 3 and the HDNA 2 subpopulation and total bacteria of microplate 2. These significant Moran's I values ranged from 0.039–0.070, with all the non-significant Moran's I values being

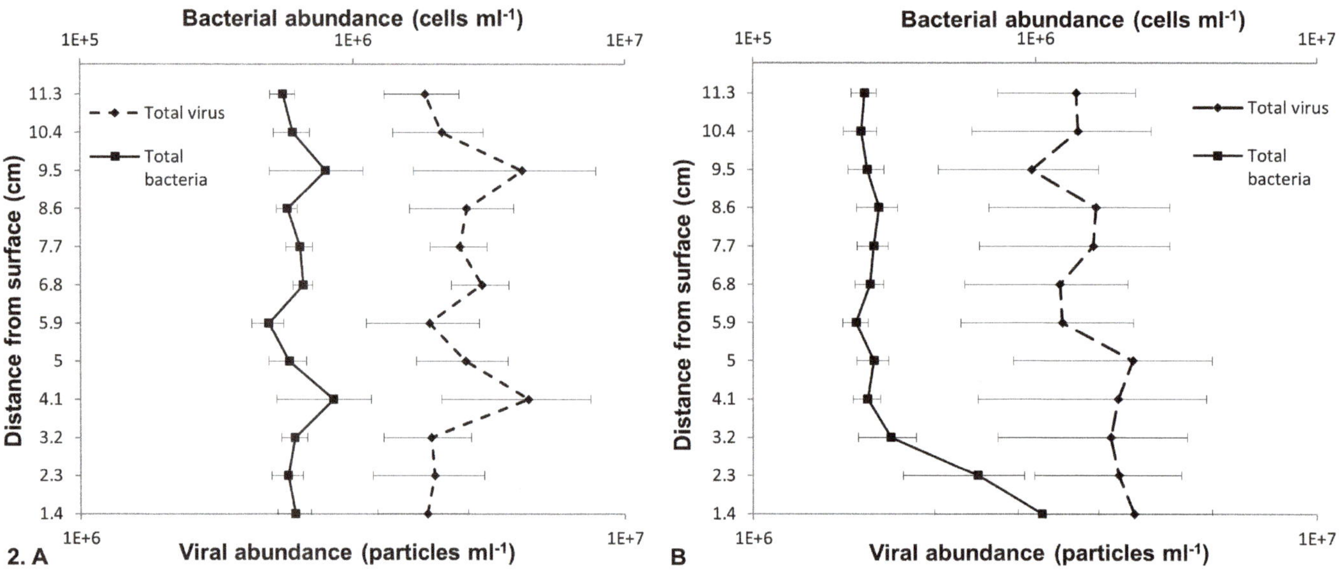


**Figure 2. Comparison of total mean bacterial and viral populations at Noarlunga via vertical depth profiles.** Total mean bacterial and total mean viral population abundance within all three microplates. **A** Noarlunga, (n = 1350). **B** St Kilda, (n = 1045). Error bars represent the 95% confidence intervals obtained from all three replicates (n = 12).

below these values (≤0.036). For these bacterial subpopulations z scores under the assumption of randomisation were much larger than the cut-off point of +1.96, with z scores ranging from 2.5 to 3.8. All the viral subpopulations at Noarlunga and St Kilda had non-significant Moran's I values.

All of the significant Moran's I values for the bacterial subpopulations were positive (Table S3 and S4). This trend is evident in the 2 dimensional contour plots of HDNA 2 and LDNA at St Kilda which show the presence of sediment-water interface surface gradients (Figure 6A–B). Statistical significance was not found for negative Moran's *I* values (Table S3 and S4). In geographic data, statistically significant positive values are more common than statistically significant negative values.

**Moran correlograms.** Moran correlograms for the bacterial subpopulations that had significant Moran's I values showed a general trend of positive to negative spatial autocorrelation as the

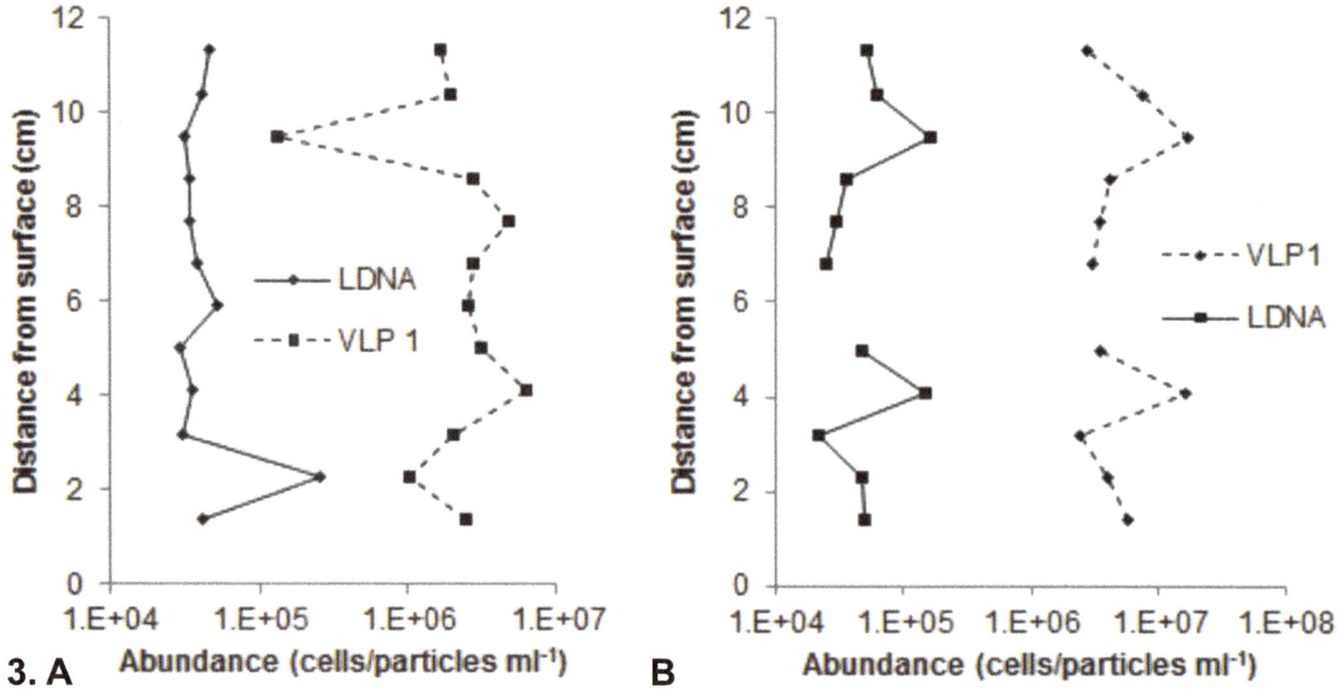


**Figure 3. Single vertical profile of VLP 1 and LDNA populations at Noarlunga. A** Microplate three showing little to no association. **B** Microplate one showing association. Gap in profile indicates a missing data point.

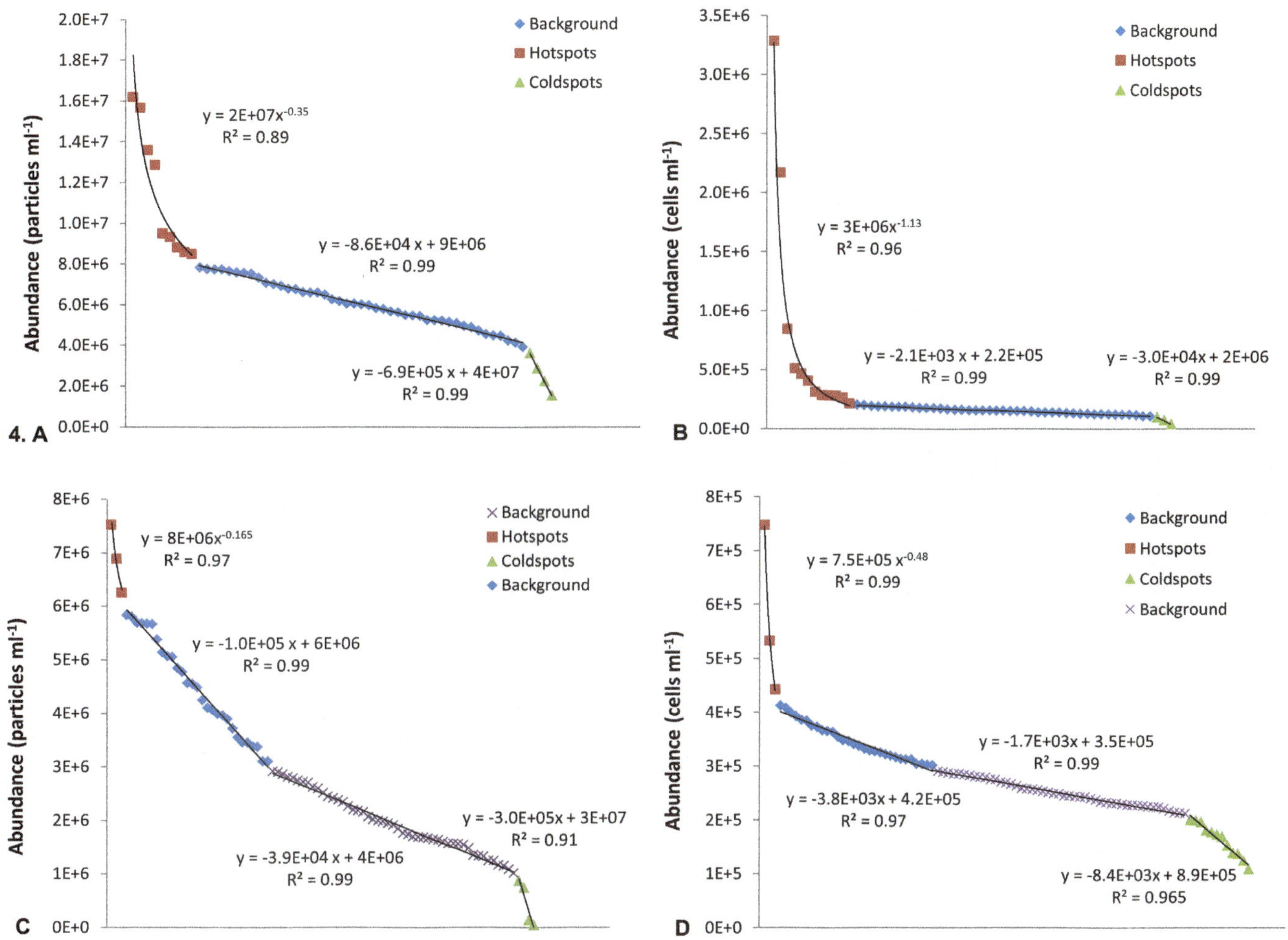


**Figure 4. Rank abundance graphs used to differentiate hotspot and coldspot values from background values.** One linear trend within the background values were characteristic of St Kilda whilst two linear trends were primarily seen at Noarlunga. **St Kilda: A** VLP 1, microplate 2. **B** HDNA 1, microplate 2. **Noarlunga: C** VLP 1, microplate 3. **D** HDNA 2, microplate 2.

distance from the sediment-water interface increased. On closer inspection, the correlograms for these bacterial subpopulations showed positive spatial correlation in the sample points located within 5.9 cm of the sediment-water interface and spatial independence as this distance from the sediment-water interface increased at >5.9 cm (Figure. 7A–D). This is indicative of surface gradients, which are shown in figures 6A–B. Moran correlograms for the LDNA, HDNA 1 and total bacteria in microplate 1, and HDNA 1 in microplate 2 at St Kilda also followed this trend despite having non-significant Moran's I values. Geary's C statistical testing was used to further analyse these subpopulations.

At Noarlunga and St Kilda, all of the viral subpopulation correlograms were non-significant and showed no general trend, with their distributions alternating between positive and negative spatial autocorrelation throughout the sampling distance indicating the presence of hotspots and coldspots.

**Geary's C.** At Noarlunga, significant Geary's C values were obtained from all of the bacterial subpopulations in microplate 1, whilst all of the bacterial subpopulations in microplate 2 and 3 were non-significant. Geary's C values were between 0.87 and 1.02; however, significance was only seen in C values of 0.87 (Table S3 and S4).

In contrast, at St Kilda significant Geary's C values were seen in 75% of the bacterial subpopulations; all of the subpopulations in microplate 1 and 2, as well as LDNA within microplate 3. Significant C values were between 0.73 and 0.92 with C values between 0.96 and 0.99 showing non-significance (Table S3 and S4).

At Noarlunga, significant Geary's C values were found in the total virus and VLP 2 subpopulation in microplate 1 whilst all other viral subpopulations were non-significant. Significance was seen in C values of 0.87, whilst C values of 0.98–1.02 showed non-significance. At St Kilda, all of the viral subpopulations had non-significant Geary's C values. C values were higher than what was seen in the bacteria at St Kilda, being between 0.96 and 1.02, which was similar to the C values obtained for the viruses at Noarlunga (Table S3 and S4).

Using Geary's C to look in more detail at the randomly distributed subpopulations that showed positive to negative spatial autocorrelation within the Moran correlograms revealed they had significant Geary's C values (Table S3 and S4). These subpopulations were the bacterial subpopulations in microplate 1, as well as the HDNA 1 subpopulation in microplate 2 at St Kilda and the bacterial subpopulations and VLP 2 and total virus populations in microplate 1 at Noarlunga.

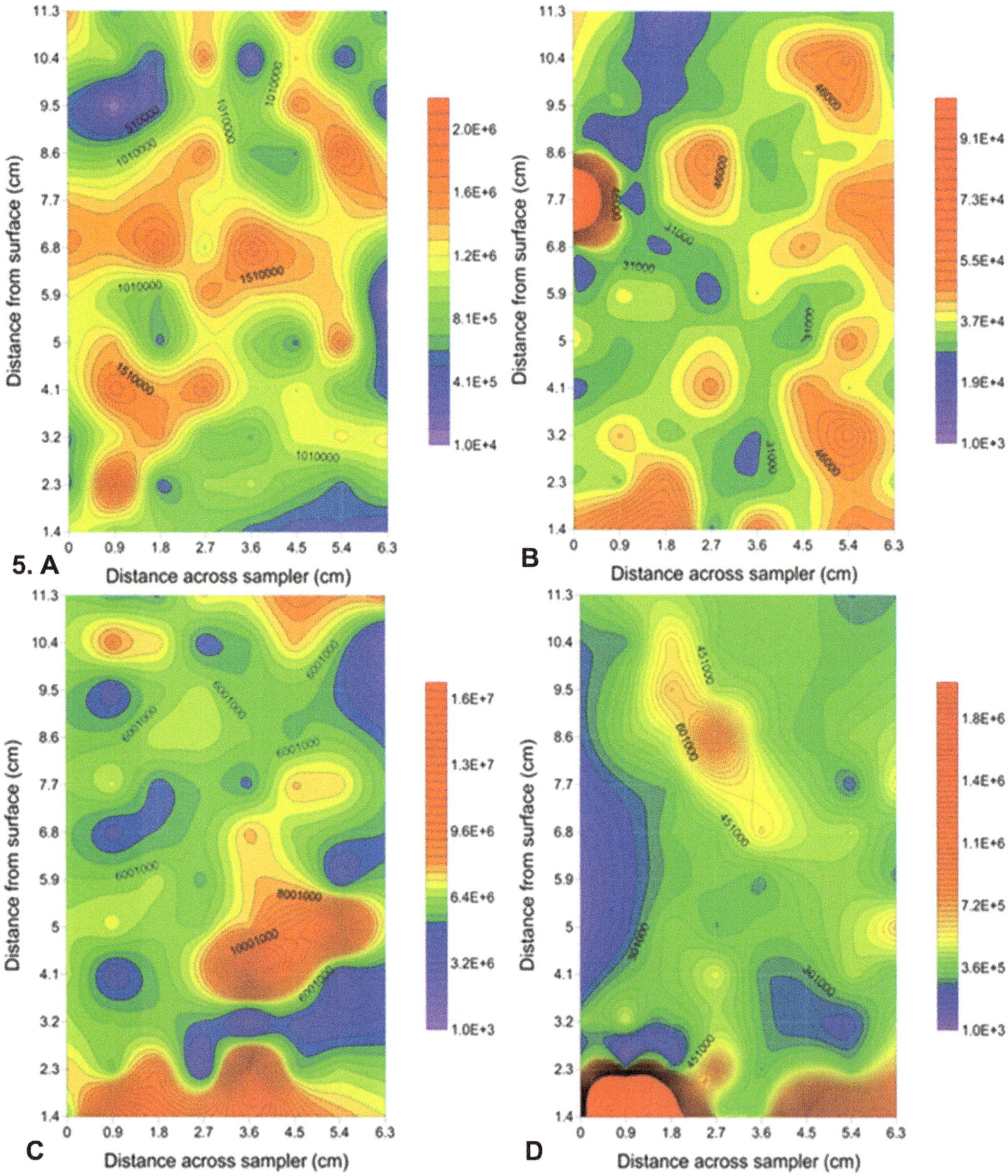


**Figure 5. Two-dimensional contour plots showing the highest change in heterogeneity due to the presence of hotspots and coldspots within bacterial and viral subpopulations.** Hotspots and coldspots seen across a distance of 6.3 cm×11.3 cm using Surfer 10 (Golden Software, Inc.). Noarlunga: A VLP 2 showing a 2585-fold change in abundance over 0.9 cm. B LDNA showing a 12.9-fold change in abundance over 0.9 cm. St Kilda: C VLP 1 showing a maximum 10.52-fold change in heterogeneity seen over 0.9 cm. D HDNA 2 showing a maximum 45.2-fold change in heterogeneity seen over 0.9 cm. There were a range of heterogeneities over 0.9 cm (Fig. S4) indicating a variety of intensities for hotspots and coldspots. Abundance levels are indicated by a colour intensity scale in units of cells/particles $ml^{-1}$. Solid red circles indicate areas of abundance higher than the maximum contour level selected. A minimum contour interval value of at least 1000 was chosen based on maximum machine error. The faint gridlines show sample interval.

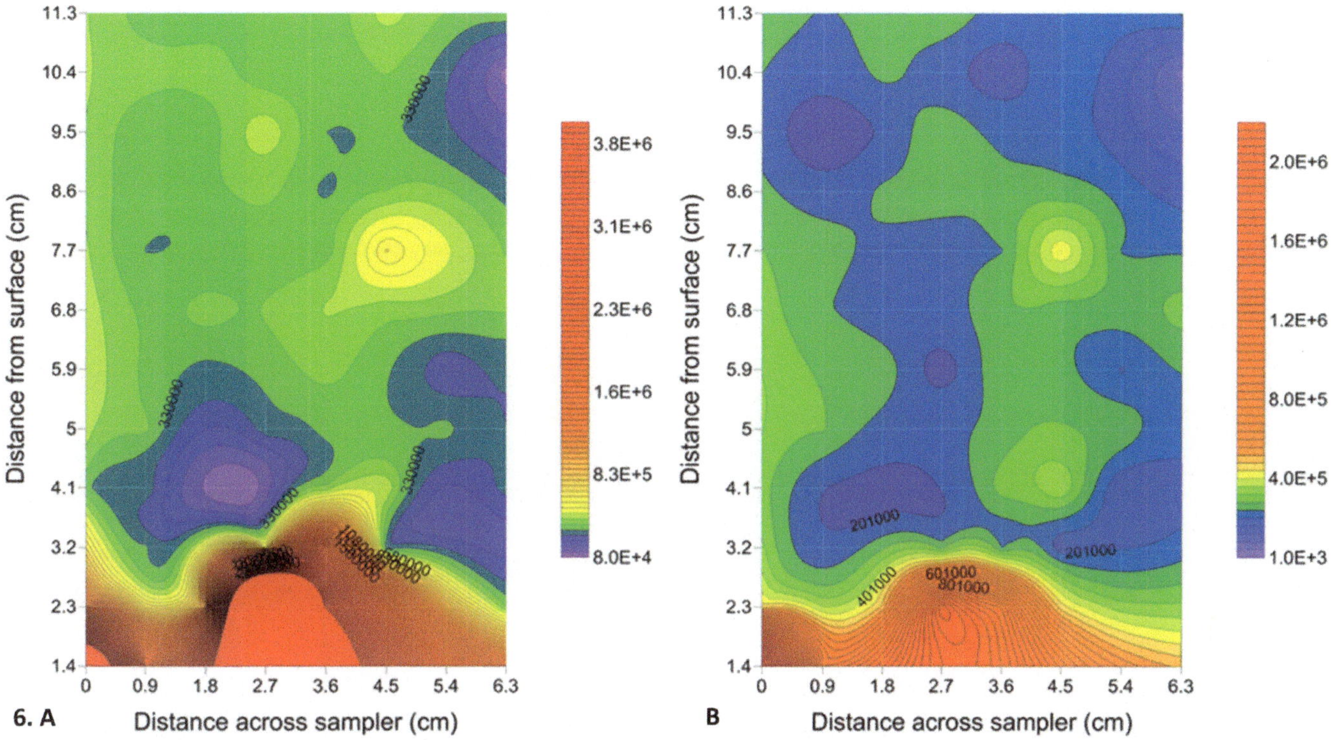


**Figure 6. Two-dimensional contour plots showing surface gradients within bacterial subpopulations at St Kilda.** Surface gradients seen across a distance of 6.3 cm × 11.3 cm using Surfer 10 (Golden Software, Inc.). Microbial abundance levels are indicated by a colour intensity scale in units of cells ml$^{-1}$. **A** HDNA 2, microplate 2. **B** LDNA, microplate 2. Solid red circles indicate areas of abundance higher than the maximum contour level selected. A minimum contour interval value of at least 1000 was chosen based on maximum machine error. The faint gridlines show sample interval.

## Discussion

### Bacterioplankton and virioplankton distributions

**Surface gradients.** Strong surface gradients were found at St Kilda, which are well established along with characteristic low dissolved oxygen levels [35]. These strong gradients rising from the sediment surface showed up to a 15-fold change in abundance over the 11.3 cm vertical sampling distance collected above the sediment-water interface, The presence of these gradients is perhaps due to the strong input of organic matter from the mangroves [46]. In support of this, Seymour *et al.* [35] found microscale microbial distributions were highly correlated to nutrient concentration gradients. Seymour *et al.* [15] found the shear velocities at St Kilda at the time of sampling were not high enough for the resuspension of microbes. Given the protected nature of the environment, resuspension may be minimal compared to exposed sites and would not play a role in the surface gradients observed.

Bacteria and viruses immediately above the sediment are exposed to a variety of biological and non-biological parameters that will affect their distribution [35]. At both sampling sites clear abundance gradients were often seen in relation to distance from the sediment-water interface. The remineralisation of organic nutrients is known to cause gradients with dissolved and particulate matter, which usually show decreasing concentrations as the distance from the sediment-water surface increases [35,59].

**Hotspots and coldspots.** Previous studies could not determine whether a high or low point in a one dimensional profile was a gradient, a multiple point hotspot/coldspot or a single point maximum/minimum [35]. In this study, two-dimensional sampling revealed the presence of single and multiple point hotspots and single point coldspots in bacterial and viral populations at Noarlunga and St Kilda. In previous studies, local increases in abundance of up to 2 times above background over a distance of 1 cm have been considered hotspots [13,14,35,43], whereas this study showed hotspots with increases of up to 2585 and 45.2 times cm$^{-1}$ in viral and bacterial subpopulations above the sediment-water interface. These hotspots were high in bacteria and viruses, consistent with bacteriophage infection being dependent on host density [15]. The correlation of these bacterial and viral hotspots suggests a repeated viral infection and lysis cycle for at least some of the bacterial species present [12,13,15], which, in turn, suggests that groups of picoplankton, at least close to the sediment interface remain close together for extended periods. Through entrainment in the overlying water these near-sediment hotspots may act to inoculate the overlying water column. With viral hotspots showing increases of up to 2585 times cm$^{-1}$ this would contribute substantially to the abundance of viral particles hence playing a role in viral production and succession. In the case of coldspots, correlations were seen with viral subpopulations but not with bacterial subpopulations. There are many possible dynamics at play here. For instance, this may indicate these coldspots are capturing the stable bacterial community numbers, which have not yet been exposed to viral communities or perhaps areas in which compatible viral particles were not present hence leading to viral decay or extensive grazing as the viruses would not have been able to protect themselves from external sources by infecting the bacteria present. Another possible explanation for these coldspots in viruses could be that they are microenvironments that favour lysogeny over lysis, such as areas of low nutrient concentration

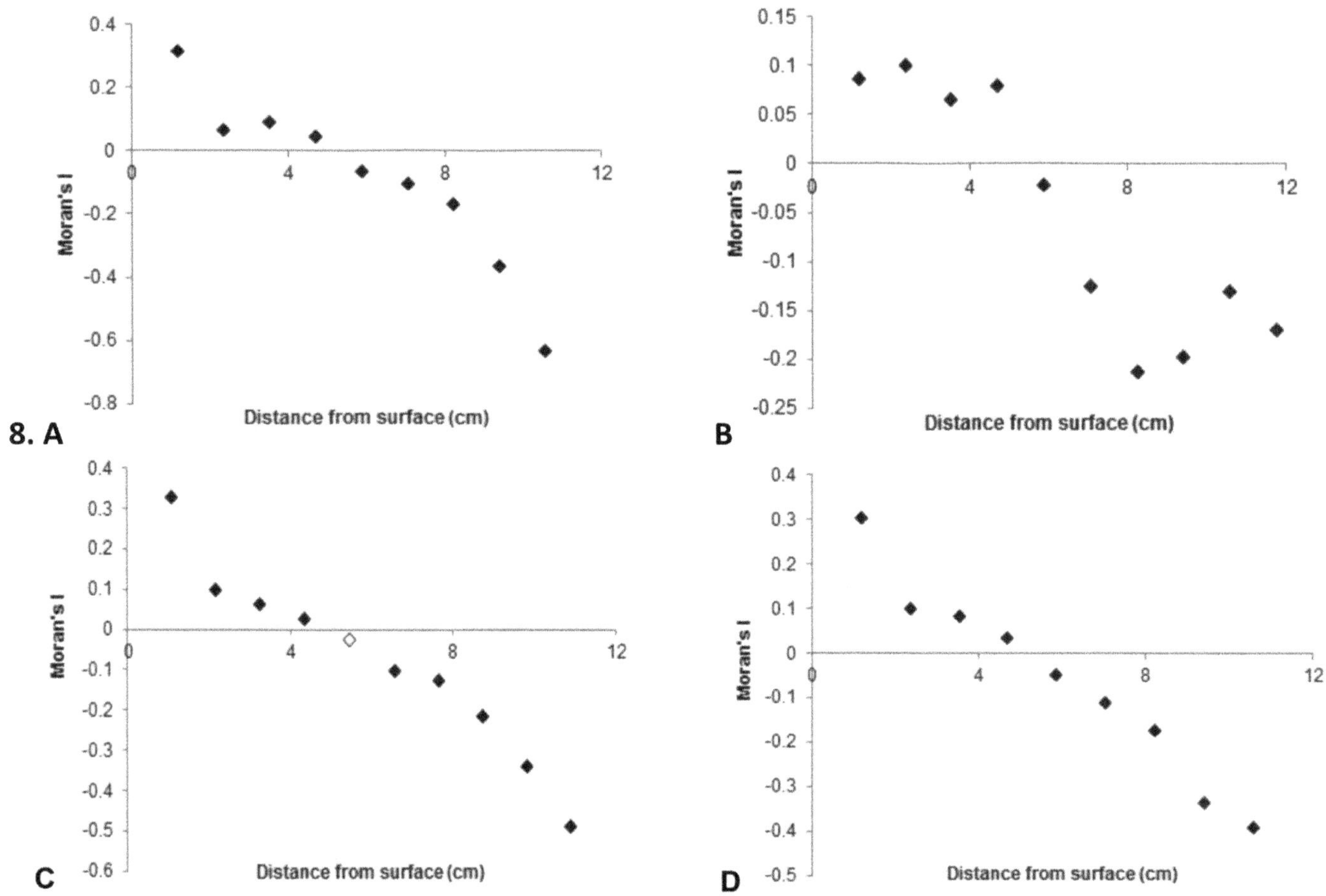


**Figure 7. Significant Moran correlograms of non-randomly distributed bacterial subpopulations at St Kilda. A** LDNA, microplate 2. **B** LDNA, microplate 3. **C** HDNA 2, microplate 2. **D** Total bacteria, microplate 2. Filled and unfilled data points indicate significant and non-significant Moran's I values ($p \leq 0.01$). Only sample points with $\geq 30$ pairs of values were included.

[60]. The wide range of hotspot values (Figure. S4) suggests that all of these and possibly more mechanisms are generating hotspots. Of course, the actual processes occurring in these hotspots and coldspots cannot be determined with biomass values alone and call for microscale nutrient and species analysis.

At St Kilda, patchy viral and bacterial hotspots and coldspots were identified and, in the case of the hotspots, often coincided. These hotspots and coldspots resulted in changes in spatial heterogeneity of 10.5-fold $cm^{-1}$ for viruses and 45-fold $cm^{-1}$ for bacteria. These results are much higher than the previously found 1.4- and 1.5-fold $cm^{-1}$ increases found in total virus and total bacteria by Seymour *et al.* [35]. As these hotspots and coldspots often occurred within single sampling points of 0.9 cm distance this highlights the importance of high sampling resolution and 2-dimensional arrays when studying the distributions of microbial communities.

The null hypothesis tested in this study was that patchiness has no effect on mean biomass estimates and therefore the mean field approach is an accurate estimate of biomass. This null hypothesis was rejected with the presence of patchy hotspots and coldspots in this study showing differences of up to four orders of magnitude in biomass. In the most extreme case, if the mean field approach had been employed without sufficient replication, estimates in microbial biomass would be off by a factor of 1,560.

Previous studies have also lacked a definition for hotspots and coldspots making comparative analysis impossible. In this study rank abundance graphs have been used as a method for determining what sample values are hotspots, coldspots or background. Background values fitted a linear trend as these values are random and are an indicator of equilibrated additive and reductive processes. Such additive processes in bacterial and viral communities could be reproduction or aggregation, whilst reductive processes could relate to grazing, decay or lysis events.

Hotspots were identified as the sample points that exceeded the linear fit and exhibited a steep power law trend, indicating their non-random nature whilst the coldspots were characterised by a linear trend with slope values much higher than those of the background values due to large differences seen between adjacent value points on the graph (Figure. 4A–D). The hotspots indicate a favouring towards additive processes and coldspots indicate a favouring towards reductive processes resulting in a non-equilibrated state when compared to the background values. Exponent values for the hotspots were similar at Noarlunga and St Kilda and ranged from $-0.1$ to $-1.4$ suggesting this is the level of structure seen within bacterial and viral communities. Exponent values at and below $-1$ indicated considerable self-organisation and strong patches [61].

The background values fit single or multiple linear trends with St Kilda being characterised primarily by single linear trends whilst Noarlunga was characterised primarily by two linear trends within its background values for bacterial and viral communities. This could be an indication of different randomisation processes present. For instance, when considering bacteria, the presence of two linear trends seen at Noarlunga could indicate equilibrium between grazing and aggregation in one linear trend whilst the second may indicate equilibrium between reproduction and lysis. By using these rank abundance graphs, not only are hotspots and coldspots identified but patterns within the background values are able to be observed. Experimental work will be required to distinguish between randomisation processes.

## Bacterial and viral abundances

The LDNA and HDNA bacterial subpopulations observed have been described previously as different phylogenetic groups present within the environment [62] or as bacteria with differing activity levels [64–66]. In the former context, the total mean abundance of HDNA 2 at St Kilda and HDNA 1 at Noarlunga being higher than the total mean abundance of the other LDNA and HDNA subpopulations, may be explained by the species within these subpopulations being more efficient at utilising resources or avoiding predatory attack [62]. The total mean abundances also indicated HDNA 1 at St Kilda and LDNA at Noarlunga had the lowest abundances of the bacterial subpopulations which could be due to the species within HDNA 1 at St Kilda and LDNA at Noarlunga being more susceptible to infection or grazing [35,62,63]. In addition, higher concentrations of predators are found within the first few centimetres above the sediment-water interface, such as heterotrophic nanoflagellates [35]. However, in the context of differing bacterial activity, where the LDNA subpopulation represents dormant bacterial cells and the HDNA subpopulations represent active cells, the higher abundances of the HDNA subpopulations seen at both sites when compared to the LDNA subpopulation would be explained by the ability of the HDNA subpopulations to replicate and grow, hence leading to their higher abundances [35,64–66]. In the context of different species, the hotspots would represent accumulation of chemotactic bacteria and the coldspots areas they had left.

Another factor controlling the abundances of the bacterial subpopulations is the viral populations present. The total mean abundance of the VLP 1 subpopulation was larger than the total mean abundance of the VLP 2 subpopulation at both sites. This has been found in previous studies, with aquatic samples being dominated by the viral population that contains small viruses between 30 and 60 nm, which corresponds to the VLP 1 subpopulation within the cytograms (Figure.1A and C) [48,67].

## Bacterial and viral coupling

Seymour *et al.* [35] did not find a correlation between total bacteria and viruses at this scale. In this study, a significant relation existed between total bacteria and total viruses at Noarlunga, whilst at St Kilda, a relation did not exist (Figure. 2A and B). This may be due to the presence of different microbial communities and the difference in environmental parameters between the two sampling sites. The dynamic nature of the environments may have led to short-lived events, such as nutrient patches, causing the aggregation and dispersion of bacteria over the course of minutes to hours. This initial increase in bacterial abundance would lead to increased viral production via lysis hence decoupling the relationship between viral and bacterial populations which may be why no relationship was seen between bacteria and viruses at St Kilda [15,38]. However, stages in between the bacterial aggregation and bacterial death via viral lysis may have been captured at Noarlunga and would explain the correlation between the bacteria and viruses in this environment.

While Seymour *et al.* [35] found no correlation between total bacteria and viruses; correlations were seen between the VLP 1, VLP 2 and LDNA subpopulations and the VLP 2 and HDNA 2 subpopulations. Therefore, showing subpopulations have independent and potentially more complex associations than what is seen in total bacteria and virus populations. These results imply that total populations reflect system biomass and the processes that may impact on this, such as grazing; whilst single subpopulations reflect the phylogenetic changes in the community from such processes as phage infection and lysis. As the individual subpopulations can be decoupled, with phylogenetic fluctuations being independent of the total system biomass, this explains why correlations were seen in the totals but not in the individual subpopulations. While the correlations are not mechanisms, they show that possible relationships exist that could be further investigated. These statements are based on biomass only. Taxonomic identification of the community will be part of a manuscript in preparation.

## Spatial autocorrelation

**Non-random vs random distribution.** At St Kilda, non-random distributions were seen in the LDNA subpopulation of microplate 2 and 3 and the HDNA 2 subpopulation and total bacteria of microplate 2 due to these subpopulations having significant Moran's I values and z scores larger than the +1.96 cut-off value. Whilst all viral subpopulations were randomly distributed due to having non-significant Moran's I values. This may be due to motile bacterial subpopulations congregating around nutrient gradients and patches or resuspension of bacteria from the sediment leading to the observed surface gradients [68]. These non-random distributions may not have been seen in the viral subpopulations as viruses are not motile and are unable to respond to chemical signals.

The significant Moran's I values seen in the bacterial subpopulations at St Kilda were positive, indicating clustering and spatial dependence. This clustering indicates that high values are located close together and low values are located close together, indicating the presence of 'hotspots' and 'coldspots' or gradients of high or low values. Moran's I values ranged from 0.04 to 0.07 for clustered distributions. These values are low when considering a Moran's I value of +1 indicates perfect clustering. These values are also low compared to Waters *et al.* [17] which had significant Moran's I values between 0.082–0.180. These low values may indicate a low level of clustering present within these distributions.

At Noarlunga, all the bacterial and viral subpopulations were randomly distributed due to having non-significant Moran's I values. This indicates that no spatial autocorrelation was present amongst the bacterial and viral subpopulation distributions and that there is no spatial dependency within the distributions of the bacterial or viral subpopulations locally or globally. On a large scale this implies that environmental processes, such as turbulence, randomize the spatial distribution of bacterial and viral subpopulations at Noarlunga.

The non-random and random bacterial distributions found at St Kilda and Noarlunga coincide with the findings of Waters *et al.* [17], which found higher spatial complexity within low-energy, high-chlorophyll environments such as St Kilda mangroves than in high-energy, low-chlorophyll environments such as Noarlunga. As St Kilda mangroves is a low-energy system with high productivity, due to the presence of dense microbial mats and mangrove forests causing stagnation in the lagoons, this may allow the development and maintenance of nutrient surface gradients [12,13,35,46].

**Correlogram autocorrelation patterns.** At St Kilda, the positive spatial autocorrelation seen within 5.9 cm of the sediment-water interface in most bacterial subpopulations indicates the presence of potential nutrient gradients that lead to bacterial clustering and the formation of surface gradients. The negative spatial autocorrelation seen beyond 5.9 cm from the sediment-water interface shows the area at which such gradients dissipate, leading to dispersed bacterial distributions. This trend may be due to eddies, Nyquist frequency limitation or a nepheloid layer. The presence of eddies could cause resuspension of particles via eddy penetration leading to a short-time mixing effect and turbulence [69–71]. Whereas, Nyquist frequency, whereby sampling frequency should be at least twice the highest frequency present in the sample data, may have contributed to this trend as if the sampling frequency was insufficient this would lead to the presence of aliases and distortions in the dataset [72]. The presence of a 5.9 cm nepheloid layer would result in high levels of suspended particulate matter being present at sediment-water interface and the presence of such layer may largely depend on the time of sampling as nepheloid layers are subject to change depths depending on the season [73,74].

Taking into account the study by Sokal [57] which showed that high Moran's I values not accompanied by high Geary's C values are representative of deterministic patterning of extreme values and the reverse implies that extreme values are randomly distributed and are hence spatially independent. Then for the bacterial subpopulations which showed positive spatial autocorrelation 5.9 cm from the sediment-water interface and then negative spatial autocorrelation beyond 5.9 cm, with relatively low corresponding Geary's C values between 0.73 and 0.83 suggests there was deterministic patterning of extreme values. Whereas, the positive to negative spatial autocorrelation trend within the Moran correlograms of the bacterial subpopulations that were found to be randomly distributed and had low Moran's I values which were less than the estimated cut-off value for clustered distributions but also had relatively low Geary's C values between 0.85 and 0.89 may indicate that globally these populations are spatially independent and hence randomly distributed, which is why the Moran's *I* value is low, but locally they show non-random distributions. In addition, the LDNA bacterial subpopulation at St Kilda in microplate 3 showed non-random distributions in the Moran's I testing however it had a higher Geary's C value than the other non-randomly distributed populations (0.92 compared to 0.73–0.83) which was more similar to the Geary's C values of the randomly distributed subpopulations that showed local association of 0.85–0.89. According to Sokal [57] this suggests that this subpopulation may be randomly distributed as the non-random result from the Moran's I was skewed by extreme values within the data set. This again highlights the importance of using Moran's I in conjunction with Geary's C as globally bacterial subpopulations may be spatially independent however locally they may be spatially dependent. As Geary's C is able to account for extreme outliers that may skew the overall result it is important to use in microbial distribution studies which have shown increases of up to 45 and 2500 times $cm^{-1}$ in bacteria and viruses.

### 'Mean field' approach limitations

The microscale spatial heterogeneity of bacterial and viral populations in this study would be underestimated, by more than 3 orders of magnitude in some cases, had the mean field approach been employed [75–77]. This is due to the presence of single point hotspots and coldspots containing more than 30% and less than 0.0008% of the total biomass within each microplate. As the distributions were random at Noarlunga but non-random at St Kilda the mean would not be representative of microbial biomass.

## Conclusion

Spatial heterogeneity in microbial abundance has been found within the 11.3 cm directly above the sediment-water interface at Noarlunga estuary and St Kilda mangroves. This heterogeneity was present in the form of single and multiple point hotspots at Noarlunga and hotspots and surface gradients at St Kilda which resulted in high levels of viral and bacterial abundance being found near the sediment-water interface. These patterns in microbial distributions have been seen in previous studies but to a lesser extent. At St Kilda, gradients in bacterial and viral abundance were much higher than previously reported with a 45- and 10.5-fold change $cm^{-1}$ found compared to the previously reported 1.5-fold change in abundance $cm^{-1}$ in bacteria [35] and 1.4-fold change $cm^{-1}$ in viruses [15, 35]. At Noarlunga, bacterial and viral abundance showed a 12.9– and 2584-fold change $cm^{-1}$. This large fold-change found $cm^{-1}$ would have been missed if lower resolution sampling was used. This high level of microscale heterogeneity highlights the importance of analysing microbial spatial distribution on the level of what individual cells experience rather than using a 'mean field' approach. In addition, this high level of microscale heterogeneity also illustrates the necessity and applicability of the novel two-dimensional high resolution sampler used in this study which was capable of collecting eight vertical profiles. The use of this sampler allowed the discrimination of single and multiple point hotspots and/or surface gradients, which was not possible in past studies as one-dimensional samplers were used.

Bacterial subpopulations were found to be randomly distributed at Noarlunga and non-randomly distributed at St Kilda with reasons for this finding relating to the low energy-nature of St Kilda allowing the formation of sediment-water interface nutrient and microbial surface gradients. Viral subpopulations only showed random distributions at both sites. Moran's *I* and Geary's C reveal that spatial autocorrelation amongst bacterial subpopulations at St Kilda show a positive to negative trend 5.9 cm from the sediment-water interface, indicating a mechanism, whether this be sampling alias or particulate resuspension, is present leading to this clustering and dispersal of bacterial populations.

Despite hotspots being ubiquitous features of microbial distributions, past studies lacked a universal definition. They qualitatively classified patches as regions of high microbial abundance that 'exceed' or are 'elevated above' background values [12,13,43]. This paper provides a new quantitative method for determining hotspots, coldspots and background values within microscale microbial distributions. It is hoped that this method will be used in future microbial microscale distributions to allow comparative analyses among studies.

## Supporting Information

**Figure S1 Collection of Vertical Profiles.** The use of an 8×12 96-well microplate allowed the two-dimensional array of 8 vertical profiles (row A, B, C, D, E, F, G and H) per microplate which consisted of 12 sampling wells per vertical profile.

**Figure S2 Comparisons of bacterial and viral subpopulations and viral subpopulations at Noarlunga via vertical depth profiles.** Bacterial and viral subpopulations of all three microplates (n = 270). **A** Total mean LDNA, HDNA 1 and HDNA 2; **B** Total mean VLP 1 and VLP 2. Error bars represent the 95% confidence intervals obtained from each subpopulation of all three replicates (n = 12).

**Figure S3 Mean vertical profile of VLP 1 and LDNA, microplate one at Noarlunga (n = 90).** Error bars represent the 95% confidence intervals obtained from one replicate (n = 12).

**Figure S4 Adjacent orthogonal gradient distributions of bacteria and viruses as rank abundance. St Kilda: A** VLP 1, **B** HDNA 2. **Noarlunga: C** VLP 2, **D** HDNA 2. Each distribution is ordered as a rank abundance. In each case the first rank was used for the 2 dimensional plots. A logarithmic trend-line was the best fit for each distribution, with the equations and $r^2$ for **A** $y = -4\times10^6 \ln(x)+2\times10^7$ ($R^2 = 0.97$), **B** $y = -7\times10^4 \ln(x)+4\times10^6$ ($R^2 = 0.99$), **C** $y = -2\times10^6 \ln(x)+8\times10^6$ ($R^2 = 0.98$), **D** $y = -8\times10^4 \ln(x)+4\times10^5$ ($R^2 = 0.92$). The p values are $<<0.05$ in all cases.

**Table S1 Mean viral abundances per microplate at Noarlunga and St Kilda.**

**Table S2 Mean bacterial abundances per microplate at Noarlunga and St Kilda.**

**Table S3 Comparison of the Moran's I values and Geary's C values obtained for each bacterial subpopulation at Noarlunga and St Kilda.**

**Table S4 Comparison of the Moran's I values and Geary's C values obtained for each viral subpopulation at Noarlunga and St Kilda.**

## Acknowledgments

We would like to extend our gratitude to Dr P MacArdle and Dr S. Bailey from the Flow Cytometry Unit of the Flinders Medical Centre, South Australia. We also thank the wildlife rangers of the Salisbury Council, South Australia for allowing access to the sampling site on the St Kilda mangrove trail.

## Author Contributions

Conceived and designed the experiments: LMD JGM PGS KN TJ JP. Performed the experiments: LMD KN JP. Analyzed the data: LMD JGM PGS KN JP. Contributed reagents/materials/analysis tools: LMD KN JP TJ. Wrote the paper: LMD JGM. Assistance with sample collection: TJ.

## References

1. Azam F, Malfatti F (2007) Microbial structuring of marine ecosystems. Nature 5: 782–791.
2. Cotner JB, Biddanda BA (2002) Small players, large Role: microbial influence on biogeochemical processes in pelagic aquatic ecosystems. Ecosystems 5: 105–121.
3. Elser JJ, Stabler BL, Hassett PR (1995) Nutrient limitation of bacterial growth and rates of bacterivory in lakes and oceans: a comparative study. Aquatic Microbial Ecology 9: 105–110.
4. Gasol JM, Del Giorgio PA (2000) Using flow cytometry for counting natural planktonic bacteria and understanding the structure of planktonic bacterial communities. Scientia Marina 64: 197–224.
5. Pace N (1997) A molecular view of microbial diversity and the biosphere. Science 276: 734–740.
6. Pontes D, Lima-Bittencourt C, Chartone-Souza E, Nascimento A (2007) Molecular approaches: advantages and artifacts in assessing bacterial diversity. Journal of industrial Microbiology & Biotechnology 34: 463–473.
7. Rodriguez-Brito B, Li L, Wegley L, Furlan M, Angly F, et al. (2010), Viral and microbial community dynamics in four aquatic environments. International Society for Microbial Ecology (ISME), 4: 739–751.
8. Venter J, Remington K, Heidelberg J, Halpern A, Rusch D, et al. (2004) Environmental genome shotgun sequencing of the Sargasso Sea. Science 304: 66–74.
9. Azam F (1998) Microbial control of oceanic carbon flux: the plot thickens. Science 280: 694–696.
10. Long RA, Azam F (2001) Microscale patchiness of bacterioplankton assemblage richness in seawater. Aquatic Microbial Ecology 26: 103–113.
11. Mitchell JG, Okubo A, Fuhrman JA (1985) Microzones surrounding phytoplankton form the basis for a stratified marine microbial ecosystem. Nature 316: 58–59.
12. Seymour JR, Mitchell JG, Seuront L (2004) Microscale heterogeneity in the activity of coastal bacterioplankton communities. Aquatic Microbial Ecology 35: 1–16.
13. Seymour JR, Seuront L, Mitchell JG (2005a) Microscale and small-scale temporal dynamics of a coastal planktonic microbial community. Marine Ecology Progress Series 300: 21–37.
14. Seymour JR, Patten N, Bourne DG, Mitchell JG (2005b) Spatial dynamics of virus-like particles and heterotrophic bacteria within a shallow coral reef system. Marine Ecology Progress Series 288: 1–8.
15. Seymour JR, Seuront L, Doubell M, Waters RL, Mitchell JG (2006) Microscale patchiness of virioplankton. Journal of the Marine Biological Association of the United Kingdom 86: 551–561.
16. Suttle CA (2007) Marine viruses – major players in the global ecosystem. Nature Reviews 5: 801–812.
17. Waters RL, Mitchell JG, Seymour J (2003) Geostatistical characterisation of centimetre-scale spatial structure of *in vivo* fluorescence. Marine Ecology Progress Series 251: 49–58.
18. Angly FE, Felts B, Breitbart M, Salamon P, Edwards RA, et al. (2006) The marine viromes of four oceanic regions. PLoS Biology 4: 2121–2131.
19. Corinaldesi C, Crevatin E, Del Negro P, Marini M, Russo A, et al. (2003) Large-scale spatial distribution of virioplankton in the Adriatic Sea: testing the trophic state control hypothesis. Applied and Environmental Microbiology 69: 2664–2673.
20. Hillmer I, Reenen PV, Imberger J, Zohary T (2008) Phytoplankton patchiness and their role in the modelled productivity of a large, seasonally stratified lake. Ecological Modelling 218: 49–59.
21. Liu H, Imai K, Suzuki K, Nojiri Y, Tsurushima N, et al. (2002) Seasonal variability of picophytoplankton and bacteria in the western subarctic Pacific Ocean at station KNOT. Deep Sea Research Part II: Topical Studies in Oceanography 49: 24–25.
22. Pan LA, Zhang LH, Zhang J, Gasol JM, Chao M (2005) On-board flow cytometric observation of picoplankton community structure in the East China Sea during the fall of different years. FEMS Microbiology Ecology 52: 243–253.
23. Pan LA, Zhang J, Zhang LH (2007) Picophytoplankton, nanophytoplankton, heterotrophic bacteria and viruses in the Changjiang Estuary and adjacent coastal waters. Journal of Plankton Research 29: 187–197.
24. Strutton PJ, Mitchell JG, Parslow JS, Greene RM (1997) Phytoplankton patchiness: quantifying the biological contribution using Fast Repetition Rate Fluorometry. Journal of Plankton Research 19: 1265–1274.
25. Curtis TP, Head IM, Lunn M, Woodcock S, Schloss PD, et al. (2006) What is the extent of prokaryotic diversity? Philosophical Transactions of the Royal Society 361: 2023–2037.
26. Grunbaum D (2002) Predicting availability to consumers of spatially and temporally variable resources. Hydrobiologia 480: 175–191.
27. Karl DM, Dore JE (2001) Microbial ecology at sea: sampling, subsampling and incubation considerations. Methods in Microbiology 30: 13–39.
28. Waters RL, Mitchell JG (2002) Centimetre-scale spatial structure of estuarine in vivo fluorescence profiles. Marine Ecology Progress Series 237: 51–63.
29. Genin A, Greene C, Haury L, Wiebe P, Gal G, et al. (1994) Zooplankton patch dynamics: daily gap formation over abrupt topography. Deep Sea Research Part I: Oceanographic Research Papers 41 (5–6): 941–951.
30. Pickett S, Cadenasso M (1995) Landscape ecology: Spatial heterogeneity in ecological systems. Science 269 (5222): 331–334.
31. Levin S, Whitfield M (1994) Patchiness in marine and terrestrial systems: from individuals to populations. Phil. Trans R. Soc 343 (1303): 99–103.
32. Grünbaum D (2012) The logic of ecological patchiness. Interface Focus 2: 150–155.
33. Brentnall SJ, Richards KJ, Brindley J, Murphy E (2003) Plankton patchiness and its effect on larger-scale productivity. Journal of Plankton Research 25: 121–140.
34. Seymour JR, Mitchell JG, Pearson L, Waters R (2000) Heterogeneity in bacterioplankton abundance from 4.5 millimetre resolution sampling. Aquatic Microbial Ecology 22: 143–153.
35. Seymour JR, Seuront L, Mitchell JG (2007b) Microscale gradients of planktonic microbial communities above the sediment surface in a mangrove estuary. Estuarine, Coastal and Shelf Sciences 73: 651–666.
36. Mitchell JG (2002) The energetic and scaling of search strategies in bacteria. American Naturalist 160: 727–740.

37. Barbara G, Mitchell JG (2003) Marine bacterial organisation around point-like sources of amino acids FEMS Microbiology Ecology 43: 99–109.
38. Blackburn N, Fenchel T, Mitchell JG (1998) Microscale nutrient patches in planktonic habitats shown by chemotactic bacteria. Science 282: 2254–2256.
39. Seymour JR, Ahmed T, Durham WM, Stocker R (2010) Chemotactic response of marine bacteria to the extracellular products of Synechococcus and Prochlorococcus. Aquatic Microbial Ecology 59: 161–168.
40. Stocker R, Seymour JR, Samadani A, Hunt DE, Polz MF (2008) Rapid chemotactic response enables marine bacteria to exploit ephemeral microscale nutrient patches. Proceedings of the National Academy of Sciences of the United States of America (PNAS) 105: 4209–4214.
41. Stocker R (2012) Marine microbes see a sea of gradients. Science Review 338: 628–633.
42. Taylor JR, Stocker R (2012) Trade-offs of chemotactic foraging in turbulent water. Science 338: 675–679.
43. Seymour JR, Seuront L, Doubell M, Mitchell JG (2008) Mesoscale and microscale spatial variability of bacteria and viruses during a Phaeocystisglobosa bloom in the Eastern English Channel. Estuarine, Coastal and Shelf Science 80: 589–597.
44. Seymour JR, Humphreys WF, Mitchell JG (2007a) Stratification of the microbial community inhabiting an anchialine sinkhole. Aquatic Microbial Ecology 50: 11–24.
45. Seuront L, Lacheze C, Doubell MJ, Seymour JR, Van Dongen-Vogels V, et al. (2007) The influence of Phaeocystisglobosa on microscale spatial patterns of chlorophyll a and bulk-phase seawater viscosity, Biogeochemistry 83: 173–188.
46. Barbara G, Mitchell JG (1996) Microlaminations of motile bacteria in microbial mats. Applied and Environmental Microbiology 62: 3985–3990.
47. Brussaard C (2004) Optimisation of procedures for counting viruses by flow cytometry. Applied and Environmental Microbiology 70: 1506–1513.
48. Marie D, Brussaard C, Thyrhaug R, Bratbak G, Vauolt D (1999) Enumeration of marine viruses in culture and natural samples by flow cytometry. Applied and Environmental Microbiology 65: 45–52.
49. Brussaard C, Marie D, Bratbak G (2000) Flow cytometric detection of viruses. Journal of Virological Methods 85 (1–2): 175–182.
50. Schapira M, Buscot MJ, Leterme SC, Pollet T, Chapperon C, et al. (2009) Distribution of heterotrophic bacteria and virus-like particles along a salinity gradient in a hypersaline coastal lagoon. Aquatic Microbial Ecology 54: 171–183.
51. Marie D, Partensky F, Jacquet S, Vaulot D (1997) Enumeration and cell cycle analysis of natural populations of marine picoplankton by flow cytometry using the nucleic acid stain SYBR Green I. Applied and Environmental Microbiology 63: 186–193.
52. Wiebe PH (1970) Small-scale spatial distribution in oceanic zooplankton. American Society of Limnology and Oceanography 15: 205–217.
53. Holm S (1979) A simple sequentially rejective multiple test procedure. Scandinavian Journal of Statistics 6: 65–70.
54. Moran PAP (1950) Notes on continuous stochastic phenomena. Biometrika 37: 17–23.
55. Rossi RE, Mulla DJ, Journel AG, Franz EH (1992) Geostatistical tools for modelling and interpreting ecological spatial dependence. Ecological Monographs 62: 277–314.
56. Geary RC (1954) The contiguity ratio and statistical mapping. The Incorporated Statistician 5: 115–145.
57. Sokal RR (1978) Spatial autocorrelation in biology: 1. Methodology. Biological Journal of the Linnean Society 10: 199–228.
58. Griffith DA (1987) Spatial autocorrelation: a primer. Resource publications in Geography. Association of American Geographers, Washington, D.C.
59. Baric A, Kuspilic G, Matijevic S (2002) Nutrient (N, P, Si) fluxes between marine sediments and water column in coastal and open Adriatic. Hydrobiologia 475/476: 151–159.
60. Weinbauer MG, Suttle CA (1999) Lysogeny and prophage induction in coastal and offshore bacterial communities. Aquatic Microbial Ecology 18: 217–225.
61. Mitchell JG (2004) Rank-size analysis and vertical phytoplankton distribution patterns. In: Handbook of Scaling Methods in Aquatic Ecology: Measurement, Analysis, Simulation (L. Seuront, P.G. Strutton eds) CRC, Boca Raton, 257–278.
62. Wang Y, Hammes F, Boon N, Chami M, Egli T (2009) Isolation and characterisation of low nucleic acid (LNA)-content bacteria. The ISME Journal 3: 889–902.
63. Bouvy M, Troussellier M, Got P, Arfi R (2004) Bacterioplankton responses to bottom-up and top-down controls in a West African reservoir (Selingue, Mali). Aquatic Microbial Ecology 34: 301–307.
64. Gasol JM, Zweifel UL, Peters F, Fuhrman JA, Hagstrom A (1999) Significance of size and nucleic acid content heterogeneity as measured by flow cytometry in natural planktonic bacteria. Applied and Environmental Microbiology 65: 4475–4483.
65. Lebaron P, Servais P, Agogue H, Courties C, Joux F (2001) Does the high nucleic acid content of individual bacterial cells allow us to discriminate between active cells and inactive cells in aquatic systems. Applied and Environmental Microbiology 67: 1775–1782.
66. Lebaron P, Servais P, Baudoux A, Bourrain M, Courties C, et al. (2002) Variations of bacterial-specific activity with cell size and nucleic acid content assessed by flow cytometry. Aquatic Microbial Ecology 28: 131–140.
67. Bergh O, Borsheim KY, Bratbak G, Heldal M (1989) High abundance of viruses found in aquatic environments. Nature 340: 467–468.
68. Thar R, Kühl M (2003) Bacteria are not too small for spatial sensing of chemical gradients: An experimental evidence. Proceedings of the National Academy of Sciences of the United States of America (PNAS) 100: 5748–5753.
69. Boegman L, Ivey GN (2009) Flow separation and resuspension beneath shoaling nonlinear internal waves. Journal of Geophysical Research. 114: 1–15.
70. Braaten DA, Paw KT, Shaw RH (1990) Particle resuspension in a turbulent boundary layer – observed and modelled. Journal of Aerosol Science 21: 613–628.
71. Nicholson KW (1988) A review of particle resuspension. Atmospheric Environment 22: 2639–2651.
72. Kester W (2009) What the Nyquist criterion means to your sampled data system design. Analog Devices 1–12.
73. Rutgers van der Loeff MM, Meyer R, Rudels B, Rachor E (2002) Resuspension and particle transport in the benthic nepheloid layer in and near Fram Strait in relation to faunal abundances and Th depletion. Deep-Sea Research 49: 1941–1958.
74. Sandilands RG, Mudroch A (1983) Nepheloid layer in Lake Ontario. Journal of Great Lakes Research 9: 190–200.
75. Alonso MC, Jimenez-Gomez F, Rodriguez J, Borrego JJ (2001) Distribution of virus-like particles in an oligotrophic marine environment (Alboran Sea, Western Mediterranean). Microbial Ecology 42: 407–415.
76. Drake LA, Choi KH, Haskell AGE, Dobbs FC (1998) Vertical profiles of virus-like particles and bacteria in the water column and sediments of Chesapeake Bay, USA. Aquatic Microbial Ecology 16: 17–25.
77. Thomson PG, Davidson AT, Enden R, Pearce I, Seuront L, et al. (2010) Distribution and abundance of marine microbes in the Southern Ocean between 30 and 80°E. Deep-Sea Research II 57: 815–827.

# Spatial Genetic Structure Patterns of Phenotype-Limited and Boundary-Limited Expanding Populations

**Qiang Dai[1,3]*, Xiangjiang Zhan[2], Bin Lu[1], Jinzhong Fu[1,4], Qian Wang[1], Dunwu Qi[3]**

**1** Chengdu Institute of Biology, Chinese Academy of Sciences, Chengdu, China, **2** Organisms and Environment Division, Cardiff School of Biosciences, Cardiff University, Cardiff, United Kingdom, **3** The Key Laboratory for Conservation Biology of Endangered Wildlife, Sichuan Province, Chengdu Research Base of Giant Panda Breeding, Chengdu, Sichuan, China, **4** Department of Integrative Biology, University of Guelph, Guelph, Ontario, Canada

## Abstract

Range expansions may create a unique spatial genetic pattern characterized by alternate genetically homogeneous domains and allele frequency clines. Previous attempts to model range expansions have mainly focused on the loss of genetic diversity during expansions. Using individual-based models, we examined spatial genetic patterns under two expansion scenarios, boundary-limited range expansions (BLRE) and phenotype-limited range expansions (PhLRE). Our simulation revealed that the genetic diversity within populations lost quickly during the range expansion, while the genetic difference accumulated between populations. Consequently, accompanying the expansions, the overall diversity featured a slow decrease. Specifically, during BLREs, high speed of boundary motion facilitated the maintenance of total genetic diversity and sharpened genetic clines. Very slight constraints on boundary motion of BLREs drastically narrowed the homogeneous domains and increased the allele frequency fluctuations from those levels exhibited by PhLREs. Even stronger constraints, however, surprisingly brought the width of homogeneous domains and the allele frequency fluctuations back to the normal levels of PhLREs. Furthermore, high migration rates maintained a higher total genetic diversity than low ones did during PhLREs. Whereas, the total genetic diversities during BLREs showed a contrary pattern: higher when migration was low than those when migration was high. Besides, the increase of migration rates helped maintain a greater number of homogeneous domains during PhLREs, but their effects on the number of homogeneous domains during BLREs were not monotonous. Previous studies have showed that the homogenous domains can merge to form a few broad domains as the expansion went on, leading to fewer homogeneous domains. Our simulations, meanwhile, revealed that the range expansions could also rebuild homogeneous domains from the clines during the range expansion. It is possible that that the number of homogeneous domains was determined by the interaction of merging and newly emerging homogeneous domains.

**Editor:** Norman Johnson, University of Massachusetts, United States of America

**Funding:** This project is supported by a National Natural Science Foundation of China grant and a National Sciences and Engineering Research Council Discovery grant to JF and Leading Foundation of the Chinese Academy of Sciences (XDA05050404) to QW. The funders had no role in study design, data collection and analysis, decision to publish, or preparation of the manuscript.

**Competing Interests:** The authors have declared that no competing interests exist.

* E-mail: daiqiang@cib.ac.cn

## Introduction

Species range expansion is an important biological process, which has shaped global biodiversity patterns. Species expand their ranges from natal sites at some point in their evolutionary history [1]. Global climatic changes have resulted in great changes in species distributions. Many Species have recolonized and/or expanded their ranges after the glaciation retreated [2,3] or during the current rapid climate change caused by global warming [4,5]. Long-distance introductions of alien species to new suitable areas produces another kind of species range expansions and are often harmful to native ecosystems [6]. Evolution in species' traits can also cause habitat adaptability change, and thus triggering a new round of range expansion [1].

The speed of expansions may vary considerably so that Nullmeier and Hallatschek [7] suggested two types of expansion, boundary-limited range expansion and phenotype-limited range expansion, which determine the speed of expansion by different mechanisms. A BLRE is typically a slow process, which happens when species expand their ranges along with the emergence of new, accessible, unoccupied, suitable habitat outside their current distribution areas. The rate at which suitable habitats become available is constrained by environmental changes over time. A classic example of the BLRE is that species expand their ranges from low-latitude to high-latitude regions in response to global warming [8,9]. In addition to environmental changes, population expansion may also be limited by the Allee effect, characterized by reductions in the population growth in small or sparse populations [10] and slower rate of range expansions [11,12]. In PhLREs, however, the speed of expansion is determined primarily by population's intrinsic characteristics, e.g. reproduction rate, migration rate and distance. When PhLREs happen, a species can claim a vast unoccupied suitable habitat once it becomes accessible when the Allee effect is either weak or absent. The successful invasion of an invasive species into a new continent likely represents a PhLRE scenario [13–15]. Moreover, a PhLRE may also occur when a new beneficial mutation expands a species' adaptive zone, hence changing the definition of its suitable habitat [16].

Several previous studies have investigated the spatial genetic consequences of PhLREs. In general, a species' range expansion is often accompanied by loss of genetic diversity because of the founder effect [2,17–19]. However, this loss is not uniform, which recently has caught much attention. For example, both petri dish experiments using microbial species [20,21] and model simulations [11,20–22] revealed a pattern characterized by the emergence of genetically homogeneous domains from genetically mixed populations, which then merged during expansion. Hallatschek and Nelson [11] suggested that this pattern was driven by strong genetic drift at the fronts of expanding populations. Meanwhile, allele frequency gradients were formed between genetically homogeneous domains as a consequence of migration leading to allele frequency clines orthogonal to the expansion direction [23].

The spatial genetic consequences of BLREs, although as important as those of PhLREs, have rarely been investigated, with the exception of simulations carried out by Nullmeier and Hallatschek [7], in which BLREs were found to maintain a higher level of genetic diversity than PhLREs were. Hallatschek and Nelson [11] and Roques et al. [12] also studied the genetic consequences of BLRE-like expansions caused by the Allee effect, and concluded that it also increased genetic diversity at expansion fronts.

In this study, we aim to explore different spatial patterns of genetic variations, specifically the number of genetically homogeneous domains and the fluctuations of allele frequencies, at the expansion front generated by the PhLREs and BLREs with varied speeds of boundary motion. We simulated the population genetic dynamics of neutrally evolved alleles at the advancing fronts using individual-based models (IBM). The impacts of migration and reproduction on the spatial genetic patterns were also examined.

## Methods

### (a) Model description

The simulations were carried out using a lattice with 75 row ×200 column grids. The populations comprised asexual, haploid individuals with discrete generations. Each individual genotype was typed by one neutral diallelic locus. We defined three population parameters: carrying capacity ($K$), reproduction rate ($r$) and migration rate ($m$). As noted by Excoffier and Ray (2008), small populations tend to produce less defined homogeneous domains and broader transition zones, therefore we set $K$ to 100 in our simulations, which was much greater than that used in Korolev et al.'s (2011) simulations ($K = 30$), and was within the range suggested by Excoffier and Ray's (2008) simulations. For reproduction, the number of offspring of each individual was drawn from independent Poisson distributions with means of $r = 1.1, 2, 3, 5$. Individuals were assumed to die immediately after reproduction. The migration rate $m$ were set to 0.01, 0.1, 0.2, and 0.4, which lied within the range employed in the previous simulations [23,24]. To examine the effect of reproduction rate, varied $r$ values ($r = 1.1, 2, 3, 5$) were tested with a constant $m$ of 0.2. To test the effect of migration rate, different $m$ values ($m = 0.01, 0.1, 0.2, 0.4$) were used with a constant $r$ of 3. The individuals migrated with a probability of $m/n$ to any of the immediately adjacent $n$ demes. The value of $n$ varied from 2 to 4, depending on the position of each deme with respect to the lattice boundaries.

Each simulation started with populations in the first 20 deme columns at the far left of the lattice; all the those populations had size of $K$ and allele frequency of 0.5 for each allele. We also simulated with initial allele frequencies of 0.01, 0.1 and 0.2 with a constant $m$ of 0.2 and $r$ of 3 to test the impact of initial frequency.

As the simulations progressed, the species range expanded from left to right of the lattice. The order of events in each generation was reproduction, migration, and finally random removal of individuals in demes where population size exceeded $K$. Each simulation was run for 8,000 generations. When a column of previously unoccupied grids reached $K$, the genetic components of those demes were recorded as a frozen record.

### (b) Boundary-limited and phenotype-limited range expansion models

Two range expansion scenarios, BLRE and PhLRE, were simulated. During PhLRE, all grids were designated as suitable habitats and range expansion occurred immediately after the simulation was initialized. During BLRE scenarios, only the first 20 columns were designated as suitable habitats at the initialization, while the others were all unsuitable habitat. The individuals can not reproduce ($r = 0$) in the unsuitable habitat. The unsuitable grids gradually turned into suitable, column by column, with the progression of generations. This process was used to imitate habitat expansion driven by environmental change. We employed six different speeds of boundary motion in BLREs, taking 600, 1000, 2000, 4000, 6000 and 7900 generations for all of the grids to change into suitable habitats. Other parameters were set as follows: $r = 3$, $m = 0.2$, and the initial allele frequency $= 0.5$. Our simulation pilots showed that it took a mean of 533.6 generations for populations to expand from left to right and all populations reach $K$ during a PhLRE, given $r = 3$ and $m = 0.2$. Therefore, 600 generations seemed to be a slight constraint on expanding populations. Additionally, the populations that do not expand make the extreme BLRE, which is usually referred as a process of isolation by distance [25–27]. We also carried out simulations with stationary habitats for non-expanded populations: only the first 20 columns were designated as suitable habitats, and the boundary was assumed to have never moved. The populations in the 20th column were considered as a "stationary front" and were compared with populations at expansion fronts during BLREs and PhLREs.

### (c) Model analysis

We counted the homogeneous domains and clines and calculated their mean widths at expansion fronts. A genetically homogeneous domain was defined as one or multiple spatially continuous homogeneous populations, and an allele frequency cline was defined as one or multiple spatially continuous genetically mixed populations.

Subtle allele frequency fluctuations orthogonal to the expansion direction were analyzed by Fast Fourier Transform (FFT) algorithm, which has been widely used in diverse fields of science to describe periodic patterns. The allele frequencies of a column of demes were considered as a wave curve $p$, and the mean frequency of the power spectrum ($\bar{f}$) was calculated:

$$\bar{f} = \frac{\sum_{k=1}^{\frac{N}{2}-1} f_k |a_k(p)|^2}{\sum_{k=1}^{\frac{N}{2}-1} |a_k(p)|^2} \tag{1}$$

Where $a_k(p)$ is the FFT of curve $p$ at index $k$, and $f_k$ is the frequency corresponding to $k$. The term of $|a_k(p)|$ is the amplitude at $f_k$. $N$ is the number of demes in a column, and the highest frequency can be identified as $\frac{N}{2}-1$ since $N$ is odd in our simulations. A higher $\bar{f}$ indicates more fluctuation of allele frequency. More homogeneous domains will also increase the $\bar{f}$.

Genetic differentiation within and among populations at an expansion front was assessed by the average number of pairwise differences within populations (PX), the average number of pairwise differences between populations (PXY), and the corrected average pairwise difference between populations (Pc) [28,29]. Pc was calculated as

$$\mathrm{Pc} = \mathrm{PXY} - \frac{\mathrm{PX}+\mathrm{PY}}{2} \tag{2}$$

Each parameter combination was simulated for 100 times. The counts and width of genetically homogeneous domain and clines,

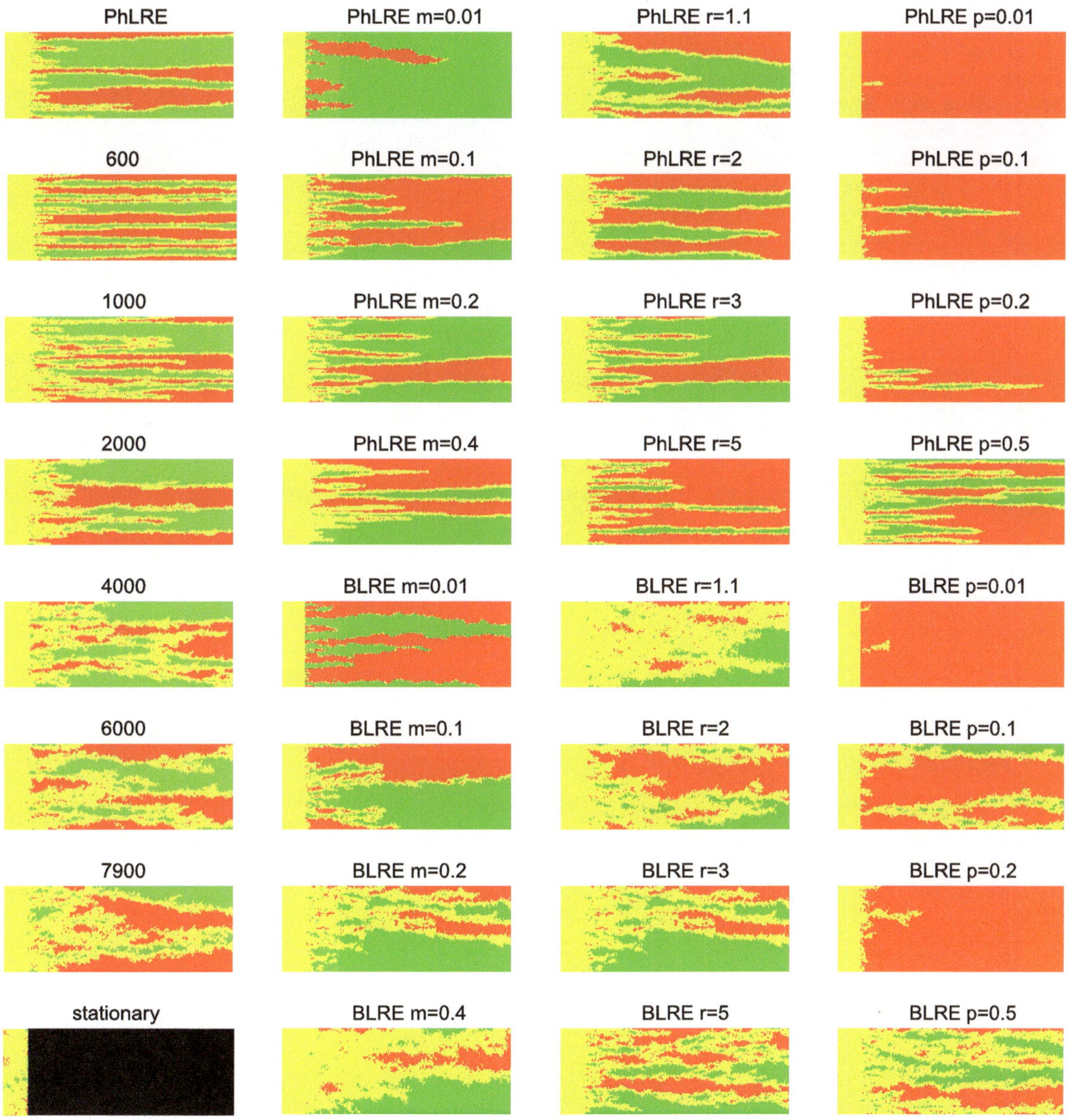


**Figure 1. Spatial distribution of allele frequencies in the expanding populations.** The allele frequencies were recorded when all populations in a column first reached their carrying capacity. The first column showed expansions with varying speeds of boundary motion (the number of generations to complete a motion), while the other parameters remained constant ($r=3$, $m=0.2$, initial frequency $=0.5$) in the first column. The 'stationary' referred to the non-expansion simulations carried out in stationary habitats, and the snapshots of genetic compositions of populations were recorded at generation 8000. The second to fourth columns showed expansions with varying migration rates, reproduction rates and initial frequencies, respectively. Column 2: $r=3$, time to complete a boundary motion was 6000 generations, initial frequency $=0.5$; Column 3: $m=0.2$, time to complete a boundary motion was 6000 generations, initial frequency $=0.5$; Column 4: $r=3$, $m=0.2$, time to complete a boundary motion was also set to 6000 generations. Carrying capacity was 100 in all the simulations. Green indicated an allele frequency of 0, red an frequency of 1, yellow an intermediate frequency, and black denoted unoccupied grids.

$\bar{f}$, and pairwise differences were computed for each simulation, and then averaged for each parameter combination. This modeling was implemented in Java, and all the other analyses were performed in R 2.15.2 [30].

## Results

### (a) Genetically homogenous domains and clines

For both the PhLRE and BLRE, many narrow homogeneous domains emerged very quickly at the beginning of the expansions (Fig. 1). Correspondingly, the number of both genetically homogeneous domains and clines increased suddenly. The homogeneous domain expanded at this stage, coupled with the quick narrowing of clines (Fig. 2). As the expansion went on, the number of homogeneous domains and clines decreased while some narrow homogenous domains merged to form a few broad domains, although the average width of the clines remained stable after the initial narrowing. However, the subtle allele frequency fluctuations $\bar{f}$ increased much earlier than number of homogeneous domains did, and kept decreasing during the expansion.

The PhLREs exhibited the narrowest clines, however slow boundary motion of BLREs broadened the clines (Fig. 2). In the non-expanded populations, genetically homogeneous domains and clines were formed in the stationary habitats, possibly due to the effect of isolation by distance [25–27]. As shown by Figure 2, the width of clines at the 20th column after 8000 generations was much wider than those at expansion fronts during BLREs or PhLREs. Additionally, the more slowly the boundary expanded, the broader the homogeneous domains were. Similarly, the slower the boundary expansions were, the higher $\bar{f}$ values. Altough PhLREs can be considered as the fastest boundary motion and the non-expansions as the slowest, the width of homogeneous domains was very broad in the former, but narrowest in the latter. Besides, the $\bar{f}$ values were the lowest during PhLREs and the highest during non-expansion process.

High migration rates broadened the clines but narrowed the homogeneous domains (Fig. 3). It also increased the $\bar{f}$, especially during PhLREs. In contrast, low migration rates increased the number of homogeneous domains more sharply than did high migrations rates at the early stage of the expansion. In the end, the higher migration rates helped maintain a greater number of homogeneous domains during PhLREs than lower ones did. However, the effect of increasing migration rate on the domain number was not monotonous during BLREs. The high reproduction rate maintained high number of homogeneous domains and clines, narrowed the clines, but broadened the homogeneous domains at least during PhLREs. The high reproduction rate also increased the $\bar{f}$, especially during PhLREs (Fig. 4).

Initial allele frequencies deviated from 0.5 accelerated the fixation and led to fewer but broader homogeneous domains and narrower clines (Fig. 5), because populations with skewed allele frequencies were more likely to lose rare alleles during the random drift and facilitated the emergence and merging of homogeneous domains. The initial skewed allele frequencies also decreased the $\bar{f}$.

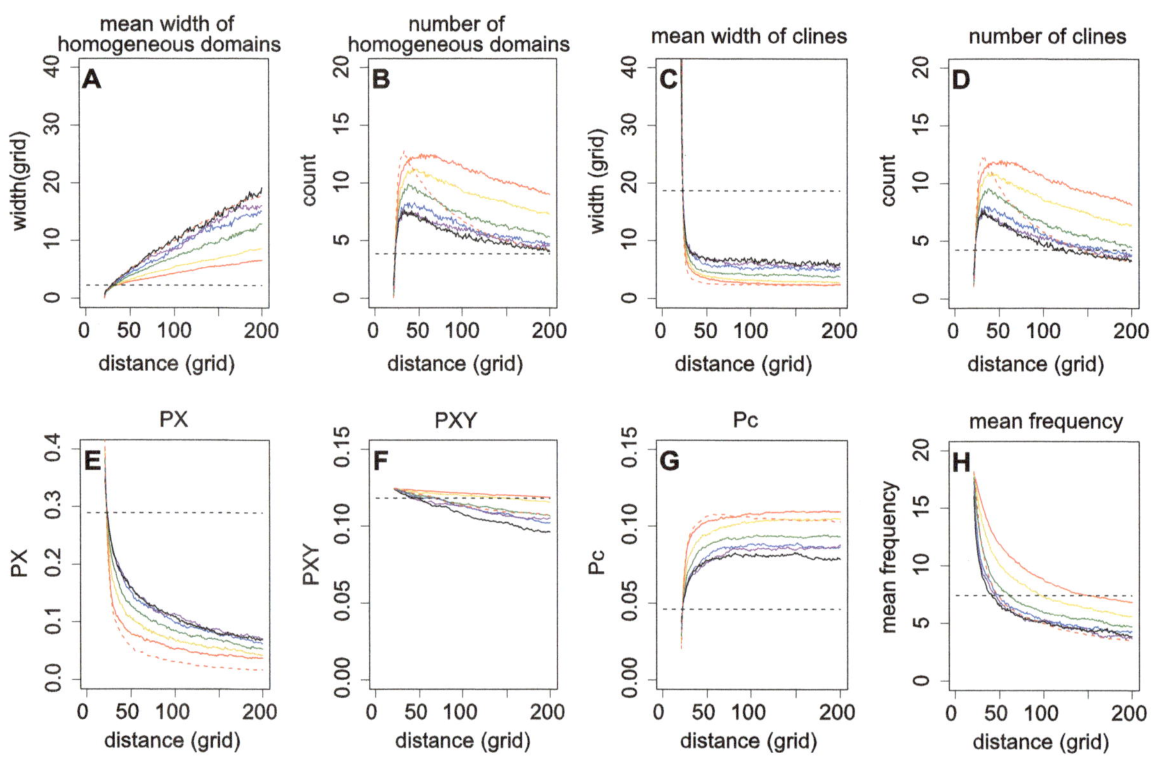


**Figure 2. The spatial patterns of genetic diversity at expansion fronts given various boundary motion speeds.** The spatial patterns of genetic diversity were described by the width and number of genetically homogeneous domains (A, B) and genetic clines (C, D), pairwise differences in PX (E), PXY (F) and Pc (G), and the mean frequency of the power spectrum (H) at expansion fronts during PhLREs (red dashed lines) and BLREs (solid lines). The speeds of boundary motion, indicated by generation to complete the boundary motion, varied from 800 to 7900 generations (red: 800, yellow: 1000, green: 2000, blue: 4000, purple: 6000, dark: 7900). The black dashed lines described the spatial genetic patterns at a distance of 20 grids and 8000 generations, which were generated in the non-expansion simulations. Other parameters were set as follows: $r = 3$, $m = 0.2$, and the initial allele frequency $= 0.5$.

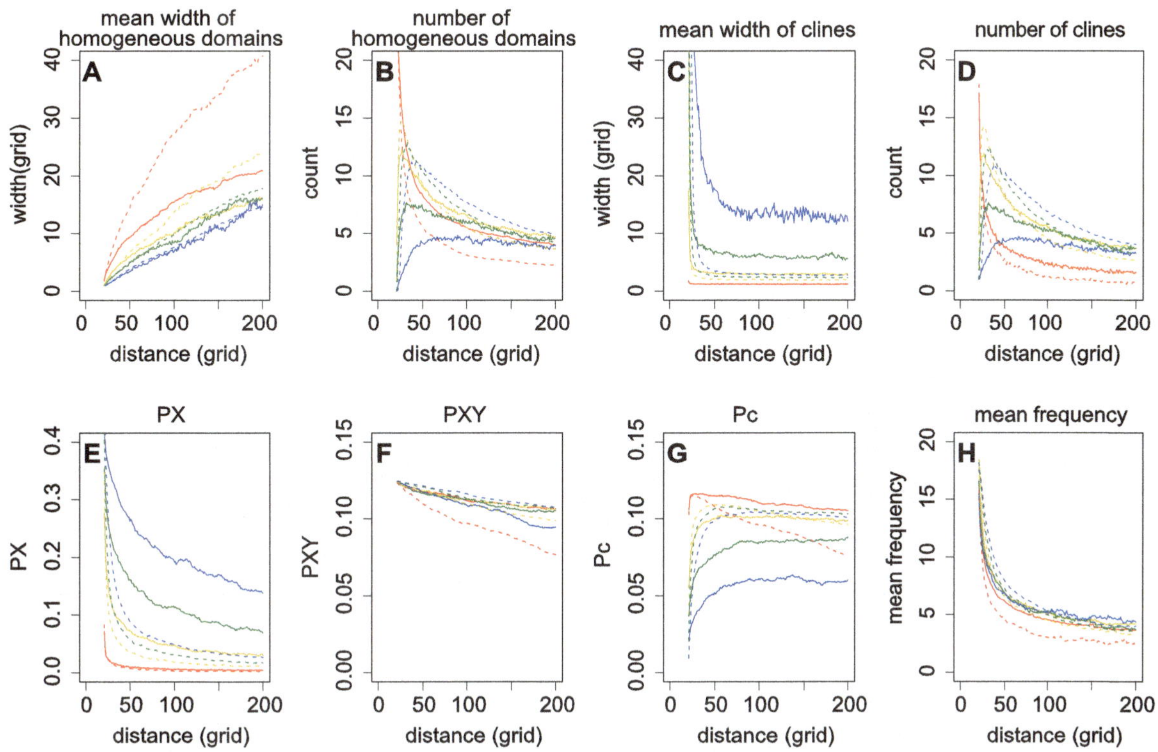


**Figure 3. The effects of migration rates on spatial patterns of genetic diversity at expansion fronts.** The migration rates varied from 0.001 to 0.4 (red: 0.01, yellow: 0.1, green: 0.2, blue: 0.4). The spatial patterns of genetic diversity were described by width and number of genetically homogeneous domains (A, B) and genetic clines (C, D), pairwise differences in PX (E), PXY (F) and Pc (G), and the mean frequency of the power spectrum (H)). The solid and dashed lines indicated BLRE and PhLRE, respectively. The value of $r$ was set as 3; initial allele frequency =0.5; and the number of generations to complete a boundary motion equal to 6000.

### (b) Population genetic heterogeneity

The simulations revealed that at the beginning of an expansion, pairwise differences within populations (PX) decreased quickly at the advancing front, while the corrected average pairwise difference between populations (Pc) increased quickly. The changing rate of PX and Pc soon decreased with the expansion (Fig. 2, 3, 4, 5). The more slowly the boundary changed, the higher the PX and lower the Pc was. And the highest PX and lowest Pc occurred in the simulations with non-expanded populations (Fig. 2). Pc during PhLREs peaked at the early stage of expansion and then declined slightly in the long run. This was because most populations at the expansion fronts became genetically homogeneous during PhLREs and thus exhibited low Pc values. The average pairwise difference between populations (PXY) decreased slowly with expansion. The higher speed of boundary motion generally maintained higher PXY than the lower speed did. However, PhLREs and simulations with stationary habitats demonstrated the exceptions again: PXY was intermediate during the former, but very high during the latter.

Generally, high migration rate increased the PX, but decreased the Pc (Fig. 3). However, there was one exception in that the Pc dropped quickly when migration rate was very low during a PhLRE. This was because most of the populations at the expansion front became genetically homogeneous under this scenario. High migration rate increased the PXY during PhLREs, but decreased it during BLREs with slow speeds of boundary motion. Reproduction rate decreased the mean genetic heterogeneity within populations, but increased Pc and PXY (Fig. 4). Initial allele frequencies deviated from 0.5 led to lower values of PX, PXY or Pc (Fig. 5).

## Discussion

### (a) The difference between PhLREs and BLREs

Our simulations showed stripe-like genetic patterns in expanded populations, which were equivalent to the genetic sectoring patterns established by previous studies [20,21,23]. Slow boundary motion, which had a strong limitation on expansion, broadened the genetic clines between homogeneous domains. PhLREs, the fast expansion, exhibited the narrowest clines and non-expansion simulations exhibited the broadest clines. This is not surprising because the expansion fronts halted for several generations before the next expansion and more migration occurred between populations at the expansion fronts during slow expansions. Similarly, Hallatschek and Nelson [11] and Roques et al. [12] also reported increased genetic diversities at expansion fronts under slow range expansions caused by the Allee effect.

The speed of boundary motion exerts great impacts on the spatial genetic pattern of the advancing front. For example, very slight constraints on boundary motion could drastically narrowed the homogeneous domains and increased the allele frequency fluctuations from those levels exhibited by PhLREs (Fig 2). In contrast, the homogeneous domains were broadened and allele frequency fluctuation decreased with even slower boundary motion. However, the strong constraint on boundary motion brought those values back to the levels exhibited by PhLREs. Simulations with non-expanded populations, though considered as the slowest boundary motion, behaved differently in that the

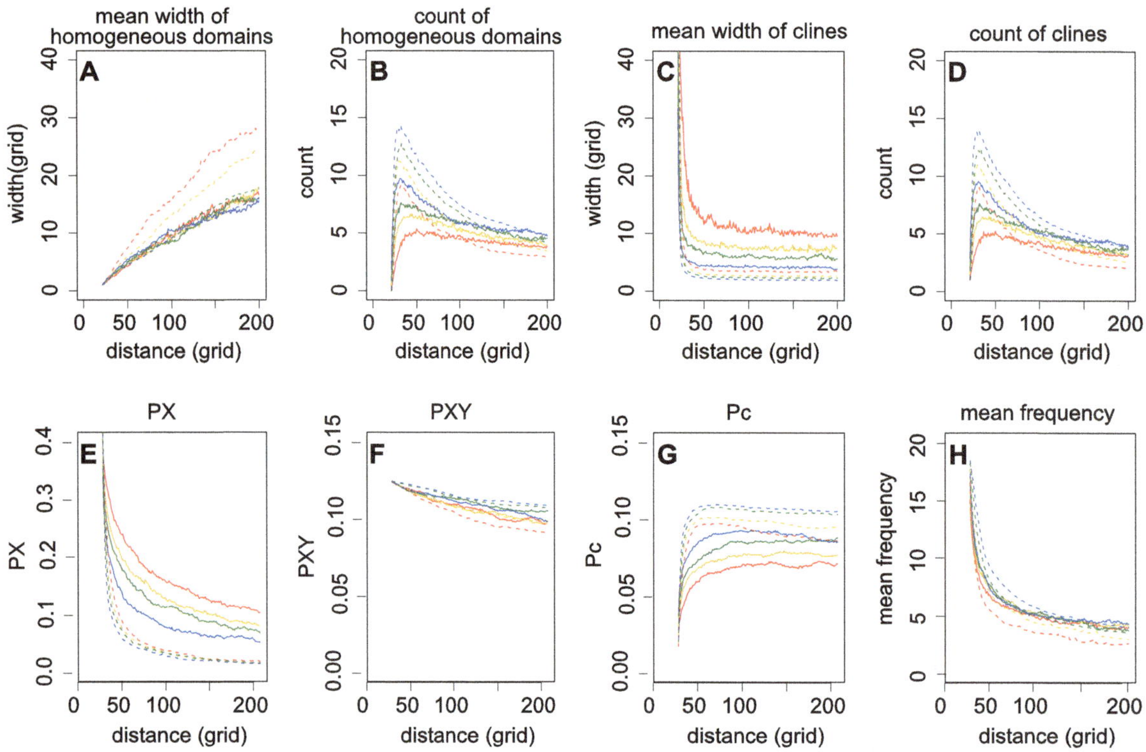


**Figure 4. The effects of reproduction rates on spatial patterns of genetic diversity at expansion front.** The spatial patterns of genetic diversity described by the width and number of genetically homogeneous domains (A, B) and genetic clines (C, D), pairwise differences in PX (E), PXY (F) and Pc (G), and the mean frequency of the power spectrum (H). The solid and dashed lines indicated BLREs and PhLREs, separately. The reproduction rates varied from 1.1 to 4 (red: 1.1, yellow: 2, green: 3, blue: 4). The value of $m$ was set as 0.2; initial allele frequency as 0.5; and the number of generations to complete a boundary motion 6000.

narrowest homogenous domains were formed and the highest level of allele frequency fluctuation reached. This maybe because the founder effect was totally absent in these simulations of stationary habitats, and it was the migration-drift balance has resulted in this spatial genetics patterns of genetic [25–27]. As for BLREs, the founder effect, however, always existed once the boundary began to expand, no matter how slow it progressed.

The average number of pairwise differences revealed a clear picture of changing spatial genetic patterns during expansions. The slow decrease in PXY, fast loss of PX, and fast increase in Pc at the expansion front suggested a process represented by slow loss of total diversity, quick loss of that within populations, and quick accumulation of genetic heterogeneity between populations corresponding to the emergence and merging of genetically homogenous domains.

The speed of boundary motion also influenced the genetic diversity at advancing fronts. Slow boundary motion increased the diversity within populations but decreased the heterogeneity between populations. Though Nullmeier and Hallatschek [7] suggested that the total loss of diversity did not depend on the speed of the range expansion, our results revealed that at least high speed of boundary motion could facilitate the maintenance of total genetic diversity during BLREs through quickly accumulating genetic heterogeneity between populations. This discrepancy maybe result from different model conditions: Nullmeier and Hallatschek [7] assumed that the front had expanded so slowly that the population density stayed at constant density up to the front. In our model, this condition could only be satisfied if the boundary motion was sufficiently slow.

Under PhLRE, though species ranges expanded most quickly, the total genetic diversity was intermediate among those during BLREs. For simulations with stationary habitats, the absence of founder effect maintained the total genetic diversity at the "non-expanding fronts". These results indicated that PhLRE and non-expansion simulations can not be simply considered as BLREs with the quickest or the slowest boundary motions.

Some previous empirical studies have implied differences in genetic patterns between BLRE and PhLRE populations. For example, examinations of several invasive species have revealed a decrease in genetic variation at expansion fronts [31,32]. However, the European larch (*Larix decidua*) maintained high genetic diversity during slow range expansion due to global warming, which has usually been attributed to the intensive mixing of genes [2]. Nevertheless, homogenous sectoring or stripe patterns are not easy to find in natural populations [32–35] because such patterns are difficult to reveal without intensive sampling. Nullmeier and Hallatschek [7] suggested that the lineage has to be sampled from a distance of more than a length $L_{\text{plateauto}} = 2Km$ to detect a range expansion, where $K$ was the carrying capacity and $m$ was the migration rate. Furthermore, many factors, e.g. landscape heterogeneity and multiple introductions [36], can further complicate patterns of spatial genetic structure.

It is notable that spatial expansions, can not only lead to allele frequency clines by eroding genetically homogeneous domains, but also rebuild genetically homogeneous domains from those clines by the ongoing demixing (Fig. 1). It is possible that the number of homogeneous domains in our study is the consequence of emergence and merging of homogeneous domains. The emergence of homogeneous domains could explain the fluctuation

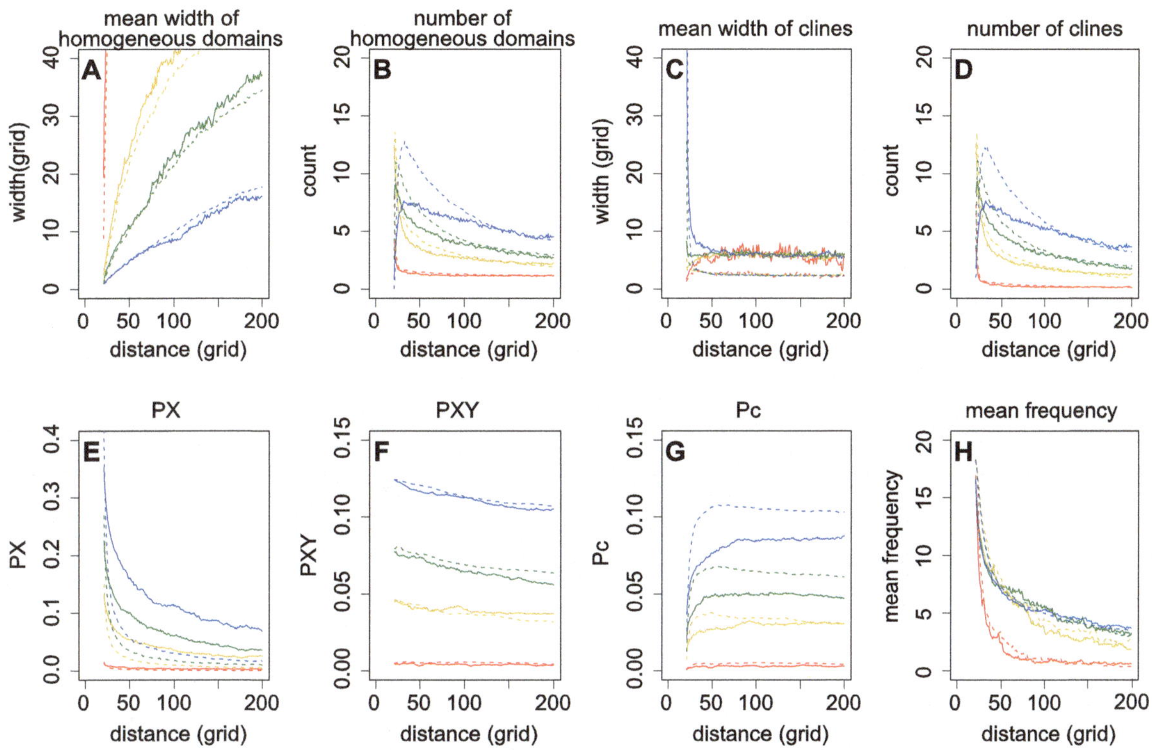


**Figure 5. The spatial patterns of genetic diversity at expansion fronts when initial allele frequencies varied.** The values of initial allele frequencies varied from 0.01 to 0.5 (red: 0.01, yellow: 0.1, green: 0.2, blue: 0.5). The solid indicated BLREs and dashed lines represented PhLREs. A to D showed the width and number of genetically homogeneous domains and genetic clines. E to F showed the pairwise differences in PX, PXY, and PC respectively. H showed the mean frequency of the power spectrum. The value of $r$ was set to 3; $m$ to 0.2; and the number of generations to complete a boundary motion 6000.

in the average numbers of genetically homogeneous domains during BLREs. However, the emergence of new homogeneous domains was rare, if any, during PhLREs. Therefore, smaller fluctuation in the average numbers of homogeneous domains was observed in PhLRES than that in BLREs (Fig 2, 3, 4, 5). Previous models [20,37,38] have shown the merging of homogeneous domains together with expansion, but failed to detect the emergence of new homogeneous domains from clines. The pattern of new homogeneous domains during BLREs was similar to those created by mutations and selection at expansion fronts [39]. However, their underlying mechanisms were totally different: the emergence of new homogeneous domains during BLREs was driven by the interaction of migration and demixing processes, as shown in our study, which differs from Kluver [40] ' s model, where the new stripes emerged from mutations.

## (b) The effects of migration rate and reproduction rate

High migration enhanced genetic diversity within populations but reduced heterogeneity between populations. Surprisingly, we found that high migration rate maintained a higher total genetic diversity than low migration rate during PhLREs, but the situation was reversed during BLREs, though Nullmeier and Hallatschek [7] suggested that the larger the migration rate, the smaller the loss of diversity. This is because high migration rates could increase both speed of range expansion and gene flow during PhLREs, but exerted little effect expansion speed during BLREs. The non-monotonous effect of increasing migration rate on the number of homogeneous domains during BLREs was also somehow counterintuitive. Specifically, the number of homogeneous domains was determined by the equilibrium of merging and emergence of homogeneous domains. High gene flow broadened the genetic clines and allowed more homogeneous domains to emerge, however, it quickly eroded existing homogeneous domains [23]. Low migration however failed to maintain broad clines and thus new homogeneous domains rarely emerge from the clines (Fig. 1,3).

Our simulations suggested that high reproduction rate reduced the genetic diversity within populations but increased the genetic heterogeneity between populations. It also narrowed the clines. It is possible that the high $r$ diluted effects of migration by reproducing more individuals from local population, and thus hindered the formation of clines. This was exactly the case in BLREs, where front populations had more time to reproduce and migrate. High reproduction also increased the number of homogeneous domains and clines during PhLREs, but not so drastically during BLREs (Fig 3). Possibly, this is because the high $r$ has another effect, i.e. lessening drift and demixing by quickly increasing population size, which will in turn impede the formation and merging of homogeneous domains. During PhLREs, species ranges kept expanding before the front populations had reached $K$, and a high $r$ helped those populations to reach larger sizes and lessen drift. During BLREs, however, front populations had enough time to obtain large sizes before further expansion took place, therefore, a high $r$ value could not guarantee a noticeable effect.

## (c) Methodological considerations

The Fast Fourier Transform (FFT) may prove to be a powerful tool for the detection of spatial genetic structure. In evolutionary and ecological studies, the period of recurrent spatial pattern was often approximately evaluated by spatial autocorrelation analysis [41]. Korolev et al. [22] presented the genetic sectorial structures by plotting the spatial heterozygosis as a function of angle, which was equivalent to a spatial autocorrelation analysis. Evaluating the allele frequency fluctuations by spatial autocorrelation analysis, nevertheless, is not realistic when the periods of fluctuation are highly variable. The spatial autocorrelation index is evaluated based on an averaged correlationship among all sites, by which the stochastic information of stripes is lost. In contrast, the FFT analysis can reveal more subtle fluctuations of allele frequency. Though the number of genetically homogeneous domains exhibited a fast increase at the early stage of expansion, $\bar{f}$ increased much earlier and showed an emergency of genetic fluctuations before the formation of homogeneous domains.

## Acknowledgments

We thank Dr. Y. Zheng for productive discussions. Thanks also to C. Hudson for editing an early version of this manuscript.

## Author Contributions

Conceived and designed the experiments: QD BL JF DQ. Performed the experiments: QD. Analyzed the data: QD QW DQ. Contributed reagents/materials/analysis tools: QD. Wrote the paper: QD XZ JF.

## References


1. Holt RD (2003) On the evolutionary ecology of species' ranges. Evol Ecol Res 5: 159–178.
2. Pluess AR (2011) Pursuing glacier retreat: genetic structure of a rapidly expanding *Larix decidua* population. Mol Ecol 20: 473–485.
3. Schmitt T, Seitz A (2002) Postglacial distribution area expansion of *Polyommatus* coridon (Lepidoptera : Lycaenidae) from its Ponto-Mediterranean glacial refugium. Heredity 89: 20–26.
4. Vandewoestijne S, Van Dyck H (2010) Population genetic differences along a latitudinal cline between original and recently colonized habitat in a butterfly. PLoS ONE 5: e13810.
5. Hill JK, Hughes CL, Dytham C, Searle JB (2006) Genetic diversity in butterflies: Interactive effects of habitat fragmentation and climate-driven range expansion. Biol Lett 2: 152–154.
6. Sakai AK, Allendorf FW, Holt JS, Lodge DM, Molofsky J, et al. (2001) The population biology of invasive specie. Annu Rev Ecol Syst 32: 305–332.
7. Nullmeier J, Hallatschek O (2013) The coalescent in boundary-limited range expansions. Evolution 67: 1307–1320.
8. Maclean IMD, Austin GE, Rehfisch MM, Blew JAN, Crowe O, et al. (2008) Climate change causes rapid changes in the distribution and site abundance of birds in winter. Global Change Biol 14: 2489–2500.
9. D'Andrea L, Broennimann O, Kozlowski G, Guisan A, Morin X, et al. (2009) Climate change, anthropogenic disturbance and the northward range expansion of Lactuca serriola (Asteraceae). J Biogeogr 36: 1573–1587.
10. Allee WC (1931) Animal aggregations, a study in general sociology. Chicago, IL: The University of Chicago Press.
11. Hallatschek O, Nelson DR (2008) Gene surfing in expanding populations. Theor Popul Biol 73: 158–170.
12. Roques L, Garnier J, Hamel F, Klein EK (2012) Allee effect promotes diversity in traveling waves of colonization. P Natl Acad Sci USA 109: 8828–8833.
13. Wang R, Wang Y-Z (2006) Invasion dynamics and potential spread of the invasive alien plant species *Ageratina adenophora (*Asteraceae) in China. Divers Distrib 12: 397–408.
14. Ficetola G, Coïc C, Detaint M, Berroneau M, Lorvelec O, et al. (2007) Pattern of distribution of the American bullfrog *Rana catesbeiana* in Europe. Biol Invasions 9: 767–772.
15. Kearney M, Phillips BL, Tracy CR, Christian KA, Betts G, et al. (2008) Modelling species distributions without using species distributions: the cane toad in Australia under current and future climates. Ecography 31: 423–434.
16. Behrman KD, Kirkpatrick M (2011) Species range expansion by beneficial mutations. J Evol Biol 24: 665–675.
17. McInerny GJ, Turner JRG, Wong HY, Travis JMJ, Benton TG (2009) How range shifts induced by climate change affect neutral evolution. P R Soc B 276: 1527–1534.
18. Hewitt G (2000) The genetic legacy of the Quaternary ice ages. Natur 405: 907–913.
19. Provan J, Maggs CA (2012) Unique genetic variation at a species' rear edge is under threat from global climate change. P R Soc B 279: 39–47.
20. Hallatschek O, Hersen P, Ramanathan S, Nelson DR (2007) Genetic drift at expanding frontiers promotes gene segregation. P Natl Acad Sci USA 104: 19926–19930.
21. Hallatschek O, Nelson DR (2010) Life at the Front of an Expanding Population. Evolution 64: 193–206.
22. Korolev KS, Xavier JB, Nelson DR, Foster KR (2011) A quantitative test of population genetics using spatiogenetic patterns in bacterial colonies. Am Nat 178: 538–552.
23. Excoffier L, Ray N (2008) Surfing during population expansions promotes genetic revolutions and structuration. Trends Ecol Evol 23: 347–351.
24. Edmonds CA, Lillie AS, Cavalli-Sforza LL (2004) Mutations arising in the wave front of an expanding population. P Natl Acad Sci USA 101: 975–979.
25. Epperson BK (1995) Spatial distributions of genotypes under isolation by distance. Genetics 140: 1431–1440.
26. Sokal RR, Jacquez GM, Wooten MC (1989) Spatial autocorrelation analysis of migration and selection. Genetics 121: 845–855.
27. Sokal RR, Wartenberg DE (1983) A test of spatial autocorrelation analysis using an Isolation-by-Distance Model. Genetics 105: 219–237.
28. Nei M, Li WH (1979) Mathematical model for studying genetic variation in terms of restriction endonucleases. P Natl Acad Sci USA 76: 5269–5273.
29. Takahata N, Nei M (1985) Gene genealogy and variance of interpopulational nucleotide differences. Genetics 110: 325–344.
30. R Core Team (2012) R: A language and environment for statistical computing. R Foundation for Statistical Computing, Vienna, Austria. ISBN 3-900051-07-0, URL http://www.R-project.org/. Vienna, Austria. http://www.R-project.org/.
31. Gasperi G, Bonizzoni M, Gomulski LM, Murelli V, Torti C, et al. (2002) Genetic differentiation, gene flow and the origin of infestations of the medfly, *Ceratitis capitata.* Gen 116: 125–135.
32. Grapputo A, Boman S, Lindstrom L, Lyytinen A, Mappes J (2005) The voyage of an invasive species across continents: genetic diversity of North American and European Colorado potato beetle populations. Mol Ecol 14: 4207–4219.
33. Armbruster P, Bradshaw WE, Holzapfel CM (1998) Effects of postglacial range expansion on allozyme and quantitative genetic variation of the pitcher-plant mosquito, *Wyeomyia smithii.* Evolution 52: 1697–1704.
34. Lihova J, Kudoh H, Marhold K (2010) Genetic Structure and Phylogeography of a Temperate-Boreal Herb, *Cardamine Scutata* (Brassicaceae), in Northeast Asia Inferred from Aflps and Cpdna Haplotypes. Am J Bot 97: 1058–1070.
35. Habel JC, Lens L, Rodder D, Schmitt T (2011) From Africa to Europe and back: refugia and range shifts cause high genetic differentiation in the Marbled White butterfly *Melanargia galathea.* BMC Evol Biol 11.
36. Bossdorf O, Auge H, Lafuma L, Rogers WE, Siemann E, et al. (2005) Phenotypic and genetic differentiation between native and introduced plant populations. Oecologia 144: 1–11.
37. Hallatschek O, Korolev KS (2009) Fisher waves in the strong noise limit. PhRvL 103: 108103.
38. Korolev KS, Avlund M, Hallatschek O, Nelson DR (2010) Genetic demixing and evolution in linear stepping stone models. Reviews of Modern Physics 82: 1691–1718.
39. Kuhr J-T, Leisner M, Frey E (2011) Range expansion with mutation and selection: dynamical phase transition in a two-species Eden model. NJPh 13: 113013.
40. Kluver J (2011) A mathematical theory of communication: meaning, information, and topology. Complexity 16: 10–26.
41. Legendre P, Legendre L (1998) Numerical ecology. Amsterdam: Elsevier Science.

# 17

# Soil Charcoal to Assess the Impacts of Past Human Disturbances on Tropical Forests

**Jason Vleminckx[1]*, Julie Morin-Rivat[2,3], Achille B. Biwolé[2], Kasso Daïnou[2,4], Jean-François Gillet[2], Jean-Louis Doucet[2], Thomas Drouet[5], Olivier J. Hardy[1]**

**1** Service d'Évolution Biologique et Écologie, Faculté des Sciences, Université Libre de Bruxelles, Brussels, Belgium, **2** Laboratoire de Foresterie des Régions tropicales et subtropicales, Unité de Gestion des Ressources forestières et des Milieux naturels, Gembloux Agro-Bio Tech, Université de Liège, Gembloux, Belgium, **3** Laboratory of Wood Biology & Xylarium, Royal Museum for Central Africa, Tervuren, Belgium, **4** Laboratory of Applied Ecology, University of Abomey-Calavi, Cotonou, Benin, **5** Laboratoire d'Écologie Végétale et Biogéochimie, Faculté des Sciences, Université Libre de Bruxelles, Brussels, Belgium

## Abstract

The canopy of many central African forests is dominated by light-demanding tree species that do not regenerate well under themselves. The prevalence of these species might result from ancient slash-and-burn agricultural activities that created large openings, while a decline of these activities since the colonial period could explain their deficit of regeneration. To verify this hypothesis, we compared soil charcoal abundance, used as a proxy for past slash-and-burn agriculture, and tree species composition assessed on 208 rainforest 0.2 ha plots located in three areas from Southern Cameroon. Species were classified in regeneration guilds (pioneer, non-pioneer light-demanding, shade-bearer) and characterized by their wood-specific gravity, assumed to reflect light requirement. We tested the correlation between soil charcoal abundance and: (i) the relative abundance of each guild, (ii) each species and family abundance and (iii) mean wood-specific gravity. Charcoal was found in 83% of the plots, indicating frequent past forest fires. Radiocarbon dating revealed two periods of fires: "recent" charcoal were on average 300 years old (up to 860 BP, n = 16) and occurred in the uppermost 20 cm soil layer, while "ancient" charcoal were on average 1900 years old (range: 1500 to 2800 BP, n = 43, excluding one sample dated 9400 BP), and found in all soil layers. While we expected a positive correlation between the relative abundance of light-demanding species and charcoal abundance in the upper soil layer, overall there was no evidence that the current heterogeneity in tree species composition can be explained by charcoal abundance in any soil layer. The absence of signal supporting our hypothesis might result from (i) a relatively uniform impact of past slash-and-burn activities, (ii) pedoturbation processes bringing ancient charcoal to the upper soil layer, blurring the signal of centuries-old Human disturbances, or (iii) the prevalence of other environmental factors on species composition.

**Editor:** Bruno Hérault, Cirad, France

**Funding:** This study is part of the FRFC project (« Fonds pour la Recherche Fondamentale et Collective», convention n ° 2.4577.10) entitled "Dynamics of tree and herbaceous communities of the tropical forests of Central Africa, in relation with past anthropogenic and climatic disturbance". J. Vleminckx and J. Morin-Rivat are funded by the "Fonds pour la formation à la Recherche dans l'Industrie et dans l'Agriculture (FRIA)". The authors gratefully acknowledge the CoForChange project funded through the ERA-net BiodivERsA (http://www.coforchange.eu/fr/). The authors declare that the funders only provided financial support for the field inventories and had no role in the study design, data collection and analysis, decision to publish, or preparation of the manuscript.

**Competing Interests:** The authors have declared that no competing interests exist.

* Email: jason.vleminckx@ulb.ac.be

## Introduction

For a long time, many tropical forests have been viewed as "virgin" or "primary" ecosystems, undisturbed by anthropogenic activities. However, increasing evidence from different continents has suggested that these forests may actually hide influences of past Human disturbances [1]. In South-East Asia for example, it has been suspected that agricultural activities developed as early as 7000 BP in Papua New Guinea [2] and 8000 BP in Thailand [3], in areas covered today by lowland uninhabited rainforests. In Northwest Belize, Ross [4] demonstrated that modern tree species composition appeared to display important differences between areas of high and low settlement of ancient Maya, because of different intensities of past forest gardening (recruitment of useful species). Similarly, in central Amazonia, "terra preta" soils covering an area of about 500 km$^2$, on which stands apparent "pristine" rainforests, have resulted from intense burning and agricultural activities occurring about 2500 BP, which have considerably enhanced the fertility of these soils and may thus have impacted floristic diversity [5].

In central African moist forests, accumulating evidences show that Human has had a profound impact on the vegetation dynamics during the three last millennia. While only sparse evidence is recorded for Human presence in central Africa during the early and middle Holocene [6,7], archaeological surveys have suggested a dramatic expansion of an ancestral Bantu population coming from the southern part of the actual Cameroon-Nigeria border during the third millennium BP [8]. Archaebotanical data have indicated that this expansion coincided with an increasing seasonality in the precipitation regime that would have generated a disruption of the forest cover and replaced it by savannas or open forest formations [9,10], and therefore facilitated the Human colonization of central Africa [11,12]. Direct evidence from pollen and diatom frequency diagrams have demonstrated increasing aridity in the region [13,14] during the second half of the third millennium BP (especially between 2400 and 2100 BP), while charred botanical remains from this period have indicated that Bantus introduced the culture of pearl millet (*Pennisetum glaucum*, Poaceae), which requires a prolonged dry season [15]. Moreover, charcoal dating and identification have revealed higher abundance of pioneer species characterizing open forest formations [16]. The dryer areas of Central Africa were then more subject to fires during the dry season, but even when more humid conditions returned, the forest did not regenerate immediately as fires continued to maintain an open vegetation. According to archeological data, these resilient fires may not have been naturally caused only but also prompted by slash-and burn agricultural activities introduced by Bantu farmers [16,17]. For some unclear reasons, a dramatic decline of Human occupation in Western Central Africa occurred about 1400 BP [7,18], and it is only around the late middle age (about 600 BP) that evidence for Human activities are found again in the region [6,19].

Nowadays, while natural openings are scarce in Central African moist forests, the upper canopy is often dominated by long-lived light-demanding tree species which seem to suffer from low regeneration rate underneath their own shade [20]. This has been explained by a decrease of forest clearing activities since Human settlements from deep forests zones have been forced to move along roadsides during the colonial period [21]. Considering these facts and the massive amounts of evidence (potsherds, charcoals, iron objects) recorded in the literature for ancient Human occupation in vast uninhabited regions of Central African forest [7,15], it has been suggested that until a recent past, the widespread practice of slash-and-burn agriculture by farmers may have favored the competitiveness of light-demanding species and profoundly influenced the dynamics of tree diversity organization.

The pioneer works of Jones [22,23] in the mahogany forest of Okomu (Nigeria) were among the firsts to link vegetation and past Human disturbances in African rainforests. Jones suggested that the abundance of old light-demanding species he observed resulted from intense past Human activities [19,22]. Mahogany forests found in Cameroon [24], Central African Republic [25,26], Republic of Congo [26] and Democratic Republic of the Congo [27] have been suspected to hide a similar history to those of Okomu [28]. More recently, Biwolé et al. [29] have emphasized coincidence between populations of light-demanding trees species and the age of the last Human disturbances in a forest of Southern Cameroon.

Therefore, if past slash-and-burn agricultural practices favored the establishment of light-demanding species and if their intensity was spatially heterogeneous, we expect a correlation between forest stand composition and signs of past Human disturbances, such as soil charcoal abundance. To our knowledge, this expectation has never been tested using an adequate statistical sampling design in tropical forests. To bridge this gap, we aimed to test the hypothesis that Human disturbance intensity varied in time and space and favored the establishment of long-lived light-demanding tree species in Central Africa. To do so, we have carried out tree inventories and soil charcoal analyses at a landscape scale in three rainforest areas located in Southern Cameroon, using soil charcoal abundance as a proxy for past slash-and-burn activities. We addressed the following specific questions: (1) Is charcoal abundance spatially structured throughout the landscape and according to soil depth? (2) What is the age distribution of charcoal and is age related to soil depth? (3) What is the spatial structure of floristic and functional properties of local tree assemblages, considering: (i) species abundance, (ii) relative abundance of three regeneration guilds (pioneer, non-pioneer light-demanding, shade-bearer), and (iii) wood-specific gravity (i.e. $g.cm^{-3}$ of wood, used as a proxy for shade-tolerance)? (4) Does soil charcoal abundance explain variation in local tree assembly properties?

## Materials and Methods

### Study area

Study areas corresponded to three forest logging concessions. Two of them, hereafter named Area 1 and Area 2, were located in South-East Cameroon while another, Area 3, was located in the South-Western part of the country (Figure 1). The vegetation of these areas consisted of transition formations between lowland evergreen and semi-evergreen moist tropical forest [24], where the upper canopy was dominated by typical long-lived light-demanding species like *Pericopsis elata* (Fabaceae) in Area 1, *Triplochiton scelroxylon* (Malvaceae) in Area 2 and *Lophira alata* (Ochnaceae) in Area 3. Topography was characterized by hilly and highly weathered plateaus displaying dense hydrographic networks, situated at an elevation of ca. 700 m a.s.l. in Areas 1 and 2, and ca. 500 m a.s.l. in Area 3. The climate within all study areas is equatorial, with mean annual temperatures of ca. 24–25°C and mean annual rainfall reaching ca. 1600–1700 mm (www.climatedata.eu). Rainfall displays a bimodal distribution, with a rainy season stretching from September to November and another from March to June, separated by two relatively less humid periods. Ferralsols constituted the most dominant type of soil, with substantial occurrence of Acrisols in Area 3 [30]. Permission to carry out our field inventories were provided by logging companies (contact details are given in Appendix S1).

### Inventory sites

A total of twelve sites were inventoried, of which six were located in Area 1, three in Area 2 and three in Area 3 (Figure 1). Each site corresponded to a linear transect along which 11 to 20 rectangular plots (40×50 m) were set up every 250 m and geo-referenced. This linear disposition of the plots within sites was designed to be able to apply torus-translation tests [31] (see details below). In Areas 1 and 2, sites were disposed along virtual transects extending over 101.5 and 66.5 km, respectively, while in Area 3, sites were irregularly disposed. Sites coordinates and the number of plots per site are given in Table S1 in File S1.

### Charcoal abundance

At each corner and at the centre of each plot, we excavated a soil pit with an auger (5 cm diameter), collecting soil samples at the following depth layers (in cm): 0–10, 10–20, 20–40, 40–60, 60–80 and 80–100. The two uppermost layers were smaller in order to analyze charcoal abundance at finer resolution, as we suspected that charcoal reflecting recent Human disturbance were mostly located in these superficial soil layers. Soil charcoal abundance is classically quantified by determining its mass after sieving the soil volume in water [32,33]. This procedure being too time-consuming in the field for the large number of measures needed, we opted for a faster yet less precise method. Therefore, for each soil volume (280 $cm^3$ for 20 cm of vertical profile), we searched manually for charcoals remains and visually attributed a "charcoal abundance index" (CAI) defined as follows: 0 = absence of charcoal; 1 = charcoal traces (scattered pieces or powder) and/or

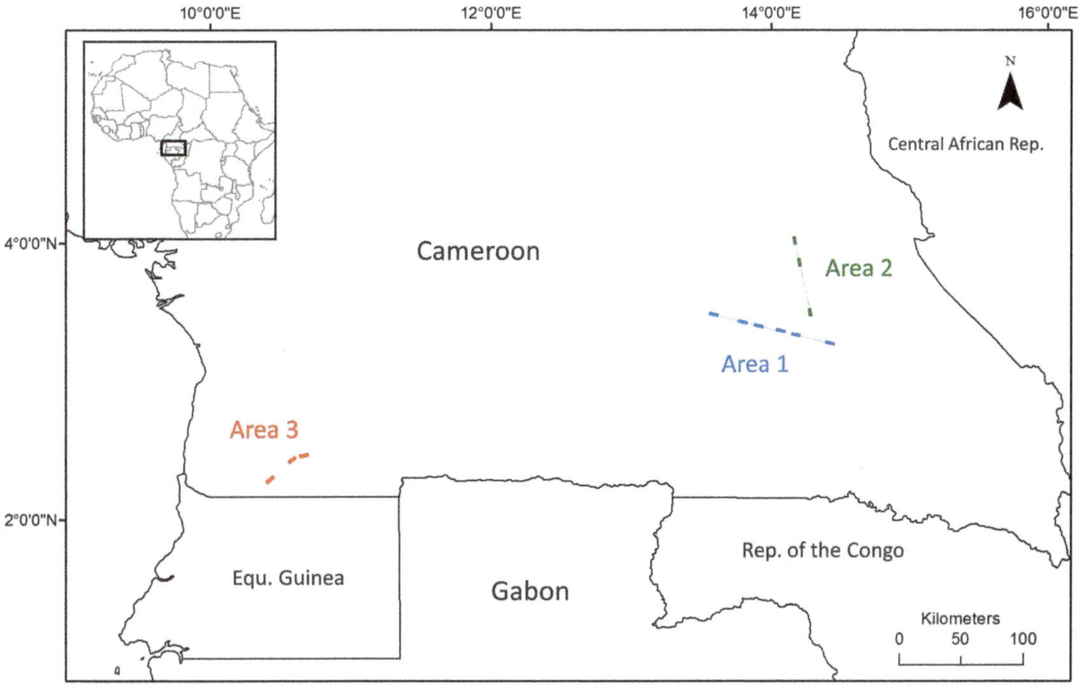


**Figure 1. Geographical location of the three study areas.** Each site is represented by a rectangle. Sites in Areas 1 and 2 are linearly disposed along a virtual transect represented by a dashed line.

≤5 pieces with a length <5 mm or 1 piece ≥5 mm; 2 = more than 5 pieces with a length <5 mm and/or more than 1 piece ≥ 5 mm. CAI could not be estimated for layers deeper than 20 cm in site 4, due to logistic constraints. Within each of the three investigated areas, CAI was estimated by the same person (JV, JMR and ABB in Area 1, 2 and 3, respectively).

We validated our CAI method in two ways. First, we tested the correlation between our CAI and the real charcoal mass for 36 soil volumes (each of 140 $cm^3$) excavated with an auger in a plot where previous digging revealed important charcoal presence (plot nr 2 of site 1, Area 1). For each volume, we first estimated the CAI, and then used a sieve of 2 mm mesh to collect all the charcoal pieces contributing significantly to the charcoal mass. We observed a significant correlation between CAI and ln(1+charcoal mass, in mg) ($r$-Pearson = 0.91, $P<0.001$) and no overlap of charcoal mass distributions between CAI values (see Appendix S2). Second, as Chabal [34] showed that the total number of charcoal particles found in a soil volume is highly correlated to the total charcoal mass, we tested the correlation between our CAI and the number of charcoal particles counted for the same soil volume on (i) 1510 soil volumes sampled in Area 2 where charcoals collected by hand were systematically counted, and (ii) 900 soil samples collected for another project (in Area 1) for which charcoals were collected using a sieve of 2 mm mesh. Pearson correlations between CAI and ln(1+charcoal particles number) reached 0.92 and 0.93 in both datasets (P<0.001, Appendix S2).

## Radiocarbon dating

We selected 60 charcoal samples found at depth ranging from 5 to 150 cm collected in 37 random pits. In eight soil pits, charcoal samples from 2 to 5 different soil layers were dated to better assess the age-depth relationship. $^{14}C$ dating was performed at the Poznań Radiocarbon Laboratory (Poland). Calibration was performed under the OxCal v4.1.7 program [35–37] with the IntCal09 atmospheric calibration curve [38] and expressed in BP.

## Species functional traits

Within each plot, we inventoried all the trees displaying a diameter at breast height (dbh) ≥20 cm. Over the three study areas, we found on average 32 individuals, 19 species, 18 genera and 13 families per plot. Table 1 summarizes abundance and diversity data for each study area. Species diversity was computed as the effective number of species expected in a random sample of $k = 2$ or $k = 100$ individuals [39], in order to give more weight to abundant and rare species, respectively. For all tree species identified in our plots, we compiled information on two key functional traits reflecting their shade-tolerance: regeneration guild (RG) and wood–specific gravity (WSG). Although well-defined classification of species in RG does not exist, three guilds could be defined, based on Hawthorne [40] and field observations in Central Africa (J. L. Doucet, A. Fayolle & J-F Gillet, pers. obs.; www.coforchange.eu): (i) pioneer species (P) require large gaps for establishment, (ii) non-pioneer light-demanding species (NPLD) can establish in shade but need a gap in the forest to grow to their full height, while (iii) shade-bearer species (SB) can be found in shade both as young and older plants. The relative abundance of each guild was computed for each of the 208 plots, after discarding unclassified species. WSG ($g/cm^3$) was used because this variable reflects the diameter growth rate (see references in Slik [41]), as fast-growing (light-demanding) species are more likely to have lighter wood than slow-growing (shade-bearer) species [41,42]. The WSG of most species could be assessed using a database [43]. When WSG data was missing for a species we assigned it the average WSG of its congeneric species, based on previous evidence showing that WSG is phylogenetically conserved [44]. For each plot, we calculated the mean WSG weighted by species relative abundances. The number of species for which the RG and

**Table 1.** Abundance and diversity data for each study area.

| | Area 1 | Area 2 | Area 3 |
|---|---|---|---|
| Nr of 0.2-ha plots | 120 (24 ha) | 53 (10.6 ha) | 35 (7 ha) |
| Nr of stems | 3799 | 1699 | 1348 |
| Total nr of identified ind. | 3531 (93%[a]) | 1693 (99.5%) | 1335 (99%) |
| Nr of species | 186 | 183 | 147 |
| Dominant species | GS (8.2%[a]) | GS (8.3%) | BW (5.5%) |
| Nr of families | 46 | 45 | 42 |
| Dominant family | Annonaceae (15.5%[a]) | Annonaceae (20.0%) | Fabaceae (11.9%) |
| ENS(2) | 51.2 | 50.5 | 54.1 |
| ENS(100) | 79.1 | 78.5 | 71.7 |
| Nr of ind. assigned to a RG | 3442 (97.5%[a]) | 1576 (93.1%) | 1263 (93.7%) |
| Nr of sp. assigned to a RG | 172 (92.0%[b]) | 155 (84.7%) | 135 (91.8%) |
| Nr of ind. with WSG | 3222 (91.2%[a]) | 1577 (93.1%) | 1217 (90.3%) |
| Nr of sp. with WSG | 170 (91.4%[b]) | 162 (88.5%) | 131 (89.1%) |
| Nr of P | 707 (20.5%[c]) | 469 (29.8%) | 261 (20.7%) |
| Nr of NPLD | 761 (22.1%[c]) | 306 (19.4%) | 261 (20.7%) |
| Nr of SB | 1974 (57.4%[c]) | 801 (50.8%) | 741 (58.7%) |

RG = regeneration guild. P = Pioneers. NPLD = Non-Pioneer Light-Demanders. SB = Shade-bearers. ENS(2) or ENS(100) = effective number of species expected for a random sample of 2 or 100 individuals. Ind. = individuals. Sp. = species. WSG = wood-specific gravity (g/cm$^3$). GS = *Greenwayodendron suaveolens* (Annonaceae). BW = *Blighia welwitschii* (Sapindaceae).
[a]Percentage calculated over the total number of stems in the study area.
[b]Percentage calculated over the total number of species in the study area.
[c]Percentage calculated over the number of individuals assigned to a RG.

WSG was known, as well as the abundance of each RG within each study area is represented in Table 1. A complete list of species, their corresponding RG and WSG, and their abundances in each study area is given in Table S2 in File S1.

## Spatial patterns

For each study area, landscape differences between sites were tested using Kruskal-Wallis tests for the following variables calculated at the plot level: (i) mean CAI in two soil layers (see next section of materials and methods), (ii) relative abundance and relative basal area of each RG, (iii) mean WSG and (iv) the abundance of each sufficiently represented species and family ($\geq$ 20 individuals). Tests were performed using function kruskal.test from R stats package [45]. Finer scale spatial patterns were assessed within each of the three areas by measuring the spatial autocorrelation of these variables using Moran's $I$ statistic [46], here defined as:

$$I_{ij} = \frac{(x_i - \bar{x})\,(x_j - \bar{x})}{Var\,(x)} + \frac{1}{(n-1)}$$

where $x_i$ and $x_j$ are the values of variable $x$ for samples $i$ and $j$, respectively; $\bar{x}$ and $Var(x)$ are, respectively, the mean and variance of variable $x$; $n$ is the total sample size. The second term is a sample bias correction, ensuring that the average $I_{ij}$ over all existing $i, j$ pairs is equal to zero. Autocorrelograms, representing the mean $I_{ij}$ values for a set of spatial distance intervals against distance, allowed a visual characterization of spatial patterns. Mantel tests [47] between the $I_{ij}$ matrix and the spatial distance matrix were used to assess the significance of spatial structure for each variable.

We analyzed the spatial autocorrelation of charcoal abundance using the CAI index obtained for each soil sample to assess patterns at contrasted spatial scales: within a soil pit (distinguishing adjacent, i.e. in contact, and non-adjacent soil layers), within a plot, within a site and among sites. At the plot level, spatial autocorrelation was also assessed for each RG relative frequency, mean WSG and the abundance of each sufficiently represented species and family. All spatial autocorrelation analyses were performed using software Torocor (Torocor 1.0, http://ebe.ulb.ac.be/ebe/Software.html) (see Appendices S3 and S4).

## Testing correlation between light-demanding species and charcoal abundance

$^{14}C$ dating revealed a bimodal distribution in charcoal age: 16 out the 60 (27%) samples were dated between 80 and 860 BP (hereafter named "recent" charcoals, relatively speaking) and were only found in the upper soil layer (up to 25 cm), while 44 of them (73%) were dated more than 1500 BP ("ancient" charcoal) and found in the whole soil profile (Figure 2). Therefore, we estimated the abundance of soil charcoals at the plot level by computing, respectively, (i) the mean CAI value over both the 0–10 and 10–20 cm soil layers (proxy for "recent" fires) and (ii) the mean CAI value over the soil layers deeper than 20 cm (proxy for "ancient" fires).

For each study area taken individually, we then tested the correlation (at the plot level) between the mean CAI above and below 20 cm depth and: (1) the relative frequencies and relative basal areas of each RG, (2) the abundance of each species and family represented by at least 20 individuals (in the study area), and (3) the mean WSG. As classical correlation test is likely to be too liberal when analyzing spatial data, we tested the correlations using (i) a classical parametric test and (ii) a non-parametric torus-translation

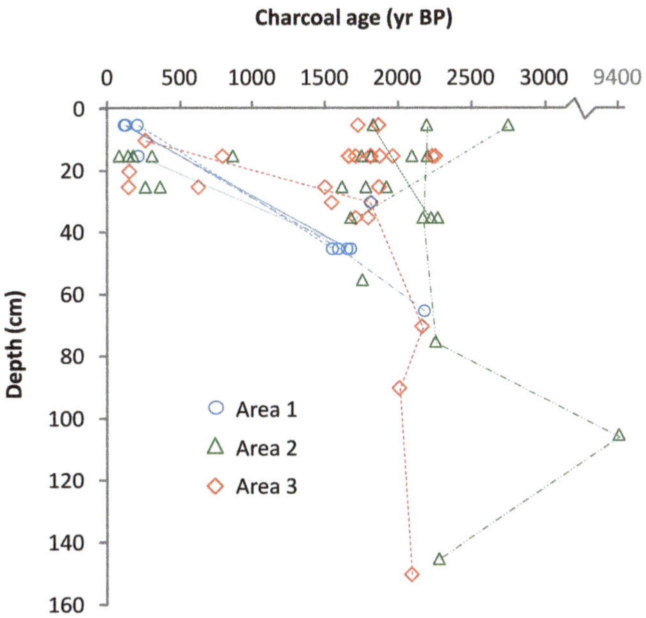


**Figure 2. Age-depth relationship for 60 charcoal fragments sampled in the three areas investigated.** Charcoals collected in a same soil pit are connected by a dashed line.

procedure taking spatial autocorrelation into account [31]. The latter consisted of generating randomized datasets where the correlation between variables was broken down while the spatial autocorrelation of each variable was kept intact. More specifically, within each site independently, the mean CAI values of the 20 plots were permuted among plots while keeping the original spatial order of plots (except at the extremities). This was like considering that the 20 plots of a site were situated along a ring and that CAI values were shifted by a random number of steps along the circle. 4999 of such permutations, independently repeated for each of the sites of a given area, were conducted to obtain null distributions of Pearson's correlation coefficients (computed using all plots). This allowed defining 95% confidence envelopes under the null hypothesis that there was no correlation between local CAI values and other variables. A test was significant when the observed correlation coefficient (i.e. without permutation) lied outside the 95% confidence envelope. By conserving the local structure (within site) of variables when permuting the objects, the torus-translation procedure tested the correlation at finer scale (i.e. within 3 to 5 km long transects) than the classic correlation test. Torus-translation analyses were performed using software Torocor (see Appendix S4).

## Results

### Charcoal ages and distribution

Charcoal was found as powder or entire fragments from 1 to 10 mm. It occurred at higher abundance between 20 and 60 cm in Area 1, between 10 and 40 cm in Area 2, and between 10 and 60 cm in Area 3 (see Appendix S3). It was present in 72, 83 and 94% of the plots between 0 and 20 cm deep in Areas 1, 2 and 3, respectively (89, 92 and 94% between 20 and 100 cm).

Radiocarbon dating revealed two distinct periods of ancient fires (Figure 2): 16 charcoal pieces were dated from 80 to 860 BP (hereafter "recent" charcoal), while 43 pieces were dated from 1495 to 2745 BP (hereafter "ancient" charcoal), and one sample was dated 9400 BP. All recent charcoal samples were found in the upper 25 cm soil layer, while ancient charcoal was found at all depths. Beside this trend between recent and ancient charcoal, there was no relationship between age and depth for ancient charcoals, even within a same soil pith (Figure 2). Calibration details for all dated samples are available in Table S3 in File S1.

For each study area, CAI values within soil pits were much more similar between adjacent soil layers than between non-adjacent ones (Figure 3, left). The correlation between mean plot CAI in soil layers above and below 20 cm depth was highly significant in Areas 1 and 2 ($r$-Pearson $=0.56$ and 0.51, respectively; $P<0.001$ under bilateral torus-translation test) but not in Area 3 ($r$-Pearson $=-0.09$; $P=0.66$). Among soil pits, there was a regular decrease of Moran's $I$ values with the logarithm of the spatial distance within each study area (Figure 3, right). In particular, Moran's $I$ values were always significantly positive between pits from a same plot, demonstrating a heterogeneity in charcoal abundance among plots within each area, although the relatively weak values found (<0.1) also indicate a high heterogeneity of charcoal abundance at a very local scale (i.e. within plot). At larger spatial scales, considering plot level mean CAI values separately in the 0–20 cm and 20–100 cm soil layers, Mantel tests between Moran's $I$ and spatial distance matrices were highly significant for both soil layers in Area 1 ($P<0.01$; see Appendix S5) while no spatial structure was found in Areas 2 and 3. Consistently, these mean CAI values differed significantly among sites only in Area 1, where charcoal was more abundant in the western part (sites 1 to 3) than in the eastern part (sites 4 to 6; $P<0.001$, see Table 2).

### Floristic and functional traits spatial patterns

In Area 1, differences between sites and spatial autocorrelations were significant ($P<0.05$) for each RG relative abundance and relative basal area, and for the mean WSG per plot, while fewer significant differences (Table 2) and spatial structures (Appendix S5) were observed within Areas 2 and 3. More specifically, the relative abundance and basal area of NPLD and P were not spatially structured in Area 2, whereas in Area 3, significant spatial autocorrelation was only observed for NPLD relative abundance and relative basal area. The proportion of species and families displaying significant spatial structure ranged between 35% and 65% within each study area (see Appendix S5). For each area, Kruskal-Wallis tests for difference of species abundances and family abundances between sites are given in Table S4 in File S1.

### Correlations between vegetation characteristics and charcoal abundance

In Area 1, the abundance of charcoal in both the 0–20 cm and 20–100 cm soil layers (dominated by "recent" and "old" charcoals, respectively) and the relative abundance and basal area of NPLD decreased from West to East (sites 1 to 6), while the relative abundance and basal area of SB displayed the reverse pattern (Figure S1). Accordingly, Pearson correlations between NPLD relative abundance/relative basal area and CAI in each soil layer (measured at the plot level) were statistically significant using classical correlation tests (Table 3). However, these correlation coefficients became non-significant when using torus-translation tests which tested for within-site correlation and corrected for spatial autocorrelation. In Areas 2 and 3, all tests of correlation between vegetation characteristics and charcoal abundance were non-significant (Table 3). Finally, no significant correlation was observed between CAI and the abundance of any species or family in any study area, regardless of the test considered (see Appendix S6).

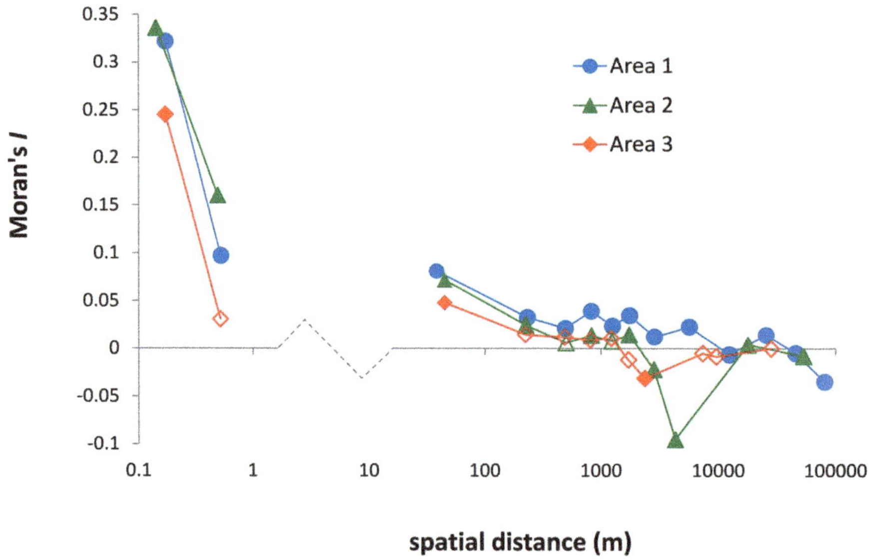


**Figure 3. Autocorrelogram of CAI values for each study area: mean Moran's *I* computed for 12 to 14 distance intervals.** On the left side (<1 m), the abscissa for the two first symbols represent the vertical distance between soil layers from a same pit, the left and right symbols distinguishing adjacent layers from non-adjacent layers, respectively. On the right side (>10 m), the abscissa corresponds to horizontal distance between soil volumes from different pits located in a same plot (first symbol between 10 m and 100 m), different plots of a same site (between 100 m and 5000 m) or different sites (>5000 m). Full symbols indicate significantly positive or negative Moran's *I* value ($P<0.05$).

## Discussion

### Origin of soil charcoal: a bimodal age distribution

Charcoal was found in the majority of the plots in each study area and in every soil layers, although at a higher frequency for the layers lying between 10 or 20 cm and 60 cm deep, which is consistent with previous studies [48]. Radiocarbon dating revealed two periods of ancient fires, with 16 samples dated between 80 to 860 BP ("recent" charcoals), and 43 samples between 1500 to 2200 BP (plus one dated 9400 BP; "ancient" charcoals). This bimodal age distribution has already been observed in previous $^{14}C$ dates distributions from other Central African countries [16,48]. It is also consistent with previous archeological evidence from Gabon and Cameroon pointing out high Human population density during the third millennium BP up to a few centuries afterwards, followed by a gap of Human occupation during the Middle Age [6,19]. This congruence between the abundances of charcoal and archeological artifacts through time supports the hypothesis that charcoal would mainly result from human-mediated fires. Ancient charcoals were found throughout the whole soil profile while recent ones were only found in the superficial soil layers (0–25 cm; Figure 2). For ancient charcoal, the absence of any age-depth relationship might be due to soil profile mixing (pedoturbation) resulting from biotic or abiotic agents. The main biotic agents in tropical soils are termites and ants [49–52] but earthworms [53], rodents, wind-fallen trees or tree uprooting can also play a role [54], while the main abiotic factor is colluvium [55] when located downhill or on a steep slope.

### Charcoal abundance displays significant spatial structure

Charcoal was found in all sites, suggesting that Human impacted the whole forest cover in every study areas, especially in Area 3, where charcoal was present in 94% of the plots. Its abundance, however, displayed significant spatial autocorrelation at different scales, reflecting differential intensity of past disturbance across each region. At the smallest spatial scale, charcoal abundance was highly correlated between adjacent layers of a same pit, but the correlation dropped sharply when non-adjacent layers were compared. Hence, charcoal tended to occur as relatively small aggregates in the soil. There was also a regular decay of spatial autocorrelation with the logarithm of the horizontal distance, indicating spatial heterogeneity among plots and sites. In Area 1, a substantial part of the spatial autocorrelation can be explained by an East-West increase in charcoal abundance among sites (Figure S1), which could reflect a landscape gradient (~100 km) in the intensity of past slash-and-burn activities along the transect. In this area, as well as in Area 2, the correlation between charcoal abundance (at the plot level) in the 0–20 cm and the 20–100 cm soil layers was highly significant, according to both classic correlation tests and torus-translation tests (the latter testing within site correlation), indicating that the similarity of charcoal abundance between both layers occurred at the landscape scale (among sites) but also locally (within a site). Unfortunately, it cannot be assessed whether this correlation indicates that slash-and-burn agriculture was concentrated in the same places over the last centuries and c. 2000 years ago because the upper soil layer contains a mixture of both recent and ancient charcoal.

### Local distribution of light-demanding species does not seem to be explained by charcoal abundance

In all study areas, most of the identified species could be categorized in a RG and characterized by a WSG. The few uncategorized species may either have a non-well known ecology or very large ecological amplitudes for light conditions. At the plot level, the relative abundance or basal area of each RG, as well as the mean WSG, displayed significant spatial autocorrelation in Area 1, while only two to three of these variables were spatially structured in Areas 2 and 3 (see Appendix S5).

Under our initial hypothesis, slash-and-burn agriculture favored the establishment of pioneer (P) species, which were then replaced by non-pioneer light demanding (NPLD) species within less than a

**Table 2.** Mean charcoal abundance index (CAI) and functional trait variables per site, and Kruskal-Wallis tests for among sites differences within each study area (using values computed at the plot level).

| | CAI[a] | | Functional traits | | | | | | |
|---|---|---|---|---|---|---|---|---|---|
| | **0–20** | **20–100** | **WSG[b]** | **P-a** | **NPLD-a** | **SB-a** | **P-b** | **NPLD-b** | **SB-b** |
| Site 1 | 0.29 | 0.37 | 0.62 | 0.29 | 0.28 | 0.43 | 0.34 | 0.36 | 0.31 |
| Site 2 | 0.29 | 0.44 | 0.64 | 0.18 | 0.32 | 0.50 | 0.24 | 0.43 | 0.35 |
| Site 3 | 0.16 | 0.27 | 0.66 | 0.15 | 0.20 | 0.65 | 0.19 | 0.35 | 0.45 |
| Site 4 | 0.18 | . | 0.62 | 0.25 | 0.27 | 0.47 | 0.31 | 0.37 | 0.31 |
| Site 5 | 0.11 | 0.30 | 0.65 | 0.16 | 0.16 | 0.68 | 0.28 | 0.21 | 0.51 |
| Site 6 | 0.14 | 0.16 | 0.62 | 0.24 | 0.14 | 0.62 | 0.41 | 0.20 | 0.41 |
| ***p*-value[c]** | *** | *** | *** | ** | *** | *** | ** | *** | ** |
| Site 7 | 0.47 | 0.35 | 0.59 | 0.32 | 0.20 | 0.48 | 0.47 | 0.23 | 0.30 |
| Site 8 | 0.35 | 0.28 | 0.58 | 0.35 | 0.22 | 0.44 | 0.43 | 0.26 | 0.31 |
| Site 9 | 0.56 | 0.41 | 0.64 | 0.24 | 0.15 | 0.61 | 0.37 | 0.22 | 0.41 |
| ***p*-value** | 0.43 | 0.39 | *** | * | 0.16 | *** | 0.31 | 0.77 | * |
| Site 10 | 0.40 | 0.38 | 0.62 | 0.19 | 0.27 | 0.55 | 0.26 | 0.42 | 0.32 |
| Site 11 | 0.42 | 0.36 | 0.65 | 0.23 | 0.15 | 0.62 | 0.28 | 0.25 | 0.47 |
| Site 12 | 0.38 | 0.55 | 0.64 | 0.22 | 0.20 | 0.58 | 0.27 | 0.32 | 0.41 |
| ***p*-value** | 0.78 | 0.19 | 0.39 | 0.53 | ** | 0.46 | 0.97 | * | 0.18 |

Sites 1 to 6, 7 to 9, and 10 to 12 correspond to Areas 1, 2 and 3, respectively.
[a]Charcoal abundance index (0–20 cm and 20–100 cm).
[b]Wood-specific gravity (g/cm$^3$).
[c]*P*-value of a Kruskal-Wallis test among sites: **p*-value <0.05, ***p*-value <0.01, *** *p*-value <0.001. P/NPLD/SB = Pioneers/Non-Pioneer Light-Demanders/Shade-Bearers. a = relative abundance. b = relative basal area.

**Table 3.** Pearson correlations between CAI in two soil layers (0–20 cm and 20–100 cm) and variables related to species functional traits.

| | Area 1 | Area 2 | Area 3 |
|---|---|---|---|
| | r[b] | r | r |
| **P** | −0.05 | −0.03 | 0.08 |
| | *−0.15* | *−0.21* | *0.00* |
| **NPLD** | 0.20* | −0.26 | 0.24 |
| | *0.28**** | *0.10* | *−0.13* |
| **SB** | −0.09 | 0.19 | −0.26 |
| | *−0.07* | *0.15* | *0.11* |
| **P relative basal area** | −0.18 | −0.07 | −0.04 |
| | *−0.20* | *−0.12* | *−0.06* |
| **NPLD relative basal area** | 0.19 | −0.08 | 0.24 |
| | *0.28*** | *0.06* | *−0.14* |
| **SB relative basal area** | 0.00 | 0.17 | −0.21 |
| | *−0.06* | *0.08* | *0.20* |
| **WSG[a]** | 0.02 | −0.01 | 0.16 |
| | *−0.10* | *0.12* | *0.08* |

P/NPLD/SB = relative abundance of Pioneers/Non-Pioneer Light-Demanders/Shade-Bearers.
[a]Wood-specific gravity (g/cm$^3$).
[b]*r*-Pearson correlation between row variable and CAI in the (i) 0–20 cm (upper line) and (ii) the 20–100 cm (bottom line; values in italics) soil layers. "*"indicates significant test with the classic correlation test: *$p<=0.05$ **$p<0.01$ ***$p<0.001$. All the tests based on toroidal translations were non-significant.

century and, eventually, shade-bearer (SB) species became dominant after a few centuries. Hence, we expected a positive correlation between CAI of the upper layer, reflecting partially the intensity of centuries-old fires, and the relative abundance of NPLD. In Area 1, the relative abundance of NPLD was indeed significantly positively correlated with the abundance of charcoals from the 0–20 cm ($r=0.2$; $P=0.03$) soil layer, but also with charcoal from the 20–100 cm ($r=0.28$; $P=0.005$) soil layer, when using a classic correlation test, not correcting for spatial autocorrelation. This is due to the parallel decay of charcoal and NPLD abundances among the six sites from West to East, which may have revealed a causal relationship if the intensity of past slash-and-burn agriculture had been stronger in the West. However, as such pattern has not been observed in the other two areas, and as toroidal translation tests also indicated that no correlation between RG relative abundance (and basal area) and soil charcoal abundance occurred within site, the among site correlation in Area 1 may well be spurious.

In Area 2 and Area 3, no significant correlations were detected. In fact, for the specific hypothesis that NPLD trees would have been favored by recent fires, we can note that while there is indeed a positive correlation between CAI in the upper soil layer and the relative abundance or basal area of NDPL in Area 1 and 3 (*r*-Pearson ranging from 0.19 to 0.24), the correlations become negative in Area 2 (*r*-Pearson ranging from −0.08 to −0.26). Hence, variation in floristic composition cannot be clearly explained by the abundance of soil charcoal.

We therefore suggest several explanations for the absence of significant signal. First, assuming a link between tree age and diameter for the entire community, we suspected that most of the trees of our dataset were too young to reflect the last Human disturbances, generating statistical noise. Nevertheless, correlation coefficients between light-demanding species frequencies and CAI did not increase after eliminating the smallest trees of our dataset (dbh <30, <40 <50, <60 and <70 cm, not shown).

Second, we assumed that charcoal from the superficial soil layer may sometimes be removed or covered by allochtonous materials due to colluvium processes, in such a way that it cannot be detected anymore. However, even after removing plots potentially subject to colluvium processes, i.e. located on a steep slope or downhill, correlation tests remained non-significant (not shown). Thus, if pedoturbation affects charcoal abundance, it is rather due to biotic agents (e.g. termites or ants). The latter could have "contaminated" the superficial soil layer (0–20 cm) with "ancient" charcoal, explaining why 59% of the dated charcoals found in the 0–25 cm soil layer were "ancient" (>1500 BP). By mixing with recent charcoals near the soil surface, ancient charcoals inevitably skewed our estimation of the abundance of "young" vs. "old" charcoals" (as recent and ancient material cannot be distinguished in the field) and therefore generated important statistical noise when testing our correlations.

Third, the absence of correlation may be due to an inappropriate study scale. Indeed, under our hypothesis, a correlation between charcoal abundance and the prevalence of light-demanding species is expected provided that there is sufficient spatial heterogeneity in the intensity of slash-and-burn agriculture to imprint a spatial heterogeneity in the abundance of light-demanding species. However, widespread and repeated slash-and-burn agriculture over the landscape may have favored light-demanding species on a large scale without causing substantial spatial heterogeneity on a more local scale.

Finally, it remains possible that the impact of past slash-and-burn agriculture on the floristic composition of forest stands is weak compared to the impact of other factors, like soil properties, climatic gradients (expected to be weak in our study areas), the history of forest establishment (e.g. forest regression ca. 3000 BP), or even stochastic processes structuring the distribution of species.

## Conclusions

Investigating the role of past Human disturbances on tropical tree communities had, to our knowledge, never been explored with an adequate statistical design based on soil charcoal. Charcoal remains were found in most of the plots in each study area, revealing frequent and widespread past forest fires in SE and SW Cameroon, presumably prompted by slash-and-burn agricultural activities. Fires occurred during two very distinct periods, one lying from ca. 80 to 850 BP and another from ca. 1500 to 2800 BP. The abundance of charcoal and non-pioneer light-demanding tree species were both spatially structured and positively correlated in only one of the three study areas. However, the reverse pattern was observed in Area 2 and correlations were non-significant at a local scale when correcting for spatial autocorrelation. Testing alternative hypotheses on the drivers of floristic composition should give more insights to confirm or not whether charcoal is an appropriate proxy to study the impacts of past Human disturbances, and whether these disturbances have significantly modified present-day tree species assemblages in Central African moist forests.

## Supporting Information

**Figure S1 Mean site values of each RG relative abundance and CAI in Area 1.** Values were computed at the plot level. Mean CAI was computed over two soil layers: 0–20 and 20–100 cm. Note that data are missing for the 20–100 cm soil layer in site 4. N-P = Non-Pioneer.

**File S1 Tables S1-S4.** Table S1 (in File S1) Sites Coordinates and number of plots per site. Table S2 (in File S1) Abundances and functional traits of all species inventoried in the three study areas. Table S3 (in File S1) Dates BP of 60 charcoal samples. Table S4 (in File S1) Mean sites values of each variable tested for their difference between sites, in each study Area.

**Appendix S1 Contact details of logging companies.**

**Appendix S2 Datasets used to calibrate the charcoal abundance index.**

**Appendix S3 Files to be used in Torocor software for testing spatial autocorrelation of charcoal abundance at multiple spatial scales.**

**Appendix S4 Plot values for spatial autocorrelation and correlation tests on Torocor.**

**Appendix S5 Results of spatial autocorrelation tests for each variable of Appendix S4.**

**Appendix S6 Pearson correlations between all variables, tested by a torus-translation procedure.**

## Acknowledgments

We gratefully acknowledge the following logging companies for their logistic support: Pallisco (M. Rougeron, L. Douaud, P. Lagoute, R. Feteke), WIJMA Cameroun S.A. and SFID (Société Forestière et Industrielle de la Doumé/Doumé Forestry Company). We also thank Adeline Fayolle (Gembloux Agro-Bio Tech) for providing data on species functional traits thanks to the European-funded CoForChange project.

## Author Contributions

Analyzed the data: JV TD OH. Contributed reagents/materials/analysis tools: JV JMR KD JFG JLD TD OH. Wrote the paper: JV JMR KD JFG JLD TD OH. Contributed to field inventories: JV JMR ABB KD JFG JLD. Organized radiocarbon dating of charcoal samples: JMR.

## References

1. Willis KJ, Gillson L, Brncic TM (2004) How "virgin" is virgin rainforest? Science 304: 402–403.
2. Denham TP, Haberle SG, Lentfer C, Fullagar R, Field J, et al. (2003) Origins of Agriculture at Kuk Swamp in the Highlands of New Guinea. Science 301: 189–193.
3. Kealhofer L (2003) Looking into the Gap: Land Use and the Tropical Forests of Southern Thailand. Asian Perspectives 42(1): 72–95.
4. Ross N (2011) Modern tree species composition reflects ancient Maya "forest gardens" in northwest Belize. Ecological Applications 21: 75–84.
5. Glaser B, Birk JJ (2012) State of the scientific knowledge on properties and genesis of Anthropogenic Dark Earths in Central Amazonia (terra preta de Indio). Geochimica et Cosmochimica Acta 82: 39–51.
6. Oslisly R (1996) The middle Ogooué valley, Gabon: Cultural changes and palaeoclimatic implications of the last four millennia. The growth of farming communities in Africa from the equator southwards, Vol. Azania XXIX-XXX (ed. by J.E.G. Sutton), pp. 324–331. The British Institute in Eastern Africa, London.
7. Oslisly R, White L, Bentaleb I, Favier C, Fontugne M, et al. (2013) Climatic and cultural changes in the west Congo Basin forests over the past 5000 years. Philosophical Transactions of the Royal Scociety, B 368, 20120304.
8. Wotzka H-P (2006) Records of activity: radiocarbon and the structure of Iron Age settlement in Central Africa. In Grundlegungen. Beiträge zur europäischen und afrikanischen Archäologie für Manfred K.H. Eggert (ed. HP . Wotzka), pp. 271–289. Tübingen, Germany.
9. Elenga H, Schwartz D, Vincens A, Bertaux J, De Namur C, et al. (1996) Diagramme pollinique holocène du lac Kitina (Congo): mise en évidence de changements paléobotaniques dans le massif forestier du Mayombe. CR Acad Sc 323: 403–410.
10. Shanahan T, Beck JW, Overpeck JT, McKay NP, Pigati JS, et al. (2012) Late Quaternary sedimentological and climate changes at Lake Bosumtwi Ghana: New constraints from laminae analysis and radiocarbon age modeling. Palaeogeography, Palaeoclimatology, Palaeoecology 361–362: 49–60.
11. Schwart D (1992) Assèchement climatique vers 3 000 B.P. et expansion Bantu en Afrique centrale atlantique: quelques réflexions. Bull. Soc. Géol. France 163(3): 353–361.
12. Maley J (1992) Mise en évidence d'une péjoration climatique entre ca 2500 et 2000 ans BP en Afrique tropicale humide. Bulletin de la Société Géologique de France 163: 363–365.
13. Vincens A, Schwartz D, Bertaux J, Elenga H, de Namur C (1998). Late Holocene Climatic Changes in Western Equatorial Africa Inferred from Pollen from Lake Sinnda, Southern Congo. Quaternary Research 50: 34–45.
14. Ngomanda A, Neumann K, Schweizer A, Maley J (2009) Seasonality change and the third millennium BP rainforest crisis in southern Cameroon (Central Africa). Quaternary Research 71: 307–318.
15. Neumann K, Bostoen K, Höhn A, Khalheber S, Ngomanda A, et al. (2011) First farmers in the Central African rainforest: A view from southern Cameroon. Quaternary International 249: 53–62.
16. Hubau W (2013) Exploring charcoal as a natural archive for palaeofire and vegetation history of the Mayumbe, Democratic Republic of the Congo. PhD thesis, Ghent University.
17. Van Gemerden BS, Olff H, Parren MPE, Bongers F (2003) The pristine rain forest? Remnants of historical Human impacts on current tree species composition and diversity. Journal of Biogeography 30: 1381–1390.
18. Morin-Rivat J, Fayolle A, Gillet J-F, Bourland N, Gourlet-Fleury S, et al. (2014). New evidence of Human activities during the Holocene in the Lowland Forests of the Northern Congo Basin. Radiocarbon 56: 209–220.
19. White LJT, Oates JF (1999) New data on the history of the plateau forest of Okomu, southern Nigeria: an insight into how Human disturbance has shaped the African rain forest. Global Ecology and Biogeography 8: 355–361.

20. Swaine MD, Hall JB (1988) The mosaic theory of forest regeneration and the determination of forest composition in Ghana. Journal of Tropical Ecology 4: 253–269.
21. Clist B (2005) Des premiers villages aux premiers Européens autour de l'estuaire du Gabon: Quatre millénaires d'interactions entre l'homme et son milieu. PhD thesis, Université Libre de Bruxelles, Brussels. 938p.
22. Jones EW (1955) Ecological studies of the rainforest of southern Nigeria IV. The plateau forest of the Okomu Forest Reserve, Part 1. The environment, the vegetation types of the forest, and the horizontal distribution of species. Journal of Ecology 43: 564–594.
23. Jones EW (1956) Ecological studies of the rain forest of southern Nigeria IV. The plateau forest of the Okomu Forest Reserve, Part 2. The reproduction and history of the forest. Journal of Ecology 44: 83–117.
24. Letouzey R (1985) Notice de la carte phytogéographique du Cameroun au 1/500.000. Institut de la Carte Internationale de la Végétation, Toulouse and Institut de la Recherche Agronomique, Yaoundé (Cameroon).
25. Boulvert Y (1986) Carte phytogéographique de la république centrafricaine au 1/1 000 000. Orstom, Paris, France.
26. Fayolle A, Engelbrecht B, Freycon V, Mortier F, Swaine M, et al. (2012) Geological Substrates Shape Tree Species and Trait Distributions in African Moist Forests. PLoS ONE 7(8): e42381. doi:10.1371/journal.pone.0042381.
27. Lebrun J, Gilbert G (1954) Une classification écologique des forêts du Congo. Publication INEAC série scientifique 63, INEAC, Bruxelles.
28. Fay JM (1997) The ecology, social organization, populations, habitat and history of the western lowland gorilla (Gorilla gorilla gorilla Savage and Wyman 1847). PhD thesis, Washington University, Washinghton D.C.
29. Biwolé AB, Morin-Rivat J, Fayolle A, Bitondo D, Dedry L, et al. (2014). New data on the recent history of the littoral forests of southern Cameroon: an insight into the role of historical human disturbances on the current forest composition. Plant Ecology and Evolution, 148 (1). In press.
30. WRB (2006) World reference base for soil resources. 2nd edition. FAO, IUSS, ISRIC (Eds.) World Soil Resources Reports No. 103. FAO, Rome, 145 p.
31. Harms KE, Condit R, Hubbell SP, Foster RB (2001) Habitat association of trees and shrubs in a 50-ha neotropical forest plot. Journal of Ecology 89: 947–959.
32. Talon B, Payette S, Filion L, Delwaide A (2005) Reconstruction of the long-term fire history of an old-growth deciduous forest in Southern Québec, Canada, from charred wood in mineral soils. Quaternary Research 64: 36–43.
33. Talon B (2010) Reconstruction of Holocene high altitude vegetation cover in the French Southern Alps: evidence from soil charcoal. The Holocene 20: 34–44.
34. Chabal L (1997) *Forêts et sociétés en Languedoc (Néolithique final, Antiquité tardive): l'anthracologie, méthode et paléoécologie*. Documents d'archéologie française, 63, éd. de la Maison des sciences de l'Homme, Paris: 189 p., 50 fig., 2 annexes, lex., bib.
35. Bronk Ramsey C (1995) Radiocarbon calibration and analysis of stratigraphy: The OxCal program. Radiocarbon 37(2): 425–430.
36. Bronk Ramsey C (2001) Development of the radiocarbon calibration program OxCal. Radiocarbon 43(2A): 355–363.
37. Bronk Ramsey C, Staff RA, Bryant CL, Brock F, Kitagawa H, et al. (2012) A Complete Terrestrial Radiocarbon Record for 11.2 to 52.8 kyr B.P. Science 338(6105): 370–374.
38. Reimer PJ, Baillie MGL, Bard E, Bayliss A, Beck JW, et al. (2009) IntCal09 and Marine09 radiocarbon age calibration curves, 0–50,000 years cal BP. Radiocarbon 51(4): 1111–1150.
39. Dauby G, Hardy OJ (2011) Sampled-based estimation of diversity sensu stricto by transforming Hurlbert diversities into effective number of species. Ecography 34: 001–012.
40. Hawthorne WD (1995) Ecological profiles of Ghanaian forest trees. Oxford Forestry Institute, Department of Plant Sciences, University of Oxford.
41. Slik JWF (2005) Assessing tropical lowland forest disturbance using plant morphological and ecological attributes. Forest Ecology and Management 205: 241–250.
42. Chave J, Coomes D, Jansen S, Lewis SL, Swenson NG, et al. (2009) Towards a worldwide wood economics spectrum. Ecology Letters 12: 351–366.
43. Zanne AE, Lopez-Gonzalez G, Coomes DA, Ilic J, Jansen S, et al. (2009) Global wood density database. Dryad. Identifier: http://hdl.handle.net/10255/dryad.235.
44. Slik JWF (2006) Estimating species-specific wood density from the genus average in Indonesian trees. Journal of Tropical Ecology 22: 481–482.
45. R Development Core Team. (2009) R: A language and environment for statistical computing. In: R Foundation for Statistical Computing (Ed.), Vienna, Austria.
46. Moran PAP (1948) The interpretation of statistical maps. Journal of the Royal Statistical Society B 10: 243–251.
47. Mantel N (1967) The detection of disease clustering and a generalized regression approach. Cancer Research 27: 209–220.
48. Gillet J-F (2013) Les forêts à Marantaceae au sein de la mosaïque forestière du nord de la république du Congo: origines et modalités de gestion. PhD thesis, Université de Liège - Gembloux Agro-bio Tech.
49. Lee KE, Wood TG (1971) Termites and soils. Academic Press, London, UK. New York, USA.
50. Miklós AAW (1992) Biodynamique d'une couverture pédologique dans la région de Botucatu (Brésil-SP). PhD thesis Université Paris-6, France.
51. Mc Brearty S (1990) Consider the humble termite: Termites as agents of post-depositional disturbance at African archaeological sites. Journal of Archaeological Science 17: 111–143.
52. Araujo AGM (2013) Bioturbation and the upward movement of sediment particles and archaeological materials: comments on Bueno, et al. Journal of Archaeological Science 40: 2124–2127.
53. Lavelle P (1983) The soil fauna of tropical savannas. Elsevier Scientific Publishing Company, Amsterdam, the Netherlands.
54. Johnson DL (1990) Biomantle evolution and the redistribution of earth materials and artifacts. Soil Science 149: 84–102.
55. Gouveia SEM, Pessenda LCR (2000) Datation par le $^{14}C$ de charbons inclus dans le sol pour l'étude du rôle de la remontée biologique de matière et du colluvionnement dans la formation de latosols de l'état de Sao Paulo, Brésil. C. R. Acad. Sci. Paris, Earth and Planetary Sciences 330: 133–138.

18

# Spatial Distribution of Oak Mistletoe as It Relates to Habits of Oak Woodland Frugivores

**Ethan A. Wilson[1]*, Patrick J. Sullivan[1], Janis L. Dickinson[1,2]**

**1** Department of Natural Resources, Cornell University, Ithaca, New York, 14850, United States of America, **2** The Cornell Lab of Ornithology, 159 Sapsucker Woods Road, Ithaca, New York, 14850, United States of America

## Abstract

This study addresses the underlying spatial distribution of oak mistletoe, *Phoradendron villosum*, a hemi-parasitic plant that provides a continuous supply of berries for frugivorous birds overwintering the oak savanna habitat of California's outer coast range. As the winter community of birds consuming oak mistletoe varies from group-living territorial species to birds that roam in flocks, we asked if mistletoe volume was spatially autocorrelated at the scale of persistent territories or whether the patterns predicted by long-term territory use by western bluebirds are overcome by seed dispersal by more mobile bird species. The abundance of mistletoe was mapped on trees within a 700 ha study site in Carmel Valley, California. Spatial autocorrelation of mistletoe *volume* was analyzed using the variogram method and spatial distribution of oak mistletoe *trees* was analyzed using Ripley's K and O-ring statistics. On a separate set of 45 trees, mistletoe volume was highly correlated with the volume of female, fruit-bearing plants, indicating that overall mistletoe volume is a good predictor of fruit availability. Variogram analysis showed that mistletoe volume was spatially autocorrelated up to approximately 250 m, a distance consistent with persistent territoriality of western bluebirds and philopatry of sons, which often breed next door to their parents and are more likely to remain home when their parents have abundant mistletoe. Using Ripley's K and O-ring analyses, we showed that mistletoe trees were aggregated for distances up to 558 m, but for distances between 558 to 724 m the O-ring analysis deviated from Ripley's K in showing repulsion rather than aggregation. While trees with mistletoe were aggregated at larger distances, mistletoe was spatially correlated at a smaller distance, consistent with what is expected based on persistent group territoriality of western bluebirds in winter and the extreme philopatry of their sons.

**Editor:** Erik Sotka, College of Charleston, United States of America

**Funding:** This work was supported by grants from National Science Foundation-IOS (0097027 and 0718416 to J.L.D.). Ethan Wilson was supported on a NSF REU Supplement to NSF grant 0718416. The funders had no role in study design, data collection and analysis, decision to publish, or preparation of the manuscript.

**Competing Interests:** The authors have declared that no competing interests exist.

* Email: ethanw31@uw.edu

## Introduction

The spatial distribution of resources is an important driver of social behavior in a range of animal species [1], [2], [3]. Where food plants and animal social systems are interdependent, they can become mutual engineers of each other's demography, dispersal, distribution, and abundance. Here we examine the spatial distribution of a key winter resource, oak mistletoe (*Phoradendron villosum*), in light of the behavior of its winter frugivores, which vary from species that hold persistent year-round territories to those that flock and adopt a mobile lifestyle in winter. Mistletoes are recognized as keystone resources throughout the world [4].

Multi-scalar assessment of landscape patterns is an important precursor to understanding the distinct processes that give rise to ecological patterns [5], but intensive data collection on keystone species is often lacking. Group-territorial frugivores like western bluebirds (*Sialia mexicana*) are likely to disperse seeds locally, producing a clumping of resources at the scale of one to two territories, whereas mobile species will tend to disperse seeds more broadly, leading to aggregation of mistletoe over larger areas. From the point of view of the plants, territorial frugivores may reduce opportunities for long distance dispersal of seeds [6]. Here we conduct spatial analysis at two different scales to look for signatures consistent with both types of seed dispersers.

Western bluebirds are common in the outer coast range of California, occupying persistent year-round territories and regularly overwintering in family groups after a brief period of juvenile dispersal in late summer and early fall [7]. The ecological driver of year-round territoriality and family-group living in winter is oak mistletoe, which is hemiparasitic on certain species of oak trees [8], [9]. Mistletoe plants produce a large and constant supply of berries from October until early spring, providing a stable winter food resource for western bluebird family groups (mistletoe seeds appear in 100% of fecal samples in winter, [10].) Western bluebirds in turn act as local dispersal agents for mistletoe plants through continuous, multi-generational occupancy of year-round territories, frequent use of favorite perches, and viscous patterns of settlement in which sons nest near their parents [11], [12].

The importance of mistletoe to delayed dispersal of western bluebirds was established at Hastings Reserve in Carmel Valley, CA with an experiment in which half the mistletoe was removed from a set of territories. Only 10% of removal territories retained sons through their first winter, compared to 87% of control territories. When parents left their territories for winter, their territories had lower volumes of mistletoe than did territories of parents remaining for winter [10]. While this experiment demonstrated that mistletoe is a causal agent for philopatry of sons [10], direct examination of the influence of western bluebirds on the dispersal and establishment of mistletoe is more challenging, because oak mistletoe germinates rarely, grows slowly, and persists on trees in excess of 60 years, based on a collection of photographs from the Hastings Reserve housed in the Museum of Vertebrate Zoology, UC Berkeley. Given that sons born on high quality territories tend to remain home and form kin neighborhoods in the vicinity of their parents, areas with abundant mistletoe tend to be used persistently. This should augment the "planting" side of the interaction producing pockets of high quality territories and thus spatial autocorrelation of mistletoe abundance, which we measure as total volume per tree. Thus spatial autocorrelation of mistletoe volume may both augment the benefits of delayed and localized dispersal in western bluebirds and also be a direct consequence of this augmentation via a positive feedback loop.

Other common winter frugivores at Hastings Reserve include American robins, *Turdus Migratorius*, and cedar waxwings, *Bombycilla cedrorens*. Where they have been studied, American robins can be territorial in winter, eat a wide variety of fruits, and engage in fruit defense against both hetero- and conspecifics [13]. Juveniles robins sometimes remain in flocks near their natal territories and these flocks remain until food supplies become insufficient [14], [15]. Mobile feeding flocks of more than 250 American robins are also seen, indicating plasticity depending on the distribution of resources and on the ability of group size to increase success competing for resources [15], [16]. Cedar waxwings, *Bombycilla cedrorum*, occur commonly at Hastings Reserve. They are nonterritorial throughout the year and, although their nests are aggregated in spring, they flock to and aggregate at fruiting sites away from the nest [17]. Their diets are comprised almost entirely of fruit in winter, and they eat a variety of fruit species as they move about in roving flocks [18]. Other frugivore species appear less important, such as band-tailed pigeons, *Patagioenas fasciata*, which are not known to be territorial in winter and commonly eat acorns in addition to other fruits [19]. Two territorial frugivores, Townsend's solitaires, *Myadestes townsendi*, and Phainopeplas, *Phainopepla nitens*, are rare in the outer coast range of California. If western bluebird territoriality concentrates dispersal of mistletoe within and around territories, mistletoe abundance will be spatially autocorrelated at the scale of territories; on the other hand, patterns of spatial autocorrelation may be disrupted by longer distance dispersal of seeds, for example, by roving flocks of species such as cedar waxwings, which are also common at Hastings Reserve.

In this paper we address the potential for multi-scalar patterns of abundance and occurrence of oak mistletoe, based on differences in seed dispersal distances for territorial versus the more mobile frugivores feeding on oak mistletoe. Using the variogram method we determine whether spatial autocorrelation of mistletoe *volume* per tree occurs over distances consistent with prolonged territory occupancy, winter family group living, territory budding, and localized dispersal of western bluebird sons. We then use Ripley's K and O-ring statistics to ask over what distance trees with mistletoe are spatially aggregated. Our expectation is that spatial autocorrelation resulting from localized dispersal of mistletoe seeds by western bluebirds will be apparent within a background of larger scale aggregation of oak trees infested with mistletoe.

## Methods

### Study areas

The mistletoe data used in this paper were collected in conjunction with a long-term (1983-present) study of box-nesting, color-banded western bluebirds breeding and wintering on Hastings Reserve and the contiguous Oak Ridge Ranch, Carmel Valley, California (herein referred to as Hastings/Oak Ridge). Identical monitoring procedures were also instigated on nearby Rana Creek Ranch from 2001–2006. Data on dispersal distances of marked natal group males were collected from Hastings/Oak Ridge and Rana Creek Ranch (2001–2006).

### Relative frequency of three main frugivores

The relative frequency of the three main frugivores was estimated using publicly available data from eBird observations entered from 2006–2014 on our study area [20]. We only considered counts conducted from September through February, eliminating redundant entries in cases where more than one individual reported data on the same birding trip.

### Mistletoe data collection

Data on mistletoe-parasitized trees were collected during the years 2001–2003 by traversing the area within 200 m of each of 363 nest boxes on the contiguous Hastings/Oak Ridge site. Our study site is well north of Arroyo Grande, CA, the dividing line above which oak mistletoe grows mostly on blue oak and valley oak trees (*Quercus douglasii* and *Quercus lobata*, respectively) and only rarely on coast live oak (*Quercus agrifolia*) [21]. Data were collected after leaf drop and before trees had fully leafed out again. The Hastings/Oak Ridge trees ($n = 2,658$) were geolocated using a differential global positioning system (Garmin), marked with uniquely numbered aluminum tree tags, and photographed next to a reference stake using a digital camera. The number of mistletoe clumps was counted and classified in the field for each tree using 6 size categories (diameter <20, 20–50, 50–75, 75–100 cm; 1–1.5 m). Mistletoe volume was calculated for the first five categories using the formula for the volume of a sphere with radius 10, 20, 35, 60, and 75 cm and summed to obtain a total mistletoe volume for each tree. Clumps measuring>1.5 m tended to be elongate and their volume was approximated as the sum of the volume for two or more 75 cm diameter spheres [10]. Data were compiled and organized within ArcGIS10 [22], and analyzed using statistical software package R [23].

In addition to collecting data on mistletoe as described above, we selected a subset of trees from another of our nearby study plots (Rana Creek Ranch) to investigate the relationship between mistletoe volume and the volume of female, fruit-bearing mistletoe plants; *P. villosum* is dioecious. In October - December of 2001 we selected and photographed a sample of 45 mistletoe-bearing blue oaks (*Quercus douglasii*) and valley oaks (*Quercus lobata*). Close inspection of trees in the field revealed that large clumps of mistletoe were often aggregates of a set of smaller clumps whose exact number was not possible to estimate. All accessible mistletoe clumps with flowers and fruits were sexed during routine, weekly phenology assessments. We used a Pearson correlation to assess the relationship between total clump volume per tree (which could be measured for all trees in our study), and total female clump volume per tree, which was feasible to measure only for a small number of trees. These measures served to test the appropriateness of using total mistletoe volume as a proxy for volume of female, fruit-bearing mistletoe available as food for western bluebirds.

### Winter territory size, dispersal distance, and spring retention of philopatric sons

Location data for assessing western bluebird territory size were collected in winter of 2004–5 from the Hastings/Oak Ridge and Rana Creek sites. Winter groups were censused for the presence of banded birds (and to band any unbanded individuals) from September through February in 2001–2006. Groups were stable by mid-October. During winter censuses, we recorded identities of all birds in the group and recorded locations to construct territory maps. We used methods similar to those described in Dickinson and McGowan [10], creating maps based upon independent data points gathered during repeat censuses of winter territories. Territories were assumed to encompass the area covered by the minimum convex polygon formed by the mapped points. In order to obtain a proxy for the bluebirds' radial territory extent, we used the total area for each territory map and computed the associated radius of a circle having the same area. The mean of these measures could then be compared to the degree of spatial dependence of mistletoe volume, as determined by the fitted variogram (see below).

To ask whether mistletoe volume was higher on territories where sons stayed home for winter and also remained on the study area in spring, we used a generalized linear model (Poisson distribution) with presence or absence in spring as a fixed factor, year as a random factor, and mistletoe volume within 100 m of the natal nest box as the response variable. The volume of mistletoe within 100 m of the natal nest box was previously shown to be positively correlated with the volume within 100 m of the central nest box on the winter territory, which in turn was positively correlated with the volume on 10 territories that were extensively mapped [9]. This data set included all sons remaining on their natal territories in winter for the five-year period where we had year-round data. Considering only the 166 males that remained in their natal groups for winter, we compared the natal territories of those that remained on the study area the following spring with natal territories of those that disappeared after surviving the winter.

In order to evaluate the dispersal distances of philopatric sons, we examined the distance dispersed by 82 males that remained on their natal territories their first winter and subsequently bred on the study area as yearlings. This analysis included winter groups monitored in 2001–2006 at Hastings Reserve and Rana Creek Ranch, and allowed us to assess the extent of localized dispersal for males that remained in their natal groups for winter [7].

## Vegetation layer describing blue oak-valley oak woodland

Over 99% of the oak mistletoe trees on our study area are blue oaks or valley oaks. Starting with vegetation classifications based on remote sensing data, we used TNTmips from 60 cm RGB and CIR aerial imagery to generate an improved vegetation map classifying the four main landcover types of interest (chaparral, blue oak-valley oak woodland, mixed oak-madrone forest, and open savanna grassland). The original remote sensing vegetation classifications more fine-grained and were found to be inaccurate at distinguishing the major vegetation types of interest; an improved classification of the vegetation layer was derived using Manifold v6.5 (Manifold Net Ltd., Carson City, NV, USA) using majority pixel analysis, which calculates the majority values over a windowed area to deal with remaining errant vegetation areas. Errors due to shadow effects were then corrected via groundtruthing and comparison to high resolution aerial photos, using the combined expertise of two individuals (J.L. Dickinson and Walter D. Koenig) who worked and lived year-round on the study area for more than 18 and 31 years, respectively.

## Spatial autocorrelation of mistletoe volume

We then examined the spatial autocorrelation of the continuous variable, mistletoe *volume*, in order to test for evidence consistent with the hypothesis that sons remaining on high mistletoe volume territories *and* establishing territories adjacent to their parents' territories will have an increased chance of acquiring a high mistletoe volume territory themselves (spatial dependence for all distances less than mean winter radius plus some distance representative of territories immediately surrounding the parents' territories). Alternatively, if no spatial autocorrelation exists, this would suggest that the mistletoe volume on a male's natal territory is a poor predictor of what he can expect to acquire by staying home for winter and dispersing next door to his parents. Spatial autocorrelation of mistletoe volume at a larger scale would not be consistent with expectations based on persistent group territoriality of western bluebirds.

In order to analyze the spatial autocorrelation of mistletoe, we used a log-transform of the volume measure. Simulation of the observed process underlying mistletoe volume was modeled by fitting a spherical variogram to the sample variogram. Due to the finite range of points, the sample variogram may contain a signal due to sampling error. To test for the significance of spatial autocorrelation via the variogram, we ran 1000 simulations in which the measurements for log-volume were randomly reassigned to the mistletoe data point locations, while the locations themselves remained the same. If the observed sample variogram falls within the 95% range of these simulated variograms, we then conclude that any autocorrelation may have occurred by chance [24]. Variogram analyses were performed in R, using the package gstat [25]. Using the fitted variogram, we created a kriged surface image portraying spatially interpolated measures of mistletoe log-volume (Fig. 1). Winter territory maps for western bluebirds were then overlaid on this image to examine territory placement with respect to mistletoe abundance.

## Spatial Point Pattern analysis for spatial aggregation of mistletoe-bearing trees

To measure the degree of spatial aggregation of the trees bearing mistletoe, we employed the Ripley's K-function and the

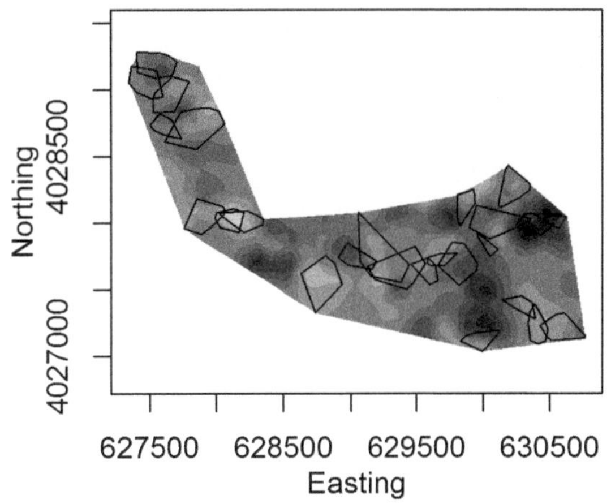


**Figure 1. Kriged surface image of predicted mistletoe log-volume, overlaid by mapped winter territories (2004–2005, from a portion of the larger study area) and the minimum bounding convex polygon for all territories.** White color represents high concentrations of mistletoe volume, while dark areas are indicative of lower intensity mistletoe volume.

O-ring statistic, complementary spatial statistical methods used to quantify aggregation in spatial point patterns [26], [27]. The Ripley's K statistic is based on the accumulation of points within a circle of increasing radius, while the O-ring statistic considers points within an annulus of increasing size. This distinction implies that as the scale of aggregation increases, strong small scale aggregation effects will continue to influence measures of large scale aggregation in the Ripley's K analysis, but not in the O-ring analysis [27].

Spatial aggregation patterns were analyzed for Hastings/Oak Ridge for the total mistletoe-bearing tree population (n = 2658 trees). Beginning with a vegetation map of blue and valley oak woodland (Fig. 2), we assumed based on our data that mistletoe growth only occurs within the real extent of blue oak-valley oak woodland. Our null model for the spatial extent of mistletoe habitat was thus restricted to that of its local host species, blue and valley oaks, as recommended in previous ecological literature [28], [29].

In both the Ripley's K and O-ring analyses, Monte Carlo simulations (199 total) were used to test the null hypothesis that the mistletoe points (trees) were distributed according to a homogeneous Poisson process. We did this by randomly reallocating 2658 points (trees), restricted to the spatial extent of oak vegetation. For each simulation, a test statistic was derived yielding a range of estimates for the expected value of the statistic under the null hypothesis. Using the results of these simulations, we derived approximate 95% confidence envelopes by using the $5^{th}$ lowest and highest of the simulated values for the statistic. These simulations were then compared to the observed test statistic for the true distribution of mistletoe points. The Ripley's K analysis was performed in R, using the package spatstat [30] and the O-ring analysis was performed using the software package Programita [27].

### Ethics Statement

The study was conducted on Hastings Natural History Reservation, a University of California Reserve associated with UC Berkeley (the UTM coordinates of Hastings office: Easting: 630032 Northing: 4027902). Queries to conduct research on this site should be communicated through the reserve director, who is also the broker for any work done on private lands adjacent to the reserve.

The work was conducted on USFWS banding permit # 23533 to Janis L. Dickinson and California Fish and Game permit # 801053 to Janis L. Dickinson. The field studies did not involve endangered or protected species. No animals were sacrificed or collected. For the purposes of this study, western bluebirds were color-banded and observed for 1-2 hours at a time to map territories. Animal use protocols are described in IACUC AUP #R212 at UC Berkeley and IACUC #2005–0137 at Cornell University.

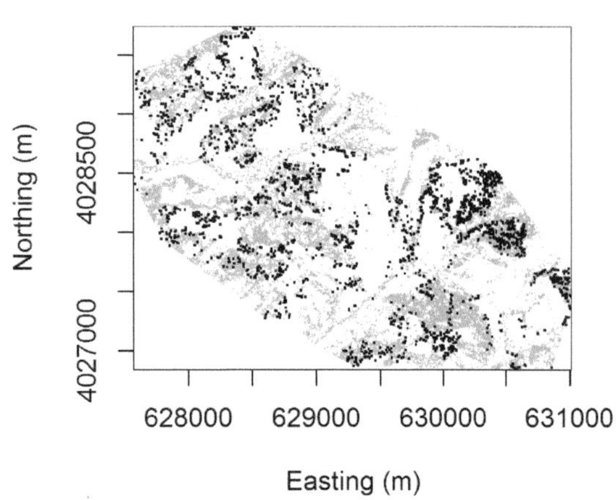


**Figure 2. Distribution of mistletoe points (black) on Hastings/Oak Ridge within the extent of our study site.** The oak background (grey) was defined spatially as the convex hull derived from the mistletoe points.

## Results

### Frequency of three main frugivores on study area

Based on 17 independent eBird counts collected over 8 winters from 2006–2014, the frequency of these three species was: 57% western bluebirds; 21% cedar waxwings; and 22% American Robin. The maximum number of individuals counted at one time for each species was: 12 western bluebirds, 15 cedar waxwings, and 12 American Robins.

### Mistletoe volume as a valid proxy for female mistletoe volume

In order to determine whether total volume per tree is a good proxy for the volume of female (fruiting) clumps per tree, we compared total and female mistletoe volume for 45 trees. Female volume increased linearly with total volume ($r^2 = 0.86$, $F_{1,43} = 265$, $P<0.001$, Fig. 3).

### Territory size for wintering western bluebirds

Extensive mapping of western bluebird territories in winter 2004–5 resulted in 25 territory maps for analysis on Hastings/Oak Ridge. The mean area of these territories was $41663 \pm 5062$ $m^2$, corresponding to a circular radius of 115 m. By finding the radius of a circle having the estimated mean area, we obtained an approximation of the distance from the center of a territory to its boundary, given that territories are not truly circular in shape. This distance was used in order to make inferences about the relationship between dispersal distance from territories and distance of spatial autocorrelation of mistletoe volume.

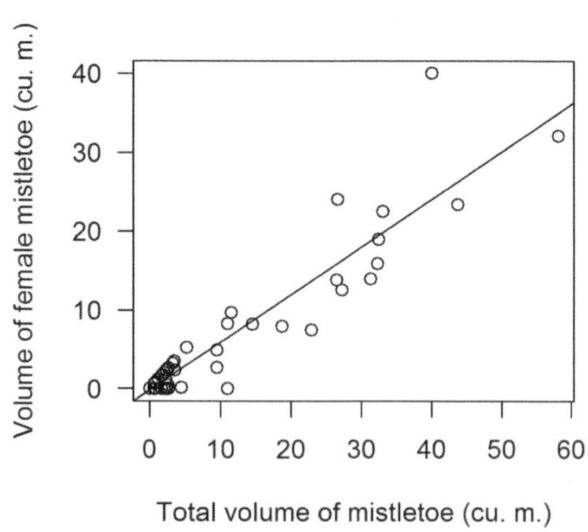


**Figure 3. Female mistletoe volume correlated with total mistletoe volume (*Phoradendron villosum*) on 45 blue oak and valley oak trees (*Quercus douglasii* and *Q. lobata*, respectively) monitored for berry production over the winter of 2001–2002 (r = 0.92, P<0.001).**

### Western bluebird sons' dispersal distances after winter

Males that settled on the study area in the breeding season came from territories with higher volumes of mistletoe than those that dispersed away from the study area after winter (lmer, mistletoe volume: $z = 13.0$, $N = 166$ males, $P<0.001$ (year as random factor), Fig. 4). Fig. 5 shows the combined within-site dispersal distribution for males that remained on their natal territories for winter at Hastings/Oak Ridge and Rana Creek Ranch sites. Approximately 99% of 82 recovered natal sons that bred settled < 900 m from their natal box and 58.5% bred within 250 m from their natal nest (Fig. 5), indicating that over half of sons that remained on their natal territory for winter bred within one territory of their natal nest. We detected one long-distance dispersal of a male about 5.6 km between the Hastings/Oak Ridge and the Rana Creek Site (not included in Fig. 5, nor in other analyses).

### Spatial autocorrelation of mistletoe volume

Mistletoe volume was analyzed using the fitted and empirical variograms. The variogram plots the semi-variance of the response variable, log-volume, as a function of distance. Plots of these, along with the parameter values for the nugget, partial sill, and range of the fitted spherical model, are represented in Fig. 6. The variogram range parameter is interpreted as evidence of spatial autocorrelation of mistletoe volume for all distances <250 meters. Fig. 6 also shows the sample variogram, compared to 1000 variograms of randomly reallocated data. The permutation test indicates that the hypothesis of absence of spatial autocorrelation of mistletoe is unlikely, as much of the empirical variogram lies outside of the major range of the permuted values [24]. Overlaying winter territory boundaries onto a kriged surface of mistletoe volume shows visually how remaining on territories with high levels of mistletoe abundance provides opportunities for acquisition of high quality territories nearby (Fig. 1).

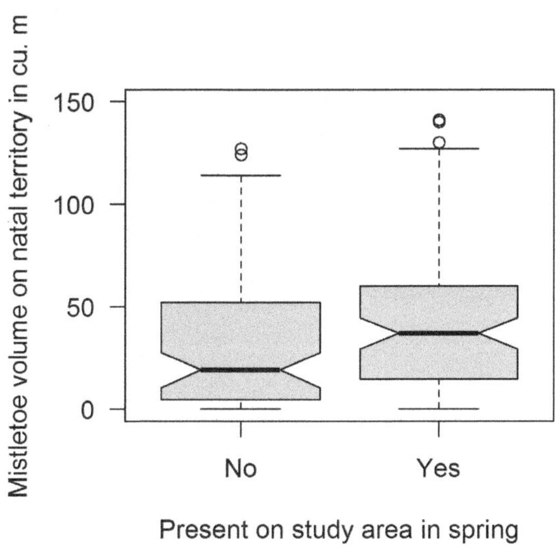


**Figure 4. Box plots demonstrating that for first-winter male western bluebirds wintering in their natal groups and surviving the winter, those present on the study area in spring had higher mistletoe volume within 100 m of their natal nest box than did those not present in spring.**

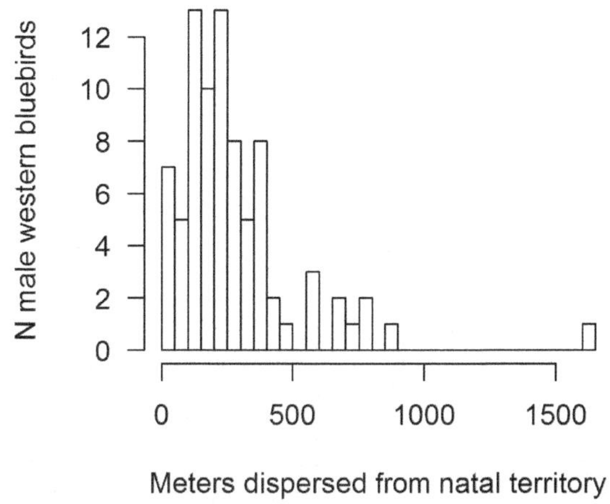


**Figure 5. Histogram of dispersal distances for western bluebird males (*Sialia Mexicana*) that wintered in their natal group in winter of 2001–2006 and bred on the study area the subsequent spring.**

### Spatial aggregation measure

Both the Ripley's *K*-function and the O-ring statistic demonstrated strong aggregation of mistletoe-bearing trees for distances up to 558 meters. However, for distances between 558 to 724 meters the Ripley's K analysis showed evidence of aggregation, while the O-ring analysis showed evidence of dispersion. The graph in Fig. 7 shows that for all distances measured, the observed K statistic lies above the confidence envelope, indicative of spatial aggregation, while the O-ring statistic transitions from above to below the confidence envelop at about 558 m. Although the graph of the observed O-ring statistic does not deviate far below the confidence envelope, it provides evidence of spatial repulsion that was not detected by the Ripley's K analysis.

## Discussion

Analysis of the variogram indicates that mistletoe volume is spatially autocorrelated for distances less than 250 meters. Given that it is only feasible for mistletoe to occur within the extent of

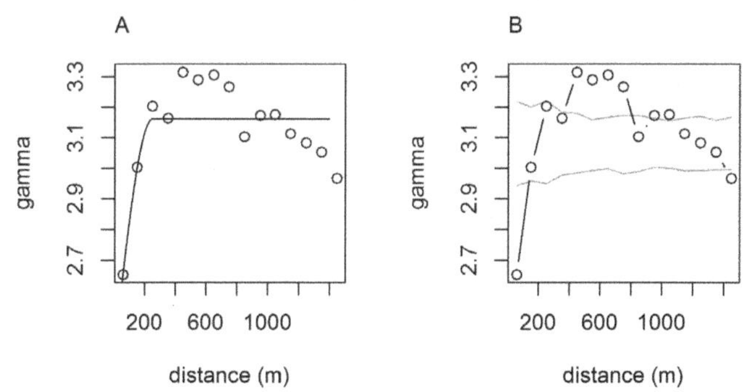


**Figure 6. Points and line representing the empirical and fitted variogram, respectively.** Parameters for the fitted spherical variogram include: partial sill = 0.79, nugget = 2.4 and range = 250.3 m (A). Permutation test of the variogram, which determines the general range of values that could occur due to random chance. The black line represents the true variogram, and the gray lines represent a 95% confidence envelope of 1000 variogram simulations of randomly permuted mistletoe volume measures; each using the same point locations as the original data (B).

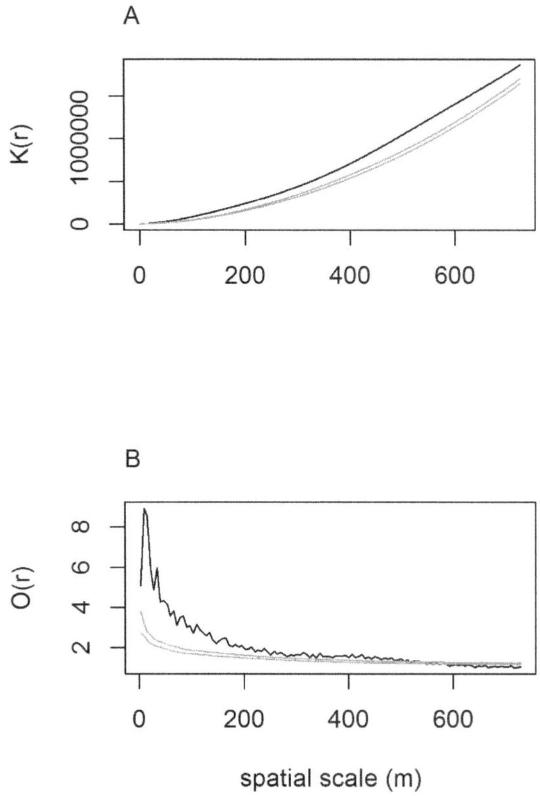


**Figure 7. Graph of the observed Ripley's K-statistic for the true distribution (black line), showing the average number of trees parasitized with mistletoe within a given distance.** The gray lines represent the 95% confidence envelopes for complete spatial randomness, based on 199 Monte Carlo simulations of 2658 random data points within the oak vegetation area (A). Graph of the observed O-ring statistic for the true distribution (black dotted line), as well as the 95% confidence envelopes (gray lines) for complete spatial randomness. This analysis was performed using a cell size of 6.2 $m^2$ and a ring width of 18.6 m. Both analyses were restricted to a maximum distance of 724 m (B).

blue and valley oak vegetation, spatial autocorrelation of volume up to 250 meters is consistent with the extent of western bluebirds' winter territories. This finding is in accordance with Aukema's (2004) suggestion that positive feedbacks can result in mistletoe augmentation at the scale of small neighborhoods by territorial bird species, although our focus here is on winter territoriality and winter fruit production [5]. As Fig. 1 indicates, from a western bluebird's point of view, the distribution of mistletoe wealth is heterogeneous at scales relevant to clustering of families on high mistletoe volume territories in winter and the dispersal of sons onto local breeding territories in spring.

Given the spatial dependence of mistletoe volume for distances up to 250 meters, the estimated mean winter territory radius of 115 meters, and data indicating that most dispersal events observed for sons fell within 0–250 meters of their natal nest, parental territories encompassing high quantities of mistletoe will typically have similarly high concentrations of mistletoe up to approximately 135 meters outside of their territory boundaries. This observed spatial autocorrelation of mistletoe volume supports the hypothesis that western bluebird sons remaining on winter territories with high mistletoe volume, as they tend to do [10], will acquire territories of similar quality when they settle next door and bud off a portion of their parents' territory in spring [9]. Specifically, using the ratio of correlated area ($A = \pi 250^2$) to estimated mean area over which mistletoe volume is spatially autocorrelated, we expect that between 2.8 and 5.2 additional territories could saturate a landscape within which spatial autocorrelation is still present. Most importantly, our results suggest that spatial autocorrelation of mistletoe volume is occurring at a scale relevant the persistent territoriality, winter family territoriality, and formation of kin neighborhoods in spring.

We cannot rule out the possibility that the underlying age structure, size or distribution of oak trees also play into the spatial aggregation of mistletoe. However, given that valley oaks are rare compared to blue oaks, it is unlikely that they generate the pattern of autocorrelation independently. Spatial autocorrelation of mistletoe volume could also be influenced disproportionately by relatively few, large plants, as opposed to many small plants, however, when large clumps of mistletoe are inspected very closely and pulled apart by cutting, it is clear that they often consist of multiple small plants that have grown together, a phenomenon that may reflect the favorite perch sites of birds living on territories.

The results of the Ripley's K and O-ring analyses provide evidence in favor of *some* seed dispersal occurring at greater scales up to 558 m. This pattern, while consistent with movements of cedar waxwings and American Robins, which are less common dispersers of mistletoe seeds on our study area, is also consistent with longer distance dispersal by western bluebirds, which make frequent trips off their territories to distant watering sites (J.L.D., unpublished observations). The observed repulsion detected by the O-ring analysis at scales beyond 558 m may be suggestive of competitive interactions, in which the likelihood of having nepotistic, related males living next door to each other decreases with distance, leading to an increased frequency of antagonistic interactions between adjacent nonrelatives.

Point data, defined loosely as presence or absence at a given location, are manifest as a non-continuous data set. However, the availability of continuous values (e.g. volume) associated with mistletoe points allowed us to examine the extent of spatial autocorrelation of resource abundance, which is more directly relevant to resource-dependent social behavior such as delayed dispersal and extreme philopatry of western bluebird sons. While the authors acknowledge that the interpretation of geostatistical methods applied to mistletoe point data should be viewed with caution, the availability of data on mistletoe volume provides an unusual opportunity to characterize some unequivocal characteristics of the underlying pattern of spatial autocorrelation.

Two complementary processes may be important to maintaining spatial autocorrelation: 1) long-term territory occupancy, in combination with territorial budding and local proliferation of dynasties from wealthy territories can increase local deposition and germination of mistletoe seeds around those territories, leading to increased spatial autocorrelation of mistletoe volume at scales of one to two territories and 2) local augmentation of mistletoe volume due to continuous occupancy can, in turn, increase the benefits of living in family groups, the likelihood that territories will be occupied for many years or even decades, and the frequency with which sons disperse next door to parents [9].

Although these relationships are usually viewed within the context of mutualism, persistent territoriality may not be beneficial to oak mistletoe if it reduces the potential for long-distance dispersal and colonization of new habitats or if it results in increased competition among mistletoe plants for germination sites or dispersal opportunities. This line of thinking is similar to recent conclusions drawn for the relationship between mistletoe and its specialist versus generalist mistletoe-eating birds [6], however, we specifically highlight territoriality and social behavior as fitness-related behaviors that can lead to patterns of seed dispersal that favor western bluebirds at the expense of oak mistletoe. Our evidence that mistletoe volume is associated with philopatry in western bluebirds is supported by both experimental and observational data [5], [9]. The spatial analysis in this paper is consistent with the idea that the underlying distribution of mistletoe is also influenced by the persistent territoriality of western bluebirds.

## Acknowledgments

We thank Walter Koenig, David Watson, and two anonymous reviewers for their comments on the manuscript. Thorsten Wiegand provided access to and advice on use of Programita software. We also thank staff of the Museum of Vertebrate Zoology and Hastings Reserve for logistical support and the owners of Oak Ridge and Rana Creek Ranches for access to their property. Many field interns helped to map mistletoe and to keep the western bluebird population banded over the years leading to this study, as well as conducting winter censuses and territory mapping.

## Author Contributions

Conceived and designed the experiments: EW PS JD. Performed the experiments: JD. Analyzed the data: EW JD. Contributed reagents/materials/analysis tools: PS. Wrote the paper: EW JD. Conceived and designed the analysis: EW PS JD. Statistical oversight and mentoring: PS. Field data collection: JD.

## References

1. Emlen ST, Oring L (1977) Ecology, sexual selection, and the evolution of mating systems. Science. 197: 215–223.
2. Emlen ST (1982a) The evolution of helping. I. An ecological constraints model. Am. Nat. 119: 29–39.
3. Emlen ST (1982b) The evolution of helping. II. The role of behavioral conflict. Am. Nat. 119: 40–53.
4. Watson DM (2001) Mistletoe- a keystone resource in forests and woodlands. Annu. Rev. Ecol. Syst. 32: 219–249.
5. Aukema JE (2004) Distribution and dispersal of desert mistletoe is scale-dependent, hierarchically nested. Ecography. 27: 137–144.
6. Watson DM, Rawsthorne J (2013) Mistletoe specialist frugivores: latterday "johnny Appleseeds" or self-serving market gardeners? Oecologia 172: 925–932.
7. Kraaijeveld K, Dickinson JL (2001) Family-based winter territoriality in western bluebirds (Sialia mexicana): the structure and dynamics of winter groups. Anim. Behav. 61: 109–117.
8. Hollinger DY (1983) Photosynthesis and water relations of the mistletoe, Phoradendron villosum, and its host, the California valley oak. Oecologia. 60: 396–400.
9. Dickinson JL, Ferree E, Stern CA, Swift R, Zuckerberg B (2014) Delayed dispersal in western bluebirds: Teasing apart the importance of resources and parents. Behavioral Ecology. doi: 10.1093/beheco/aru042.
10. Dickinson JL, McGowan A (2005) Winter resource wealth drives delayed dispersal and family-group living in western bluebirds. Proc. R. Soc. B. 272: 2423–2428.
11. Lewontin RC (1982) Organism and environment. In: H. C . Plotkin, editors.Learning, development and culture. Wiley, New York, pp.151–170.
12. Laland KN, Odling-Smee FJ, Feldman MW (1996) On the evolutionary consequences of niche construction. J. Evolution Biol. 9: 293–316.
13. Sallabanks R (1993) Fruit defenders vs. fruit thieves: winter foraging behavior in American Robins. J. Field Ornithol. 64: 42–48.
14. Hirth DH (1969) Dispersal and flocking of marked young robins (Turdus migratorius) after fledgling. J. Field Ornithol. 40: 208–215.
15. Sallabanks R, James FC (1999) American Robin (Turdus migratorius). The Birds of North America Online (A. Poole, Ed.). Ithaca, NY: Cornell Laboratory of Ornithology; The Birds of North America Online database. Available: http://bna.birds.cornell.edu/bna/. Accessed 2008 Mar 25. doi: 10.2173/bna.462.
16. Chavez-Ramirez F, Slack RD (1994) Effects of avian foraging and post-foraging behavior on seed dispersal patterns of Ashe Juniper. Oikos 71: 40–46.
17. Putnam LS (1949) The life history of the Cedar Waxwing. Wilson Bull. 61: 141–182.
18. Witmer MC, Mountjoy DJ, Elliot L (1997) Cedar Waxwing (Bombycilla cedrorum). The Birds of North America Online (A. Poole, Ed.). Ithaca: Cornell Lab of Ornithology; the Birds of North America Online database. Available: http://bna.birds.cornell.edu.proxy.library.cornell.edu/bna/species/309 doi: 10.2173/bna.309.
19. Keppie DM, Braun CE (2000) Band-tailed pigeon (Patagioenas fasciata). The Birds of North America Online (A. Poole, Ed.). Ithaca: Cornell Lab of Ornithology; the Birds of North America Online. Available: http://bna.birds.cornell.edu.proxy.library.cornell.edu/bna/species/530 doi: 10.2173/bna.530.
20. eBird. 2012. eBird: An online database of bird distribution and abundance [web application]. eBird, Cornell Lab of Ornithology, Ithaca, New York. Available: http://www.ebird.org. Accessed 2 February 2012.
21. Thomson VE, Mahall BE (1983) Host specificity by a mistletoe, Phoradendron villosum Nutt, Subsp. villosum, on three oak species in California. Bot. Gaz. 144: 124–131.
22. ESRI (2011) ArcGIS Desktop: Release 10. Redlands, CA: Environmental Systems Research Inst.
23. R Development Core Team (2011) R: A language and environment for statistical computing. R Foundation for Statistical Computing, Vienna, Austria. ISBN 3-900051-07-0, URL Available: http://www.R-project.org/.
24. Bivand RS, Pebesma E, Gómez-Rubio V (2008) Interpolation and Geostatistics. In: Gentleman R, Hornik K, Parmigiani G, editors. Applied Spatial Data Analysis with R. Springer Science and Business Media, LLC, pp.191–236.
25. Pebesma EJ (2004) Multivariable geostatistics in S: the gstat package. Comput. Geosci. 30: 683–691.
26. Ripley BD (1981) Spatial statistics. John Wiley and Sons, New York, New York, USA.
27. Wiegand T, Moloney KA (2004) Rings, circles, and null-models for point pattern analysis in ecology. Oikos 104: 209–229.
28. Kraft CE, Sullivan PJ, Karatayev AY, Burlakova LE, Nekola JC, et al. (2002) Landscape patterns of an aquatic invader: Assessing dispersal extent from spatial distributions. Ecol. Appl. 12: 749–759.
29. Jolles AE, Sullivan PJ, Alker AP, Harvell CD (2002) Disease transmission of aspergillosis in sea fans: inferring process from spatial pattern. Ecology. 83: 2373–2378.
30. Baddeley A, Turner R (2005) Spatstat: an R package for analyzing spatial point patterns. J. Stat. Softw. 12: 1–42. Available: http://www.jstatsoft.org/v12/i06/paper.

# Spatial Structure and Distribution of Small Pelagic Fish in the Northwestern Mediterranean Sea

**Claire Saraux[1]*, Jean-Marc Fromentin[1], Jean-Louis Bigot[1], Jean-Hervé Bourdeix[1], Marie Morfin[1], David Roos[1], Elisabeth Van Beveren[1], Nicolas Bez[2]**

**1** IFREMER (Institut Français de Recherche pour l'Exploitation de la MER), Research Unit EME (UMR 212), Sète, France, **2** IRD (Institut de Recherche pour le Développement), Research Unit EME (UMR 212), Sète, France

## Abstract

Understanding the ecological and anthropogenic drivers of population dynamics requires detailed studies on habitat selection and spatial distribution. Although small pelagic fish aggregate in large shoals and usually exhibit important spatial structure, their dynamics in time and space remain unpredictable and challenging. In the Gulf of Lions (north-western Mediterranean), sardine and anchovy biomasses have declined over the past 5 years causing an important fishery crisis while sprat abundance rose. Applying geostatistical tools on scientific acoustic surveys conducted in the Gulf of Lions, we investigated anchovy, sardine and sprat spatial distributions and structures over 10 years. Our results show that sardines and sprats were more coastal than anchovies. The spatial structure of the three species was fairly stable over time according to variogram outputs, while year-to-year variations in kriged maps highlighted substantial changes in their location. Support for the McCall's basin hypothesis (covariation of both population density and presence area with biomass) was found only in sprats, the most variable of the three species. An innovative method to investigate species collocation at different scales revealed that globally the three species strongly overlap. Although species often co-occurred in terms of presence/absence, their biomass density differed at local scale, suggesting potential interspecific avoidance or different sensitivity to local environmental characteristics. Persistent favourable areas were finally detected, but their environmental characteristics remain to be determined.

**Editor:** Konstantinos I. Stergiou, Aristotle University of Thessaloniki, Greece

**Funding:** Surveys were cofunded by IFREMER and the European Union through the Data Collection Framework (DCF). This study is part of the program EcoPelGol (Study of the Pelagic ecosystem in the Gulf of Lions), funded by France Filière Pêche (FFP). The funders had no role in study design, data collection and analysis, decision to publish, or preparation of the manuscript.

**Competing Interests:** The present study is part of a more general program investigating changes in small pelagic populations in the Gulf of Lions, named EcoPelGol. This program is partly funded by the interprofessional association France Filière Pêche. The authors are free to fully develop and publish their research without consulting France Filière Pêche, as per previous signed agreement with the association. France Filière Pêche has not yet seen these results, which will be communicated to them once published only.

* Email: claire.saraux@ifremer.fr

## Introduction

Because animal spatial distribution is often strongly associated with population dynamics, spatial indices may provide valuable tools for assessing the status of these populations, and in particular the status of endangered or exploited species [1–4]. At the population scale, spatial distribution can be seen, in the absence of substantial anthropogenic impacts, as the emergent property of habitat selection. In addition to social motivation (presence *vs.* absence of conspecifics; [5]), spatio-temporal aggregation patterns in animal populations may be explained by individuals sharing similar needs and dealing with similar biotic and abiotic pressures, such as prey abundance *vs.* predation risk [6], propitious *vs.* detrimental environmental conditions [7]. Nonetheless, as population density increases, so does intra-specific competition [8], and individuals are expected to spread towards less suitable habitats when a certain threshold is reached ('Basin hypothesis'; [9]). Both population density and its occupation area should then vary with its abundance [10–12]. If their density-dependent and density-independent drivers are difficult to disentangle (but see [13]), spatial distributions in themselves offer valuable information for both ecological understanding and management.

Indeed, besides an obvious fundamental interest in terms of population dynamics and marine ecosystem functioning [14], information on fish biomass location is also of crucial importance for stock management. For instance, the effective implementation of Marine Protected Areas (MPAs) requires detailed knowledge on species spatio-temporal dynamics, *i.e.* on temporal variability in fish spatial distributions [15]. In addition, knowledge on interspecific interactions, such as the co-occurrence or repulsion of species based on their spatial distributions, may greatly help scientists to understand competition or predator-prey processes but also policy makers to implement ecosystemic rather than single species management measures, as presently done.

In this study, we investigated the spatio-temporal distribution of small pelagic fish in the Gulf of Lions using a unique dataset of 10 years of acoustic surveys. Because small pelagic fish are key species of the pelagic ecosystem due to their central place in the food web, transferring energy from the lowest trophic levels (plankton) towards top-predators [16], and because their life-history traits (short lifespan, large fecundity) make them strongly dependent on the abiotic environment [17], changes in their distribution and abundance have been a major source of concern for scientists and managers worldwide ([18]; Global Ocean Ecosystem Dynamics program - GLOBEC;).

Over the past years, important changes in the two main target species (the European sardine, *Sardina pilchardus* and the European anchovy, *Engraulis encrasicolus*) biomass [19], along with a shift in the size-distribution of these species towards smaller individuals [20] have been observed in the Northwestern Mediterranean Sea (Gulf of Lions), resulting in important economic losses for fisheries. Both stocks are now considered to have a low biomass and a low fishing mortality by General Fisheries Commission for the Mediterranean [19]. In parallel, a third small pelagic species, the sprat (*Sprattus sprattus*), which is not commercially exploited in the Western Mediterranean has appeared in the system with a steadily increasing biomass since 2007 [19]. This unexpected situation offered us the opportunity to describe and compare simultaneous spatio-temporal distributions of three species sharing the same trophic level, with similar feeding behaviours [21–22], but with completely different trends in biomass, through the analysis of 10 years of scientific acoustic surveys. First, we studied the aggregation patterns and spatial structures of these species separately. Second, we investigated species spatial dynamics by considering spatio-temporal changes in biomass, and defined optimal recurrent areas based on a combination of biomass levels and their variability. Finally, we focused on interspecific relationships by studying the overlap between species at different spatio-temporal scales.

## Methods

### Ethic statement

The study was conducted in the Gulf of Lions (Longitude in [3.05°; 5.20°] and Latitude in [42.44°; 43.44°]), a public sea area. No sampling was operated from private land and field studies did not involve endangered or protected species. All data used in this study came from PELMED acoustic surveys, which comply with the MEDIAS (Mediterranean Acoustic Survey) protocol. The sampling has been performed under repeated international standardized surveys where the research vessel had full permission to sample from all relevant national public authorities (governments). Acoustic data were collected at a distance, which does not require any particular ethic approval. Further data used in this study came from scientific trawls conducted during these same PELMED surveys. Again, no approval by an ethic committee was required as the targeted species are exploited species and trawling methods done according to international standard trawl surveying.

### Data collection and survey design

Every July since 2003, the French Research Institute IFREMER has been carrying out acoustic surveys of the pelagic resources present in the Gulf of Lions, Mediterranean Sea. The summer period of the survey corresponds to contrasted biological periods for our 3 species. Indeed, this is the peak of reproduction for anchovies, while sprats and sardines reproduce in winter [23]. Due to the biological cycle of these species, it is important not to extrapolate our results outside the summer period. Sampling was performed along 9 parallel transects, regularly spaced by 12 nautical miles (nm) (see Figure S1 in File S1). Acoustic data were recorded every 1 nm using multi-frequency echosounders (Simrad EK500 and ER60), while travelling at a constant speed of 8 $nm.h^{-1}$. All 4 frequencies were visualized during sampling to help deciding when to trawl for species identification. However, only energies from the 38 kHz (typical frequency used for fish) channel were used to estimate fish density. Acoustic data analyses, such as bottom correction, were later performed using *Movies+* [24] and *FishView* IFREMER softwares. Species discrimination and echo-partitioning were performed by the combination of echo trace classification and trawl outputs [25]. Species biomass and abundance were finally estimated from species energy using specific target strength ($TS = 20 \log(L) - 71.2$, where L is the length of the fish for all 3 species, see [26] for more details on acoustic surveys and analyses). Main survey features are summarised in Table 1.

### Analyses

We defined and used different spatial indicators and geostatistical methods described in details below. An overview of these indicators, including their calculation formula, representative scale and biological meaning are summarised in Table 2. A key concept in statistics in general and in spatial statistics in particular is the support of the information, i.e. the geographical area over which measures are recorded [27–28]. In this study, the support size was the size of the Elementary Sampling Distance Unit, that is 1 nm. All statistics derived from these data are thus associated to the 1 nm sampling support and, for all of them, their values would have been different if computed at another support. Even though some metrics are labelled like spatial statistics (e.g. space selectivity index, local index of collocation), they are not sensitive to the location of the points in space (i.e. any exchange between two data does not change the result). They are sensitive to the spatial patterns that exist at scales smaller than the support size and that are integrated in the observations. For some of them (e.g. Empirical Orthogonal Functions), it remains possible to represent them as geographical distributions making the confusion even worth.

**Aggregation patterns and spatial structure.** Presence Area (PA) was calculated as the percentage of sampled points at which the species was found, independently of its abundance.

A space selectivity index was defined on the basis of annual concentration curves [29–30]. Like Lorenz curves (as used in [31]), concentration curves represent the maximum proportion of biomass as a function of the proportion of samples [30]. For instance, a proportion of biomass of 0.6 for a proportion of sampled area of 0.2 would indicate that 60% of population total biomass was found in only 20% of the total area sampled. For a homogeneously distributed population, each proportion of biomass should be found in the same proportion of total area (i.e. 10% of biomass in 10% of area, etc.), so that the annual concentration curve of a homogeneously distributed population equals the first diagonal. It follows that the more concentrated a population, the further the curve falls from the first diagonal, *i.e.* a higher proportion of biomass is situated in a given proportion of the total area. Consequently, the space selectivity index is defined as twice the area between the concentration curve and the first diagonal [29–30], so that the higher the space selectivity index, the more concentrated the spatial distribution of the fish.

Spatial structure was defined as the spatial autocorrelation between values at different locations. In other words, spatial structure considers whether the correlation between biomass in two locations depends on the distance separating them. It was investigated using variograms [27] that calculate the value of the spatial autocorrelation at different lags (distance intervals). As is often the case, the frequency distributions of species biomass were highly skewed with a large proportion of zeros or small values, and few extremely large values contributing importantly to total biomass (coefficients of variation CV were 2.6, 3.8 and 4.2 for anchovies, sprats and sardines, respectively). Besides, the variance

**Table 1.** Main PELMED survey features from 2003 to 2012.

| Year | Sampling dates | Nb of nm | Nb of EDSU | Nb of trawls | Anchovy biomass | Sardine biomass | Sprat biomass |
|---|---|---|---|---|---|---|---|
| 2003 | 07/07/03-06/08/03 | 1274 | 284 | 27 | 27 860 | 126 120 | 685 |
| 2004 | 05/07/04-04/08/04 | 1174 | 285 | 29 | 25 953 | 215 560 | 786 |
| 2005 | 08/07/05-07/08/05 | 1462 | 294 | 33 | 15 962 | 264 024 | 1 955 |
| 2006 | 09/07/06-08/08/06 | 1473 | 288 | 39 | 25 658 | 102 276 | 772 |
| 2007 | 11/07/07-10/08/07 | 1500 | 290 | 43 | 13 654 | 88 297 | 15 |
| 2008 | 19/06/08-30/07/08 | 2151 | 273 | 60 | 23 395 | 91 546 | 5 002 |
| 2009 | 24/06/09-29/07/09 | 2173 | 284 | 43 | 30 424 | 52 977 | 7 845 |
| 2010 | 24/06/10-29/07/10 | 2000 | 276 | 39 | 23 514 | 51 819 | 15 760 |
| 2011 | 27/06/11-31/07/11 | 1704 | 282 | 42 | 25 906 | 44 926 | 26 638 |
| 2012 | 27/06/12-31/07/12 | 1172 | 279 | 37 | 39 061 | 80 537 | 70 263 |

Biomasses are indicated in tons, nm stands for nautical mile.

in biomass increased proportionally with mean biomass in our data. This led us to log-transform the studied variables (the optimal Box-Cox transformation was $\mathbf{Y = ln\ (B{+}c)}$, with $\boldsymbol{c}$ a positive constant added to insure positive values). This constant was set to the smallest biomass observed in our data, *i.e.* 0.05 tons. While distributions of the log-biomass were still skewed, the proportionality effect disappeared. The log-transformed data was used to calculate empirical variograms for all three species for each year of survey. To account for the anisotropy of the sampling (only along parallel transects) and for a potential anisotropic structure due to a bathymetry gradient, bidirectional variograms (along transects and perpendicularly to them) were calculated and modelled by automatic fitting using a least square method [32]. The temporal variability of the spatial structure was assessed on annually standardised data by comparing annual variograms to the mean variogram (average across all years of survey, pairs being retained only if the two sampling locations belong to the same year) and its 95%-confidence interval (adaptation from [12] for bidirectional variograms). The confidence interval was obtained by simulating 1000 random fields according to the mean modelled variogram using the turning-bands method [33] and estimating the 1000 variograms associated with the sampling points extracted from these random fields.

**Species spatial distributions.** To capture the spatial patterns of the 3 populations as simply as possible and to investigate year-to-year variations, we calculated the centres of gravity of their biomass along with their associated inertia [34–35]. The centre of gravity represents the mean location of the population, while inertia describes the dispersion of the population around its centre of gravity. Because fish populations are often aggregated in a given area and therefore spatial distributions heterogeneous, we also investigated the presence of spatial patches of high biomass densities, by adapting the recursive algorithm developed by [35]. This was adapted to take into account the anisotropy of our sampling by setting two different threshold distances (one parallel to the transect and another one perpendicular). The sensitivity to the threshold distances was tested using different values. Very small distances did not enable us to identify patches with more than 10% of biomass, while increasing the distance reduced the number of patches to one which comprised the whole presence area. The final distances (i.e. 6 nm in the direction of the transect and 24 perpendicular to it) were set in the interval where the resulting number of patches was stable and were kept constant both across species and years. The number of patches, centres of gravity and inertias of all patches were then used to study year-to-year and interspecific variations.

Kriging maps were produced for each species on a 1 nm*1 nm-grid over the study area, using the modelled mean or annual variograms depending on the results of interannual variability tests. In order to account for anisotropic sampling, we used an anisotropic search of neighbours when conducting kriging, *i.e.* the distance at which neighbours were looked for was smaller in the transect direction than perpendicular to it to make sure that each point comprised neighbours in two different transects.

Two different methods were used to quantify the temporal stability/variability of spatial distributions at 2 different scales (Table 2). First, to assess the stability/variability of the entire distribution, we calculated the Empirical Orthogonal Functions (EOF) on the raw acoustic data for each species. EOF analysis is a decomposition of a spatio-temporal dataset in terms of orthogonal basis functions. It is thus very similar to a Principal Component Analysis, except that it is applied to spatio-temporal data, using time as a descriptor and space (i.e. sampling locations) as objects [36]. EOF was performed on standardized data to compare

**Table 2.** Spatial indicators used in the study.

| Indicator name | Formula | Spatial scale | Time scale | Biological meaning |
|---|---|---|---|---|
| *Presence area* | | | | |
| **Presence area** | $PA = \frac{\sum 1_i}{n} * 100$ | Global | Annual | Area of presence(in %) of the species over the studied area |
| *Aggregation and spatial structure* | | | | |
| **Space selectivity index** | | Global | Annual | Spatial compactness |
| **Annual variograms** | | | Annual | Spatial autocorrelation |
| **Mean variogram** | | | Decadal | Mean Spatial autocorrelation |
| *Spatial distribution* | | | | |
| **Centre of gravity** | $CG = \frac{\int x\, z(x)\, dx}{\int z(x)\, dx}$ | Global | Annual & decadal | Mean location of the species |
| **Inertia** | $I = \frac{\int (x - CG)^2\, z(x)\, dx}{\int z(x)\, dx}$ | Global | Annual& decadal | Spatial dispersion around CG |
| **Patches** | | Intermediate | Annual | Concentration of biomass |
| **EOF** | | Global | decadal | Spatio-temporal variability |
| **CV** | $CV_x = \frac{\sigma_{z_a(x)}}{Mean(z_{a(x)})}$ | Local | decadal | Temporal variability |
| *Interspecific relationships/co-occurence* | | | | |
| **Overlap** | $O_{i,j} = \frac{A_{E_i \cap E_j}}{A_{E_i \cup E_j}}$ | Global | Annual decadal | Overlap between the spatial distribution envelops of two species |
| **Overlap of patches** | $O_{i,j} = \frac{\sum_{p_i} \sum_{p_j} A_{Ep_i \cap Ep_j}}{A_{\bigcup_{p_i, p_j} E_{p_i} E_{p_j}}}$ | Intermediate | Annual decadal | Proportion of patches where both species present |
| **Co-occurrence index** | $Co_{i,j} = \frac{\sum 1_{i,j}}{\sum 1_i + \sum 1_j}$ | Local | Annual decadal | Level of local co-occurrence between 2 species based on presence/absence |
| **Local index of collocation** | $LIC_{ij} = \frac{\int z_i(x)\, z_j(x)\, dx}{\sqrt{\int z_i^2(x)\, dx}\sqrt{\int z_j^2(x)\, dx}}$ | Local | Annual decadal | Level of local cooccurence between species based on densities |
| **Dominance index** | $Dom_{i,j} = \frac{z_i(x) - z_j(x)}{z_i(x) + z_j(x)}$ | Local | Annual decadal | Relative biomass in each point |
| *Mapping* | | | | |
| **Kriging** | Anisotropic neighbourhood | Global | Annual | Distribution maps |
| *Area classification* | | | | |
| | Variability map * Average map | Global | Decadal | Identification of recurrent, occasional & unfavourable areas |

n: Number of sampling points, z(x): biomass density in x, $z_a$ (x): annual biomass density in x.

annual spatial distributions independently of annual total levels of biomass. The first axis (eigen vector) of the EOF is the linear combination of years which maximizes the percentage of the interannual variance of spatial distributions. When all yearly contributions to this first axis share the same sign, the greater the percentage of variance explained by the first EOF, the more persistent the spatial distributions [37;12]. In order to have a local estimation of temporal variability/stability and see whether some areas were more stable than others, we also calculated the Coefficient of Variation (CV) of the 10 annual values in each sampling point. As biomass varied substantially between years, we used relative biomass (i.e absolute biomass divided by the total annual biomass) instead of absolute biomass, so that the observed CV corresponded to a change in geographical location rather than a change in overall biomass.

Finally, we defined recurrent, occasional and unfavourable areas by combining average and variability maps [38]. To produce average and variability maps, we calculated the mean and the standard deviation of the 10 annual kriged maps node by node (1 nm*1 nm). Then, each 1 nm*1 nm pixel was assigned one of the 3 categories (recurrent, occasional and unfavourable) depending on its value in the average and variability maps. Pixels with both low mean and variability (i.e. inferior to the median) were considered unfavourable; those with high mean and low variability recurrent, while pixels with high variability were defined as occasional.

**Interspecific indices: collocation index and overlap between species.** We quantified the collocation at three different spatial scales in a coherent manner: from a global scale comparing the distribution over the whole study area, to a local scale comparing species at each sampling point, passing by an intermediate scale comparing previously defined patches.

Global overlap was defined based on centres of gravity and associated inertia ellipses (see above), as follows:

$$O_{i,j} = \frac{A_{E_i \cap E_j}}{A_{E_i \cup E_j}}$$

where $E_i$ and $E_j$ are the inertia ellipses of species i and j respectively and A is the area. The index varies from 0 when the two ellipses are totally separated to 1 when the two ellipses are identical.

At the intermediate scale, we used the same index but calculated from the inertia ellipses associated with the patches of the monospecific distributions:

$$O_{i,j} = \frac{\sum_{P_i} \sum_{P_j} A_{Ep_i \cap Ep_j}}{A_{\bigcup_{P_i, P_j} Ep_i\, Ep_j}}$$

where $E_{pi}$ and $E_{pj}$ are the ellipses associated with the patches of species i and j respectively.

To characterize the co-occurence of two species at the local scale, we used two different indices: (*i*) an index of co-occurrence defined as a proportion of sampling points where both species co-occur (taking into account only the sampling points where at least one of the two species is present), and (*ii*) a local index of collocation (LIC; [39]) defined as:

$$LIC_{ij} = \frac{\int z_i(x)\, z_j(x)\, dx}{\sqrt{\int z_i^2(x)\, dx}\sqrt{\int z_j^2(x)\, dx}}$$

where $z_i$ (x) and $z_j$ (x) are the biomass density of the species i and j in the location x. Both indices vary between 0 (no location where the 2 species are found simultaneously) and 1 (the two species are situated in exactly the same locations). Yet, they differ in the fact that the former is based only on presence/absence while the second is based on biomass density.

To evaluate the deviation of these indices from random expectation, we used a randomization test by non-parametric bootstrap of 1000 values. For species of strong local overlap (i.e. anchovies and sardines), we also calculated a dominance index based on the relative difference in biomass of the two species in each point (Table 2). This index varies between -1, where only the species j is present and 1 where only the presence i is present, 0 corresponding to a situation where species i and j are equally abundant [40]. This index enabled us to better represent which species dominate where.

**Softwares.** Statistics were conducted using R v. 2.15.0 [41]. Geostatistical analyses were performed with the package RGeoS [42]. Spatial data are given in WGS84 coordinate system. Analyses were conducted on both raw and kriged data. As both methods yielded very similar results, we only presented results on raw data for clarity purposes. Kriging was produced only for mapping purposes.

## Results

### Aggregation patterns and spatial structure

Species were absent from several sampled locations, regardless of whether anchovies (absent from 9.3% of sampled locations), sardines (12.9%) or sprats (51.0%) were considered. However, these absences were not spatially consistent across years. When pooling all data, anchovies and sardines could be observed at least once in all sampled locations; sprats in contrast were never found after the 200 m isobath (Fig. S1 in File S1). During each of the 10 study-years, sardines and anchovies occupied most of the Gulf of Lions, and presence area only varied slightly for these two species (PA: 85.0 to 94.7% and 67.2 to 99.6% for anchovies and sardines, respectively). Variability was far greater for sprats, which were almost completely absent in 2007 (PA = 0.3%), but covered most of the Gulf of Lions in 2012 (PA = 91.8%). Presence area was positively correlated with total log-biomass index for sprats (LM: $R^2_{adj} = 0.87$, $p<0.001$), but not for anchovies ($p = 0.68$) or sardines ($p = 0.10$) (Figure 1), *i.e.* the area occupied by the population expanded with total logbiomass in sprats but not in the other two species. Log-biomass and presence area were far less variable for sardines and anchovies than for sprats (CVs being at least 4 to 7 fold smaller for PA and biomass, respectively). The mean density in presence area also increased with log-biomass for anchovies ($R^2_{adj} = 0.71$, $P = 0.001$) and sprats ($R^2_{adj} = 0.69$, $P = 0.002$). The same trend was observed in sardines, but the probability was lower ($R^2_{adj} = 0.27$, $P = 0.07$).

The degree of aggregation appeared to be different in anchovies compared to sprats or sardines (Figure 1; Figure S2 in File S1). Anchovies were, on average, less concentrated than the two other species. This was confirmed by the space selectivity index, which was lower in anchovies (0.722) than in sardines and sprats (0.850 and 0.867 respectively). Over the 10 study-years, yearly space selectivity indices decreased significantly with increasing log-biomass in anchovies (LM: $R^2_{adj} = 0.43$, $P = 0.02$,) and in sprats ($R^2_{adj} = 0.94$, $P<0.001$), but not in sardines ($P = 0.41$, Figure 1).

Mean variograms of log biomass exhibited spatial structure for the three species in both directions (i.e, increasing variance when the distance between sampling points increased before levelling off; Figure 2), meaning that the spatial distribution was not random and that two close points had a higher probability to have similar values than distant points. While sardine and anchovy variograms shared similar main characteristics, sprat mean variogram differed slightly. In particular, it differed in the range at which variance stabilized (Figure 2). Variance stabilized rapidly in sprats (range = 4 and 12 nm respectively for the first and second spherical component), while the first structure appeared around 15 nm for sardines and anchovies and the second was not apparent at the scale of our study, *i.e.* variance did not stabilize completely within the 50 nm limit that we fixed. Further, the mean structure seemed similar between the two perpendicular directions in sprats, while it differed slightly in anchovies and sardines, though the mean variogram of each direction was mostly included in the confidence interval around the mean variogram of the other direction.

Annual variograms were computed for each of the three species (Figure 2). All exhibited clear spatial structure and were included in the 95% confidence interval around the mean variogram for anchovies and sardines (see Figure 2). This revealed the absence of a year effect on the spatial structure of these 2 species. In sprats however, some variograms (especially along transect) were not included in the confidence interval, suggesting interannual variability in spatial structure.

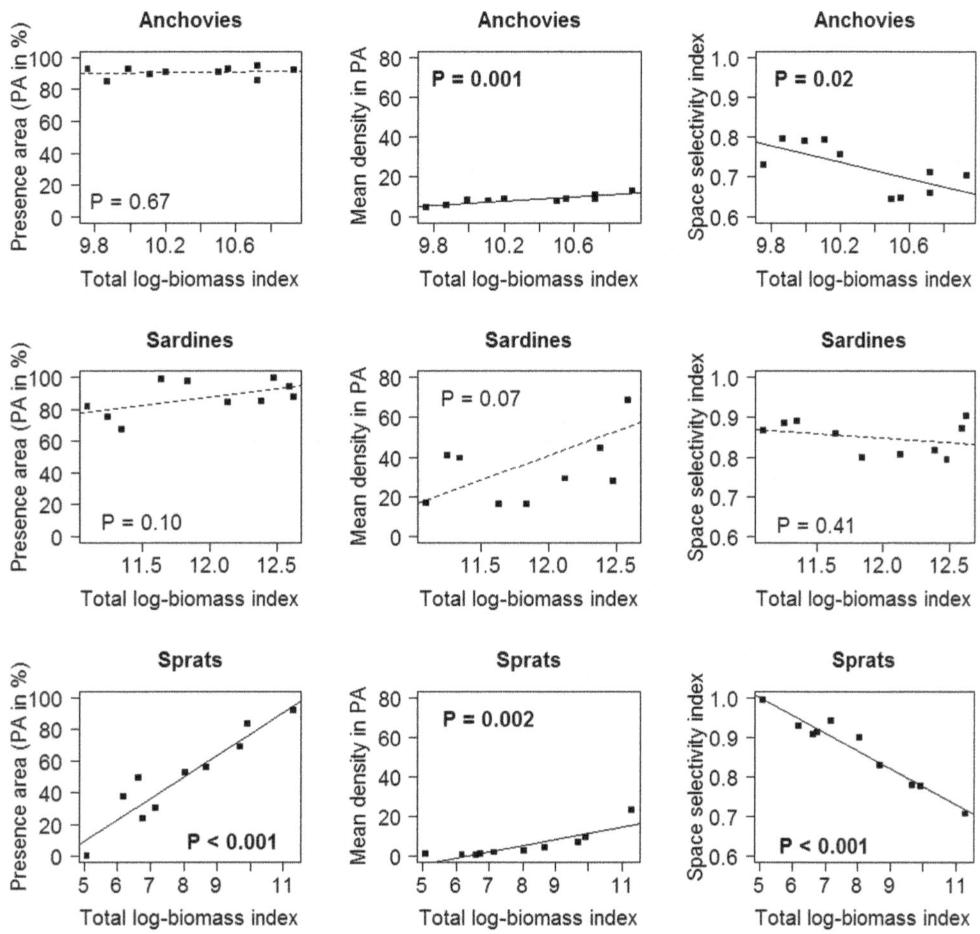


**Figure 1. Presence areas, biomass densities and space selectivity indices relatively to total log-biomass indices in anchovies, sardines and sprats.** Lines represent the linear regressions. Significant linear regressions are represented by plain lines and Pvalues are indicated in bold. For non-significant relationships, the trend is shown by a dotted line.

## Species spatial distribution

**Log-biomass distributions.** The annual spatial distributions exhibited similarities between species in some years (*e.g.* in 2011 a large area in the middle of the shelf was unoccupied both for anchovies and sardines), but not in others (Figures S3, S4, S5 in File S1). Some consistent differences between species also appeared. For instance, anchovies occupied the centre of the continental shelf, while sardines were more coastal. For the three species, kriged annual distributions also revealed the important interannual variability of these populations both in terms of biomass levels and their spatial repartitions.

Centres of gravity of spatial distributions did not vary much according to species or year and were situated close to the geometrical centre of gravity of the sampled area (8.2±0.8 nm). Interestingly, the associated inertia was high for anchovies and sardines (inertia = 1152±70 nm$^2$), meaning that the mean distance between a presence point and the centre of gravity of its population was 34 nm. This distance was remarkably close to that between a sampling point and the geometrical centre of gravity of the total area (37 nm). As a consequence, the ellipse associated with this inertia covered most of the sampled area for these two species (Figure S6 in File S1), *i.e.* the populations were dispersed in the entire Gulf of Lions. In sprats, while inertia was quite low in the first 6 years (447±101 nm$^2$; Figure S6 in File S1) due to their low presence, it reached values almost as high as the ones in sardines and anchovies, in the last 4 years (991±75 nm$^2$).

The number of patches present per year and per species varied between 1 and 4 (Figure 3). As inferred by the square root of the inertia associated with the patch, the mean distance between a presence point in a patch and its centre of gravity ranged from 6 nm to 17 nm (excluding 2007 in sprats for which the species had only been observed in 1 location). The centres of gravity of anchovy patches were slightly more offshore than those of sardines (14.9±0.8 vs. 11.1±1.0 nm, Wilcoxon test: W = 662, P = 0.001), confirming the tendency of sardines to be more coastal than anchovies.

**Instability of the spatial distribution.** EOF of the 10 annual maps of each species confirmed the visual inspection highlighting high temporal variations in the spatial distributions. The first component of the EOF explained no more than 20% of the variance (14%, 20% and 17% for anchovies, sardines and sprats, respectively). Further, no trend could be detected from the contributions of each year to the first 2 axes of the EOF. The sign of the correlations between annual maps and the two first axes varied depending on the year for the three species (Figure S7 in File S1).

Coefficients of variation (calculated in each sampling point) were higher in sprats than in sardines and anchovies. Most coefficients of variation were higher than 1 (100%, 92.3% and

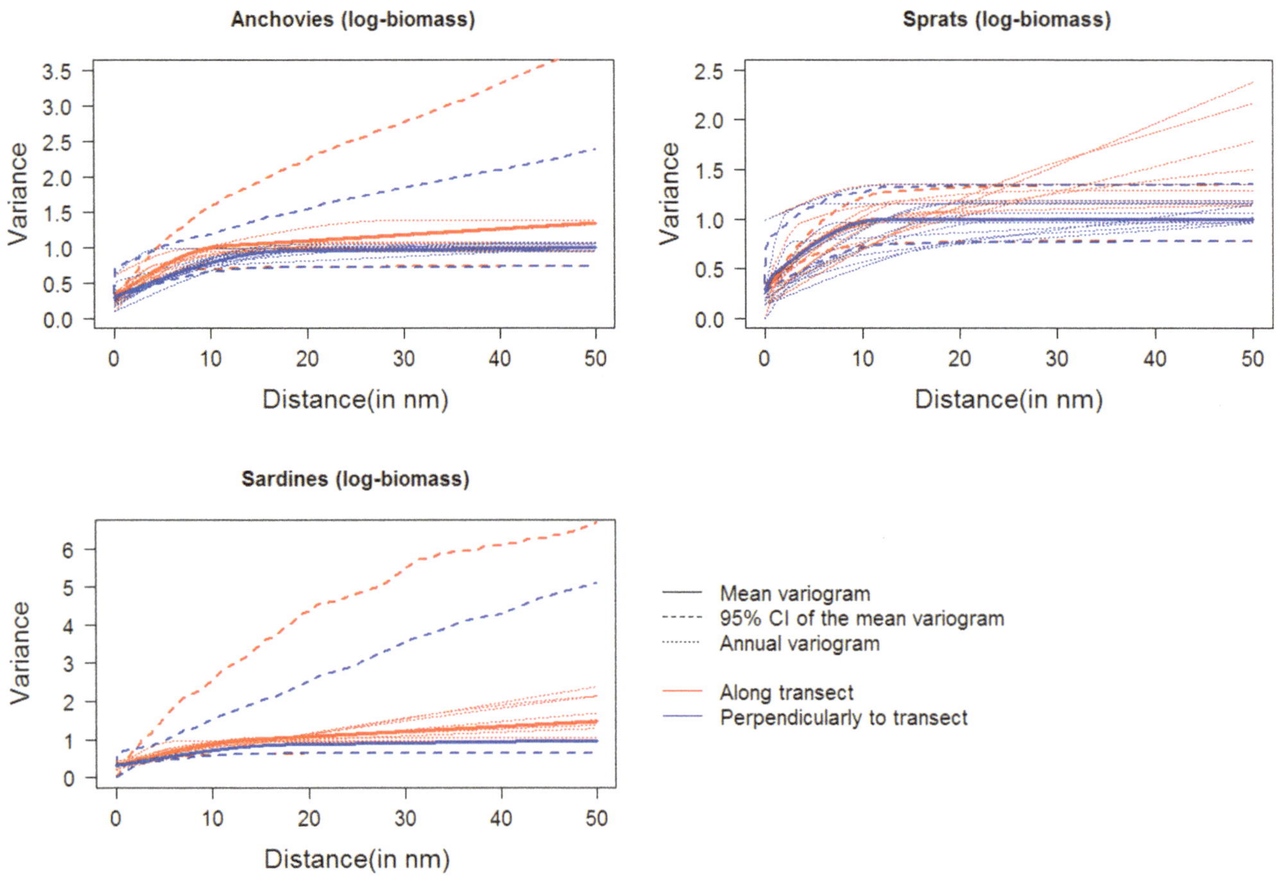


**Figure 2. Annual (black) and mean (red) modelled variograms of anchovies, sardines and sprats.** The red dotted lines correspond to the 95% confidence interval of the mean variogram deduced from 500 simulations.

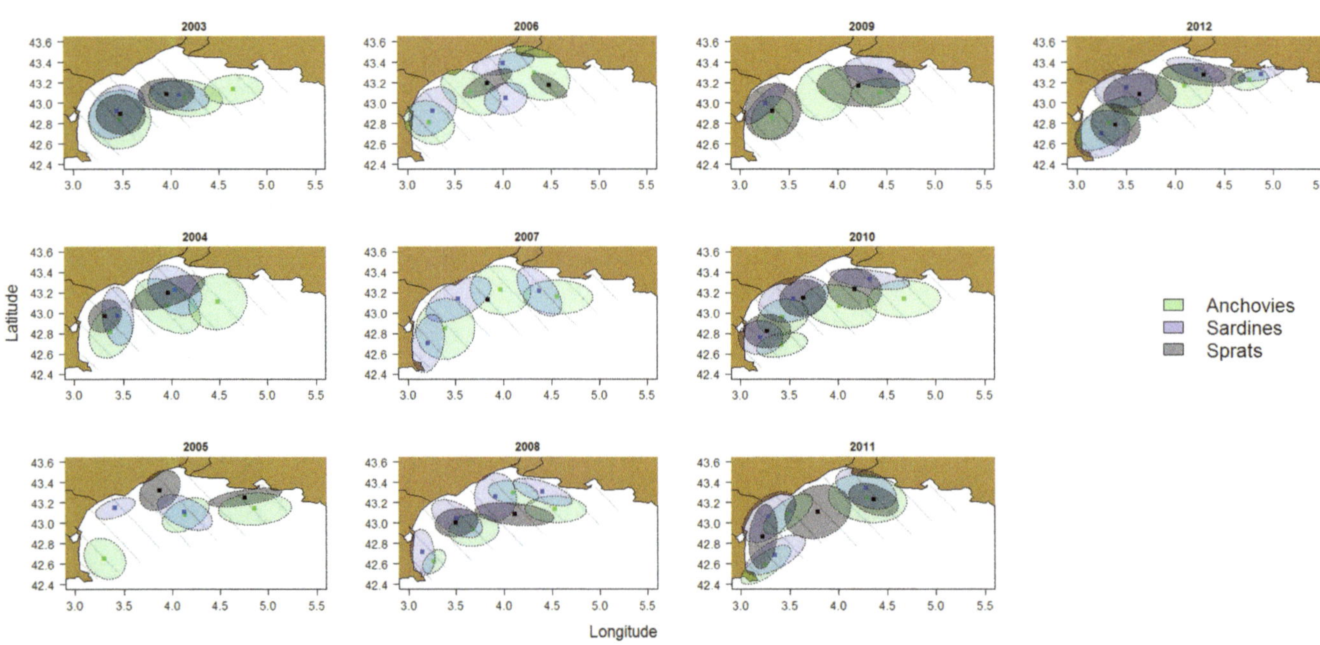


**Figure 3. Annual maps of centres of gravity and inertia of patches for anchovies (in green), sardines (in blue) and sprats (in black).**

81.6% of CV values were higher than 1 in sprats, sardines and anchovies, respectively), meaning that the distributions were overdispersed and again suggesting high temporal variability of the spatial distributions for all three species.

**Recurrent, occasional and unfavourable areas.** Average maps confirmed the tendency for sardines to be quite coastal, while anchovies occupied most of the shelf (Figure 4). Sprats seemed in an intermediate position where high biomass areas were situated in the centre of the Gulf of Lions, neither too coastal nor too offshore, but, as noted above, sprat biomass was also more variable. The combination of average and variability maps enabled us to detect a recurrent (persistent) area for anchovies in the centre of the Gulf, slightly West of the Rhone estuary, and, for sardines, in western coastal areas. Recurrent areas were minimal for sprats as the period encompassed some years with a quasi-absence of this species. Finally, deep waters represented unfavourable areas both for sardines and sprats.

## Interspecific indices: collocation index and overlap between species

Regardless of the scale at which we investigated it, collocation indices were higher between sardines and anchovies than between any of these 2 species and sprats (Table S1 in File S1), meaning that sardines and anchovies co-occurred more often than they did with sprats. Additionally, collocation indices were higher, on average, at the global scale than at the intermediate scale (*e.g.* between anchovies and sardines $0.53 \pm 0.04$ *vs.* $0.30 \pm 0.04$ for the global and intermediate scales, respectively), suggesting that species globally lived in the same areas, but were not always found together in each given sampling location. Collocation indices, whether global, local or intermediate, varied substantially from year-to-year (Table S1 in File S1). Collocation indices at the three different scales were not correlated with each other, except for the global and intermediate indices between sprats and anchovies. For instance in 2004, the index of collocation between sprats and sardines was fairly high at the global scale ($2^{nd}$ highest observed), while the associated LIC was low and not significantly different from random expectation. The two local collocation indices were not correlated for anchovy and sardine association or anchovy and sprat association ($\rho = -0.61$, $P = 0.07$; $\rho = 0.44$, $P = 0.20$ respectively), suggesting that the co-occurrence of species (i.e. species found in the same location), did not necessarily concur with a co-occurrence of their hotspots or peaks of biomass. They were however slightly correlated when looking at the association between sardines and sprats ($\rho = 0.72$, $P = 0.02$).

Finally, although the co-occurrence between sardines and anchovies was high (Table S1 in File S1), in most sampled points one of the two species clearly dominated the other (Figure 5). The

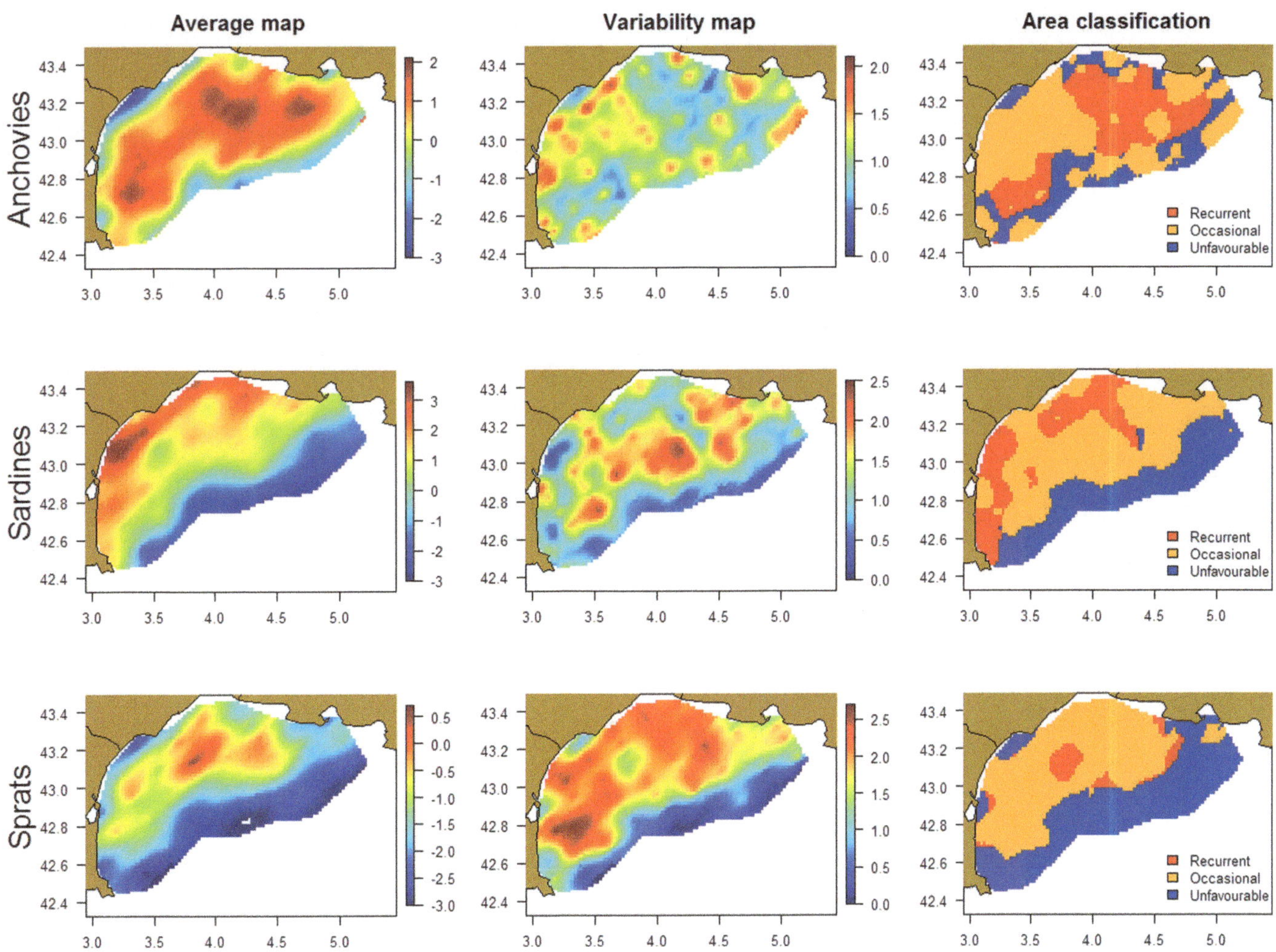


**Figure 4. Average and variability maps and area classification for anchovies, sardines and sprats.**

dominance index increased with depth, *i.e.* anchovies dominated sardines in deeper offshore waters, while sardines were dominant in shallow waters close to the coast (see also statistics in Figure S8 in File S1). Besides, the dominance index also increased with longitude, anchovies dominating towards the East (see statistics in Figure S8 in File S1).

## Discussion

The strong increase in sprat biomass (from almost 0 to 70,000 tons from 2007 to 2012) that has paralleled a strong decline in anchovy and sardine biomass over the past few years [19], highlighted the urgent need to better understand the cross and mutual dynamics of those three species. This study documents the spatio-temporal patterns observed in these three species in the Gulf of Lions. However, it should be noted that surveys were conducted only in summer, corresponding to peak reproduction of anchovies but resting period for sardines and sprats. The results presented here thus correspond to different biological phases for the three species and should not be extrapolated to other non-monitored seasons.

Our main objective was to investigate the temporal variability of biomass and spatial distribution in the three species. Biomass estimates obtained by direct acoustic stock assessment substantially varied during the 10 years of survey. However, sprat was the only species in which a clear trend in biomass could be identified over the study period. For anchovies and sardines, biomass variability corresponded, respectively, either to fluctuations around a low value or a succession of an increasing and decreasing trends. The increase in sprat biomass resulted in both an increase of fish local densities and an expansion of fish towards new areas, in agreement with the McCall basin hypothesis [9]. On the contrary, the Gulf of Lions being a favourable area for small pelagics [43], sardine and anchovy presence area was large and stable over the entire study period and fluctuations in biomass only resulted in fluctuations of local densities. Changes in local densities could be a buffer for fluctuating biomasses of already well implanted species. However, if a species continuously increase or decrease its biomass, this might well not be enough anymore. For instance, sprats which were almost absent from the area at the start of the study had to expand their presence area to cope with their increasing biomass. Inversely, a declining population would see local areas disappear from its presence area as these species are gregarious and have to stay in shoals. Our results may thus support the hypothesis that interannual variability may not be enough to observe contraction/ expansion, and that a directional trend would be needed to investigate the link between abundance and distribution [44–45]. Still, using space selectivity indices to investigate aggregation patterns in more details, we found that sprats and anchovies became less selective with increasing biomass, suggesting that they selected less suitable areas rather than increased their local density. This is in contrast with sardines for which the space selectivity index did not vary according to total biomass and densities 5 fold higher could be reached. This could suggest different social constraints/benefits, sardines being able to aggregate in much denser shoals than the other 2 species.

Similarly to what was found in demersal species [12], we highlighted the stability of the spatial structure (i.e. spatial autocorrelation) of small pelagic fish in the Gulf of Lions over time. Only for sprats did some interannual variability appear, but this is probably due to a reduced confidence interval around the mean variogram due to low and distance-independent variance between sampling points in the first half of the study, when sprats were quasi-absent. Spatial structure is defined as the spatial

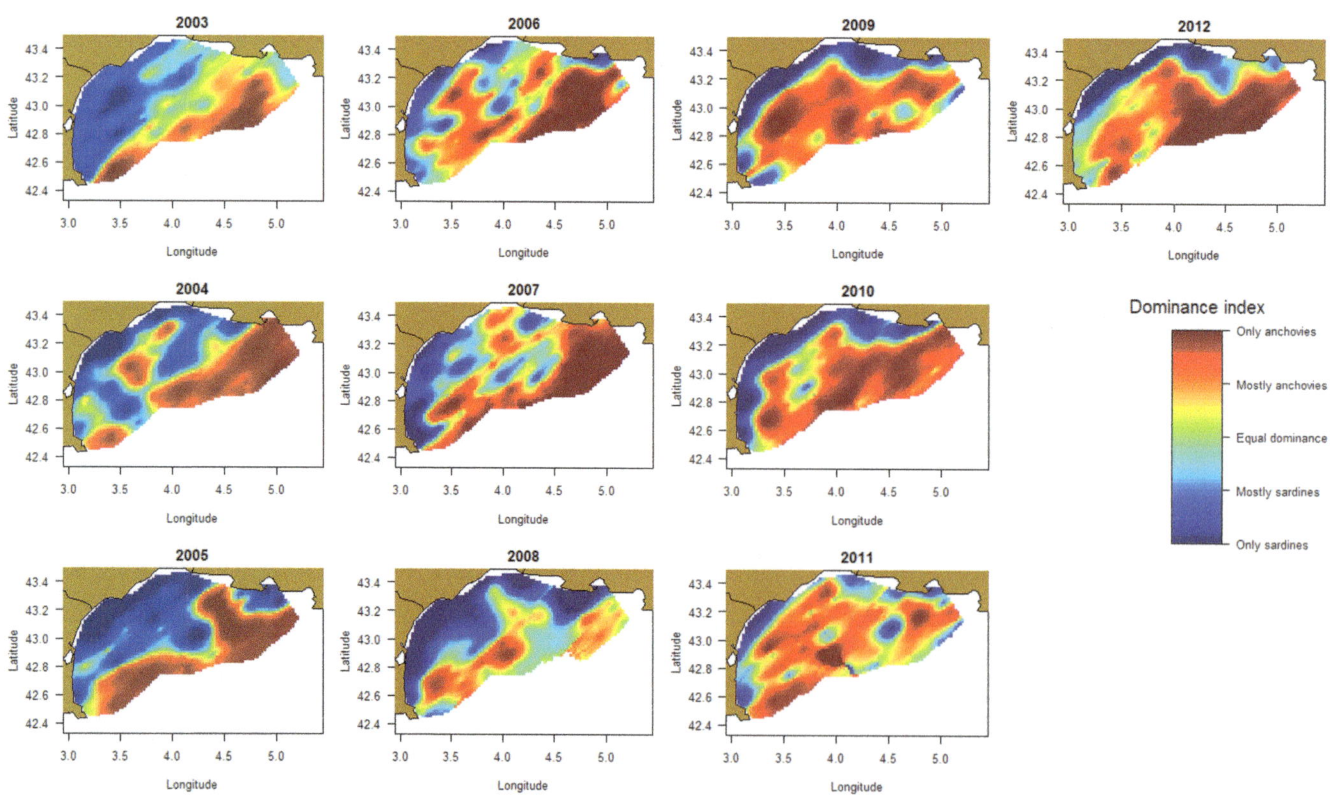


**Figure 5. Annual maps of relative biomass between anchovies and sardines.**

autocorrelation between values at different locations, i.e. in our case whether biomasses taken at 2 close points were more correlated than biomasses of points situated far apart. From an ecological point of view, spatial structure could thus indicate the occurrence of concentration in certain areas due to similar needs (e.g. habitat preferences) or aggregation patterns due to species social constraints (formation of shoals for instance here). Such stability of the spatial structure could result from important aggregative behaviours of small pelagic species, giving them intrinsic aggregation properties, independently from their abiotic environment.

Though necessary, the study of the spatial structure offers no insight on the locations where these aggregations took place or where the population was absent. In a second step, we thus investigated temporal changes in the geographic locations of biomass. In contrast to the stable spatial structure, the geographical distributions of small pelagics were highly variable from year-to-year. Although such variability is clearly visible when comparing the maps, it is more difficult to quantify it. Therefore, we resorted to spatial indicators (CVs and empirical orthogonal functions) for summarizing annual information and exploring possible spatial shifts in population distribution [35;46]. These indicators confirmed the visual inspection of the annual maps (high interannual variability) and thus the expected mobility of small pelagic fish, which contrasted to the stable and area-specific spatial distributions of demersal species of the same area [12]. Such result was to be expected, as demersal species are restricted to the bottom and very dependent on substrates or bathymetry. By contrast, small pelagic fish live in the water column, are much more mobile and should be affected mostly by dynamic environmental factors, which change from year to year. Previous studies on small pelagic fish habitats conducted in the Mediterranean [43] have identified the Gulf of Lions as a potential favourable habitat for sardines and anchovies. In this smaller-scale study, we tried to identify whether there could be within Gulf variations in habitat quality for small pelagic fish. From the presence/absence data pooled over the entire study period, we saw that the three species could use the entire Gulf of Lions (except past the 200 m isobath for sprat). Nonetheless, it was clear that spatial distributions were not uniform over the Gulf and that they exhibited more than one patch of high biomass. The location of patches varied across years and between species, suggesting that they reacted to some external variable factors. In the future, linking spatio-temporal distributions to environmental variables at a minute scale while accounting for aggregation structures should help in understanding the environmental drivers affecting the population dynamics of those three pelagic species. Previous studies on environmental drivers of small pelagic fish habitats have already been conducted at a large scale in the Mediterranean, showing the importance of temperature and chlorophyll concentration as forcing factors of the spatial distribution (*e.g.* [43]). However, processes that drive the small pelagic populations may be different at smaller scales (i.e. within the Gulf of Lions). In particular, the relative importance of density-dependent *versus* density-independent (environmental) processes on spatial distributions is likely to be dependent on the scale of the study [47]. A study accounting for both density-dependent and independent variables is thus of necessity to understand the identified areas [13].

Despite the important variability in spatial distributions, we defined unfavourable areas close to the coast for anchovies and on the 150–200 m-deep stratum for sprats and sardines. In contrast, we could assess recurrent areas (high average biomass and low variability) for sardines and anchovies in the Gulf of Lions. For sardines, recurrent areas were situated near the coast and in the Western part of the Gulf of Lions. For anchovies, two recurrent areas were identified further from the coast (isobaths of ~70 to 100 m). One was situated on the west part while the other one could be associated with the Rhone river and its plume on the east. As the surveys occurred during peak reproduction for anchovies [23], it is likely these areas corresponded to the spawning grounds of anchovies in this region. Interestingly, these 2 areas are very similar to the spawning areas detected from egg surveys in the 60 s [48]. This gives us important insights on their spawning grounds, exhibiting their stability despite several changes in external variables between these 2 periods (e.g. fishing effort and captures, environment). Also, such information is important for the elaboration of spatially explicit management plan (e.g. MPA, etc.). Indeed, it shows that despite the high mobility of small pelagic fish and high interannual variability in peak biomass locations, some areas consistently offer favourable spawning grounds for this species, making the task easier if one wanted to protect them.

Finally, we investigated interspecific relationships between 3 species sharing similar trophic level (i.e. zooplankton and especially copepods constitute the bulk of their diet at that period, though phytoplankton is also consumed in particular by sardines [21–22]). Different results were obtained according to the scale at which it was studied. Indeed, the index of overlap that we proposed decreased from the global (the population envelop) to the intermediate scale, indicating that if the three species co-habited in the Gulf of Lions, their hotspots were not always situated in the same area. This suggests that global environmental factors may have driven the three species to inhabit the Gulf of Lions, but that local environmental factors and/or inter-specific competition may have resulted in segregation within the Gulf. This was confirmed by the difference obtained between the two local collocation indices used. The collocation index translating species co-occurrence was surprisingly high (especially for anchovies and sardines), revealing that at a 2-nm sampling scale, species almost always co-occurred. The LIC index (based on density rather than presence/absence) had lower values, suggesting that for most sampled locations a given species predominantly occurred in each location (as also indicated by the dominance index). In particular, sardines clearly dominated the first depth strata (0–50 m) and the western part of the Gulf of Lions, while anchovies were more abundant in deeper waters and towards the central and eastern part of the Gulf. Collocation analyses give us interesting information on potential competition of these three planktivorous species, as their co-occurrence is a potential source of competition if/when food becomes limiting.

In summary, this study exhibited the relatively stable aggregation structure of small pelagic species, probably due to their inherent social structure (shoals, etc.). In contrast, it confirmed that the three species were highly mobile and could inhabit every stratum of the Gulf of Lions, though some preferred and unfavourable habitats could be highlighted for each species. The environmental drivers of both these habitats and the variability in locations of peak biomass remain to be investigated to better understand population dynamics. Finally, differences between species aggregative behaviour and habitat preferences were also highlighted, exhibiting potential competition avoidance.

## Supporting Information

**File S1 Supplementary material. Figure S1.** Presence of anchovies, sardines and sprats in the Gulf of Lions. **Figure S2.** Annual aggregation curves for each species. **Figure S3.** Annual maps of log-biomass for anchovies. **Figure S4**. Annual maps of log-biomass for sardines. **Figure S5**. Annual maps of log-biomass for sprats. **Figure S6**. Centres of gravity and inertia. **Figure S7**. Empirical Orthogonal Function analysis. **Figure S8**. Dominance index depending on depth or longitude strata. **Table S1**. Yearly and global collocation indices at three different time scales.

## Acknowledgments

We are very grateful to the crew of the Ifremer RV *L'Europe* and everyone who participated to the PELMED surveys between 2003 and 2012. PELMED surveys are cofunded by IFREMER and the European Union through the Data Collection Framework (DCF). This study is part of the program EcoPelGol (Study of the Pelagic ecosystem in the Gulf of Lions), funded by France Filière Pêche (FFP). We also wish to thank Vincent A. Viblanc for his helpful suggestions and his English-proofing of the manuscript. Finally, we thank 2 anonymous reviewers for their helpful comments.

## Author Contributions

Conceived and designed the experiments: CS JLB DR. Performed the experiments: JLB JHB DR EVB CS. Analyzed the data: CS NB JMF. Contributed reagents/materials/analysis tools: CS MM. Wrote the paper: CS JMF NB. Made useful suggestions on the manuscript: MM EVB.

## References

1. Woillez M, Petitgas P, Rivoirard J, Fernandes P, Hoftstede Rter, et al. (2006) Relationships between population spatial occupation and population dynamics. ICES CM 2006/O:05
2. Agostini VN, Hendrix AN, Hollowed AB, Wilson CD, Pierce SD, et al. (2008) Climate-ocean variability and Pacific hake: a geostatistical modeling approach. Journal of marine systems 71: 237–248.
3. Bacheler NM, Bailey KM, Ciannelli L, Bartolino V, Chan KS (2009) Density-dependent, landscape and climate effects on spawning distribution of walleye Pollock Theragra chalcogramma. Marine Ecology Progress Series 391: 1–12.
4. Cardinale M, Hagberg J, Svedang H, Bartolino V, Gedamke T, et al. (2010) Fishing through time: population dynamics of plaice (Pleuronectes platessa) in the Kattegat-Skagerrak over a century. Population Ecology 52: 251–261.
5. Boulinier T, Danchin E (1997) The use of conspecific reproductive success for breeding patch selection in terrestrial migratory species. Evolutionary Ecology 11: 505–517.
6. Hixon MA, Anderson TW, Buch KL, Johnson DW, McLeod JB, et al. (2012) Density dependence and population regulation in marine fish: a large-scale, long-term field manipulation. Ecological Monographs 82: 467–489.
7. Petitgas P, Alheit J, Peck MA, Raab K, Irigoien X, et al. (2012) Anchovy population expansion in the North Sea. Marine Ecology Progress Series 444: 1–13.
8. Lewis S, Sheratt TN, Hamer KC, Wanless S (2001) Evidence of intra-specific competition for food in pelagic seabird. Nature 412: 816–819.
9. MacCall AD (1990) Dynamic Geography of Marine Fish Populations. Seattle: University of Washington Pr. 153 p.
10. Paloheimo JE, Dickie LM (1964) Abundance and fishing success. Rapport et Proces Verbaux des Reunions du Conseil International pour l'Exploration de la Mer 155: 152–163.
11. Bertrand A, Segura M, Gutiérrez M, Vásquez L (2004) From small-scale habitat loopholes to decadal cycles: a habitat-based hypothesis explaining fluctuation in pelagic fish populations off Peru. Fish and Fisheries 5: 296–316.
12. Morfin M, Fromentin JM, Jadaud A, Bez N (2012) Spatio-temporal Patterns of Key Exploited Marine Species in the Northwestern Mediterranean Sea. PlosOne 7 (5): e37907.
13. Ciannelli L, Bartolino V, Chan KS (2012) Non-additive and non-stationary properties in the spatial distribution of a large marine fish population. Proceedings of the Royal Society B 279: 3635–3642.
14. Ciannelli L, Fisher JAD, Skern-Mauritzen M, Hunsicker ME, Hidalgo M, et al. (2013) Theory, consequences and evidence of eroding population spatial structure in harvested marine fishes: a review. Marine Ecology Progress Series 480: 227–243.
15. Apostolaki P, Milner-Gulland EJ, McAllister MK, Kirkwood GP (2002) Modelling the effects of establishing a marine reserve for mobile fish species. Canadian Journal of Fisheries and Aquatic Sciences 59: 405–415.
16. Cury P, Bakun A, Crawford RJM, Jarre-Teichmann A, Quinones R, et al. (2000) Small pelagics in upwelling systems: patterns of interaction and structural changes in "wasp-waist" ecosystems. ICES J. Mar. Sci. 57: 603–618.
17. Bakun A (1996) Patterns in the ocean: ocean processes and marine population dynamics. University of California Sea Grant, San Diego, California, USA in cooperation with Centro de Investigaciones Biologicas de Noroeste, La Paz, Baja California Sur, Mexico. 323 pp.
18. Checkley DM, Ayon P, Baumgartner TR, Bernal M, Coetzee JC, et al. (2009) Habitats. In: Climate Change and Small Pelagic Fish. A. Checkley, J. Alheit, Y. Oozeki & C. Roy (eds) New York: Cambridge University Press
19. General Fisheries Commission for the Mediterranean (2012) Report of the working group on stock assessment of small pelagic species. Split, Coratia.
20. Van Beveren E, Bonhommeau S, Fromentin JM, Bigot JL, Bourdeix JH, et al. (2014) Rapid changes in growth, condition, size and age of small pelagic fish in the Mediterranean. Marine Biology 161: 1809–1822.
21. Costalago D, Navarro J, Álvarez-Calleja I, Palomera I (2012) Ontogenetic and seasonal changes in the feeding habits and trophic levels of two small pelagic fish species. Marine Ecology Progress Series 460: 169–181
22. Nikolioudakis N, Isari S, Somarakis S (2014) Trophodynamics of anchovy in a non-upwelling system: direct comparison with sardine. Marine Ecology Progress Series 500: 215–229.
23. Palomera I, Olivar MP, Salat J, Sabates A, Coll M, et al. (2007) Small pelagic fish in the NW Mediterranean Sea: An ecological review. Progress in Oceanography 74: 377–396.
24. Weill A, Scalabrin C, Diner N (1993) MOVIES-B: An acoustic detection description software. Application to shoal species classification. Aquatic Living Resources, 6: 255–267
25. Simmonds J, MacLennan D (2005) Fisheries Acoustics, Theory and Practice, 2nd edn. Oxford: Blackwell Publishing, 437 pp.
26. Doray M, Masse J, Petitgas P (2010) Pelagic fish stock assessment by acoustic methods at Ifremer, Internal report, pp.18, *epic.awi.de/33739/1/ifremer-acoustic-methods.pdf*.
27. Matheron G (1963) Principles of geostatistics. Economic Geology 58: 1246–1266.
28. Dungan JL, Perry JN, Dale MRT, Lengendre P, Citron-Pousty S, et al. (2002) A balanced view of scale in spatial statistical analysis. Ecography 25: 626–640.
29. Gini C (1921) Measurement of inequality of income. Economic Journal 31: 22–43.
30. Petitgas P (1998) Biomass-dependent dynamics of fish spatial distributions characterized by geostatistical aggregation curves. ICES Journal of Marine Science 55: 443–453.
31. Myers RA, Cadigan NG (1995) Was an increase in natural mortality responsible for the collapse of northern cod? Canadian Journal of Fish and Aquatic Science 52: 1274–1285.
32. Rivoirard J, Simmonds J, Foote KG, Fernandes P, Bez N (2000) Gesotatistics for estimating fish abundance. Blackwell Publishing, Oxford.
33. Chilès J-P, Delfiner P (1999) Geostatistics: Modeling Spatial Uncertainty. 1st ed. New York: Wiley-Interscience. 720 p.
34. Bez N, Rivoirard J (2001) Transitive geostatistics to characterise spatial aggregations with diffuse limits: an application on mackerel ichtyoplankton. Fisheries Research 50: 41–58.
35. Woillez M, Poulard J-C, Rivoirard J, Petitgas P, Bez N (2007) Indices for capturing spatial patterns and their evolution in time, with application to European hake (Merluccius merluccius) in the Bay of Biscay. ICES Journal of Marine Science 64: 537–550.
36. Lorenz EN (1956) Empirical orthogonal functions and statistical weather prediction. Department of Meteorology, MIT, Scientific Report n°1.
37. Korres G, Pinardi N, Lascaratos A (2000) The Ocean Response to Low-Frequency Interannual Atmospheric Variability in the Mediterranean Sea. Part II: Empirical Orthogonal Functions Analysis. Journal of Climate 13: 732–745.
38. Bellier E, Planque B, Petitgas P (2007) Historical fluctuations in spawning location of anchovy (Engraulis encrasicolus) and sardine (Sardina pilchardus) in the Bay of Biscay during 1967–73 and 2000–2004. Fisheries Oceanography 16: 1–15.
39. Bez N, Rivoirard J (2000) Indices of collocation between populations. In: Checkley DM, Hunter JR, Motos L, von der Lingen CD (editors).Workshop on the Use of Continuous Underway Fish Egg Sampler (CUFES) for mapping spawning habitat of pelagic fish. GLOBEC Rep., pp. 48–52.
40. Barange M, Coetzee JC, Twatwa NM (2005) Strategies of space occupation in anchovy and sardine in the southern Benguela: the role of stock size and intra-species competition. ICES Journal of Marine Science 62: 645e654.
41. R Development Core Team (2012) R: A language and environment for statistical computing. R Foundation for Statistical Computing, Vienna, Austria.
42. Renard D, Bez N, Desassis N, Beucher H, Ors F (2013) RGeoS: Geostatistical Package. R package version 9.1.6. (Ecoles des Mines website, available at http://cg.ensmp.fr/rgeos, accessed 2014 October 10$^{th}$).
43. Giannoulaki M, Iglesias M, Tugores MP, Bonanno A, Patti B, et al. (2013) Characterizing the potential habitat of European anchovy Engraulis encrasicolus in the Mediterranean Sea, at different life-stages. Fisheries Oceanography 22: 69–89.

44. Gaston KJ, Blackburn TM, Gregory RD (1999) Intraspecific abundance–occupancy relationships: case studies of six bird species in Britain. Div. Distr. 5: 197–212. (doi:10.1046/j.1472-4642.1999.00054.x)
45. Fisher JAD, Frank KT (2004) Abundance–distribution relationships and conservation of exploited marine fishes. Mar. Ecol. Prog. Ser. 279: 201–213. (doi:10.3354/meps279201)
46. Atkinson DB, Rose GA, Murphy EF, Bishop CA (1997) Distribution changes and abundance of northern cod (*Gadus morhua*), 1981–1993. Canadian Journal of Fisheries and Aquatic Sciences 54: 132–138.
47. Shepherd T, Litvak M (2004) Density-dependent habitat selection and the ideal free distribution in marine fish spatial dynamics: considerations and cautions. Fish and Fisheries 5: 141–152.
48. Aldebert Y, Tournier H (1971) La reproduction de la sardine et de l'anchois dans le Golfe du Lion. Revue de Travail de l'Institut des Pêches maritimes 35: 57–75.

# Range Expansion and Population Dynamics of an Invasive Species: The Eurasian Collared-Dove (*Streptopelia decaocto*)

**Spencer N. Scheidt*, Allen H. Hurlbert**

Department of Biology, University of North Carolina, Chapel Hill, North Carolina, United States of America

## Abstract

Invasive species offer ecologists the opportunity to study the factors governing species distributions and population growth. The Eurasian Collared-Dove (*Streptopelia decaocto*) serves as a model organism for invasive spread because of the wealth of abundance records and the recent development of the invasion. We tested whether a set of environmental variables were related to the carrying capacities and growth rates of individual populations by modeling the growth trajectories of individual populations of the Collared-Dove using Breeding Bird Survey (BBS) and Christmas Bird Count (CBC) data. Depending on the fit of our growth models, carrying capacity and growth rate parameters were extracted and modeled using historical, geographical, land cover and climatic predictors. Model averaging and individual variable importance weights were used to assess the strength of these predictors. The specific variables with the greatest support in our models differed between data sets, which may be the result of temporal and spatial differences between the BBS and CBC. However, our results indicate that both carrying capacity and population growth rates are related to developed land cover and temperature, while growth rates may also be influenced by dispersal patterns along the invasion front. Model averaged multivariate models explained 35–48% and 41–46% of the variation in carrying capacities and population growth rates, respectively. Our results suggest that widespread species invasions can be evaluated within a predictable population ecology framework. Land cover and climate both have important effects on population growth rates and carrying capacities of Collared-Dove populations. Efforts to model aspects of population growth of this invasive species were more successful than attempts to model static abundance patterns, pointing to a potentially fruitful avenue for the development of improved invasive distribution models.

**Editor:** Csaba Moskát, Hungarian Academy of Sciences, Hungary

**Funding:** The authors have no support or funding to report.

**Competing Interests:** The authors have declared that no competing interests exist.

* Email: spencer.scheidt@gmail.com

## Introduction

Invasive species, though considered by many to represent a significant threat to global biodiversity [1,2], can provide a unique opportunity to study ecological and evolutionary processes on a scale that is otherwise infeasible and potentially unethical [3–5]. In particular, studying a species as it spreads across a broad geographic region can provide a unique context for examining the primary factors influencing distribution, abundance, and population dynamics. In cases where the invasion and spread are ongoing, such analyses can be used to make predictions about future distributions and the environmental features that may be facilitating population spread and establishment [6]. However, despite several classical examples [7–9], ecologists still lack a comprehensive predictive model of invasion dynamics [4,10].

Even though species invasions are relatively common and widespread [11,12], few invaders are ideal for ecological study. First, most invasions either occur without human knowledge or are not noticed until well after initial colonization and establishment. Second, because many invasive species are only detectable by scientists or natural resource managers, the number of records may be few and patchily distributed, especially during the early stages of invasion. Avian invaders, however, are much less likely to escape the notice of the legions of amateur birdwatchers constantly on the lookout for rarities, and semi-standardized surveys such as the North American Breeding Bird Survey and the Christmas Bird Count provide quantitative indices of abundance in addition to the more commonly available information on presence-absence over a broad geographical extent. It is thanks to the efforts of these types of surveys that the invasion of North America by one species, the Eurasian Collared-Dove (*Streptopelia decaocto*), has been documented in detail over the past forty years [13,14].

Originally a native of India, the Eurasian Collared-Dove (hereafter the ECD) was introduced into the Bahamas in the mid-1970s and has spread rapidly throughout the continental United States in the past four decades in a similar manner to its invasion of Europe earlier in the 20th century [13,15–17]. A known grain forager, the ECD has been linked to human settlements and non-intensive agricultural zones in several large scale ecological studies in both the U.S. and Europe [16,18–20]. Researchers have also found negative relationships with forest cover and areas of intensive agriculture [16,21,22]. With respect to rates of spread, Hooten and Wikle (2008) used a hierarchical Bayesian diffusion model and found higher rates of diffusion in the western U.S. compared to regions closer to the point of origin of the invasion. Higher rates of spread could be due either to landscape features that facilitate dispersal, or to local environmental conditions that allow greater population growth and the production of more propagules. However, because the ECD exhibits a characteristic jump dispersal pattern, often dispersing great distances and 'backfilling' areas in-between [21], and inhabits a wide climatic range, identifying clear patterns in its invasive spread has proven difficult [19,20,23,24]. In addition, local sites vary in the length of time that ECDs have been present, from sites that have only recently been colonized to areas that have been continuously occupied for decades. The fact that the abundance surface of the ECD is inherently non-equilibrial may be part of the reason why studies that have attempted to explain spatial variation in abundance from a snapshot in time have done so with limited predictive power [20].

Here, we make use of historical information on abundance from several hundred locations throughout North America to evaluate how local population trajectories of an invasive species vary across its expanding range. By fitting simple population growth models (exponential, quadratic and logistic), we are able to estimate parameters such as population growth rate and carrying capacity (for sites that are no longer increasing) for many local sites. We are then able to assess the importance of climate, land cover, and historical or geographic variables for explaining these two parameters, and make the following predictions. First, sites close to the point of invasion (i.e. south Florida) with older populations are most likely to have reached carrying capacity, while sites along the range expansion front are more likely to still be in a phase of exponential growth. Second, while distance from the point of invasion and population age should be predictors of ECD abundance in any one year [19,20], these historical variables should be less important in explaining local carrying capacities at sites that show signs of having reached an equilibrium. Third, based on previous literature [18,21,22], it is predicted that areas associated with developed and agricultural land cover will be positively correlated with growth rate and carrying capacity, while forested land cover is predicted to be negatively correlated with these parameters.

## Methods

Abundance data for the ECD up through 2010 were obtained from two publicly available datasets: the North American Breeding Bird Survey (BBS) [25], and the Christmas Bird Count (CBC) [26]. Each BBS route consists of 50 3-minute point counts spaced along a 39.4 km roadside route that are conducted on a single day during the breeding season (typically in June), over which a single observer records all birds seen or heard within 400 m of each point. Data that did not meet BBS quality standards (due to inclement weather, inappropriate survey times, etc.) were removed from our analysis. Abundance counts from all 50 stops at each individual BBS route were combined into a single metric of abundance for that particular route. Each CBC survey consists of a circular region 24 km in diameter in which variable numbers of observers count all of the birds seen or heard over the course of the entire day in late December or early January. To account for variation in survey effort through time and across count circles, we divided these raw counts by the number of party hours (i.e. the total number of hours independent birding parties spent counting birds on that day). These two surveys are conducted at distinct times of year and at very different grain sizes, and as a result may provide different insight into the ECD invasion.

Traditionally, studies have attempted to explain static abundance patterns across the range, and so we analyzed the abundance surface of the ECD for a single year, 2010, across 792 CBC sites that reported the presence of the ECD. Our main focus, however, was on characterizing how abundance has varied through time at local sites across the range, an approach we hypothesized would have more explanatory power than the static approach. Logistic, exponential, and quadratic models were fit to population trajectories for each route (BBS) or count circle (CBC) with at least 8 years of abundance data; 107 BBS routes and 292 CBC count circles fit these criteria. Quadratic functions were fit because a number of empirical patterns were hump-shaped, and the quadratic allows an estimate of an upper limit to carrying capacity. Logistic curves were fit using maximum likelihood, although models failed to reach convergence for 10 BBS and 31 CBC sites for which a logistic was a poor empirical fit. Because the ECD was known to be absent from all sites initially, sites that were best fit by monotonically decreasing functions (negative exponential, negative linear, or negative logistic; a total of 14 BBS and 36 CBC sites) were presumed to have missed the dynamics of initial colonization and population growth and were discarded from further analyses. These negative growth rate model fits could be due to birders not recording the presence of the ECD during the early years of the invasion. The similarity of the ECD to the Ringed Turtle-Dove (*Streptopelia risoria*) combined with its rapid spread has been a point of confusion for many birders, at least until they became aware of the ECD's presence and appearance [23].

Growth model fit was assessed using the small sample-corrected Akaike Information Criterion ($AIC_c$) [27], for each model in each location. The model with the lowest $AIC_c$ score was presumed to be the best fit, although sites that had $\Delta AIC_c$ scores of less than 2 indicate that other models also had some measure of support [27]. Exponential growth rates were estimated using linear regression on log-transformed abundance values for all sites where $\Delta AIC_c$ for the exponential model was less than 2. Local carrying capacity was estimated in one of two ways. For sites that were best fit by quadratic models, the maximum abundance value where the derivative equaled 0 was identified as an upper limit on carrying capacity. For population trajectories best fit by a logistic model, carrying capacity was taken as the asymptote parameter. Analyzing these two types of carrying capacity separately had little effect on model results, and so only the combined analysis is presented and the term "carrying capacity" is used to refer to both. Sites that were best fit by an exponential model but that had a quadratic or logistic $\Delta AIC_c$ score of less than 2 were included in the carrying capacity analysis using the value of carrying capacity from whichever of the two models had the lowest $\Delta AIC_c$ score. After imposing these filters, 50 BBS routes and 136 CBC circles were used in the analysis of carrying capacity, and 58 BBS routes and 187 CBC circles were used in the analysis of population growth rate.

We examined three categories of predictors of carrying capacity or growth rate: historical/geographical, land cover, and climate variables. Historical/geographical variables include invasion distance and population age. Invasion distance was estimated as the great circle distance between each local site and the initial point of introduction for each respective data set (Florida route 36 in 1986 for the BBS and Bahamas count circles BASA and BASC in 1985 for the CBC) [6,20]. It should be noted that this metric ignores natural boundaries to dispersal (such as the Gulf of Mexico separating south Florida from south Texas), but is expected to be a reasonable approximation for total land-based distance. Population age was estimated as the number of years since ECDs were first observed at a local site. This metric assumes that the ECD was immediately detected after its initial colonization.

Land cover data were extracted from the National Land Cover Dataset (NLCD) 2006 raster, which covers the conterminous U.S. with 16 land cover classes at a resolution of 30 m, where each pixel is assigned to only one land cover class [28]. BBS routes were clipped with a 400 m buffer along their entire length (39.4 km), which corresponds to the survey area of the route. CBC sites were clipped with a 24 km diameter buffer. Land cover percentage values were calculated by dividing the number of pixels in each land cover class by the total number of pixels in the buffer and multiplying by 100. Because the ECD is known to associate with human settlement [16,18–20], variables reflecting different types of developed and agricultural land cover were selected for further analysis. In addition, researchers have documented negative relationships between ECD prevalence and forest cover [16,21,22], and so an aggregated measure of forested cover was also included in the analysis. Specifically, we calculated the percentage of land cover surrounding focal BBS routes and CBC circles in each of the following NLCD categories: open (<20% impervious cover), low intensity developed (20–49% impervious cover), medium intensity developed (50–79% impervious cover), high intensity developed (>80% impervious cover), pasture (pasture or hay accounts for ≥20% of vegetation), cropland (crops account for ≥20% of vegetation), and forest (≥20% tree cover).

Climate data were extracted from the WorldClim BioClim data set (20 arc minute resolution) [29], over the same spatial buffers around BBS and CBC sites. Preliminary univariate analyses allowed us to limit the number of climate variables to two strong predictors that would affect primary productivity, average habitat suitability and environmental tolerance: mean annual precipitation (mm) and mean annual temperature (°C), each averaged over the appropriate buffer. The distributions of land cover and climate variables encompassed by the BBS and CBC datasets are shown in Figure S1, and differences between the distributions were assessed with Mann-Whitney two-sample rank sum tests.

Our goal was to develop predictive models of ECD population carrying capacities and growth rates using these three classes of predictor variables, and to assess the relative importance of those predictors. As a preliminary analysis, we utilized regression trees to identify potentially important interactions between predictor variables [30]. No strong interactions were found, with the exception of CBC invasion distance and CBC low intensity percent developed cover in our growth rate analysis, which we ignored because it increased the AIC score of our CBC growth model. Because all of the predictor variables in our three categories had *a priori* support, we considered the set of linear models representing all possible combinations of those eleven variables as main effects for five distinct analyses: 1) predicting carrying capacity using BBS data, 2) predicting carrying capacity using CBC data, 3) predicting 2010 abundance using CBC data from 2010, 4) predicting population growth rate using BBS data, and 5) predicting growth rate using CBC data. All variables and responses for each dataset were independently mean centered at 0 and scaled so that the standard deviation was equal to 1. This normalization facilitates the comparison of parameter estimates across predictors that vary in units and scale. Because the distributions of environmental variables differed slightly between BBS and CBC datasets (Figure S1), we also conducted the above analyses on jointly standardized variables in which values from the BBS and CBC sites were combined prior to normalization. Models were evaluated in an information theoretic context, and for each linear model in a set, the $AIC_c$ score, $R^2$ and Akaike weight were computed. For predictive purposes, we used model averaging in which parameter estimates for each variable are averaged across all models in which that variable occurs, weighted by each model's Akaike weight. Model averaging provides a more precise and less biased inference about the predicted effects of a set of variables, especially when no one model has overwhelming support [27]. Finally, we calculated individual variable relative importance weights ($w_{i+}$), which are a measure of how important a variable is relative to the other variables considered, by summing the Akaike weights of all models containing that particular variable (Burnham and Anderson 2002). A variable relative importance weight close to 1 indicates that that variable tends to be included in the models that collectively have the highest levels of support.

For each data set and response variable (population growth rate or carrying capacity), a linear model containing all of the weighted parameters (see Table 1) was evaluated for potential over-fitting using Leave-One-Out cross validation. Cross validation is a common model evaluation technique that divides a data set into training and validation segments [31], and can be used to assess how well the training data predict the validation data [32–34]. Leave-One-Out-cross-validation was applied to each model, where $n$-1 data points are used to predict the remaining datum $n$ times to create an unbiased estimate of average prediction error. This prediction error was used to calculate R-squared inflation due to overfitting of the data set for each model. Finally, because cross-validation may overestimate the goodness of fit of models when model residuals are spatially autocorrelated [35], we plotted Moran's I correlograms using the ncf package in R [35] for our four main dependent variables (growth rate and carrying capacity for each of the two datasets), and for their respective average model residuals.

All derived data used both in model selection and cross validation are archived in the Dryad Digital Depository.

## Results

In both BBS and CBC data sets, the type of population model (exponential, quadratic, or logistic) receiving the greatest support at a particular site varied as a function of the distance to the original invasion point (BBS: $p = 0.003$; CBC: $p << 0.001$) and of the time since colonization (BBS: $p = 0.007$; CBC: $p << 0.001$; Figure 1, Figure 2). Hump-shaped population trajectories occurred at sites with the oldest populations and that were located closest to the point of invasion, while most of the younger populations out on the expansion front exhibited exponential growth. Sites best fit by logistic growth are intermediate on average in both age and distance.

### Carrying capacity

The variables most important for explaining spatial variation in carrying capacity differed for the BBS and CBC data sets (Table 1, Figure 3). Although the BBS model with the lowest $AIC_c$ score described a moderate amount of variance ($R^2 = 0.46$), its Akaike weight was only 0.03, and several other models had roughly equivalent support. As such, we evaluated overall variable importance weights rather than simply identifying which variables were included in the "best model", and we interpret model-averaged parameters, which collectively explained 48% of the variance in carrying capacity. An examination of variable weights across all models indicated that medium intensity developed cover (a positive relationship, $w_{i+} = 0.99$) and high intensity developed cover (a negative relationship, $w_{i+} = 0.93$) were the variables with the strongest effects on carrying capacity in the BBS data set (Table 1).

**Table 1.** Variable relative importance weights and weighted parameter estimates.

| | *K*BBS | | *K*CBC | | *NCBC: 2010* | | *r*BBC | | *r*CBC | |
|---|---|---|---|---|---|---|---|---|---|---|
| | ***wi+*** | $\hat{\beta}$ | ***wi+*** | $\hat{\beta}$ | ***wi+*** | $\hat{\beta}$ | ***wi+*** | $\hat{\beta}$ | ***wi+*** | $\hat{\beta}$ |
| *Historical/geographical variables* | 0.29 | −0.02 | 0.28 | −0.03 | 0.35 | 0.02 | **0.98** | **0.59** | **1.00** | **0.59** |
| Distance to invasion origin | 0.23 | −0.01 | **0.99** | **0.40** | **1.00** | **0.29** | 0.31 | −0.04 | 0.26 | 0.00 |
| Time since colonization | | | | | | | | | | |
| *Land cover variables* | 0.29 | −0.03 | 0.63 | −0.12 | **0.95** | (0.14 | 0.24 | 0.01 | 0.29 | (0.01 |
| Developed: open cover | 0.55 | (0.19 | 0.61 | 0.16 | 0.91 | **0.15** | 0.30 | 0.01 | **0.76** | **0.15** |
| Developed: low intensity | **0.99** | **1.27** | 0.33 | 0.01 | **0.82** | **−0.11** | 0.48 | 0.18 | 0.37 | 0.03 |
| Developed: med. Intensity | **0.93** | **−0.84** | 0.48 | 0.08 | 0.37 | 0.01 | 0.36 | −0.05 | 0.38 | 0.04 |
| Developed: high intensity | 0.42 | −0.08 | 0.37 | −0.03 | 0.49 | −0.03 | 0.25 | 0.01 | 0.59 | −0.06 |
| Forest | 0.39 | −0.06 | 0.49 | −0.05 | 0.35 | −0.01 | **0.93** | **−0.29** | 0.26 | 0.00 |
| Pasture | 0.22 | 0.00 | **0.99** | **0.27** | 0.43 | 0.02 | 0.34 | −0.04 | 0.32 | 0.01 |
| Crops | | | | | | | | | | |
| *Climate variables* | 0.35 | 0.05 | **1.00** | **−0.54** | **1.00** | **−0.34** | 0.25 | 0.02 | 0.35 | −0.04 |
| Avg. Precipitation | 0.51 | 0.13 | **0.76** | (0.18 | 0.65 | (0.05 | 0.60 | 0.16 | 0.57 | (0.07 |
| Avg. Temperature | | | | | | | | | | |
| Model Averaged R2 | 0.48 | | 0.35 | | 0.17 | | 0.41 | | 0.46 | |

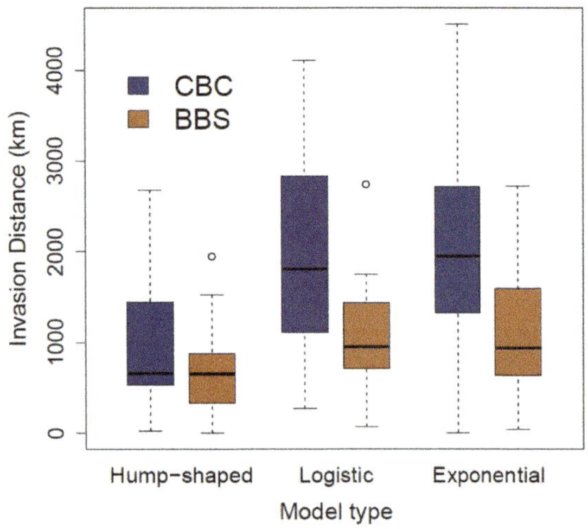


**Figure 1. Invasion distance and model type.** Boxplots of distance of a site from the original introduction point grouped by population trajectory type for the Christmas Bird Count (blue) and the Breeding Bird Survey (orange).

The CBC model using model averaged parameter estimates explained 35% of the variance in carrying capacity. Based on variable relative importance weights, the strongest positive predictors were time since colonization ($w_{i+} = 0.99$) and cropland cover ($w_{i+} = 0.99$), while the strongest negative predictor was precipitation ($w_{i+} = 1.00$), with a moderate negative effect of average temperature ($w_{i+} = 0.76$, Table 1). For comparison, we also modeled spatial variation in observed abundance from a single year, as opposed to carrying capacity estimated from the available time series, for 792 count circles in 2010. As in the carrying capacity analysis, time since colonization and average precipitation were two of the most consistent predictors of single-year abundance, with differences in which land cover variables were identified as most important (Table 1). Most importantly, however, the variance in single-year abundance was inherently noisier ($R^2 = 0.17$), with the model averaged parameters accounting for less than half of the variation explained in carrying capacity.

### Population Growth Rate

Similar to the carrying capacity results, the collective model averaged BBS and CBC models incorporated different predictors (Table 1). The model averaged BBS growth rate model explained 41% of the variance in growth rate. Variable relative importance weights point to a strong positive relationship with distance to the point of invasion origin ($w_{i+} = 0.98$), and a negative relationship with the amount of pastureland ($w_{i+} = 0.93$, Table 1). Total explained variation in population growth rate based on CBC data was similar to the BBS analysis (model averaged $R^2 = 0.46$), and distance to invasion origin was again the strongest predictor ($w_{i+} = 1.00$). However, for the CBC, low intensity developed cover was the most supported land cover variable ($w_{i+} = 0.76$, positive relationship) while pastureland appeared to be unimportant (Table 1).

### Cross Validation, Spatial Autocorrelation and Variable Standardization

As expected, Leave-One-Out-cross-validation detected model over-fitting, although the extent of the average prediction error and the degree of reduction in R-squared varied from model to model (Table 2). In general, cross validation suggested that our models still accounted for a large percentage of the variation in both carrying capacity and growth rate. Both the BBS and CBC growth rate models experience R-squared decreases of less than 10%, and still explained between 37 and 45% of the variance (Table 2). Cross validation of carrying capacity models exhibited larger decreases in R-squared (between 29 and 36%), although both still explained 150–200% of the variation in carrying capacity compared to the cross-validated single year abundance model (Table 2).

Raw values of carrying capacity and population growth rate exhibited varying degrees of spatial autocorrelation (especially growth rates). However, this was accounted for primarily by spatial autocorrelation in the predictors, and the residuals of our averaged models exhibited very little autocorrelation even at the closest lag distances (Figure S2). As such, spatial autocorrelation should not influence our interpretation of goodness-of-fit from the cross-validation results.

In our carrying capacity and population growth rate analyses, independent standardization of the BBS and CBC datasets had minimal impact on our model-averaged results compared to the analysis in which both datasets were combined prior to standardization. No parameter estimates changed sign and all variable relative importance weights were identical to within 0.01 (see Table 1 and Table S1).

## Discussion

Examining the invasive spread of the ECD by modeling population trajectories at individual sites across the range reveals novel aspects of the range expansion of this species. With both data sets we found that the best fitting growth models varied in a similar fashion with invasion distance and population age: hump-shaped distributions tended to fit older populations close to the initial introduction point, logistic distributions fit middle-aged populations at moderate invasion distances, and exponential models fit younger populations closer to the invasion front. This repeated pattern suggests that populations of the ECD may frequently go through three phases during colonization and establishment: (1) exponential increase, (2) a leveling off at or above local carrying capacity, and (3) a subsequent decline. Although the reasons for this decline are beyond the scope of this study, this may be evidence that populations of the ECD routinely overshoot the carrying capacity of local habitats. Another possibility is that these population trajectories could be depicting a traveling invasion wave that is initially fed by dispersing birds along the invasion front; once these dispersers pass through, declines in local abundance result. Regardless of why such a decline may occur, it seems reasonable to expect that populations of the ECD are more likely to show evidence of a carrying capacity as they age.

While the overall shape of population trajectories was related to historical and geographical variables, local carrying capacities and population growth rates varied in a predictable manner with land cover and climate variables across the ECD's geographic range. As confirmed by cross validation analyses, our main effects models were able to explain at least 23–45% of the variance in our two main response variables (Table 2). The predictive ability of models for these population parameters was much greater than for the model attempting to predict static single year abundance values across the range. The fact that the strongest predictor of abundance in 2010 was population age confirms our prediction that abundance in any given year is more influenced by invasion

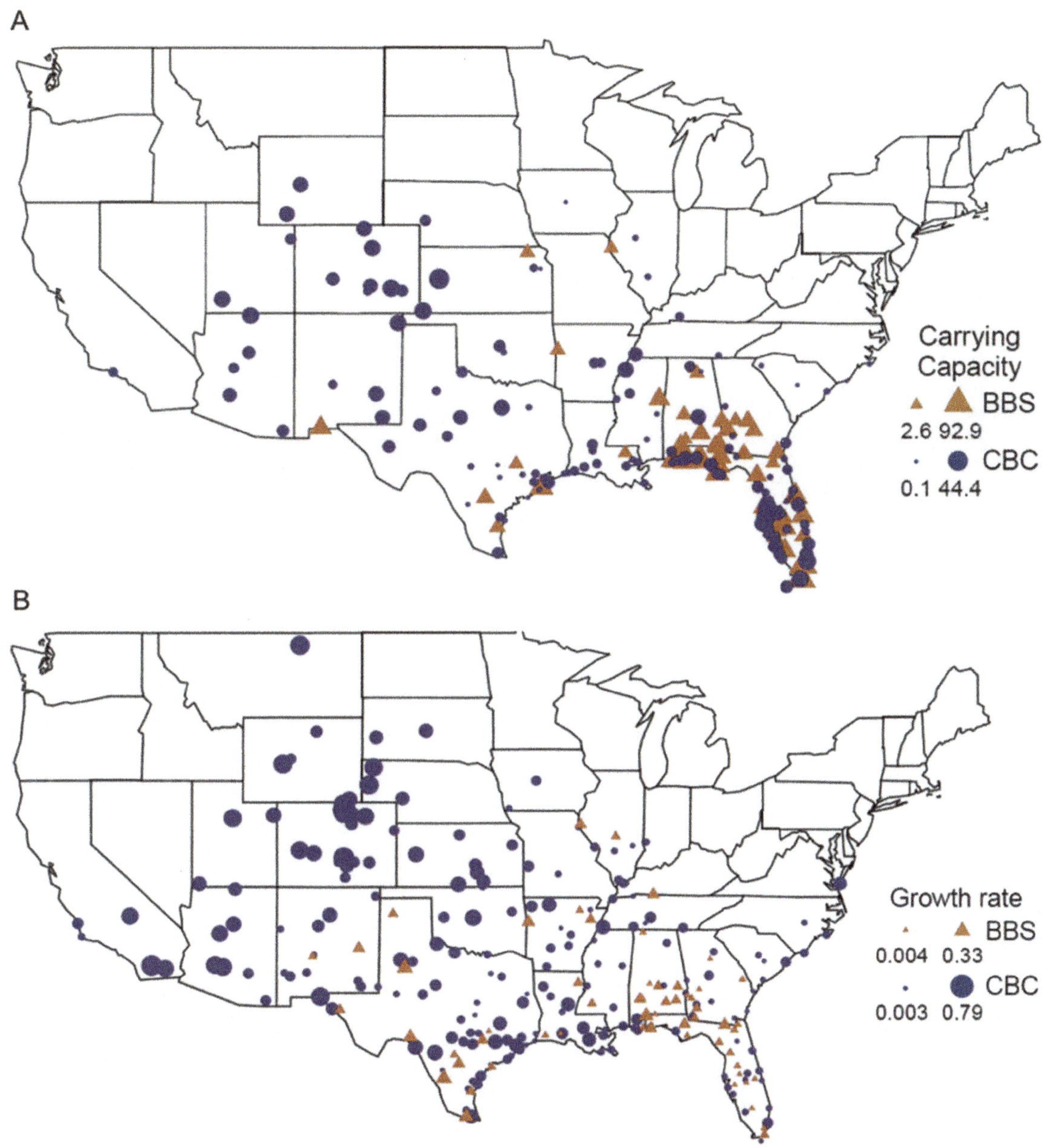


**Figure 2. Carrying capacities and growth rates.** Map of all (a) Christmas Bird Count circles (blue) and Breeding Bird Survey routes (orange) used in carrying capacity analysis scaled by estimated carrying capacity, *k* (b) Christmas Bird Count circles (blue) and Breeding Bird Survey routes (orange) used in growth rate analyses with the sizes of each location's point scaled by estimated population growth rate, *r*. Minimum and maximum symbol sizes and their meanings are given in the legend of each figure panel.

history than is a site's estimated carrying capacity. Given the non-equilibrial nature of this system, the estimation of population parameters from modeling growth trajectories therefore facilitates the development of more predictive models that can link the occurrence of ECD to characteristics of the local environment.

Our results show that both land cover and climate are important factors for explaining the observed patterns. Developed land cover in different forms (BBS: medium and high intensity developed; CBC: open and low intensity developed) was a consistently strong predictor across both data sets for predicting variation in carrying capacity and growth rate. Given the previously known propensity of the ECD to exploit bird-feeders and grain as sources of food and use telephone poles, suburban trees, and buildings for roosting [17], its seems likely that the ECD is numerically responding to these basic habitat and energetic constraints. Many of our analyses found negative relationships with both open and high intensity developed cover, indicating that neither golf courses and large lawns, nor heavily urban centers offer the ECD the resources necessary to achieve high abundances or growth rates. In contrast, both low and medium intensity developed cover (20–80% impervious surface) typified by single family housing areas enhanced carrying capacity and population growth rates, depending on the data set. These results parallel observations that link the ECD to suburban land cover but find it is not present in dense urban areas [21,23]. We also found that both forested cover and pastureland had weak negative effects on

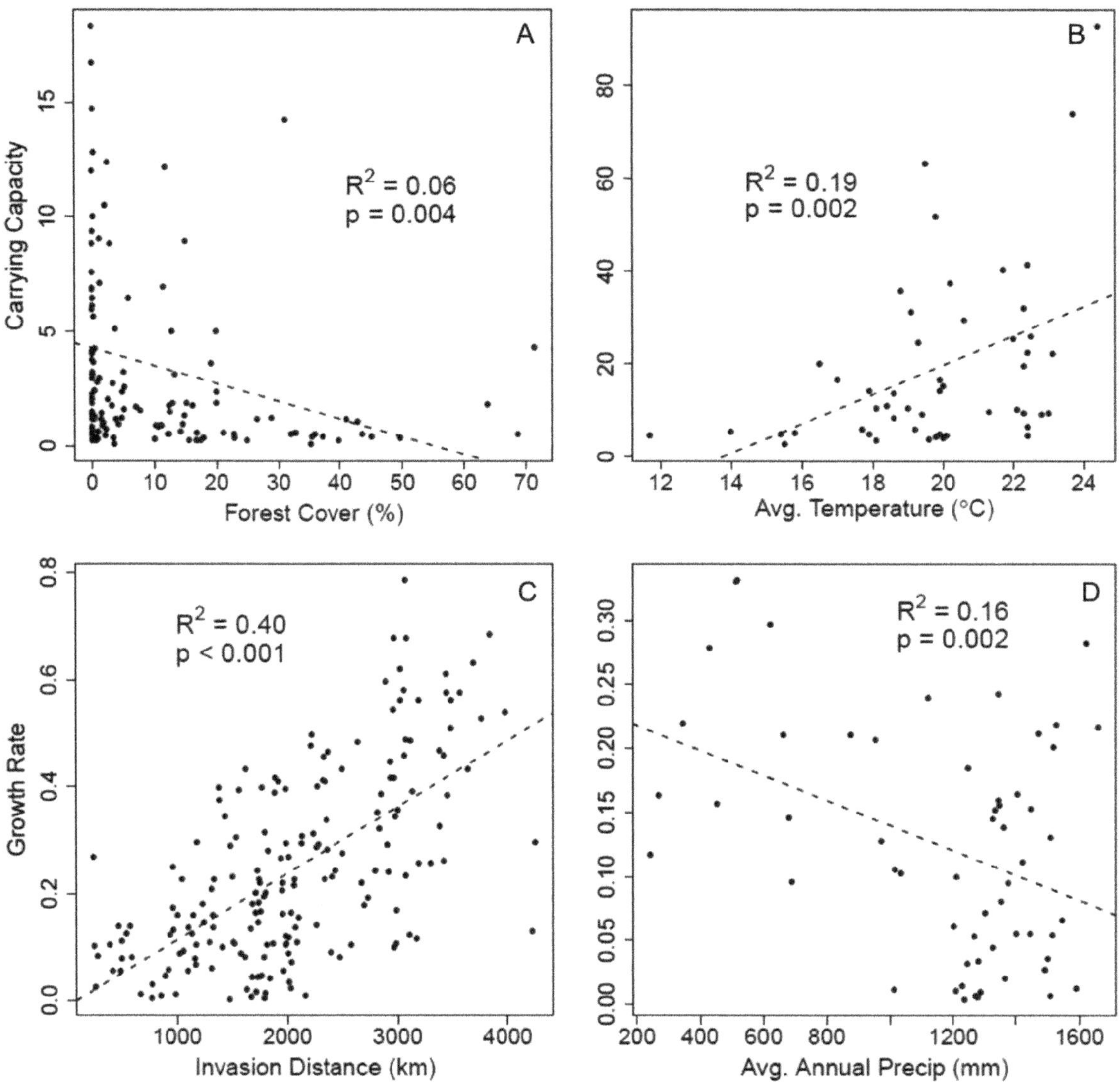


**Figure 3. Linear predictors of carrying capacities and growth rates.** Univariate plots of the strongest predictor variables for: (a) Christmas Bird Count (CBC) carrying capacity versus percent developed low intensity cover, (b) Breeding Bird Survey (BBS) carrying capacity versus minimum temperature of the coldest month (°C); and population growth rate: (c) CBC population growth rate versus invasion distance, (d) BBS population growth rate versus average annual temperature (°C).

**Table 2.** Cross Validation Results.

| | **Original Results** | | **Leave-One-Out cross validation** | | **% $R^2$ decrease (LOOCV)** |
|---|---|---|---|---|---|
| | **MSE** | **$R^2$** | **MSE** | **$R^2$** | |
| *Model* | | | | | |
| $K_{BBS}$ | 0.67 | 0.48 | 0.83 | 0.34 | 29.1 |
| $K_{CBC}$ | 0.70 | 0.35 | 0.84 | 0.23 | 36.1 |
| $N_{CBC: 2010}$ | 0.84 | 0.17 | 0.85 | 0.16 | 6.0 |
| $r_{BBS}$ | 0.73 | 0.41 | 0.76 | 0.37 | 9.5 |
| $r_{CBC}$ | 0.57 | 0.46 | 0.58 | 0.45 | 3.0 |

Leave-One Out-cross-validation (LOOCV) results for the model averaged main effects models predicting carrying capacities and growth rates of the BBS and CBC data sets. The mean square error (MSE) and $R^2$ values are provided for the original, uncorrected models and the corrected LOOCV models, with their associated R-squared shrinkage values.

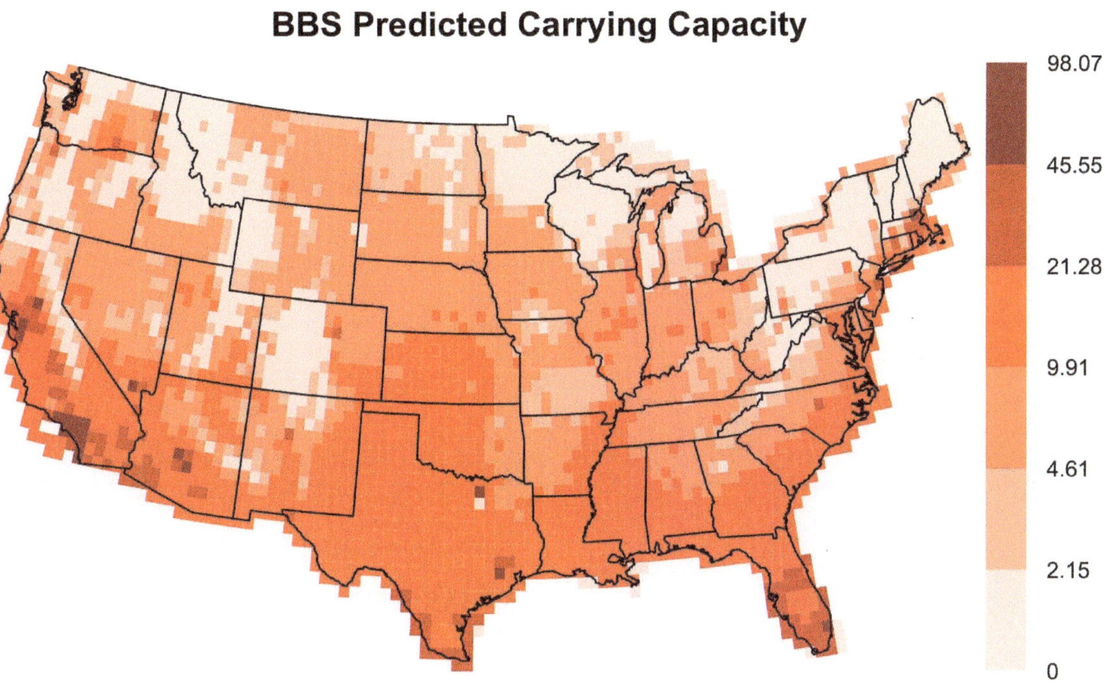


**Figure 4. Carrying capacity prediction map.** Map of predicted carrying capacity of the Eurasian Collared-Dove across the contiguous United States (Albers Equal Area Conic projection) based on the modeled relationships between Breeding Bird Survey population trajectories and climate and land cover data (see Table 1). Individual cells are 0.5° in each direction, and values represent the expected number of individuals observed on a BBS route within each cell for populations that have reached their carrying capacity.

carrying capacity and population growth rate, consistent with previous studies [13,21,22], while cropland may enhance ECD populations in some cases (Table 1). This suggests that agricultural land cannot be aggregated when studying the ECD – pastureland does not offer the ECD much in the way of food or resources [17] – and may explain why other studies have not found stronger coarse scale relationships between agricultural cover and the ECD [19,20].

Climate variables were also some of the strongest components of the best main effects models. In general, the ECD responded differently to climate depending on the data set being examined. Both of our BBS analyses found positive relationships with both precipitation and temperature, while all of our CBC analyses found negative relationships with the same variables. This could be due to differences in grain, spatial extent and sampling time, as discussed below.

Both carrying capacity and population growth rates were strongly related to land cover and climatic variables overall, but population growth rates were also affected by invasion distance to an equal extent in both data sets (Table 1). Population growth rates peaked at the expansion front (Figure 2), a result that parallels findings by Hooten and Wikle (2008) showing that the rate of spatial diffusion was greatest in the western U.S. We believe that populations at the invasion front could be experiencing higher growth rates for several reasons. First, rates of dispersal may be higher along the invasion front due to a lack of barriers to dispersal, especially dense forests. Second, sites closer to the original invasion point may be more likely to have begun approaching a population asymptote. Two sites might be on identical logistic trajectories, but the fitted estimate of population growth rate will be lower for sites that are farther along that trajectory. However, the correlation coefficient between growth rates calculated over a fixed time window (the first 8 years of data) versus growth rates calculated from the full time-series was quite high ($r = 0.84$), indicating that variation in time series length is not the primary source of variation in growth rates. Third, sites along the invasion front could offer more of the resources and habitats that facilitate the maintenance and growth of ECD populations. Regardless, populations of the ECD are increasing at a rapid rate far from the original point of introduction, which implies that the ECD is still on the path toward rapid range expansion.

Although the BBS and CBC datasets were similar with respect to the relative importance of broad classes of variables, the specific variables with the greatest support in our models differed depending on the data set. Several differences between the BBS and the CBC seem pertinent for explaining some of this variation. One obvious difference is that CBC surveys are conducted in winter, while the BBS is conducted in May or June. The ECD may exhibit seasonal behaviors that make individuals more or less likely to be detected in different seasons [36], and this might explain some of the discrepancies over the directionality of climate relationships between the two data sets. However, the ECD is capable of breeding year round throughout much of its North American range, and at this point no data suggest seasonal flocking or aggregation in the United States [17]. More importantly, the grain size of the two surveys differs by more than an order of magnitude: a CBC circle encompasses a total area of over 450 $km^2$ while a BBS route covers only 25 $km^2$. The larger sampling area of CBC surveys means that CBCs will be more likely to record ECD presence compared to BBS routes surveyed in the same location, especially when ECD densities are low. This explains the greater geographic coverage of the CBC relative to the BBS (and hence the greater overall distances from the invasion origin, Figure 1). This does not imply that the ECD is absent throughout the western U.S. during the breeding season, but rather that the number of sites on which it has been observed for a minimum of 8 years on BBS routes lags behind the number of CBC circles. This difference in spatial extent means that the CBC analysis covers a broader range of environmental conditions and reflects a different distribution of land cover values (Figures S1). For example, CBC sites tended to have less developed open land cover, pasture, and forest compared to BBS sites (Figure S1) which could potentially obscure or enhance the strength and direction of population parameter-environment relationships. The difference in CBC and BBS survey areas also means that environmental relationships are being evaluated at two distinct grain sizes, and it has long been appreciated that species-habitat relationships are strongly scale-dependent [37,38]. The fact that growth rate models had higher predictive power in the CBC while carrying capacity models had higher predictive power in the BBS might suggest that population growth rate is intrinsically determined at a coarser grain size than carrying capacity. This could be the case if population growth rate were driven in part by immigration from adjoining areas as might be expected with an actively invading species, where coarse-scale landscape configuration and connectivity may be important. Unfortunately, no strong conclusions may be made about the differences in results between BBS and CBC from these analyses alone, but we hope to have highlighted several important avenues of future research.

The models developed here can be used to make predictions about how this invasion will continue to play out across the North American continent (Figure 4). Since low and medium intensity developed cover has been shown to positively affect both carrying capacity and population growth rate, we expect that the ECD will quickly achieve high abundances in suburban areas, especially where agricultural cropland is close at hand (e.g. greater Los Angeles, southern Florida, California Central Valley, Figure 4). In contrast, we expect urban centers, dense forests, mountain ranges and pastureland will limit both abundance and growth rate of individual populations (e.g. Appalachian and Rocky Mountains, and the Dallas metropolitan center, Figure 4). Up to the present, the ECD has been largely absent from the northeastern U.S. [23], despite a seemingly suitable level of development. Greater forest cover throughout this region may be limiting rates of dispersal and colonization. While our model predicts that the ECD will eventually colonize the region more completely, it also predicts that the ECD will not achieve high local abundances there (Figure 4). More targeted research is needed to fully understand this absence, and the continued collection of ECD observations through the BBS, CBC and other programs will eventually put these predictions to the test.

Our examination of population trajectories rather than a single year's abundance pattern provides additional insight into future dynamics over the course of range expansion. Many of the older populations in the southeastern U.S. have seen declines in numbers after initial periods of exponential growth. We expect that many of the sites experiencing rapid population growth at the invasion front may eventually exhibit subsequent declines as well. Another important avenue of future research is to identify the environmental features that determine whether ECDs asymptotically approach local carrying capacity or overshoot it.

Our analysis did not take into account the presence or abundance of other competitors or predators that might impact ECD populations. However, several studies that have examined habitat and dietary overlap of the ECD with native dove species have found only positive associations [19,39], and more generally some have argued that invasive species only rarely cause the complete competitive exclusion of native species [5,11,40]. Thus, invasive species may not interact as strongly with natives as previously thought. Nevertheless, the simultaneous incorporation of both biotic and abiotic variables presents an interesting area of future study with important implications for the modeling and management of range-expanding species.

## Conclusions

Our geographic analysis of population time series across the range of a rapidly expanding invasive species illustrates two points. First, species invasions can be modeled using a predictable population ecology framework that is primarily structured by dispersal, distance and time. Second, landscape features and environment may play a large role in structuring local carrying capacities and population growth rates of an invasive bird species. Future studies of the ECD should attempt to determine the extent to which these population parameters are structured by predation and intraspecific interactions, barriers to dispersal, and food availability. Examining whether these patterns apply to other invasive species will also be important for creating more integrated and informative models of range expansion and invasive spread.

## Supporting Information

**Figure S1 Environmental variable kernel density plots.** Kernel density plot for each environmental variable used in both our carrying capacity and population growth rate analyses for both BBS (orange) and CBC (blue) datasets. The displayed p-values are the result of Mann-Whitney two sample ranked sum tests.

**Figure S2 Spatial autocorrelation in response variables.** Correlograms of 4 different response variables showing Moran's I as a function of lag distance (km). Black lines represent raw response values (either carrying capacity or population growth rate), while green lines represent our model averaged residuals and their respective confidence intervals (if they are large enough to be plotted). Zero spatial autocorrelation is represented by the dashed line in each plot.

**Table S1 Variable relative importance weights and weighted parameter estimates for collectively standardized dataset.** Variable relative importance weights ($w_{i+}$) and weighted parameter estimates () based on all multivariate models predicting carrying capacity ($K$) and growth rate ($r$) for both the BBS and CBC datasets. Variables with a relative importance weight greater than 0.7 and their corresponding parameter estimates are highlighted in bold. Coefficient of determination ($R^2$) and Akaike weight for the model with the lowest $AIC_c$ score are provided for each dependent variable and dataset. Data for the models was drawn from a merged, scaled dataset containing both BBS and CBC data.

## Acknowledgments

We would like to thank the thousands of Breeding Bird Survey and Christmas Bird Count volunteers, without whom this research would not have been possible. We would also like to thank Robert Hijmans, Jack Weiss, and James Umbanhowar for their advice on analytical methods, and Wesley Hochachka and an anonymous reviewer for constructive comments that improved the paper.

## Author Contributions

Conceived and designed the experiments: SNS AHH. Performed the experiments: SNS AHH. Analyzed the data: SNS AHH. Contributed reagents/materials/analysis tools: SNS AHH. Wrote the paper: SNS AHH.

## References


1. Sala OE, Chapin, Armesto JJ, Berlow E, Bloomfield J, et al. (2000) Global biodiversity scenarios for the year 2100. Science 287: 1770–1774.
2. McKinney ML, La Sorte FA (2007) Invasiveness and homogenization: Synergism of wide dispersal and high local abundance. Global Ecology and Biogeography 16: 394–400.
3. Bruno JF, Fridley JD, Bromberg KD, Bertness MD (2005) Insights into biotic interactions from studies of species invasions. Species invasions: Insights into ecology, evolution, and biogeography. Sunderland, MA: Sinauer Associates Inc. pp. 13–40.
4. Kinlan BP, Hastings A (2005) Rates of population spread and geographic range expansion: What exotic species tell us. Species invasions: Insights into ecology, evolution, and biogeography. Sunderland, MA: Sinauer Associates Inc. pp. 381–419.
5. Sax DF, Stachowicz JJ, Brown JH, Bruno JF, Dawson MN, et al. (2007) Ecological and evolutionary insights from species invasions. Trends in Ecology & Evolution 22: 465–471.
6. Urban MC, Phillips BL, Skelly DK, Shine R (2008) A toad more traveled: The heterogeneous invasion dynamics of Cane Toads in Australia. The American Naturalist 171: E134–E148.
7. Elton CS (1958) The ecology of invasions by plants and animals. London: Methuen.
8. Fritts TH, Rodda GH (1998) The role of introduced species in the degradation of island ecosystems: A case history of Guam. Annual review of Ecology and Systematics: 113–140.
9. Holway DA (1999) Competitive mechanisms underlying the displacement of native ants by the invasive Argentine ant. Ecology 80: 238–251.
10. Ricklefs RE (2005) Taxon cycles: Insights from invasive species. Species invasions: Insights into ecology, evolution, and biogeography. Sunderland, MA: Sinauer Associates Inc. pp. 165–199.
11. Sax DF, Gaines SD, Brown JH (2002) Species invasions exceed extinctions on islands worldwide: A comparative study of plants and birds. The American Naturalist 160: 766–783.
12. Stachowicz JJ, Tilman D (2005) Species invasions and the relationships between species diversity, community saturation, and ecosystem functioning. Species invasions: Insights into ecology, evolution, and biogeography. Sunderland, MA: Sinauer Associates Inc. 41–64. Available: http://www.cedarcreek.umn.edu/biblio/fulltext/t1921.pdf. Accessed 19 September 2012.
13. Hengeveld R (1993) What to do about the North American invasion by the Collared-Dove? Journal of Field Ornithology: 477–489.
14. Hooten MB, Wikle CK (2008) A hierarchical Bayesian non-linear spatio-temporal model for the spread of invasive species with application to the Eurasian Collared-Dove. Environmental and Ecological Statistics 15: 59–70.
15. Fisher J (1953) The collared turtle dove in Europe. British Birds 46: 153–181.
16. Hengeveld R (1988) Mechanisms of biological invasions. Journal of Biogeography: 819–828.
17. Romagosa CM (2012) Eurasian Collared-Dove (*Streptopelia decaocto*). The Birds of North America Online. Available: http://bna.birds.cornell.edu/bna/species/630. Accessed 10 October 2011.
18. Coombs CFB, Isaacson AJ, Murton RK, Thearle RJP, Westwood NJ (1981) Collared-Doves (*Streptopelia decaocto*) in urban habitats. Journal of Applied Ecology: 41–62.
19. Bonter DN, Zuckerberg B, Dickinson JL (2010) Invasive birds in a novel landscape: habitat associations and effects on established species. Ecography 33: 494–502.
20. Fujisaki I, Pearlstine EV, Mazzotti FJ (2010) The rapid spread of invasive Eurasian Collared-Doves (*Streptopelia decaocto*) in the continental USA follows human-altered habitats. Ibis 152: 622–632.
21. Hudson R (1972) Collared-Doves in Britain and Ireland during 1965–70. British Birds 65: 139–155.
22. Eraud C, Boutin JM, Roux D, Faivre B (2007) Spatial dynamics of an invasive bird species assessed using robust design occupancy analysis: the case of the Eurasian Collared-Dove (Streptopelia decaocto) in France. Journal of Biogeography 34: 1077–1086.
23. Romagosa CM, Labisky RF (2000) Establishment and dispersal of the Eurasian Collared-Dove in Florida. Journal of Field Ornithology 71: 159–166.
24. Bled F, Royle JA, Cam E (2011) Hierarchical modeling of an invasive spread: the Eurasian Collared-Dove (*Streptopelia decaocto*) in the United States. Ecological Applications 21: 290–302.
25. Bystrak D (1981) The North American breeding bird survey. Studies in Avian Biology 6: 34–41.
26. National Audobon Society (2010) The Christmas Bird Count Historical Results [Online]. Available: http://www.christmasbirdcount.org. Accessed 2011 Oct 10.
27. Burnham KP, Anderson DR (2002) Model selection and multimodel inference: a practical information-theoretic approach. 2nd ed. Springer Verlag. Available: http://books.google.com/books?hl=en&lr=&id=fT1Iu-h6E-oC&oi=fnd&pg=PR7&dq=burnham+anderson&ots=tdvt32UGj1&sig=lV00Wj5lQFxqk01gQcnLB_15yWU. Accessed 2012 Sep 19.
28. Fry JA, Xian G, Jin S, Dewitz JA, Homer CG, et al. (2011) Completion of the 2006 National Land Cover Database for the Conterminous United States. Photogrammetric Engineering and Remote Sensing 77: 858–864.
29. Hijmans RJ, Cameron SE, Parra JL, Jones PG, Jarvis A (2005) Very high resolution interpolated climate surfaces for global land areas. International Journal of Climatology 25: 1965–1978.
30. Breiman L, Friedman J, Stone CJ, Olshen RA (1984) Classification and regression trees. Chapman & Hall/CRC.
31. Refaeilzadeh P, Tang L, Liu H (2009) Cross-validation. Encyclopedia of Database Systems 5.
32. Kohavi R (1995) A study of cross-validation and bootstrap for accuracy estimation and model selection. International Joint Conference on Artificial Intelligence 14: 1137–1145.
33. Arlot S, Celisse A (2010) A survey of cross-validation procedures for model selection. Statistics Surveys 4: 40–79.
34. Stone M (1974) Cross-validatory choice and assessment of statistical predictions. Journal of the Royal Statistical Society Series B (Methodological) 36: 111–147.
35. Telford RJ, Birks HJB (2005) The secret assumption of transfer functions: problems with spatial autocorrelation in evaluating model performance. Quaternary Science Reviews 24: 2173–2179.
36. Best LB (1981) Seasonal changes in detection of individual bird species. Studies in Avian Biology 6: 252–261.
37. Wiens JA (1989) Spatial scaling in ecology. Functional Ecology 3: 385–397.
38. Cushman SA, McGarigal K (2002) Hierarchical, multi-scale decomposition of species-environment relationships. Landscape Ecology 17: 637–646.
39. Poling TD, Hayslette SE (2006) Dietary overlap and foraging competition between Mourning Doves and Eurasian Collared-Doves. Journal of Wildlife Management 70: 998–1004.
40. Davis MA, Chew MK, Hobbs RJ, Lugo AE, Ewel JJ, et al. (2011) Don't judge species on their origins. Nature 474: 153–154.

# Similar Processes but Different Environmental Filters for Soil Bacterial and Fungal Community Composition Turnover on a Broad Spatial Scale

**Nicolas Chemidlin Prévost-Bouré[1]*, Samuel Dequiedt[2], Jean Thioulouse[3], Mélanie Lelièvre[2], Nicolas P. A. Saby[4], Claudy Jolivet[4], Dominique Arrouays[4], Pierre Plassart[2], Philippe Lemanceau[1], Lionel Ranjard[1,2]**

**1** Unité Mixte de Recherche 1347 Agroécologie, Institut National de la Recherche Agronomique-AgroSup Dijon-Université de Bourgogne, Dijon, France, **2** Unité Mixte de Recherche 1347 Agroécologie-Plateforme GenoSol, Institut National de la Recherche Agronomique-AgroSup Dijon-Université de Bourgogne, Dijon, France, **3** Unité Mixte de Recherche 555 Laboratoire de Biométrie et Biologie Evolutive, Université Lyon 1-Centre National de la Recherche Scientifique, Villeurbanne, France, **4** Unité de Services 1106 InfoSol, Institut National de la Recherche Agronomique, Orléans, France

## Abstract

Spatial scaling of microorganisms has been demonstrated over the last decade. However, the processes and environmental filters shaping soil microbial community structure on a broad spatial scale still need to be refined and ranked. Here, we compared bacterial and fungal community composition turnovers through a biogeographical approach on the same soil sampling design at a broad spatial scale (area range: 13300 to 31000 km$^2$): i) to examine their spatial structuring; ii) to investigate the relative importance of environmental selection and spatial autocorrelation in determining their community composition turnover; and iii) to identify and rank the relevant environmental filters and scales involved in their spatial variations. Molecular fingerprinting of soil bacterial and fungal communities was performed on 413 soils from four French regions of contrasting environmental heterogeneity (Landes<Burgundy≤Brittany<<South-East) using the systematic grid of French Soil Quality Monitoring Network to evaluate the communities' composition turnovers. The relative importance of processes and filters was assessed by distance-based redundancy analysis. This study demonstrates significant community composition turnover rates for soil bacteria and fungi, which were dependent on the region. Bacterial and fungal community composition turnovers were mainly driven by environmental selection explaining from 10% to 20% of community composition variations, but spatial variables also explained 3% to 9% of total variance. These variables highlighted significant spatial autocorrelation of both communities unexplained by the environmental variables measured and could partly be explained by dispersal limitations. Although the identified filters and their hierarchy were dependent on the region and organism, selection was systematically based on a common group of environmental variables: pH, trophic resources, texture and land use. Spatial autocorrelation was also important at coarse (80 to 120 km radius) and/or medium (40 to 65 km radius) spatial scales, suggesting dispersal limitations at these scales.

**Editor:** Maarja Öpik, University of Tartu, Estonia

**Funding:** This study was granted by ADEME (Energy and Environment Management Agency) and by the French National Research Agency (ANR). RMQS soil sampling and physico-chemical analyses were supported by a French Scientific Group of Interest on soils: the "GIS Sol", involving the French Ministry for Ecology and Sustainable Development (MEDAD), the French Ministry of Agriculture (MAP), the French Institute for Environment (IFEN), the French Agency for Energy and Environment (ADEME), the French Institute for Research and Development (IRD) and the National Institute for Agronomic Research (INRA). This work, through the involvement of technical facilities of the GenoSol platform of the infrastructure ANAEE378 Services, received a grant from the French state via the National Agency for Research under the program "Investments for the Future" (reference ANR-11-INBS-0001), as well as a grant from the Regional Council of Burgundy. The funders had no role in study design, data collection and analysis, decision to publish, or preparation of the manuscript.

* Email: n.chemidlin@agrosupdijon.fr

## Introduction

For over two centuries, biogeographical studies have been carried out on macroorganisms and have provided a better understanding of species distribution, extinction and interactions [1–2]. For microorganisms, the first biogeographic postulate was developed by Baas Becking in 1934 [3]: "Everything is everywhere, *but*, the environment selects" suggesting that microbial "species" may be everywhere due to huge dispersal potentials, but that their abundances are constrained by contemporary environmental context, which may be especially true at broad spatial scales (spatial scales larger than 100 km$^2$ are considered as broad in this study). The number of studies in microbial biogeography has increased exponentially over the past decade thanks to new molecular tools applicable in routine on wide scale sampling networks constituted of several hundreds of samples [4–5]. These studies revealed that soil microorganisms are not strictly cosmopolitan since their distributions are systematically heterogeneous and structured into biogeographical patterns [2], [6–9].

One way to discriminate the spatial processing of microbial diversity is to evaluate either the Taxa-Area Relationship (TAR), *i.e.* the accumulation of new taxa with increasing sampling area, or the Distance-Decay Relationship (DDR), *i.e.* the rate of change in compositional similarity with increasing distance. [10–11]. Although significant TAR and DDR have recently been demonstrated for both soil fungal [12] and bacterial [8], [11], [13–14] communities, the relative importance of the ecological processes shaping these communities is still under debate. Therefore, it needs to be more deeply considered at the community level. According to Vellend [15], four processes are involved in shaping microbial community composition: selection, dispersal, ecological drift and speciation. Speciation is difficult to consider at the community level because the molecular markers used to discriminate microbial taxa mainly target highly conserved regions (*e.g.* ribosomal genes) with low mutation rates. The stochastic demographic processes underlying ecological drift are also difficult to consider since it remains a challenge to fully characterize demographic evolutions within complex microbial communities in environmental samples. Consequently, most biogeographical studies have focused on environmental selection and dispersal limitations, the later leading to a spatial autocorrelation between sites independently of environmental factors. Numerous studies have identified environmental selection as relevant in shaping soil bacterial community composition [8–9], [16–24]. Conversely, dispersal limitation is still under debate regarding the high dispersal potentials of microorganisms and because some environmental variables always remain unmeasured. Nevertheless, recent publications also suggest that bacteria may be dispersal limited [9], [21–23] or that part of soil bacterial communities is endemic [17]. As regards soil fungi, the relevance of environmental selection and dispersal limitations has been demonstrated at the community level and for ectomycorrhizal groups [16], [19], [22], [24–29]. Nevertheless, most of these studies were performed on different sampling designs with different molecular techniques. Only few studies have investigated such processes for both soil fungal and bacterial communities simultaneously to compare their biodiversity turnover. Most of them were performed in particular ecosystems and lead to diverging conclusions: Pasternak et al [18] concluded that bacterial and fungal communities were primarily shaped by environmental selection rather than dispersal limitations at the scale of the Israeli desert. On the contrary, Talbot et al. [27] highlighted a strong endemism for fungi in pine forests and Hovatter [22] suggested that the ecological processes shaping soil bacterial community could differ at a local scale due to the presence/absence of a particular plant. Altogether, this suggests that environmental heterogeneity may determine the relative importance of the ecological processes at work and therefore affect the distance-decay relationship for both soil bacteria and fungi [9–11], also suggested by other macrobial studies [30–31]. The comparison of different microbial communities along different levels of environmental heterogeneity may therefore help to reach a consensus.

Both selection and dispersal are based on the various ecological attributes of soil bacteria and fungi in terms of soil colonization, dispersal forms, trophic requirements, biological interactions and adaptation to environmental conditions, together with stochastic factors. Consequently, studies focusing particularly on soil bacteria or soil fungi have identified numerous environmental filters involved in shaping these particular communities but no consensus could be reached regarding microbial community as a whole on broad spatial scales. The filter most frequently identified for bacteria is soil pH [5], [17], [20], [23–24], [32–34] and it is commonly assumed that this is an important driver for fungal communities [18], [24]. Soil texture and carbon content have also been identified as important filters for bacteria [6], [22–23], [33–34]. Similarly, the quality of soil organic matter, represented by the C:N ratio, and the amount of N were shown to have a significant effect on the abundance and composition of bacterial and fungal communities [24], [34–35]. These edaphic factors are often considered as the main determinants of bacterial diversity since the importance of climate may vary across biomes at a continental scale [5], [12], [16]. Land-use, agricultural practices, and plant community composition are also important filters for both bacteria and fungi on a wide scale [18], [19], [22–24], [36–38]. Therefore, this suggests that common filters determine the composition of both bacterial and fungal communities, but they still need to be ranked according to their relative importance to reach a consensus. This may be achieved by comparing bacterial and fungal communities over regions contrasted in terms of habitat heterogeneity but also with wide ranges of variations for the identified filters.

The objectives of this study were: i) to examine the spatial structuring of bacterial and fungal communities on a broad spatial scale; ii) to investigate the relative importance of environmental selection and spatial autocorrelation in determining the community composition turnover of these communities; and iii) to identify and rank the relevant environmental filters and scales involved in their spatial variations. To attain these objectives, four regions in the RMQS data set ("Réseau de Mesures de la Qualité des Sols" = French Monitoring Network for Soil Quality, recovering 2,200 soils over the whole of France) were selected along a gradient of environmental heterogeneity, representing a total of 413 soils. This gradient was chosen in order to confront the community composition turnover rates of bacterial and fungal communities to soil habitat heterogeneity [9]. Bacterial and fungal communities were characterized by Automated RISA fingerprinting of soil DNA. Community composition turnover (z) was estimated by means of a similarity DDR using an exponential model as suggested by Harte et al [39–40] for microorganisms. Together with this measure of community composition turnover over broad spatial scales, the initial similarity of communities was evaluated [30]. It represents the variability of community composition at finer spatial scales. High initial similarity corresponds to low local variability. The relative influence of environmental selection and spatial autocorrelation was investigated through a variance partitioning approach involving pedoclimatic characteristics and land-use and spatial variables (geographic coordinates and Principal Coordinates of Neighbour Matrices; PCNM), respectively.

## Methods

### Soil samples

Soil samples were provided by the Soil Genetic Resource Conservatory (platform GenoSol, http://www.dijon.inra.fr/plateforme_genosol, [41]) and obtained from the soil storage facility of the RMQS ("Réseau de Mesures de la Qualité des Sols" = French Monitoring Network for Soil Quality). The RMQS database consists of observations of soil properties on a 16-km regular grid across the 550000 $km^2$ French metropolitan territory and was designed to monitor soil properties [42]. The baseline survey consisting of 2,200 sites (each corresponding to a composite soil sample constituted of 25 soil cores) was completed in 2009. The sites were selected at the centre of each 16×16-km cell. In this study, we focused on a subset of 413 sites from the RMQS data set. The samples were organized into four regions: Brittany (131 sites), Burgundy (109 sites), Landes (52 sites) and South-East (121 sites, Fig. 1A) which are contrasted in terms of soil type, land-use (coarse level of the CORINE Land Cover classification; IFEN, http://

www.ifen.fr; 7 classes: forest, crop systems, grasslands, particular natural ecosystems, vineyards/orchards, parkland and wild land), climate and geomorphology (Table S1). Within a region, sites were separated by 16 km at least. For each soil, the pedo-climatic characteristics considered were particle-size distribution, pH in water ($pH_{water}$), organic carbon content ($C_{org}$), N content, C:N ratio, soluble P contents, $CaCO_3$ and exchangeable cations (Ca, K, Mg), sum of annual temperature (°C) and annual rainfall (mm). Physical and chemical analyses were performed by the Soil Analysis Laboratory of INRA (Arras, France) which is accredited for such analyses by the French Ministry of Agriculture.

## Bacterial and fungal community fingerprinting

**Soil DNA extraction.** For each soil sample, the equivalent of 1.5 g of dry soil was used for DNA extraction, following the procedure optimized by platform GenoSol [35]. Briefly, extraction buffer (100 mM Tris pH 8.0, 100 mM EDTA pH 8.0, 100 mM NaCl and 2% (w/v) SDS) was added to the sample in the proportion 3:1 (v/w), with two grams of glass beads (106 µm diameter) and eight glass beads (2 mm diameter) in a bead-beater tube. All beads were acid washed and sterilized. The samples were homogenized for 30 s at 1600 rpm in a mini bead-beater cell disruptor (Mikro-dismembrator, S. B. Braun Biotech International), incubated for 30 min at 70°C in a water bath and centrifuged for 5 min at 7000 g and room temperature. The supernatant was collected, incubated on ice with 1/10 volume of 3 M potassium acetate (pH 5.5) and centrifuged for 5 min at 14000 g. DNA was precipitated with one volume of ice-cold isopropanol and centrifuged for 30 min at 13000 rpm. The DNA pellet was washed with ice-cold 70% ethanol and dissolved in 100 µl of ultra pure water. For purification, aliquots (100 µL) of crude DNA extracts were loaded onto PVPP (polyvinyl polypyrrolidone) minicolumns (BIORAD, Marne la Coquette, France) and centrifuged for 4 min at 1000 g and 10°C. This step was repeated if the eluate was opaque. The eluate was then collected and purified for residual impurities using the Geneclean Turbo kit as recommended by the manufacturer (Q Biogene, France).

**PCR conditions.** The bacterial ribosomal IGS was amplified using the PCR protocol described in Ranjard et al [43]. 12.5 ng of DNA was used as the template for PCR volumes of 25 µl. The fungal ribosomal ITS was amplified using the primer set ITS1F/ITS4-IRD800 (5′- CTTGGTCATTTAGAGGAAGTAA -3′/5′-IRD800-TCCTCCGCTTATTGATATGC -3′). 20 ng of DNA was used as the template for PCR volumes of 25 µl with the following PCR conditions: denaturation at 95°C for 3 min, 35 cycles of 30 s at 95°C, 45 s at 55°C and 1 min at 72°C, and a final elongation of 7 min at 72°C. The primer Tm was the same for bacterial IGS and fungal ITS. Every PCR products were purified using the MinElute Kit (QIAGEN, Courtaboeuf, France) and quantified using a calf thymus DNA standard curve.

**ARISA fingerprinting conditions.** 2 µL of the PCR product was added to deionized formamide and denatured at 90°C for 2 min. Bacterial and Fungal ARISA fragments were resolved on 3.7% polyacrylamide gels under denaturing conditions as

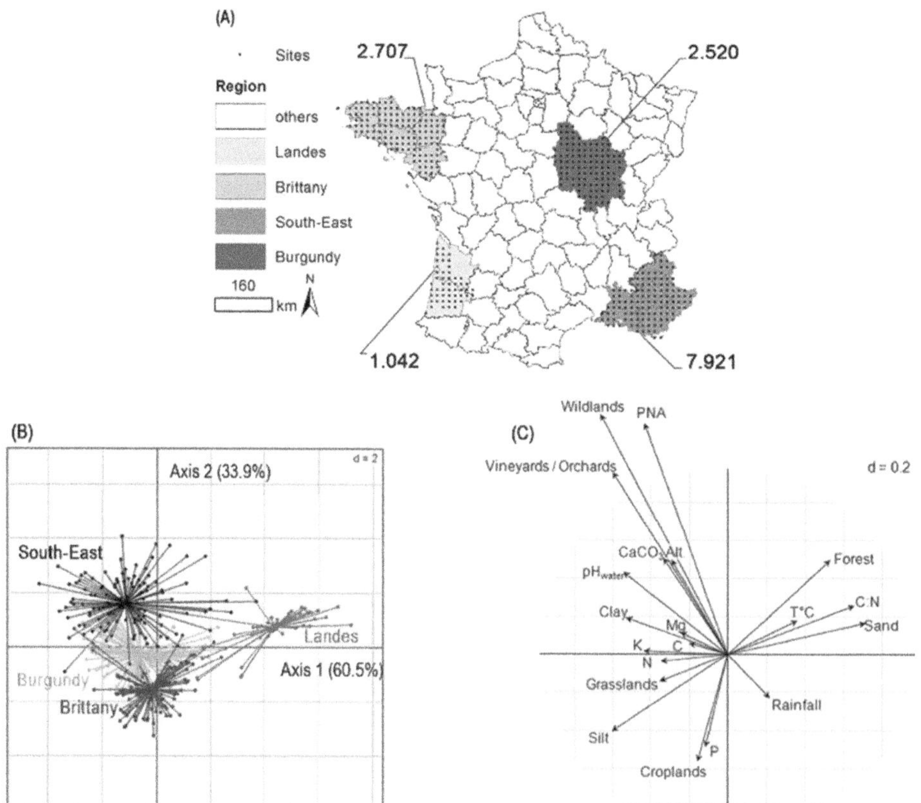


**Figure 1. Comparison of the regions considered on the basis of their soil habitats.** A. Soil average dissimilarity of soil habitat for the different regions (number linked to the corresponding region) and position of sites; B. Between group analysis of soil habitats according to the region; C. Correlation circle of the variables defining soil habitat in the between group analysis. The length of the arrow corresponds to the Pearson's correlation coefficient for quantitative variables and to the correlation ratio for qualitative variables. Symbols: Alt.: Elevation; T°C: Sum of annual temperatures; $P_{ass}$: Assimilable P; C:N: Carbon to Nitrogen ratio; $C_{org}$: Organic Carbon content.

described in Ranjard et al [9] on a LiCor DNA sequencer (ScienceTec).

**Image analysis.** The data were analyzed using the 1D-Scan software (ScienceTec), converting fluorescence data into electrophoregrams, where peaks represented PCR fragments (100 to 110 peaks retained per sample, the resolution limit to avoid considering background noise). The height of the peaks was calculated in conjunction with the median filter option and the Gaussian integration in 1D-Scan, and represented the relative proportion of the fragments in the total products. Lengths (in base pairs) were calculated by using a size standard with bands ranging from 200 to 1659 bp. The data were then converted into a contingency table with prepRISA package in R.

## Statistical analyses

**Characterization of habitat variability and average dissimilarity across regions.** Habitats were compared between regions in a Hill & Smith multivariate analysis [44] using the ade4 package in R [45–46]. The analysis was applied to pedoclimatic characteristics, land-use and geomorphology, by centering and scaling the quantitative variables, and converting the qualitative ones into weighted binary variables (weight equal to 1/n; n is the number of classes for the qualitative variables). Differences between these regions were examined by between group analysis and tested by applying a Monte-Carlo permutation test (1000 permutations). The average dissimilarity between soil habitats was determined by transposing a method based on the dissimilarity matrix for communities [47] to soil habitat. The dissimilarity matrix for soil habitat was derived from the site coordinates in the Hill & Smith analysis, following equation 1 [9]:

$$D_{i,j} = \frac{ED_{i,j}}{ED\max} \quad (1)$$

Where $D_{i,j}$ and $ED_{i,j}$ are the dissimilarity and the Euclidean distance between sites i and j, respectively. $ED_{max}$ is the maximum Euclidean distance observed between sites. The average dissimilarity between soil habitats was then calculated as follows [47]:

$$\overline{D}_{habitat} = \frac{1}{n}\sum_{h=1}^{n-1}\sum_{i=h+1}^{n} D_{h,i}^{2} \quad (2)$$

$\overline{D}_{habitat}$ is the average dissimilarity between soil habitats of soil habitat and n the number of sites in the region.

**Evaluation of the similarity distance-decay relationship and initial similarity of Bacterial and Fungal community composition.** The *similarity distance-decay* relationship was estimated as proposed by Harte et al [39] for organisms with large populations per taxa. From this relationship, the community composition turnover rates (z) for bacterial and fungal communities composition were derived as described in Ranjard *et al.* [9] following the method and the exponential model (equation 3) proposed by Harte et al [39–40] for microorganisms.

$$\log_{10}(\chi_d) = (-2 * z) * \log_{10}(d) + b \quad (3)$$

Where $\chi_d$ is the observed Sørensen's similarity between two soil samples that are d meters apart from each other; b is the intercept of the linear relationship and z the turnover rate of the community composition. The z estimate and its 95% confidence interval were derived from the slope (–2*z) of the relationship between similarity and distance by weighted linear regression. The overlap of the 95% confidence intervals was used to test for significant differences in community composition turnover rates between regions or between bacteria and fungi. The initial similarity of community composition was taken as the average similarity between sites 16 km apart and the 95% confidence intervals of the mean were determined [30].

**Variance partitioning of community composition variations according to environmental filters and space.** The relative importance of spatial variables, pedoclimatic characteristics and land-use in determining community composition turnover was tested by db-RDA [47–48]. Quantitative data were centered and scaled. Spatial variables were constructed from site coordinates (x, y, elevation) to reveal potential spatial trends at scales larger than the region, and of Principle Coordinates of Neighbour Matrices [49] in each region. The PCNM approach creates independent spatial descriptors that can be introduced in canonical analysis models to consider the spatial autocorrelation between sites in the model [49]. PCNMs with a significant Moran index ($P<0.001$) were selected. Land-use corresponded to the Corinne Land Cover classes recoded into dummy variables. Pedo-climatic characteristics consisted of climate and all the physico-chemical variables except sand. The most parsimonious model was obtained by forward selection from null to full model in two steps: a first step for selecting environmental variables and a second step for selecting the relevant PCNMs. Then, the pure effects of each set of filters or each individual filter were tested with an anova-like permutation test for canonical analyses (anova.cca function in vegan package, [50]). The PCNMs approach does not provide directly the range of the spatial descriptor. Nevertheless, Bellier et al [51] demonstrated that kriging approach could be applied to PCNMs to estimate their spatial range. Consequently, when PCNMs were selected in the most parsimonious model, their ranges were determined by standard kriging techniques (ordinary kriging with a Gaussian model). The hierarchy of these filters must nevertheless be considered with caution due to the small amounts of variance explained by each one. Land-use was not included in the filter-ranking since it corresponded to a set of categories and a global "land-use" category was already taken into account in the processes section. Maps of soil fungal community structure variations are provided as (Fig. S1, mapping methodology is described in the legend).

# Results

## Heterogeneity of soil habitat

The four regions were selected for their contrasting environmental heterogeneity as demonstrated by the between group analysis (Fig. 1B) and comparison of the calculated average dissimilarity between soil habitats ($\overline{D}_{habitat}$) (Fig. 1A and C). Multivariate analysis revealed a clear discrimination of the four regions on the first and second axes (Monte-Carlo permutation test, $P<0.001$): Landes was significantly discriminated from Brittany, Burgundy and South-east on the first axis and these three regions were discriminated from each other on the second axis. In addition, the environmental variability strongly differed between the four regions as demonstrated by the dispersal of sites in the factorial map. Sites from Landes were less dispersed on the factorial map than the sites from Brittany or Burgundy, which were less dispersed than sites from South-East. The calculated $\overline{D}_{habitat}$ ranged from 1.042 in the Landes to 7.921 in the South-East, with intermediate values for Burgundy and Brittany (2.520 and 2.707, respectively; Fig. 1A). It provided the same discrimination between regions. Fig. 1C shows that the four regions could mainly be distinguished according to land-use (*e.g.*: 86% of the

Landes sites are forest sites), a restricted set of soil physico-chemical characteristics (sand and silt contents, $pH_{water}$ and $CaCO_3$ content, assimilable P content and organic matter quality as measured by C:N ratio) and by differences in elevation. Climatic conditions did not play a significant role in regional discrimination.

## Distance-Decay Relationship for bacterial and fungal soil communities

Bacterial and fungal community similarity decreased with increasing distance in each region (Fig. 2). Community similarity was systematically higher for bacteria than for fungi in all regions at small or large distances. For bacteria, the linear regression model was highly significant in each region ($P<0.001$, Fig. 2) except in Landes where it was just below the significance threshold of 5% ($P<0.02$). For fungi, the linear regression model was highly significant in each region ($P<0.001$) except Landes (Fig. 2).

Community composition turnover rates were derived from the parameters of the linear regression. The community composition turnover rates for the bacterial and fungal communities ranged from 0.006 to 0.013 (Table 1). No significant differences were highlighted between these organisms when the community composition turnover rates were compared within each region. When the community composition turnover of bacterial or fungal communities was compared between regions, a significant difference was only found between Brittany and the South-East ($P<0.05$) for bacteria.

The initial similarity was always higher for bacterial communities than for fungal communities within each region, ranging respectively from 54.4% to 61.4% and from 39.7% to 43.1%. The initial similarity of the bacterial community in Landes was significantly lower than in the other regions, which did not differ from each other. For fungi, the initial similarities in Landes and Brittany were similar and significantly lower than those in Burgundy and the South-East.

These results were confirmed by a covariance analysis comparing the models between organisms within a region and between regions for a given organism (data not shown).

## Variance partitioning of community composition variations

The relative importance of the sets of spatial variables, land-use and pedo-climatic characteristics on variations in bacterial and fungal community composition was tested by db-RDA using the Sørensen index (Fig. 3). The amount of variance in bacterial and fungal community composition explained by the three sets of filters

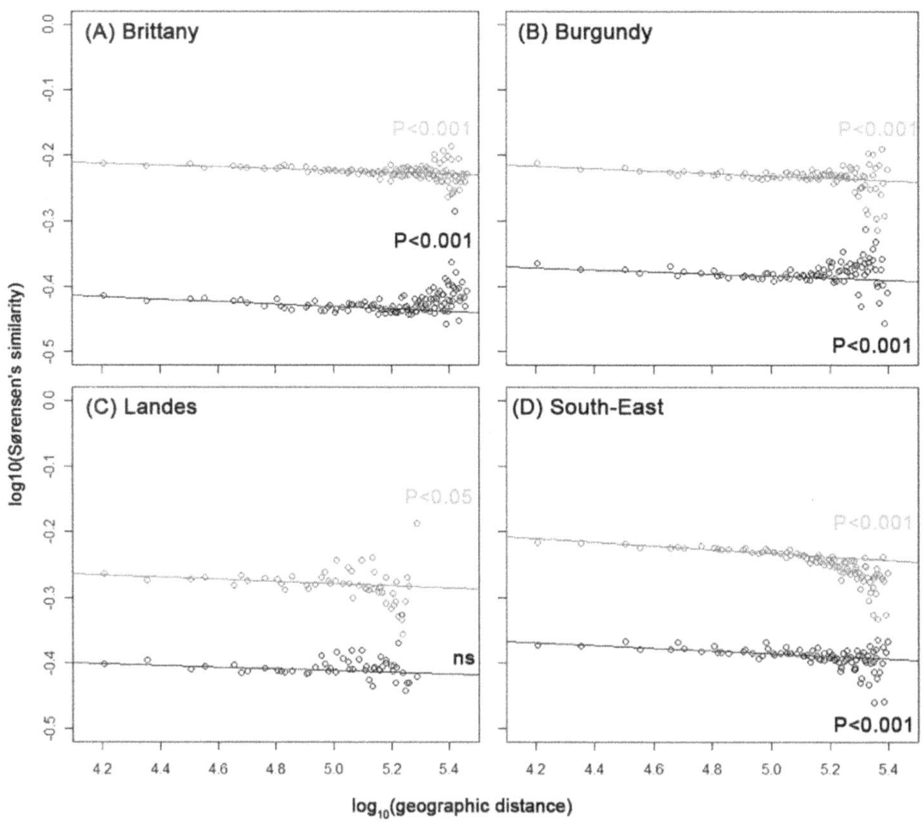


**Figure 2. Distance-Decay Relationships for bacteria and fungi.** Each panel correspond to a region: Brittany (A), Burgundy (B), Landes (C) and South-East (D) Points the average Sørensen's similarity between sites for each distance class. Lines represent the regression model based on the whole set of paired comparisons; for bacteria (grey) and fungi (black). The equations for the regression models were as follows: (**A**) Brittany: Bacteria***: log10(Sørensen's similarity) = −0.014×log10(geographic distance)−0.156; Fungi***: log10(Sørensen's similarity) = −0.017×log10(geographic distance)−0.350; (**B**) Burgundy: Bacteria*** log10(Sørensen's similarity) = −0.018×log10(geographic distance)−0.144; Fungi***: log10(Sørensen's similarity) = −0.015×log10(geographic distance)−0.316; (**C**) Landes: Bacteria*: log10(Sørensen's similarity) = −0.017×log10(geographic distance)−0.198; Fungi $^{ns}$: log10(Sørensen's similarity) = −0.012×log10(geographic distance)−0.357; (**D**) South-East: Bacteria***: log10(Sørensen's similarity) = −0.027×log10(geographic distance)−0.101; Fungi***: log10(Sørensen's similarity) = −0.019×log10(geographic distance)−0.298. A graph with points representing all paired-comparisions between sites as points can be found in Figure S2. Significance of the model is indicated as an exponent for each organism: ns: not significant; P<0.05: *; P<0.01: **, P<0.001: ***.

**Table 1.** Regression parameters of the Distance-Decay Relationships for Bacteria and Fungi.

| Region | Parameter | Organism | Estimate | 95% Confidence interval |
|---|---|---|---|---|
| Brittany (131) | Z | Bacteria | 0.007 | [0.005; 0.009] |
| | | Fungi | 0.009 | [0.006; 0.011] |
| | Initial similarity | Bacteria | 61.4% | [60.4%; 62.3%] |
| | | Fungi | 38.5% | [37.7%; 39.4%] |
| Burgundy (109) | z | Bacteria | 0.009 | [0.006; 0.012] |
| | | Fungi | 0.008 | [0.003; 0.012] |
| | Initial similarity | Bacteria | 61.4% | [60.4%; 62.5%] |
| | | Fungi | 43.1% | [42.0%; 44.3%] |
| Landes (52) | z | Bacteria | 0.009 | [0.001; 0.016] |
| | | Fungi | 0.006 | [−0.001; 0.013] |
| | Initial similarity | Bacteria | 54.4% | [52.9%; 56.0%] |
| | | Fungi | 39.7% | [38.4%; 40.9%] |
| South-East (121) | z | Bacteria | 0.013 | [0.011; 0.016] |
| | | Fungi | 0.009 | [0.006; 0.013] |
| | Initial similarity | Bacteria | 60.9% | [59.9%; 61.9%] |
| | | Fungi | 42.4% | [41.5%; 43.3%] |

The number of observations per region is provided in brackets beside the name of the region. The community composition turnover rate (z) and the initial similarity are derived from the slope of the regression (−2z) and the mean of similarity at 16 km; respectively. The statistical comparison between region and organism was performed by examining the overlap of the 95% confidence intervals of turnover rates or initial similarities.

ranged from 17% to 32%. The significance of the interactions between the three sets of filters could not be tested but always explained a small amount of the total variance (from 1.2% to 8.4%).

Among the sets of filters, soil pedo-climatic characteristics were the main contributor to variations in bacterial and fungal community composition. Spatial variables systematically explained a lower amount of variance than pedo-climatic characteristics, but a higher amount than land-use for both bacteria and fungi except for bacteria in Landes region where neither spatial variables nor land-use were significant.

The amount of variance explained by pedo-climatic characteristics for bacteria or fungi was always significant, ranging from 5% to 15%, and was similar between regions. Nevertheless, within each region, the amount of variance explained by soil pedo-climatic characteristics was always higher for bacteria than for fungi (Fig. 3).

Spatial variables significantly explained part of the community composition variations in all regions except for bacteria in Landes region. When significant, spatial variables represented from 3% to 9% of the total variance and were of the same order of magnitude both between bacteria and fungi and between regions (Fig. 3).

Similarly, land-use explained a significant amount of community composition variations in all regions except for bacteria in Landes region. The amount of explained variance ranged from 2.6% to 6.5% of the total variance. Within each region, land-use explained similar amounts of bacterial and fungal community variations when significant. Between regions, amounts of variance explained by land-use were similar for both bacteria and fungi (Fig. 3).

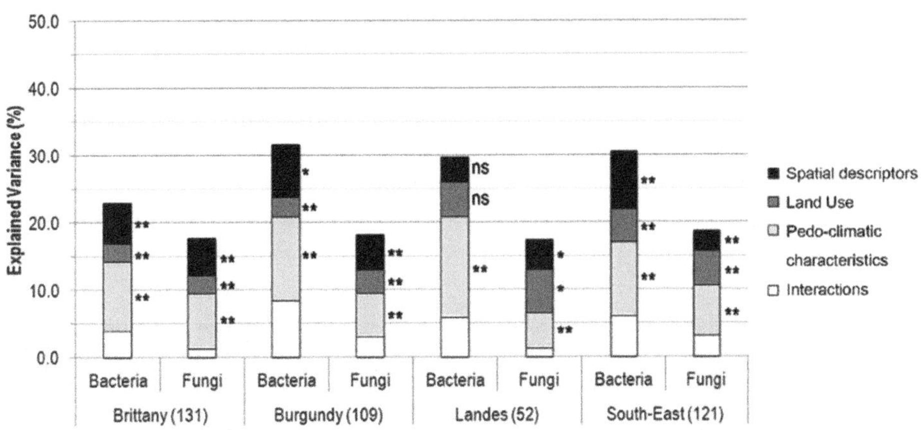


**Figure 3. Variance partitioning of bacterial and fungal community composition.** The number indicated in brackets corresponds to the number of samples for the region. Significance levels: ns: not significant; P<0.05: *; P<0.01: **, P<0.001: ***.

### Hierarchy of environmental filters

Within the sets of soil pedo-climatic characteristics and spatial variables, the pure effects of individual filters on bacterial and fungal community structure are presented in Figure 4. These pure effects account for relatively small proportions of the total variance (from 0.7% to 6.5%) because of the large number of filters explaining the total variance of bacterial or fungal community compositions. Regarding the pedo-climatic characteristics, the significant filters for soil bacterial communities in Brittany were first pH and secondly N content. For fungi, these were the quality of organic matter resource as indicated by the selection of N content, $C_{org}$ content and C:N ratio. The least significant filters corresponded to soil texture and other soil nutrients for bacteria (clay and silt contents, Mg and $CaCO_3$ concentrations) and fungi (clay content, K and P concentrations, Figure 4). In Burgundy, as in Brittany, the soil bacterial and fungal communities were principally affected by pH. Beside pH, bacterial community composition was shaped by the quality of organic matter resource (N content, $C_{org}$ content, C:N ratio) followed by clay content and annual rainfall. The fungal community composition was shaped by C:N ratio only. The only filter that had a significant effect in the Landes region, on both soil bacteria and soil fungi, was the C:N ratio. The pH was the most important filter, followed by clay content and K concentration, for bacteria in the South-East. In this region, N content and K concentration were more important filters for fungi than pH and clay content.

The spatial variables corresponded to the sites coordinates and 16 significant PCNM eigenfunctions, each representing a different spatial scale of analysis (coarse, medium and fine scales, Figure 4). Longitude, latitude or altitude coordinates did not influence community composition except latitude for fungi in Burgundy region (Figure 4). PCNMs representing spatial structures of 80–120 km radius explained significant amounts of variance in the composition of bacterial and fungal communities in Brittany and South-East. This highlighted that these communities were spatially structured at a coarse spatial scale. Similarly, PCNMs representing spatial structures of 40–65 km radius explained significant amounts of variation in the composition of bacterial community in Brittany and of fungal community in Burgundy, South-East and Landes. This showed the spatial structuration of these communities at medium spatial scales. PCNMs accounting for fine scale variables were neither significant for bacteria nor fungi in any region.

## Discussion

The four regions were selected to challenge the hypothesis that different levels of environmental heterogeneity, *i.e.* different habitat diversity and fragmentation for soil bacteria and fungi, results in different community composition turnovers [9], [14] and to compare their determinism. The multivariate analysis and the calculated $\overline{D}_{habitat}$ both highlighted a significant gradient in environmental heterogeneity following the sequence: Landes< Burgundy≤Brittany<<South-East. The four regions were mainly discriminated by environmental variables already demonstrated to influence soil microbial community abundance and diversity, such as land-use [52] and soil characteristics (texture, $pH_{water}$, P content and C:N ratio; [6], [14], [33], [35]). Among these four regions, Landes and South-East represent two extremes of environmental heterogeneity. Landes was distinct in having sites with conifer forest on acidic soils and low altitudinal variations. This specificity may turn Landes into an outlier for its environmental variability. On the other hand, South-East is characterized by a strong altitudinal gradient and a mosaic of land-use. In comparison, Burgundy and Brittany were also characterized by a mosaic of land-use but were more marked by croplands than South-East and less by forests. In addition, these two regions presented intermediate soil types regarding Landes and South-East: silty-clay calcic soils and silty acidic soils, respectively; associated to intermediate levels of organic Carbon content. Altogether, these observations highlight that these four regions allow the consideration of a large range of environmental conditions associated to different levels of environmental heterogeneity. Therefore, community similarity turnover rates can be confronted to environmental heterogeneity and their comparison may lead to a consensus regarding the environmental filters shaping soil microbial communities.

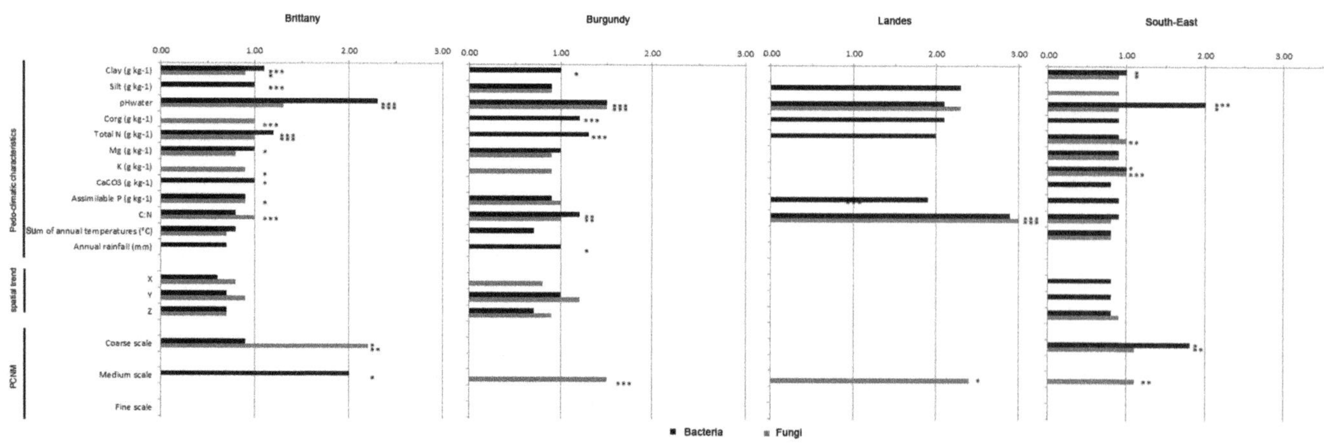


**Figure 4. Variations of microbial communities partitionned according to edaphic variables and space.** For each organism and region, only variables retained in the most parsimonious model are presented and their pure effect is tested by a permutation test. Significance levels are: P< 0.05: *; P<0.01: **, P<0.001: ***. Missing values or variables indicate that the variable was not retained in the model. Sand was removed prior to model evaluation since it was represented by the opposite of the sum of silt and clay content. Rainfall: Sum of annual rainfall (mm). Temperature: Sum of annual temperature (°C). Spatial components were summarized according to the scale considered: trend (x, y and z coordinates), coarse, medium or fine. The interval in brackets indicates the numbers of PCNMs retained in the model for each scale. The proportion of variance for each scale was determined as the sum of the pure effects of each PCNM when these were significant. Coarse, medium and fine scales correspond to PCNM with a spatial range of 80 to 120 km, 40 to 65 km and less than 40 km; respectively.

In all regions, except Landes, the soil bacterial and fungal communities were spatially structured as indicated by the significant DDR. This supports that the concept of DDR for soil microorganisms may be generalized, as suggested by several other studies [1], [9], [12], [24], [53]. The estimated bacterial and fungal community composition turnover rates ranged from 0.006 to 0.013. This is in agreement with the recent community composition turnover rates observed for soil microorganisms [10], [14], [24]. However, the turnover rates estimates were low. This could be due to technical limitations; particularly the low taxonomic resolution of DNA fingerprinting. Species variations were aggregated by the DNA fingerprints into few dominant bands, which precluded the accumulation of new minor species with increasing distance [54–55]. This would be supported by the higher community composition turnovers observed in Zinger [24] with the high resolution level provided be the pyrosequencing approach. The sampling design could also have led to low estimates of the community composition turnover rate. Indeed, lower composition turnover rates are commonly observed in regions composed of contiguous habitats [13] with gradual variations of habitat characteristics across sites and higher rates of community composition turnover have been observed at finer taxonomic levels and spatial scales [56].

Beside these technical points, community composition turnover rates were very close within each region for bacteria and fungi. This observation was in agreement with maps of bacterial and fungal communities' structures which revealed large patches of *ca.* 100–140 km radius (see Fig. 2 in [7] for bacteria and Fig. S1 for fungi). Altogether, this indicated a low-level of aggregation of both communities on a broad spatial scale [57]. This observation would suggest that differences in terms of biological and ecological features (habitat characteristics, colonization modes, trophic requirements, biotic interaction, dormancy; [6], [58–59]) would not lead to different rates of community composition turnover on a broad spatial scale. This would be supported by strong biotic interactions between bacterial and fungal communities highlithed by a significant correlation between bacterial and fungal communities' compositions. Nevertheless, this correlation was much lower than that between each community composition and environmental variables (data not shown). This strong dependency to environmental conditions resulted in a trend towards higher turnover rates in regions with higher $\overline{D}_{habitat}$. This would corroborate our first hypothesis that the microbial DDR is positively related to environmental heterogeneity. This observation was more marked and significant for bacterial communities but only a tendency for fungal communities. Although this conclusion is based on analyses of only 4 regions, it is in agreement with Ranjard *et al.* [9] who demonstrated a correlation between habitat heterogeneity and turnover rate for soil bacterial communities on a broad spatial scale, and with other microbial and macrobial studies [24], [30–31], [60].

The differences between the sensitivities of soil bacteria and fungi to environmental heterogeneity could be due to the ecology of soil fungi [58] at the spatial scale of this study. Nevertheless, according to Peay et al [25–26], ectomycorhizal fungi are spatially structured at very fine spatial scales. This would suggest that these organisms are strongly dependent on environmental conditions but that the grain size of this study (252 $km^2$) did not allow perceiving this dependence. This hypothesis would be supported by the estimated similarity of the microbial communities at a local spatial scale (initial similarity, [30]). Indeed, the initial similarity was systematically higher for bacterial than for fungal communities. According to Morlon et al [57], this would indicate that fungal communities are more aggregated than bacterial communities, *i.e.* more variable than bacterial communities, at a local spatial scale. This would lead to the conclusion that fungi are somehow more dependent on environmental conditions than bacteria at a local scale, in agreement with Peay et al [25–26] and the numerous biological interactions they are involved in.

The observed patterns and significant turnover rates result from different ecological processes [8] which may shape differently bacterial and fungal community composition on a broad spatial scale according to their relative importance. Recent studies have demonstrated that the main processes involved in increasing or decreasing community composition turnover rate are environmental selection and dispersal, respectively [8–9]. Dispersal is commonly supposed, but not empirically demonstrated to be high for microbes [61]. The maintenance of significant turnover rates, as recorded in our study, thus requires environmental selection to be high and dispersal not infinite. This hypothesis was tested by partitioning the β-diversity variations of microbial communities within regions. This partition was made according to filters involved in environmental selection: soil pedo-climatic characteristics and land-use; and to spatial variables characterizing spatial autocorrelation unexplained by the environmental variables. The residual spatial autocorrelation could result from dispersal limitations, but also from unmeasured spatially autocorrelated environmental variables despite the extensive number of environmental variables considered in this study. The amount of explained variance ranged from 23% to 32% for bacteria and from 17.5% to 18.8% for fungi. These values are within the range reported in the literature for the whole communities of bacteria [8], [14] and fungi [25–26], [62].

Soil characteristics accounted for higher amounts of variance, for both bacteria and fungi, than spatial variables. This supports the idea that soil microbial communities are primarily affected by environmental selection and secondly by other processes leading to spatial structuration independently of the environment [2], [7–8], [38]. More precisely, environmental selection was mainly driven by pedo-climatic characteristics, which accounted for larger amounts of variance than land-use. This is in agreement with several studies conducted on a broad spatial scale evidencing the higher dependency of bacterial and fungal communities on soil physico-chemical characteristics [18], [38], [63] than on land-use. Nevertheless, the amount of variance explained by pedo-climatic characteristics was higher for bacteria than for fungi, highlighting the greater effect of soil habitat on shaping the bacterial community. This observation is supported by the higher diversity and reactivity of the soil bacterial community to changes in surrounding conditions which, in turn, leads to a community structure that is better fitted to the habitat characteristics [13], [58]. On the contrary, land-use explained a higher amount of fungal community variance than of bacterial community variance. This latter could be due to plant-soil microbe interactions resulting from the type of vegetative cover [19], [38], [52], [64] as well as from human activities, especially agricultural or industrial practices [35]; [37] potentially affecting fungi more strongly than bacteria. This difference would be in agreement with a highly patchy distribution of soil fungi at a local scale as suggested above by the low level of initial similarity. Altogether, these results support the postulate of Baas-Becking "the environment selects" in summarizing microbial biogeography.

Nevertheless, variance partitioning also highlighted that bacterial and fungal communities are spatially structured independently of environmental characteristics since spatial variables accounted for a significant amount of community variance. A hypothesis in explaining this spatial autocorrelation is that soil bacterial and fungal communities might be dispersal limited, even if this result may also be related to unmeasured, spatially structured, environmental variables. This hypothesis would be supported by the study of Ranjard et al [9] demonstrating significant turnover rates of bacterial diversity in fully homogenous regions and by the consideration of several soil physico-chemical characteristics in the variance partitioning approach of the present study. Nevertheless, this must be confirmed by more in-depth studies since few evidence remains to date on the shaping of bacterial and fungal communities by limited dispersal [8], [17], [26–27], [65]. Comparisons have indicated that fungi tend to be less dependent on spatial variables than bacteria. This result is surprising since fungi are demonstrated to be dispersal limited [27] and bacteria are generally expected to disperse over larger distances, even wider than our regional scale [14], [66]. It was estimated in the literature that $10^{18}$ viable bacteria were transported annually in the atmosphere between continents [61]. Moreover, particular fungal populations have been demonstrated to disperse over short distances, *e.g.* ectomycorrhiza at the "plant island scale" [25–26], or to be endemic [27]. On the other hand, variance explained by spatial variables was similar in each region for fungi, whereas it followed the environmental heterogeneity gradient: Landes$<$ Brittany$\leq$Burgundy$<$South-East for bacteria. This observation supports the hypothesis that bacteria are more sensitive to geomorphology and that the presence of natural barriers (*e.g.* mountains), or a higher fragmentation of landscape in the South-East, due to a mosaic of agricultural and natural plots [9] may lead to significant spatial variations.

Our last objective was to identify and rank the environmental filters and the spatial variables involved in environmental selection process and spatial autocorrelation of bacteria and fungi. To do so, the environmental and spatial filters selected in the variance partitioning approach were ranked according to their pure effect on bacterial and fungal community composition turnover. First, environmental selection was mainly based on soil characteristics but not on climatic conditions. This is in accordance with previous studies highlighting the weak influence of climate on soil microbial diversity on a broad spatial scale [6–7], [35], [37], [67], [68]. Nevertheless, recent studies also demonstrated strong differences between bacterial communities across different biomes, especially cold deserts *versus* temperate biomes and hot deserts, suggesting that climate may play a role [32] at very large spatial scales. Overall, the main soil characteristics identified as important filters for soil bacterial and fungal community composition were pH, trophic resources (N, C, K, P contents; C:N ratio) and texture (clay, silt content). These findings are consistent with the literature where i) pH is regularly identified as the main filter for both soil bacteria [6], [16–17] and fungi [25], [62], ii) soil organic C and N contents and the C:N ratio constitute important components of microbial niches [18], [32], [35], [58], and iii) texture determines the size and stability of soil micro-habitats [18], [69], [70]. The hierarchy of these filters depended on the type of organisms and the regions. The above-described hierarchical sequence (pH$>$ trophic resources$\geq$texture) was observed for three of the four regions, but not for Landes. In the Landes region, the filter variability is low and environmental selection is basically driven by organic matter quality (C:N ratio). This is related to the large number of conifer forests sites leading to soil organic matter with a high C:N ratio and strong recalcitrance to microbial decomposition. A strong selection of particular populations with enzymatic ability to transform this organic matter is occurring in such soils [58], which is deeply influencing the corresponding community structure. Comparison of the overall filter hierarchies for bacteria and fungi in a given region did not reveal any discrepancy for the primary drivers. The only differences were observed for secondary filters, including C:N ratio and mineral nutrients (such as K and P) for fungi, and clay and N content for bacteria. This is consistent with the well-known dependency of the soil fungal community on soil P content [34] and soil organic matter quality [58]. As regards bacteria, clay content is positively correlated with the biotic capacity of soil as well as its indigenous bacterial diversity by enhancing the level of protection of the soil habitat and the retention of nutrients [69–70]. Second, regarding spatial autocorrelation, different scales were derived from the hierarchy of spatial variables and their range. The main scale identified was the coarse scale (80 to 120 km radius), for both bacteria and fungi in Brittany and South-east regions. This spatial scale is smaller than the one at which soil habitat changed on the RMQS Network (150 to 470 km, [9]), whereas it is in agreement with the large patches obtained by mapping the bacterial and fungal community structure over these regions [7] and Fig. S1). In addition, finer spatial scales (medium scales; 40 to 65 km radius) were also identified as significant in Burgundy, Landes and South-East for fungi and in Brittany for bacteria. Altogether, this scale dependency would support a hypothesis for the dispersal limitation of bacterial and fungal communities [33]. This scale dependency is in agreement with Martiny et al [14] for bacteria and is supported by the observations of Peay et al [25–26] on soil fungi. As in other studies at continental scales [14], this highlights the importance of considering multiple scales to better understand microbial ecology.

Altogether, our study demonstrated the spatial structuring of soil bacterial and fungal communities on local to coarse scales, which was based on environmental selection and on an unexplained spatial autocorrelation that could be related to limited dispersal. Selection and spatial autocorrelation were shown to have a similar influence on soil bacteria and soil fungi but the filters involved could differ depending on the environmental heterogeneity. Nevertheless, the comparison of bacterial and fungal communities helped to propose a primary consensus regarding the environmental filters shaping soil microbial community composition as a whole: Land-Use and pH are the primary filters, followed by trophic resources quantity (organic carbon content and nitrogen content) and then quality (C:N ratio). The results of this study increase our knowledge on the effects of soil habitat and provide insights in the scale at which dispersal may occur according to the ecological attributes of bacteria and fungi. However further investigations, based on up-scaling approaches, are now required to: i) provide a direct measurement of bacterial cells dispersal which is now crucial to demonstrate limited dispersal of bacteria; ii) identify the filters operating at each spatial scale. Especially, these upscaling approaches could be associated to high-throughput sequencing to achieve a finer resolution on the communities. This would improve our ability to sustainably manage soil biodiversity.

## Supporting Information

**Figure S1 Maps of interpolated MULTISPATI scores for the first three MULTISPATI axes (columns) and for the four geographical regions (rows).** Each map was generated as described in Dequiedt et al (2009) and corresponds to the spatial synthesis of the F-ARISA genetic structure of indigenous fungal communities from the corresponding soils sampled in the four regions of France. Colours on the map are proportional to the score of each soil sample on each MULTISPATI axis following the scale provided at the bottom of the figure. Below each column, the empirical variogram is provided for each MULTISPATI axis.

**Figure S2 Distance-Decay Relationship for bacteria and fungi.** Each panel correspond to: **(A–D): Bacteria** in Brittany,

Burgundy, Landes and South-East; **(E–H): Fungi** in Brittany, Burgundy, Landes and South-East. Points represent paired-comparisons between sites and line the linear model. The equations for the regression models were as follows: **(A)** log10(Sørensen's similarity) = −0.014×log10(geographic distance)−0.156; **(B)** log10(Sørensen's similarity) = −0.018×log10(geographic distance)−0.144; **(C)** log10(Sørensen's similarity) = −0.017×log10(geographic distance)−0.198; **(D)** (Sørensen's similarity) = −0.027×log10(geographic distance)−0.101; **(E)** log10(Sørensen's similarity) = −0.017×log10(geographic distance)−0.350; **(F)** log10(Sørensen's similarity) = −0.015×log10(geographic distance)−0.316; **(G)** log10(Sørensen's similarity) = −0.012×log10(geographic distance)−0.357; **(H)** log10(Sørensen's similarity) = −0.019×log10(geographic distance)−0.298. Significance of the model is indicated as an exponent for each organism: ns: not significant; $P<0.05$: *; $P<0.01$: **, $P<0.001$: ***.

**Table S1 Summary statistics of regions characteristics.** PNE: Particular Natural Ecosystems, SE: standard error of the mean.

## Author Contributions

Conceived and designed the experiments: NCPB SD NPAS CJ DA LR PL. Performed the experiments: NCPB ML. Analyzed the data: NCPB SD JT PP. Contributed reagents/materials/analysis tools: NCPB SD JT ML. Wrote the paper: NCPB LR.

## References


1. Horner-Devine MC, Lage M, Hughes JB, Bohannan BJM (2004) A taxa-area relationship for bacteria. Nature 432: 750–753.
2. Martiny JBH, Bohannan BJM, Brown JH, Colwell RK, Fuhrman JA, et al. (2006) Microbial biogeography: putting microorganisms on the map. Nat Rev Microbiol 4: 102–112.
3. Baas Becking LGM (1934) *Geobiologie of inleiding tot de milieukunde*, The Hague, The Netherlands.
4. Maron P-A, Mougel C, Ranjard L (2011) Soil microbial diversity: Methodological strategy, spatial overview and functional interest. C R Biol 334: 403–411.
5. Gilbert J, Meyer F, Jansson J, Gordon J, Pace N, et al. (2010) *The Earth Microbiome Project: Meeting report of the "1st EMP meeting on sample selection and acquisition"* at Argonne National Laboratory October 6th 2010.
6. Fierer N, Jackson RB (2006) The diversity and biogeography of soil bacterial communities. Proc Natl Acad Sci U 103: 626–631.
7. Dequiedt S, Thioulouse J, Jolivet C, Saby NPA, Lelievre M, et al. (2009) Biogeographical patterns of soil bacterial communities. Environ Microbiol Rep 1: 251–255.
8. Hanson CA, Fuhrman JA, Horner-Devine MC, Martiny JBH (2012) Beyond biogeographic patterns: processes shaping the microbial landscape. Nat Rev Microbiol 10: 497–506.
9. Ranjard L, Dequiedt S, Chemidlin Prévost-Bouré N, Thioulouse J, Saby NPA, et al. (2013) Turnover of soil bacterial diversity driven by wide-scale environmental heterogeneity. Nat Commun 4: 1434.
10. Gleason AH (1922) On the relationship between species and area. Ecology (N. Y.), 3: 158–162.
11. Zinger L, Boetius A, Ramette A (2014) Bacterial taxa–area and distance–decay relationships in marine environments. Mol Ecol 23: 954–964.
12. Green JL, Holmes AJ, Westoby M, Oliver I, Briscoe D, et al. (2004) Spatial scaling of microbial eukaryote diversity. Nature 432: 747–750.
13. Prosser JI, Bohannan BJM, Curtis TP, Ellis RJ, Firestone MK, et al. (2007) Essay - The role of ecological theory in microbial ecology. Nat Rev Microbiol 5: 384–392.
14. Martiny JBH, Eisen JA, Penn K, Allison SD, Horner-Devine MC (2011) Drivers of bacterial β-diversity depend on spatial scale. P Natl Acad Sci USA 108: 7850–7854.
15. Vellend M (2010) Conceptual synthesis in community ecology. Q Rev Biol 85: 183–206.
16. Green JL, Bohannan BJM (2006) Spatial scaling of microbial biodiversity. Trends Ecol Evol 21: 501–507.
17. Chu H, Fierer N, Lauber CL, Caporaso JG, Knight R, et al. (2010) Soil bacterial diversity in the Arctic is not fundamentally different from that found in other biomes. Environ Microbiol DOI: 0.1111/j.1462-2920.2010.02277.x.
18. Pasternak Z, Al-Ashhab A, Gatica J, Gafny R, Avraham S, et al. (2013) Spatial and Temporal Biogeography of Soil Microbial Communities in Arid and Semiarid Regions. PLoS ONE 8: e69705. doi:10.1371/journal.pone.0069705.
19. Brodie E, Edwards S, Clipson N (2002) Bacterial Community Dynamics across a Floristic Gradient in a Temperate Upland Grassland Ecosystem. Microb Ecol 44: 260–270.
20. Rousk J, Baath E, Brookes PC, Lauber CL, Lozupone C, et al. (2010) Soil bacterial and fungal communities across a pH gradient in an arable soil. ISME J 4: 1340–1351.
21. Langenheder S, Szekely AJ (2011) Species sorting and neutral processes are both important during the initial assembly of bacterial communities. ISME J 5: 1086–1094.
22. Hovatter SR, Dejelo C, Case AL, Blackwood CB (2010) Metacommunity organization of soil microorganisms depends on habitat defined by presence of Lobelia siphilitica plants. Ecology 92: 57–65.
23. Yergeau E, Bezemer TM, Hedlund K, Mortimer SR, Kowalchuk GA, et al. (2010) Influences of space, soil, nematodes and plants on microbial community composition of chalk grassland soils. Environ Microbiol 12: 2096–2106.
24. Zinger L, Lejon DPH, Baptist F, Bouasria A, Aubert S, et al. (2011) Contrasting Diversity Patterns of Crenarchaeal, Bacterial and Fungal Soil Communities in an Alpine Landscape. PLoS ONE 6: e19950.
25. Peay KG, Bruns TD, Kennedy PG, Bergemann SE, Garbelotto M (2007) A strong species–area relationship for eukaryotic soil microbes: island size matters for ectomycorrhizal fungi. Ecol Lett 10: 470–480.
26. Peay KG, Schubert MG, Nguyen NH, Bruns TD (2012) Measuring ectomycorrhizal fungal dispersal: macroecological patterns driven by microscopic propagules. Mol Ecol 21: 4122–4136.
27. Talbot JM, Bruns TD, Taylor JW, Smith DP, Branco S, et al. (2014) Endemism and functional convergence across the North American soil mycobiome. P Natl Acad Sci USA 111: 6341–6346.
28. Taylor DL, Hollingsworth TN, McFarland JW, Lennon NJ, Nusbaum C, et al. (2013) A first comprehensive census of fungi in soil reveals both hyperdiversity and fine-scale niche partitioning. Ecol Monogr 84: 3–20.
29. Taylor DL, Herriott IC, Stone KE, McFarland JW, Booth MG, et al. (2010) Structure and resilience of fungal communities in Alaskan boreal forest soilsThis article is one of a selection of papers from The Dynamics of Change in Alaska's Boreal Forests: Resilience and Vulnerability in Response to Climate Warming. Can J For Res 40: 1288–1301.
30. Soininen J, McDonald R, Hillebrand H (2007) The distance decay of similarity in ecological communities. Ecography 30: 3–12.
31. Drakare S, Lennon JJ, Hillebrand H (2006) The imprint of the geographical, evolutionary and ecological context on species–area relationships. Ecol Lett 9: 215–227.
32. Fierer N, Leff JW, Adams BJ, Nielsen UN, Bates ST, et al. (2012) Cross-biome metagenomic analyses of soil microbial communities and their functional attributes. Proc Natl Acad Sci USA 109: 21390–21395.
33. Ramette A, Tiedje JM (2007) Multiscale responses of microbial life to spatial distance and environmental heterogeneity in a patchy ecosystem. P Natl Acad Sci USA 104: 2761–2766.
34. Lauber CL, Strickland MS, Bradford MA, Fierer N (2008) The influence of soil properties on the structure of bacterial and fungal communities across land-use types. Soil Biol Biochem 40: 2407–2415.
35. Dequiedt S, Saby NPA, Lelievre M, Jolivet C, Thioulouse J, et al. (2011) Biogeographical patterns of soil molecular microbial biomass as influenced by soil characteristics and management. Global Ecol Biogeog 20: 641–652.
36. Brodie E, Edwards S, Clipson N (2003) Soil fungal community structure in a temperate upland grassland soil. FEMS Microbiol Ecol 45: 105–114.
37. Lienhard P, Tivet F, Chabanne A, Dequiedt S, Lelièvre M, et al. (2013) No-till and cover crops shift soil microbial abundance and diversity in Laos tropical grasslands. Agron Sustain Dev 33: 1–10.
38. Griffiths RI, Thomson BC, James P, Bell T, Bailey M, et al. (2011) The bacterial biogeography of British soils. Environ Microbiol 13: 1642–1654.
39. Harte J, Kinzig A, Green J (1999) Self-Similarity in the Distribution and Abundance of Species. Science 284: 334–336.
40. Harte J, Smith AB, Storch D (2009) Biodiversity scales from plots to biomes with a universal species–area curve. Ecol Lett 12: 789–797.
41. Ranjard L, Dequiedt S, Lelievre M, Maron PA, Mougel C, et al. (2009) Platform GenoSol: a new tool for conserving and exploring soil microbial diversity. Environ Microbiol Rep 1: 97–99.
42. Arrouays D, Jolivet C, Boulonne L, Bodineau G, Saby NPA, et al. (2002) A new projection in France: a multi-institutional soil quality monitoring networkUne initiative nouvelle en France : la mise en place d'un réseau multi-institutionnel de mesure de la qualité des sols (RMQS). Comptes Rendus de l'Académie d'Agriculture de France 88: 93–105.
43. Ranjard L, Poly F, Nazaret S (2000) Monitoring complex bacterial communities using culture-independent molecular techniques: application to soil environment. Res Microbiol 151: 167–177.
44. Hill MO, Smith AEJ (1976) Principal Component Analysis of Taxonomic Data with Multi-State Discrete Characters. Taxon 25: 249–255.

45. Thioulouse J, Dray S (2007) Interactive multivariate data analysisin R with the ade4 and ade4TkGUI packages. J stat softw 22: 1–14.
46. R Development Core Team (2011) *R: A Language and Environment for Statistical Computing*, Vienna, Austria.
47. Legendre P, Borcard D, Peres-Neto PR (2005) Analyzing beta diversity: Partitioning the spatial variation of community composition data. Ecol Monogr 75: 435–450.
48. Legendre P, Fortin MJ (2010) Comparison of the Mantel test and alternative approaches for detecting complex multivariate relationships in the spatial analysis of genetic data. Mol Ecol Resour 10: 831–844.
49. Dray SP, Legendre P, Peres-Neto PR (2006) Spatial modelling: a comprehensive framework for principal coordinate analysis of neighbour matrices (PCNM) Ecol Model 196: 483–493.
50. Oksanen J, Blanchet FG, Kindt R, Legendre P, Minchin PR, et al. (2011) vegan: Community Ecology Package.
51. Bellier E, Monestiez P, Durbec J-P, Candau J-N (2007) Identifying spatial relationships at multiple scales: principal coordinates of neighbour matrices (PCNM) and geostatistical approaches. Ecography 30: 385–399.
52. Drenovsky RE, Steenwerth KL, Jackson LE, Scow KM (2010) Land use and climatic factors structure regional patterns in soil microbial communities. Global Ecol Biogeog 19: 27–39.
53. Bell T, Ager D, Song J-I, Newman JA, Thompson IP, et al. (2005) Larger Islands House More Bacterial Taxa. Science 308: 1884.
54. Woodcock S, Curtis TP, Head IM, Lunn M, Sloan WT (2006) Taxa–area relationships for microbes: the unsampled and the unseen. Ecol Lett 9: 805–812.
55. MacArthur RH, Wilson EO (1967) *The theory of island biogeography/Robert H. MacArthur and Edward O. Wilson*. Princeton University Press, Princeton, N.J.
56. Noguez AM, Arita HT, Escalante AE, Forney LJ, Garcia-Oliva F, et al. (2005) Microbial macroecology: highly structured prokaryotic soil assemblages in a tropical deciduous forest. Global Ecol Biogeog 14: 241–248.
57. Morlon H, Chuyong G, Condit R, Hubbell S, Kenfack D, et al. (2008) A general framework for the distance–decay of similarity in ecological communities. Ecol Lett 11: 904–917.
58. de Boer W, Folman LB, Summerbell RC, Boddy L (2005) Living in a fungal world: impact of fungi on soil bacterial niche development. FEMS Microbiol Rev 29: 795–811.
59. Lennon JT, Jones SE (2011) Microbial seed banks: the ecological and evolutionary implications of dormancy. Nat Rev Micro 9: 119–130.
60. Jobe RT (2008) Estimating ladscape-scale species richness: reconciling frequency- and turnover-based approaches. Ecology 89: 174–182.
61. Fenchel T (2003) Biogeography for Bacteria. Science 301: 925–926.
62. Dumbrell AJ, Nelson M, Helgason T, Dytham C, Fitter AH (2010) Relative roles of niche and neutral processes in structuring a soil microbial community. ISME J 4: 337–345.
63. Hossain Z, Sugiyama S-I (2011) Geographical structure of soil microbial communities in northern Japan: Effects of distance, land use type and soil properties. Eur J Soil Biol 47: 88–94.
64. Bryant JA, Lamanna C, Morlon H, Kerkhoff AJ, Enquist BJ, et al. (2008) Microbes on mountainsides: Contrasting elevational patterns of bacterial and plant diversity. P Natl Acad Sci USA 105: 11505–11511.
65. Monroy F, van der Putten WH, Yergeau E, Mortimer SR, Duyts H, et al. (2012) Community patterns of soil bacteria and nematodes in relation to geographic distance. Soil Biol Biochem 45: 1–7.
66. Cho JC, Tiedje JM (2000) Biogeography and Degree of Endemicity of Fluorescent Pseudomonas Strains in Soil. Appl Environ Microb 66: 5448–5456.
67. Lozupone CA, Knight R (2007) Global patterns in bacterial diversity. Proceedings of the National Academy of Sciences 104: 11436–11440.
68. Auguet J-C, Barberan A, Casamayor EO (2009) Global ecological patterns in uncultured Archaea. ISME J 4: 182–190.
69. Ranjard L, Richaume AS (2001) Quantitative and qualitative microscale distribution of bacteria in soil. Res Microbiol 152: 707–716.
70. Kong AYY, Scow KM, Cordova-Kreylos AL, Holmes WE, Six J (2011) Microbial community composition and carbon cycling within soil microenvironments of conventional, low-input, and organic cropping systems. Soil Biol Biochem 43: 20–30.

# Relative Roles of Grey Squirrels, Supplementary Feeding, and Habitat in Shaping Urban Bird Assemblages

**Colin Bonnington[1], Kevin J. Gaston[2], Karl L. Evans[1]***

**1** Department of Animal and Plant Sciences, University of Sheffield, Sheffield, United Kingdom, **2** Environment and Sustainability Institute, University of Exeter, Penryn, Cornwall, United Kingdom

## Abstract

Non-native species are frequently considered to influence urban assemblages. The grey squirrel *Sciurus carolinensis* is one such species that is widespread in the UK and is starting to spread across Europe; it predates birds' nests and can compete with birds for supplementary food. Using distance sampling across the urbanisation intensity gradient in Sheffield (UK) we test whether urban grey squirrels influence avian species richness and density through nest predation and competition for supplementary food sources. We also assess how urban bird assemblages respond to supplementary feeding. We find that grey squirrels slightly reduced the abundance of breeding bird species most sensitive to squirrel nest predation by reducing the beneficial impact of woodland cover. There was no evidence that grey squirrel presence altered relationships between supplementary feeding and avian assemblage structure. This may be because, somewhat surprisingly, supplementary feeding was not associated with the richness or density of wintering bird assemblages. These associations were positive during the summer, supporting advocacy to feed birds during the breeding season and not just winter, but explanatory capacity was limited. The amount of green space and its quality, assessed as canopy cover, had a stronger influence on avian species richness and population size than the presence of grey squirrels and supplementary feeding stations. Urban bird populations are thus more likely to benefit from investment in improving the availability of high quality habitats than controlling squirrel populations or increased investment in supplementary feeding.

**Editor:** Antoni Margalida, University of Lleida, Spain

**Funding:** This work was funded by a Natural Environment Research Council studentship. The funders had no role in study design, data collection and analysis, decision to publish, or preparation of the manuscript.

**Competing Interests:** The authors have declared that no competing interests exist.

* Email: karl.evans@sheffield.ac.uk

## Introduction

Urbanisation is one of the fastest growing land uses, and generates environments with very different selection pressures than the rural ones which it replaces [1–3]. The resultant species assemblages in towns and cities thus also differ markedly in their structure and composition than equivalent assemblages in more rural environments [4–6], but can include considerable populations of some species of conservation concern [7]. Numerous factors can drive this divergence in assemblage structure, with changes in habitat quality and the introduction of non-native species frequently considered important [8–9]. Much of the research on urban assemblage structure has focused on avian assemblages, which can also be influenced by predation risk and provision of supplementary food [10–12].

In the UK, the grey squirrel *Sciurus carolinensis* is an exotic species that is widespread in urban areas [13]. Grey squirrels sometimes predate birds' nests [14], and there is evidence that they can out-compete birds for supplementary food [15,16]. As a consequence of these observations concern has been expressed that grey squirrels may limit avian population size, but empirical evidence is very limited. In rural areas the populations of a small number of bird species are slightly reduced when grey squirrels are present with the assumption being that this is due to nest predation [17–18], but empirical data are lacking from urban areas.

The provision of supplementary food for wild birds is a widespread activity in many developed countries, with almost half of UK households providing food for birds [19]. This activity is associated with grey squirrel occurrence [13]. Positive associations have also been documented between supplementary feeding and the size of wintering and breeding bird populations [10,11]. Such positive associations may arise through improved survival rates or increased productivity [20,21]. Some recent studies have, however, found that supplementary feeding is associated with reduced reproductive success [22], and that provision of low quality food can reduce maternal investment in egg quality [23]. Moreover, there is much spatial variation in the effects of supplementary feeding on avian population size, with positive effects in northern Europe (Finland, UK) but negligible effects in central Europe (France), and not all species that use feeders exhibit increased population size [10–11;24–25]. One possible mechanism for this is that supplementary feeding promotes interference competition with a small number of dominant aggressive species monopolising resources which could reduce the benefits they provide to native species [26,27]. Indeed, experimental evidence demonstrates that grey squirrel presence at supplementary feeding stations severely limits food intake rates (lowered by over 90%) of a wide range of passerines demonstrating the potential for strong interference competition [16]. It is unknown, however, if such competition influences avian population size, for example this would not be the case if feeding stations are sufficiently abundant that displaced birds can readily find alternative stations that lack squirrels or if the availability of such stations does not regulate bird population size.

Our primary objective is to assess if grey squirrels influence the structure of urban avian assemblages through either nest predation or competition at supplementary feeding stations. We do so by comparing breeding and wintering avian assemblage structure across multiple urban locations, in which grey squirrels are both absent and present, whilst taking the amount of green space, its quality, and provision of supplementary food into account. In so doing we also provide an additional assessment of the association between supplementary feeding and avian assemblage structure. We use Sheffield (UK), the fifth largest urban municipality in the UK, as a case study.

## Materials and Methods

No specific permissions were required to conduct this work. Research did not involve endangered or protected species or collection of biological material.

### Sampling approach

This study was conducted in urban Sheffield, which contains c. 555,500 people [28]. Urban areas were defined as 1 km×1 km squares with at least 25% coverage of hard surface. This definition has been used in numerous other urban ecology studies conducted in the Sheffield region, and is appropriate as it excludes areas of the countryside which fall within the city's administrative boundaries, but retains all parks and other green-spaces surrounded by built up land within the urban landscape. This resulted in 143 1 km×1 km squares, each of which was split into 16 cells of 250 m×250 m. The green space cover (%) of each of these smaller cells was determined using OS Mastermap digital maps (EDINA Digimap, Edinburgh, UK) in ArcMap 9.3 (ESRI Corporation, Redlands, CA, USA). Each cell was assigned to one of ten categories, based on its green space cover (%) ranging from category 1 (0–10% green space) to category 10 (91–100% green space). Ten sampling points were then selected using a random stratified approach, in each category of green space, resulting in 100 sampling points. An additional 40 sampling points were selected from cells with intermediate amounts of green space (41–80% green space) to increase sampling effort in those areas most likely to have variation in squirrel occurrence.

### Field observations

Visual surveys and counting dreys in the winter are the two most economical methods for surveying squirrels [29]. Drey counts require data on the mean number of dreys used by an individual squirrel, but such data are unavailable for urban areas. Moreover, in towns and cities grey squirrel dreys can be very difficult to detect as they are frequently built inside old corvid nests, the roof space of buildings and evergreen trees [30], which reduces the method's value. We thus assess grey squirrel distributions using a visual survey method. We used a fixed duration point count methodology which has previously been used to estimate densities of other squirrel species [31–33]. We followed reference [32] and used a ten minute point count; we used a fixed radius of 100 m [34]. The size of the point count survey radius is thus approximately 3 ha, which incorporates the typical home range size (0.5 to 3 hectares) of grey squirrels occupying patches of fragmented habitat [35]. Our methodology is described in full in [13]. These point counts were also used to sample the avian assemblage; such methods have previously been frequently used to sample urban avian assemblages in both the breeding and non-breeding season [36,37].

Each sampling point was visited four times in 2010 on dry calm days (less than 4 on the Beaufort scale), once every season (winter: 4 February –2 March; spring: 8–27 April; summer: 5–21 July; autumn: 25 October –10 November). Surveys were not conducted during very cold days (<3°C) as these can reduce squirrel activity. Sampling points were located in the field using a handheld GPS receiver and a map of each location from Google Earth. The exact sampling point was accessible in 102 out of 140 cases (73%); when it was not accessible, the observer (C. Bonnington) stood at the nearest accessible point within the same cell. Ten-minute avian point counts of a fixed radius of 100 m were conducted in daylight hours within 5 hours of sunrise or 5 hours of sunset, during each season. Observations began immediately on arrival at the point location. For each detected individual, the species, radial distance from the observer (within the 100 m point count radius) and detection type (whether seen or heard) were recorded. Distances were recorded using a range finder (Bushnell Laser range finder Sport 450, Overland Park, KS, USA). Grey squirrels and actively stocked supplementary feeding stations were recorded in the same manner as the avifauna, but if no squirrel or feeding station was recorded during the point count period the survey area was searched for a maximum of 10 minutes to confirm the absence of squirrels and actively stocked feeding stations.

The habitat characteristics of each survey area were recorded: the height of the 20 closest trees (>2 m high and ≥20 cm diameter at breast height in survey area) to the sampling point; canopy cover (%) of the survey area estimated from aerial Google Earth maps (using imagery recorded in summer 2008); and green space (%) of the survey area from OS Mastermap (see above). Ground-truthing during the surveys confirmed the estimates of canopy cover and green space.

### Density estimates

Distance software (v.6, [38]) was used to calculate avian densities (number/hectare) at each sampling point. We excluded a small number of rarely detected species that exclusively use habitats that are not used by grey squirrels: wetlands (mallard *Anas platyrhynchos*, 2 detections; coot *Fulica atra*, 1 detection; moorhen *Gallinula chloropus*, 4 detections; sand martin *Riparia riparia*, 20 detections; sedge warbler *Acrocephalus schoenobaenus*, 2 detections) and open fields (skylark *Alauda arvensis*, 1 detection). Following standard distance sampling protocols [34] we fitted half-normal and hazard-rate distributions to all bird datasets to model how detection declined with distance from bird observations, across the four seasons. Urban form (green space % of the survey area), detection type (visual or audial) and season were included as covariates in models of detectability functions. The grouping of appropriate distance bands was explored until a good fit between each model and the data was obtained, by comparing the modelled detection function against the observed distance data, the goodness-of-fit statistics and Akaike's Information Criterion values for alternative grouping (and covariate combinations) and detection function options. For those species with fewer than 32 observations detectability functions were constructed by including observations of a surrogate species following reference [39]. The best fitting models were used to generate density estimates for each avian species, at all sampling points, in each season (Table S1). The same approach was used to calculate densities of grey squirrels and actively stocked supplementary feeding stations.

There was, however, limited variation between point counts in the resultant density estimates so subsequent statistical analyses were restricted to using the presence/absence of grey squirrels and feeding stations as predictors (respectively termed squirrel occurrence and supplementary feeding stations).

## Species classifications

Species were classified as using supplementary food if they were recorded as doing so in at least 75% of gardens in the BTO's Garden Bird Feeding Survey (GBFS) in the last four years (http://www.bto.org/volunteer-surveys/gbfs/results). Data are not available from this survey for a small number of our recorded species (feral pigeon *Columba livia*, carrion crow *Corvus corone* and most summer migrants). The feeder use of these species was classified following Fuller *et al.* (2008). The resultant classification (Table S1) is identical to reference [11] except that we also classify coal tit *Periparus ater* and long-tailed tit *Aegithalos caudatus* as using supplementary feeders, these species take supplementary food in respectively 90% and 75% of gardens in the GBFS. Species that were present during the winter were classified in two categories of sensitivity to interference competition, i.e. most and least sensitive (Table S1). The most sensitive species comprised those that met all of the following criteria: i) they used supplementary feeders (as defined above); ii) obtained most of their supplementary food from raised or hanging feeding stations, rather than feeding on the ground, as these are the types of feeders typically used by grey squirrels; and iii) were unaggressive small bodied species. These criteria are derived from the results of a food competition experiment which demonstrated that small bodied avian passerines, with the exception of the aggressive robin *Erithacus rubecula*), that used hanging feeders were susceptible to interference competition from grey squirrels [16]. All other species were classified as least sensitive to food competition from grey squirrels.

Species detected during the breeding season were classified as most or least sensitive to grey squirrel predation (Table S1). Species were classified as most sensitive if they nested in habitats frequently used by grey squirrels, and their daily nest failure rates were greater than 1% nest/day as calculated from the British Trust for Ornithology's nest record card scheme [40] as the vast majority of nest failures are caused by predation in our focal species. Data for pheasant *Phasianus colchicus* were obtained from the Game and Wildlife Conservation Trust [41]. Using this methodology cavity nesters and aggressive large bodied species were typically classified as least sensitive to predation, whilst other species were classified as being most sensitive to predation. There was a clear gap in the distributions of daily nest predation rates between the two groups of species (most sensitive species: 1.1 to 4.27%; least sensitive species: 0.08 to 0.74%).

## Statistical analysis

Analyses were conducted to assess how grey squirrels and supplementary feeding influenced the structure of urban avian assemblages during the breeding and wintering seasons. Although surveys were conducted in all four seasons, we focus on assessing how the breeding and wintering avifaunas were influenced by grey squirrels as these are likely to be most susceptible respectively to nest predation and food competition by grey squirrels. Data on the composition of the avifauna during the breeding season used observations from the spring point counts for the vast majority of species, but for a small number of late arriving summer migrants (Table S1) the summer survey data were used as some individuals would not have arrived at their breeding sites by the time of the spring survey period.

R v. 2.15.1 [42] was used for all statistical analyses. The species richness and density of each avian assemblage (i.e. species grouped by their sensitivity categories) was modelled, using generalised linear models, as a function of squirrel occurrence, mean tree height (m), canopy cover (%), green space in the 250 m×250 m cell (%), and presence of supplementary feeding stations. Interaction terms between squirrel occurrence and other predictors were included to assess whether squirrels altered the slope of the relationships between assemblage structure (i.e. species richness and density) and key predictor variables, i.e. presence of supplementary feeding stations, and measures of habitat quality. This is important because positive relationships between predators and their avian prey have been reported by other studies [12,17,18], probably reflecting a mutual preference for similar habitats, which may mask any suppression of population densities by grey squirrels.

The above analyses provide some information on the influence of supplementary feeding stations on assemblage structure, but to assess this further additional analyses were conducted that modelled the species richness and density of species that use and do not use supplementary feeders as a function of the presence of these feeders, mean tree height, canopy cover and green space. These analyses build on previous analyses of this issue by using empirical data on active feeder occurrence at the focal survey site and considering a wider range of habitat indicators than only the amount of green space.

Preliminary exploration indicated that all relationships with continuous predictors were linear and thus square terms were not included as predictors. Tolerance values of all predictors were sufficiently above the threshold (0.1; minimum 0.48 for canopy cover) below which correlations between predictors can bias the results of multiple regressions [43]. Model selection for all generalised linear models adopted an information theoretic approach; we constructed all possible models given the suite of our predictor variables using the MuMIn package, and the 95% confidence set of models comprised those whose cumulative weights summed to 0.95. Model averaging was conducted across this set of models to assess the influence of all predictors on avian species richness and density for each category of species. We used the spdep package to test all response variables for spatial autocorrelation following the methodology of Dormann *et al.* [44]. Moran's I tests demonstrated that spatial autocorrelation was extremely limited (maximum Moran's I = 0.037, for the breeding species richness of species most sensitive to nest predation), and non-significant for the majority of response variables. For those response variables with statistically significant Moran's I values comparison of full models constructed with and without taking spatial autocorrelation into account indicated that spatial autocorrelation had limited influence on parameter estimates and explanatory power (Tables S2 & S3). We thus only report the results from non-spatial models.

# Results

## Breeding season - sensitivity to nest predation

Multiple regression models explained between one sixth and just over a third of the variation in the richness and density, respectively, of breeding avian assemblages (Table 1). There was no evidence that the presence of grey squirrels, as a main effect, was negatively associated with avian species richness or density; indeed, squirrel occurrence was positively associated with the richness and density of those bird species most sensitive to grey squirrel nest predation, but explanatory capacity was limited (respective model averaged partial $r^2 = 0.06$ and 0.07; Table 1).

The richness and density of these species was negatively influenced by the interaction between squirrel occurrence and canopy cover, indicating that grey squirrel presence reduced the benefits of increasing woodland cover for these species, but explanatory capacity was again limited (respective model averaged partial $r^2 = 0.05$ and 0.04; Table 1). For the richness and density of those least sensitive bird species, interaction terms between squirrel occurrence and other predictors had negligible explanatory capacity (model averaged partial $r^2 < 0.02$; Table 1).

Predictors other than grey squirrel occurrence, and its interactions, typically explained more of the variation in the response variables for both the most and least sensitive species to grey squirrel nest predation. Canopy cover was the habitat predictor that explained most of the variation in species richness and density (respective model averaged partial $r^2 = 0.12$ and 0.1; Table 1) of species most sensitive to grey squirrel nest predation. The presence of supplementary feeding stations explained the most variation in species richness and density (respective model averaged partial $r^2 = 0.06$ and 0.1; Table 1) of species least sensitive to grey squirrel nest predation. The proportion of green space influenced the richness and density of the most and least sensitive bird species to nest predation, although the explanatory capacity was limited (model averaged partial $r^2 < 0.05$; Table 1).

## Winter - food competition

Multiple regression models explained between one sixth and a third of the variation in the richness and density, respectively, of wintering avian assemblages (Table 2). There was no evidence that the presence of grey squirrels, as a main effect, was negatively associated with avian species richness or density; indeed, squirrel occurrence was positively associated with the richness of those bird species most sensitive to food competition with grey squirrels, but explanatory capacity was limited (model averaged partial $r^2 = 0.03$; Table 2). There was very limited evidence that the interaction between squirrel occurrence and green space influenced the richness of the most sensitive bird species (model averaged partial $r^2 = 0.02$; Table 2). For the density of those most sensitive bird species, and the richness and density of those bird species least sensitive to food competition, interaction terms between squirrel occurrence and other predictors had negligible explanatory capacity (model averaged partial $r^2 < 0.01$; Table 2).

Predictors other than grey squirrel occurrence, and its interactions, typically explained more of the variation in the response variables for both the most and least sensitive species to food competition with grey squirrels. Canopy cover was the habitat predictor that explained most of the variation in species richness and density (respective model averaged partial $r^2 = 0.14$ and 0.16; Table 2) of species most sensitive to food competition. The proportion of green space and the presence of supplementary feeding stations explained most of the respective variation in species richness and density (respective model averaged partial $r^2 = 0.05$ and 0.07; Table 2) of species least sensitive to food competition.

## Supplementary feeding effects during the breeding season and winter

When taking habitat factors into account the presence of supplementary feeding stations had very limited influence on the species richness (model averaged partial $r^2 = 0.03$) and density (model averaged partial $r^2 = 0.04$) of wintering avian assemblages of species that regularly use supplementary feeders (Table 3). We found no evidence to suggest that supplementary feeding stations explained any of the variation in the winter species richness or density of species that do not regularly use supplementary feeders

**Table 1.** Generalised linear models of relationships between avian species richness and density and grey squirrel occurrence during the breeding season, whilst taking habitat and occurrence of supplementary feeders into account.

| Avian sensitivity category response variable | Model $R^2$ | Canopy cover: Partial $R^2$ | Canopy cover: Mean ± s.e. | Mean tree height: Partial $R^2$ | Mean tree height: Mean ± s.e. | Supplementary feeding stations: Partial $R^2$ | Supplementary feeding stations: Mean ± s.e. | Green space: Partial $R^2$ | Green space: Mean ± s.e. | Squirrel occurrence: Partial $R^2$ | Squirrel occurrence: Mean ± s.e. | Interactions with squirrel occurrence: Partial $R^2$ | Interactions with squirrel occurrence: Mean ± s.e. |
|---|---|---|---|---|---|---|---|---|---|---|---|---|---|
| Most sensitive species richness | 0.352 | 0.115 | 0.056±0.012 | 0.008 | −0.046±0.057 | 0.001 | −0.098±0.212 | 0.042 | 0.021±0.008 | 0.064 | 3.343±1.172 | 0.049 | −0.063±0.022[d] |
| Most sensitive density | 0.326 | 0.097 | 0.135±0.035 | 0.001 | −0.037±0.083 | 0.007 | 0.654±0.950 | 0.023 | 0.032±0.027 | 0.068 | 11.587±4.544 | 0.042 | −0.090±0.078[d] |
| Least sensitive species richness | 0.163 | 0.005 | 0.004±0.006 | 0.003 | −0.012±0.023 | 0.064 | 0.705±0.229 | 0.040 | 0.011±0.005 | 0.005 | 0.448±0.695 | 0.003 | −0.004±0.007[c] |
| Least sensitive density | 0.162 | 0.002 | −0.010±0.024 | 0.005 | −0.101±0.167 | 0.102 | 5.405±1.696 | 0.021 | 0.044±0.036 | 0.021 | 0.126±3.499 | 0.014 | −3.352±4.070[a] |

Species are classified by their relative sensitivity to grey squirrel nest predation. An information theoretic approach to model selection was adopted and data reported are model averaged values; standard errors are unconditional. In the final column, the partial $R^2$ is for all relevant interactions combined, and the mean ± s.e. are the model averaged values for the interaction with the highest explanatory power (and thus contributing the most to the combined partial $R^2$ in this column). Superscript letters are given to represent which interaction term has the highest explanatory power: [a] supplementary feeding stations*squirrel occurrence; [b] mean tree height*squirrel occurrence; [c] green space* squirrel occurrence, [d] canopy cover*squirrel occurrence.

**Table 2.** Generalised linear models of relationships between avian species richness and density and grey squirrel occurrence during the winter, whilst taking habitat and occurrence of supplementary feeders into account.

| Avian sensitivity category response variable | Model $R^2$ | Canopy cover | | Mean tree height | | Supplementary feeding stations | | Green space | | Squirrel occurrence | | Interactions with squirrel occurrence | |
|---|---|---|---|---|---|---|---|---|---|---|---|---|---|
| | | Partial $R^2$ | Mean ± s.e. | Partial $R^2$ | Mean ± s.e. | Partial $R^2$ | Mean ± s.e. | Partial $R^2$ | Mean ± s.e. | Partial $R^2$ | Mean ± s.e. | Partial $R^2$ | Mean ± s.e. |
| Most sensitive species richness | 0.306 | 0.137 | 0.032±0.008 | 0.017 | −0.051±0.040 | 0.018 | −0.229±0.255 | 0.027 | 0.010±0.006 | 0.025 | 1.462±1.363 | 0.017 | −0.016±0.021[c] |
| Most sensitive density | 0.285 | 0.161 | 0.196±0.043 | 0.003 | −0.097±0.158 | 0.004 | −0.803±1.225 | 0.016 | 0.033±0.034 | 0.007 | 3.307±4.768 | 0.002 | −0.029±0.055[c] |
| Least sensitive species richness | 0.197 | 0.005 | 0.006±0.010 | 0.018 | 0.079±0.070 | 0.039 | 1.064±0.462 | 0.052 | 0.024±0.009 | 0.012 | 2.216±2.472 | 0.009 | −0.018±0.028[c] |
| Least sensitive density | 0.129 | 0.001 | −0.001±0.021 | 0.002 | 0.088±0.185 | 0.072 | 7.922±2.490 | 0.026 | 0.088±0.059 | 0.004 | 4.483±8.101 | 0.004 | −1.863±3.246[a] |

Species are classified by their relative sensitivity to food competition from grey squirrel. An information theoretic approach to model selection was adopted and data reported are model averaged values; standard errors are unconditional. In the final column, the partial $R^2$ is for all relevant interactions combined, and the mean ± s.e. are the model averaged values for the interaction with the highest explanatory power (and thus contributing the most to the combined partial $R^2$ in this column). Superscript letters are given to represent which interaction term has the highest explanatory power: [a] supplementary feeding stations*squirrel occurrence; [b] mean tree height*squirrel occurrence; [c] green space* squirrel occurrence, [d] canopy cover*squirrel occurrence.

(model averaged partial $r^2<0.01$; Table 3). In breeding assemblages, the richness, and density, of species that regularly use feeders was positively associated with the presence of supplementary feeding stations (model averaged partial $r^2=0.07$), and canopy cover explained comparable variation in the richness of these species (model averaged partial $r^2=0.07$). We found no evidence that the presence of supplementary feeding stations influenced the breeding richness, and density, of the species that do not regularly use feeders (model averaged partial $r^2<0.01$), and the richness, and density, of these species were respectively influenced most by the proportion of green space and mean tree height (respective model averaged partial $r^2=0.08$ and 0.1; Table 3).

## Discussion

Grey squirrels predate birds' nests [14]. It has been widely suggested that this predation can alter the composition of avian assemblages and reduce the abundance of those species that are sensitive to nest predation [15]. Empirical data are limited to rural environments with some evidence that at the 1 km × 1 km spatial scale grey squirrels may slightly reduce the population size of a small number of bird species [17]. The strongest evidence that we found for such impacts in urban assemblages was that the presence of grey squirrels reduced the slope of the relationship between canopy cover and the density and richness of species most sensitive to nest predation by grey squirrels. Such relationships were not detected amongst species that were less sensitive to nest predation. Therefore, and whilst explanatory capacity was limited, the presence of grey squirrels appears to be having some influence on the structure of urban bird assemblages. All the species in our study that are sensitive to grey squirrel nest predation occur in woodland environments, and notably measures related to the abundance or type of green space, including canopy cover, had consistently stronger impacts on the size and species richness of urban breeding bird assemblages.

Grey squirrel population size and distribution in urban areas are positively influenced by supplementary feeding stations [13,45], at which they can outcompete numerous bird species, restricting their food intake rates by over 90%, and generating potential for interference competition [16]. Despite this we find no evidence, when taking habitat availability and type into account, that grey squirrels reduced the abundance or species richness of wintering avian assemblages in a highly urbanised region. This remains the case even when assemblage composition was restricted to those species most sensitive to food competition with grey squirrels. This may be a consequence of the high abundance of supplementary feeding stations, with approximately half of UK households feeding garden birds [19], as this will increase the probability of a bird displaced from one feeder by a grey squirrel rapidly finding an alternative food source, thus limiting the impact of competition on food intake rates. The apparent lack of adverse impacts of grey squirrels on birds using supplementary feeding stations could also arise from relatively low densities of grey squirrels. Our results thus do not preclude the possibility that grey squirrels could adversely influence wintering bird populations in situations with either a lower density of supplementary feeding stations or higher squirrel densities.

We find negligible evidence that the species richness or density of bird species is influenced by supplementary feeders, which may also contribute to the lack of adverse impacts of interference competition from grey squirrels on the structure of wintering avian assemblages. In contrast, the density and species richness of breeding bird species were positively associated with the presence

**Table 3.** The results of the generalised linear model analyses of avian assemblages categorised based on whether the species is a supplementary feeding species (suppl. feeders) or non-supplementary feeding species (non-suppl. feeders), as a function of metrics of habitat quality and the occurrence of supplementary feeding stations.

| | | Canopy cover | | Mean tree height | | Supplementary feeding stations | | Green space | |
|---|---|---|---|---|---|---|---|---|---|
| **Avian response variable** | **Model average $R^2$** | **Partial $R^2$** | **Mean ± s.e.** | **Partial $R^2$** | **Mean ± s.e.** | **Partial $R^2$** | **Mean ± s.e.** | **Partial $R^2$** | **Mean ± s.e.** |
| Winter richness (suppl. feeders) | 0.184 | 0.070 | 0.039±0.015 | <0.001 | −0.003±0.018 | 0.030 | 0.920±0.540 | 0.021 | 0.016±0.012 |
| Winter richness (non-suppl. feeders) | 0.245 | 0.021 | 0.006±0.006 | 0.027 | 0.037±0.027 | 0.001 | −0.026±0.061 | 0.077 | 0.011±0.003 |
| Breeding season richness (suppl. feeders) | 0.203 | 0.071 | 0.039±0.012 | <0.001 | −0.003±0.020 | 0.074 | 1.293±0.458 | 0.017 | 0.014±0.012 |
| Breeding season richness (non-suppl. feeders) | 0.246 | 0.021 | 0.006±0.005 | 0.027 | 0.038±0.027 | 0.002 | −0.044±0.084 | 0.078 | 0.011±0.003 |
| Winter density (suppl. feeders) | 0.127 | 0.022 | 0.111±0.100 | 0.001 | −0.064±0.181 | 0.041 | 6.844±3.171 | 0.031 | 0.110±0.080 |
| Winter density (non-suppl. feeders) | 0.173 | 0.001 | 0.003±0.010 | 0.093 | 0.398±0.107 | 0.004 | −0.386±0.626 | 0.020 | 0.027±0.020 |
| Breeding season density (suppl. feeders) | 0.158 | 0.024 | 0.124±0.094 | 0.001 | −0.081±0.181 | 0.073 | 9.457±2.865 | 0.023 | 0.094±0.076 |
| Breeding season density (non-suppl. feeders) | 0.172 | 0.001 | 0.003±0.010 | 0.095 | 0.401±0.107 | 0.003 | −0.313±0.546 | 0.019 | 0.026±0.020 |

An information theoretic approach to model selection was adopted and data reported are model averaged values; standard errors are unconditional.

of feeders, supporting the current advocacy that supplementary feeding should occur in spring and summer. The explanatory capacity of the associations between feeders and breeding species richness and density were, however, rather limited, especially in comparison to previous work suggesting that supplementary feeding explains nearly half of the variation in the densities of breeding avian assemblages in Sheffield [11]. The two studies were conducted five years apart, and it is possible that changes in the nature or extent of supplementary feeding by the public, or avian responses to it, may have changed over that time period. Supplementary feeding may, for example, have become sufficiently common that the occurrence of feeders now has less of a role in regulating avian population size than was previously the case. Methodological differences, such as the use of direct observations of feeder presence (this study) in comparison to estimated densities of feeders based on socio-economic variables [11], are though perhaps more likely to contribute to the variation in results of the two studies. Although both studies indicate that breeding bird assemblages benefit from supplementary feeding the magnitude of this effect may be less marked than previously thought.

## Conclusions and management implications

We find negligible evidence that interference competition from grey squirrels at supplementary feeding stations influences the structure of urban bird assemblages during the winter, which may in part be a consequence of negligible benefits of supplementary feeders to these assemblages. During the breeding season avian densities and species richness, of those species that used feeders, responded positively to the presence of supplementary feeders. Grey squirrel occurrence reduced the densities and species richness, of those species sensitive to nest predation, by limiting the beneficial impacts of woodland cover. The positive effects of feeders and the negative effects of grey squirrels were, however, rather limited in their explanatory capacity. The amount of green space and its quality, i.e. canopy cover, exerted a stronger influence on urban bird assemblages during both the winter and breeding seasons. Conservation management activities in the focal region targeting urban bird populations should, given the current densities of grey squirrels, focus on improving the availability of high quality habitat rather than controlling grey squirrels or further increasing supplementary feeding.

## Supporting Information

**Table S1 Species classifications and attributes of their detectability functions obtained from distance sampling.**

**Table S2 Assessments of how taking spatial autocorrelation into account influences full models that assess the responses of avian assemblages to grey squirrels.**

**Table S3 Assessments of how taking spatial autocorrelation into account influences full models that assess the responses of avian assemblages to supplementary feeders.**

## Acknowledgments

This research was funded by the Natural Environment Research Council. We would like to thank Sheffield City Council and the numerous households who gave permission to work on their land, G. Benassi for assisting with fieldwork and F. Eigenbrod, M. Dallimer and Z. Davies for assistance with ArcMap 9.3.

## Author Contributions

Conceived and designed the experiments: KLE KJG CB. Analyzed the data: CB. Contributed to the writing of the manuscript: CB KLE KJG. Conducted the fieldwork: CB.

## References


1. Faeth SH, Warren PS, Shochat E, Marussich WA (2005) Trophic dynamics in urban communities. Bioscience 55: 399–407.
2. Grimm NB, Faeth SH, Golubiewski NE, Redman CL, Wu JG, et al. (2008) Global change and the ecology of cities. Science 319: 756–760.
3. Shochat E, Lerman SB, Anderies JM, Warren PS, Faeth SH, Nilon CH (2010) Invasion, competition, and biodiversity loss in urban ecosystems. BioScience 60: 199–208.
4. Blair RB (1996) Land use and avian species diversity along an urban gradient. Ecol Appl 6: 506–519.
5. Tratalos J, Fuller RA, Evans KL, Davies RG, Newson SE, et al. (2007) Bird densities are associated with household densities. Glob Change Biol 13: 1685–1695.
6. Evans KL, Chamberlain DE, Hatchwell BJ, Gregory RD et al. (2011) What makes an urban bird? Global Change Biol 17: 32–44.
7. Gregory RD, Baillie SR (1998) Large-scale habitat use of some declining British birds. J Appl Ecol 35: 785–799.
8. McKinney ML (2006) Urbanization as a major cause of biotic homogenization. Biol Conserv 127: 247–260.
9. Evans KL, Newson SE, Gaston KJ (2009) Habitat influences on urban avian assemblages. Ibis 151: 19–39.
10. Jokimäki J, Clergeau P, Kaisanlahti-Jokimäki ML (2002) Winter bird communities in urban habitats: a comparative study between central and northern Europe. J Biogeogr 29: 69–79.
11. Fuller RA, Warren PH, Armsworth PR, Barbosa O, Gaston KJ (2008) Garden bird feeding predicts the structure of urban avian assemblages. Divers Distrib 14: 131–137.
12. Sims V, Evans KL, Newson SE, Tratalos JA, Gaston KJ (2008) Avian assemblage structure and domestic cat densities in urban environments. Divers Distrib 14: 387–399.
13. Bonnington C, Gaston KJ & Evans KL (2014) Squirrels in suburbia: influence of urbanisation on the occurrence and distribution of a common exotic mammal. Urban Ecosyst 17: 533–546.
14. Møller H (1983) Foods and foraging behaviour of Red (*Sciurus vulgaris*) and Grey (*Sciurus carolinensis*) squirrels. Mamm Rev 13: 81–98.
15. Hewson CM, Fuller RJ, Mayle B, Smith KW (2004) Possible impacts of Grey Squirrel on birds and other wildlife. Br Wildl 15: 183–191.
16. Bonnington C, Gaston KJ & Evans KL (2014) Assessing the potential for Grey Squirrels *Sciurus carolinensis* to compete with birds at supplementary feeding stations Ibis 156: 220–226.
17. Newson SE, Leech DI, Hewson CM, Crick HQP, Grice PV (2010a) Potential impact of grey squirrels *Sciurus carolinensis* on woodland bird populations in England. J Ornithol 151: 211–218.
18. Newson SE, Rexstad EA, Baillie SR, Buckland ST, Aebischer NJ (2010b) Population change of avian predators and grey squirrels in England: is there evidence for an impact on avian prey populations? J Appl Ecol 47: 244–252.
19. Davies ZG, Fuller RA, Loram A, Irvine KN, Sims V, et al. (2009) A national scale inventory of resource provision for biodiversity within domestic gardens. Biol Conserv 142: 761–771.
20. Jones DN, Reynolds SJ (2008) Feeding birds in our towns and cities: a global research opportunity. J Avian Biol 39: 265–271.
21. Robb GN, McDonald RA, Chamberlain DE, Reynolds SJ, Harrison TJE (2008a) Winter feeding of birds increases productivity in the subsequent breeding season. Biol Lett 4: 220–223.
22. Harrison TJE, Smith JA, Martin GR, Chamberlain DE, Bearhop S, et al. (2010) Does food supplementation really enhance productivity of bird nestlings? Oecologia 164: 311–320.
23. Plummer KE, Bearhop S, Leech DI, Chamberlain DE, Blount JD (2013) Fat provisioning in winter impairs egg production during the following spring: a landscape-scale study of blue tits. J Anim Ecol 82: 673–682.
24. Jokimäki J, Kaisanlahti-Jokimäki ML (2003) Spatial similarity of urban bird communities: a multiscale approach. J Biogeogr 30: 1183–1193.
25. Parsons H, Major RE, French K (2006) Species interactions and habitat associations of birds inhabiting urban areas of Sydney, Australia. Austral Ecol 31: 217–227.
26. Daniels GD, Kirkpatrick JB (2006) Does variation in garden characteristics influence the conservation of birds in suburbia? Biol Conserv 133: 326–335.
27. Robb GN, McDonald RA, Chamberlain DE, Bearhop S (2008b) Food for thought: supplementary feeding as a driver of ecological change in avian populations. Front Ecol Environ 6: 476–484.
28. Office for National Statistics (2010) Population estimates for UK, England and Wales, Scotland and Northern Ireland, Mid-2010. http://www.ons.gov.uk/ons/publications/Accessed 2011 May 20.
29. Gurnell J, Lurz PWW, Shirley MDF, Cartmel S, Garson PJ, et al. (2004) Monitoring red squirrels *Sciurus vulgaris* and grey squirrels *Sciurus carolinensis* in Britain. Mamm Rev 34: 51–74.
30. Harris S, Yalden DW (2008) Mammals of the British Isles: handbook 4th edition. The Mammal Society, Southampton.
31. Sieving KE, Willson MF (1998) Nest predation and avian species diversity in north-western forest understory. Ecology 79: 2391–2402.
32. Bayne E, Hobson K (2000) Relative use of contiguous and fragmented boreal forest by red squirrels (*Tamiasciurus hudsonicus*). Can J Zool 78: 359–365.
33. Russell RE, Lehmkuhl JF, Buckland ST, Saab VA (2010) The short-term responses of red squirrels to prescribed burning in the interior Pacific. Northwest J Wildl Manag 74: 12–17.
34. Buckland ST, Anderson DR, Burnham KP, Laake JL, Borchers DL et al. (2004) Advanced distance sampling. Oxford University Press, Oxford, England.
35. Koprowski JL (2005) The response of tree squirrels to fragmentation: a review and synthesis. Anim Conserv 8: 369–376.
36. Fuller RA, Tratalos J, Gaston KJ (2009) How many birds are there in a city of half a million people? Divers Distrib 15: 328–337.
37. Zhou DQ, Chu LM (2012) How would size, age, human disturbance, and vegetation structure affect bird communities of urban parks in different seasons? J Ornithol 153: 1101–1112.
38. Thomas L, Laake JL, Rexstad E, Strindberg S, Marques FFC, et al. (2009) Distance 6.0. Release "6.0". Research Unit for Wildlife Population Assessment, University of St. Andrews, St. Andrews, UK. http://www.ruwpa.st-and.ac.uk/distance/.
39. Newson SE, Evans KL, Noble DG, Greenwood JJD, Gaston KJ (2008) Use of distance sampling to improve estimates of national population sizes for common and widespread breeding birds in the UK. J Appl Ecol 45: 1330–1338.
40. Baillie SR, Marchant JH, Leech DI, Massimino D, Eglington SM et al. (2013) BirdTrends 2012: trends in numbers, breeding success and survival for UK breeding birds. BTO Research Report No. 644. BTO, Thetford. http://www.bto.org/birdtrends.
41. Game and Wildlife Conservancy Trust (2013) Pheasant nest predation. Available at https://www.gwct.org.uk/game/research/species/pheasant/pheasant-nest-predation/. Accessed 2014 January.
42. R Development Core Team (2012) R: A Language and Environment for Statistical Computing. Vienna, Austria: R Foundation for Statistical Computing.
43. Dormann CF, Elith J, Bacher S, Buchmann C, Carl G, et al. (2013) Collinearity: a review of methods to deal with it and a simulation study evaluating their performance. Ecography 36: 27–46.
44. Dormann CF, McPherson JM, Araújo MB, Bivand R, Bolliger J, et al. (2007) Methods to account for spatial autocorrelation in the analysis of species distributional data: a review. Ecography 30: 609–628.
45. Parker TS, Nilon CH (2008) Gray squirrel density, habitat suitability, and behavior in urban parks. Urban Ecosyst 11: 243–255.

# PERMISSIONS

All chapters in this book were first published in PLOS ONE, by The Public Library of Science; hereby published with permission under the Creative Commons Attribution License or equivalent. Every chapter published in this book has been scrutinized by our experts. Their significance has been extensively debated. The topics covered herein carry significant findings which will fuel the growth of the discipline. They may even be implemented as practical applications or may be referred to as a beginning point for another development.

The contributors of this book come from diverse backgrounds, making this book a truly international effort. This book will bring forth new frontiers with its revolutionizing research information and detailed analysis of the nascent developments around the world.

We would like to thank all the contributing authors for lending their expertise to make the book truly unique. They have played a crucial role in the development of this book. Without their invaluable contributions this book wouldn't have been possible. They have made vital efforts to compile up to date information on the varied aspects of this subject to make this book a valuable addition to the collection of many professionals and students.

This book was conceptualized with the vision of imparting up-to-date information and advanced data in this field. To ensure the same, a matchless editorial board was set up. Every individual on the board went through rigorous rounds of assessment to prove their worth. After which they invested a large part of their time researching and compiling the most relevant data for our readers.

The editorial board has been involved in producing this book since its inception. They have spent rigorous hours researching and exploring the diverse topics which have resulted in the successful publishing of this book. They have passed on their knowledge of decades through this book. To expedite this challenging task, the publisher supported the team at every step. A small team of assistant editors was also appointed to further simplify the editing procedure and attain best results for the readers.

Apart from the editorial board, the designing team has also invested a significant amount of their time in understanding the subject and creating the most relevant covers. They scrutinized every image to scout for the most suitable representation of the subject and create an appropriate cover for the book.

The publishing team has been an ardent support to the editorial, designing and production team. Their endless efforts to recruit the best for this project, has resulted in the accomplishment of this book. They are a veteran in the field of academics and their pool of knowledge is as vast as their experience in printing. Their expertise and guidance has proved useful at every step. Their uncompromising quality standards have made this book an exceptional effort. Their encouragement from time to time has been an inspiration for everyone.

The publisher and the editorial board hope that this book will prove to be a valuable piece of knowledge for researchers, students, practitioners and scholars across the globe.

# LIST OF CONTRIBUTORS

**Andrew J. Gregory**
School of Earth, Environment and Society, Bowling Green State University, Bowling Green, Ohio, United States of America

**Jeffrey L. Beck**
Department of Ecosystem Science and Management, University of Wyoming, Laramie, Wyoming, United States of America

**Pedro Aragón**
Departamento de Biogeografía y Cambio Global, Museo Nacional de Ciencias Naturales, Madrid, Spain

**Patrick S. Fitze**
Departamento de Biodiversidad y Biología Evolutiva, Museo Nacional de Ciencias Naturales, Madrid, Spain
Université de Lausanne, Department of Ecology and Evolution (DEE), Biophore, Lausanne, Switzerland
Instituto Pirenaico de Ecología (IPE-CSIC), Avenida Regimiento de Galicia s/n, Jaca, Spain
Fundación Araid, Edificio Pignatelli, Zaragoza, Spain

**Esperanza Manzano-Piedras and F. Xavier Picó**
Departamento de Ecología Integrativa, Estación Bioló gica de Doñana (EBD), Consejo Superior de Investigaciones Científicas (CSIC), Sevilla, Spain

**Arnald Marcer**
CREAF, Cerdanyola del Vallés, Spain
Universitat Autónoma de Barcelona, Cerdanyola del Vallés, Spain

**Carlos Alonso-Blanco**
Departamento de Genética Molecular de Plantas, Centro Nacional de Biotecnología (CNB), Consejo Superior de Investigaciones Científicas (CSIC), Madrid, Spain

**James J. Bell and Danielle Hannan**
School of Biological Sciences, Victoria University of Wellington, Wellington, New Zealand

**David Smith**
Coral Reef Research Unit, University of Essex, Colchester, United Kingdom

**Abdul Haris and Jamaludin Jompa**
Research and Development Center on Marine, Coastal and Small Islands, Hasanuddin University, Makassar, Indonesia

**Luke Thomas**
The Oceans Institute, School of Plant Biolog University of Western Australia, Australia

**Ju-yu Lian, Hong-lin Cao, Wan-hui Ye, Lin-fang Wu and Yue Bin**
Key Laboratory of Vegetation Restoration and Management of Degraded Ecosystems, South China Botanical Garden, Chinese Academy of Sciences, Guangzhou, China

**Jian-Xiong Huang and Yong Shen**
Key Laboratory of Vegetation Restoration and Management of Degraded Ecosystems, South China Botanical Garden, Chinese Academy of Sciences, Guangzhou, China
University of Chinese Academy of Sciences, Beijing, China

**Jian Zhang**
Department of Renewable Resources, University of Alberta, Edmonton, Alberta, Canada

**Sze Hang Fu**
Centre for Global Health Research, Li Ka Shing Knowledge Institute, St. Michael's Hospital, University of Toronto, Toronto, Ontario, Canada

**Prabhat Jha**
Centre for Global Health Research, Li Ka Shing Knowledge Institute, St. Michael's Hospital, University of Toronto, Toronto, Ontario, Canada
Dalla Lana School of Public Health, University of Toronto, Toronto, Ontario, Canada

**Prakash C. Gupta**
Healis-Sekhsaria Institute of Public Health, Navi Mumbai, India

**Rajesh Kumar**
School of Public Health, Post Graduate Institute of Medical Education and Research, Chandigarh, India

**Rajesh Dikshit**
Department of Epidemiology, Tata Memorial Hospital, Mumbai, India

**Dhirendra Sinha**
South East Asia Regional Office, World Health Organization, New Delhi, India

**Rémi M. Daigle and Anna Metaxas**
Department of Oceanography, Dalhousie University, Halifax, Nova Scotia, Canada

**Brad deYoung**
Department of Physics and Physical Oceanography, Memorial University, St. John's, Newfoundland, Canada

**Antonio Gazol, MaarjaÖpik, Mari Moora, Ülle Saks, Annika Uibopuu and Martin Zobel**
Department of Botany, Institute of Ecology and Earth Sciences, University of Tartu, Tartu, Estonia

**Kadri Koorem**
Department of Botany, Institute of Ecology and Earth Sciences, University of Tartu, Tartu, Estonia
Department of Terrestrial Ecology, Netherlands Institute of Ecology, NIOO-KNAW, Wageningen, The Netherlands

**Virve Sõber**
Department of Zoology, Institute of Ecology and Earth Sciences, University of Tartu, Tartu, Estonia

**Roger Jovani, David Serrano, José L. Tella, Jordi Figuerola, Ismael Galván and Eduardo Aguilera**
Estación Bioló gica de Doñana (CSIC), Sevilla, Spain

**Javier Diaz-Real**
Estación Bioló gica de Doñana (CSIC), Sevilla, Spain
Departamento de Ecoloxía e Bioloxía Animal. Universidade de Vigo, Campus As Lagoas Marconsende, Vigo, Pontevedra, Spain

**Sofía Fernández-González**
Departamento de Zoología y Antropología Física. Universidad Complutense de Madrid, Madrid, Spain

**Javier Pérez-Tris**
Departamento de Zoología y Antropología Física. Universidad Complutense de Madrid, Madrid, Spain
Grupo Ornitoló gico SEO-Monticola. Unidad de Zoología. Universidad Autónoma de Madrid, Madrid, Spain

**Javier De la Puente, Diana De Palacio, José L. Martínez, Rubén Moreno-Opo and Ana Bermejo**
Grupo Ornitoló gico SEO-Monticola. Unidad de Zoología. Universidad Autónoma de Madrid, Madrid, Spain

**Juan A. Calleja**
Grupo Ornitoló gico SEO-Monticola. Unidad de Zoología. Universidad Autónoma de Madrid, Madrid, Spain
Departamento Biología Animal, Vegetal y Ecología, Universidad Autónoma de Barcelona, Bellaterra, Barcelona, Spain

**Óscar Frías and Guillermo Blanco**
Department of Evolutionary Ecology, Museo Nacional de Ciencias Naturales (CSIC), Jose Gutiérrez Abascal, Madrid, Spain

**Carlos Ponce**
Grupo Ornitoló gico SEO-Monticola. Unidad de Zoología. Universidad Autónoma de Madrid, Madrid, Spain
Department of Evolutionary Ecology, Museo Nacional de Ciencias Naturales (CSIC), Jose Gutiérrez Abascal, Madrid, Spain

**Anders P. Møller**
Laboratoire d'Ecologie, Systématique et Evolution, CNRS UMR 8079, Université Paris-Sud 11, Bâtiment 362, Orsay, France

**Csongor I. Vágási**
Evolutionary Ecology Group, Hungarian Department of Biology and Ecology, Babeş-Bolyai University, Cluj Napoca, Romania

**Péter L. Pap**
Evolutionary Ecology Group, Hungarian Department of Biology and Ecology, Babeş-Bolyai University, Cluj Napoca, Romania
MTA-DE "Lendület" Behavioural Ecology Research Group, Department of Evolutionary Zoology and Human Biology, University of Debrecen, Debrecen, Hungary

**István Kovács**
'Milvus Group' Bird and Nature Protection Association, Târgu Mureş, Romania

**Leandro Meléndez**
Unidad Mixta de Investigacionen Biodiversidad. Instituto Cantábrico de Biodiversidad (CSIC Universidad de Oviedo), Oviedo, Spain

**Juan Carlos Senar and Antoni Borràs**
Unidad Asociada de Ecología Evolutiva y del Comportamiento, Museo de Ciencias Naturales de Barcelona (CSIC), Barcelona, Spain

**Francisco Atiénzar, Emilio Barba, Verónica Cortés, Juan S. Monrós and Rubén Piculo**
Cavanilles Institute of Biodiversity and Evolutionary Ecology, University of Valencia, Paterna, Spain

**José L. Cantó**
Parque Natural del Carrascal de la Font Roja, Alcoi, Spain

**Matthias Vögeli**
Federal Office for the Environment FOEN, Species, Ecosystems, Landscape Division, Bern, Switzerland

**Carlos Navarro**
Profesor Gonzalo Sanchez Vazquez, Sevilla, Spain

**Alexandre Mestre**
Department of Microbiology and Ecology, University of Valencia, Burjassot, Spain

**Stephanie G. Schuttler, Jessica A. Philbrick and Lori S. Eggert**
Division of Biological Sciences, University of Missouri, Columbia, Missouri, United States of America

**Kathryn J. Jeffery**
Agence Nationale des Parcs Nationaux, Libreville, Gabon
School of Natural Sciences, University of Stirling, Stirling, Scotland
Institut de Recherche en Ecologie Tropicale, Libreville, Gabon

**Craig Smith and Nina Kung**
Queensland Centre for Emerging Infectious Diseases, Department of Agriculture, Fisheries and Forestry, Brisbane, Queensland, Australia

**Billie Roberts**
Griffith School of Environment, Griffith University, Brisbane, Queensland, Australia

**Hume Field**
Queensland Centre for Emerging Infectious Diseases, Department of Agriculture, Fisheries and Forestry, Brisbane, Queensland, Australia
EcoHealth Alliance, New York, New York, United States of America

**Chris Skelly**
Biosecurity Intelligence Unit, Department of Agriculture, Fisheries and Forestry, Brisbane, Queensland, Australia
GIS People Pty Ltd, Brisbane, Queensland, Australia

**Andrew K. Skidmore and Tiejun Wang**
Department of Natural Resources, Faculty of Geo-Information Science and Earth Observation (ITC), University of Twente, Enschede, The Netherlands

**Xinping Ye**
Department of Natural Resources, Faculty of Geo-Information Science and Earth Observation (ITC), University of Twente, Enschede, The Netherlands
Foping National Nature Reserve, Shaanxi, China

**Jennifer A. Moore**
Biology Department, Grand Valley State University, Allendale, Michigan, United States of America
Department of Fisheries and Wildlife, Michigan State University, East Lansing, Michigan, United States of America

**Hope M. Draheim**
Department of Zoology, Michigan State University, East Lansing, Michigan, United States of America

**Dwayne Etter**
Wildlife Division, Michigan Department of Natural Resources, East Lansing, Michigan, United States of America

**Scott Winterstein and Kim T. Scribner**
Department of Fisheries and Wildlife, Michigan State University, East Lansing, Michigan, United States of America

**Wanwan Liu, Yuejia Cheng, Rongbin Yu, Hongbing Shen, Feng Chen and Zhihang Peng**
Department of Epidemiology & Biostatistics, School of Public Health, Nanjing Medical University, Nanjing, Jiangsu, China

**Changjun Bao, Yefei Zhu, Wendong Liu, Jianli Hu, Qi Liang, Ying Wu, Minghao Zhou and Fenyang Tang**
Jiangsu Province Center for Disease Control and Prevention, Nanjing, Jiangsu, China

**Lisa M. Dann, James G. Mitchell, Peter G. Speck, Kelly Newton and James Paterson**
School of Biological Sciences, Flinders University, Adelaide, South Australia, Australia

**Thomas Jeffries**
School of the Environment, University of Technology Sydney, Sydney, New South Wales, Australia

**Bin Lu and Qian Wang**
Chengdu Institute of Biology, Chinese Academy of Sciences, Chengdu, China

**Xiangjiang Zhan**
Organisms and Environment Division, Cardiff School of Biosciences, Cardiff University, Cardiff, United Kingdom

**Qiang Dai**
Chengdu Institute of Biology, Chinese Academy of Sciences, Chengdu, China
The Key Laboratory for Conservation Biology of Endangered Wildlife, Sichuan Province, Chengdu Research Base of Giant Panda Breeding, Chengdu, Sichuan, China

**Jinzhong Fu**
Chengdu Institute of Biology, Chinese Academy of Sciences, Chengdu, China
Department of Integrative Biology, University of Guelph, Guelph, Ontario, Canada

**Dunwu Qi**
The Key Laboratory for Conservation Biology of Endangered Wildlife, Sichuan Province, Chengdu Research Base of Giant Panda Breeding, Chengdu, Sichuan, China

**Jason Vleminckx and Olivier J. Hardy**
Service d'Évolution Biologique et Écologie, Faculté des Sciences, Université Libre de Bruxelles, Brussels, Belgium

**Achille B. Biwolé, Jean-François Gillet and Jean-Louis Doucet**
Laboratoire de Foresterie des Régions tropicales et subtropicales, Unité de Gestion des Ressources forestières et des Milieux naturels, Gembloux Agro-Bio Tech, Université de Liège, Gembloux, Belgium

**Julie Morin-Rivat**
Laboratoire de Foresterie des Régions tropicales et subtropicales, Unité de Gestion des Ressources forestières et des Milieux naturels, Gembloux Agro-Bio Tech, Université de Liège, Gembloux, Belgium
Laboratory of Wood Biology & Xylarium, Royal Museum for Central Africa, Tervuren, Belgium

**Kasso Daïnou**
Laboratoire de Foresterie des Régions tropicales et subtropicales, Unité de Gestion des Ressources forestières et des Milieux naturels, Gembloux Agro-Bio Tech, Université de Liège, Gembloux, Belgium
Laboratory of Applied Ecology, University of Abomey-Calavi, Cotonou, Benin

**Thomas Drouet**
Laboratoire d'Écologie Végétale et Biogéochimie, Faculté des Sciences, Université Libre de Bruxelles, Brussels, Belgium

**Ethan A. Wilson and Patrick J. Sullivan**
Department of Natural Resources, Cornell University, Ithaca, New York, 14850, United States of America

**Janis L. Dickinson**
Department of Natural Resources, Cornell University, Ithaca, New York, 14850, United States of America
The Cornell Lab of Ornithology, 159 Sapsucker Woods Road, Ithaca, New York, 14850, United States of America

**Claire Saraux, Jean-Marc Fromentin, Jean-Louis Bigot, Jean-Hervé Bourdeix, Marie Morfin, David Roos and Elisabeth Van Beveren**
IFREMER (Institut Franc¸ais de Recherche pour l'Exploitation de la MER), Research Unit EME (UMR 212), Séte, France

**Nicolas Bez**
IRD (Institut de Recherche pour le Développement), Research Unit EME (UMR 212), Séte, France

**Spencer N. Scheidt and Allen H. Hurlbert**
Department of Biology, University of North Carolina, Chapel Hill, North Carolina, United States of America

**Nicolas Chemidlin Prévost-Bouré and Philippe Lemanceau**
Unité Mixte de Recherche 1347 Agroécologie, Institut National de la Recherche Agronomique-AgroSup Dijon-Université de Bourgogne, Dijon, France

**Lionel Ranjard**
Unité Mixte de Recherche 1347 Agroécologie, Institut National de la Recherche Agronomique-AgroSup Dijon-Université de Bourgogne, Dijon, France
Unité Mixte de Recherche 1347 Agroécologie-Plateforme GenoSol, Institut National de la Recherche Agronomique-AgroSup Dijon-Université de Bourgogne, Dijon, France

**Samuel Dequiedt, Mélanie Lelièvre and Pierre Plassart**
Unité Mixte de Recherche 1347 Agroécologie-Plateforme GenoSol, Institut National de la Recherche Agronomique-AgroSup Dijon-Université de Bourgogne, Dijon, France

**Jean Thioulouse**
Unité Mixte de Recherche 555 Laboratoire de Biomé trie et Biologie Evolutive, Université Lyon 1-Centre National de la Recherche Scientifique, Villeurbanne, France

**Nicolas P. A. Saby, Claudy Jolivet and Dominique Arrouays**
Unité de Services 1106 InfoSol, Institut National de la Recherche Agronomique, Orléans, France

**Colin Bonnington and Karl L. Evans**
Department of Animal and Plant Sciences, University of Sheffield, Sheffield, United Kingdom

**Kevin J. Gaston**
Environment and Sustainability Institute, University of Exeter, Penryn, Cornwall, United Kingdom

# Index